D0592814

WIRELESS TECHNOLOGY

Protocols, Standards, and Techniques

WIRELESS TECHNOLOGY

Protocols, Standards, and Techniques

Michel Daoud Yacoub

CRC PRESS

Boca Raton London New York Washington, D.C.

Library of Congress Cataloging-in-Publication Data

Yacoub, Michel Daoud.
Wireless technology : protocols, standards, and techniques / by Michel Daoud Yacoub.
p. cm.
Includes bibliographical references and index.
ISBN 0-8493-0969-7 (alk. paper)
1. Wireless communication systems. 2. Mobile communication systems. I. Title.

TK6570.M6 Y35 2001
621.382--dc21 2001052586

This book contains information obtained from authentic and highly regarded sources. Reprinted material is quoted with permission, and sources are indicated. A wide variety of references are listed. Reasonable efforts have been made to publish reliable data and information, but the author and the publisher cannot assume responsibility for the validity of all materials or for the consequences of their use.

Neither this book nor any part may be reproduced or transmitted in any form or by any means, electronic or mechanical, including photocopying, microfilming, and recording, or by any information storage or retrieval system, without prior permission in writing from the publisher.

The consent of CRC Press LLC does not extend to copying for general distribution, for promotion, for creating new works, or for resale. Specific permission must be obtained in writing from CRC Press LLC for such copying.

Direct all inquiries to CRC Press LLC, 2000 N.W. Corporate Blvd., Boca Raton, Florida 33431.

Trademark Notice: Product or corporate names may be trademarks or registered trademarks, and are used only for identification and explanation, without intent to infringe.

Visit the CRC Press Web site at www.crcpress.com

© 2002 by CRC Press LLC

No claim to original U.S. Government works
International Standard Book Number 0-8493-0969-7
Library of Congress Card Number 2001052586
Printed in the United States of America 1 2 3 4 5 6 7 8 9 0
Printed on acid-free paper

Dedication

Technology has its time; knowledge is timeless.

Technowledge boosts *Technowledge.*

To those who idealize, conceive, standardize, implement, test, operate, maintain, upgrade; to those professionals, for whose knowledge and work we owe the technology and all it conveys—to those *technowledgers*—I pay my most sincere tribute.

I dedicate this book to my beloved family. Thank you, Maria Nídia, my precious wife, and thank you, Alexandre, Helena, Carolina, Ricardo, Vinícius, and Elisa, my wonderful children, for your love, patience, and unconditional support.

Preface

We can always wait a bit longer to write a better book on technology. We can always wait . . .

In this ever-changing technological scenario, keeping pace with the rapid evolution of wireless technology is a formidable, exciting, and indispensable task more than a challenge. The work is indeed herculean and often discouraging, for technology is vast, the number of topics to be approached is immense, the documentation on standards and recommendations comprises piles of uncountable pages, and we often find we are leaving something important behind when selecting the appropriate subject matter to explore. The consolation, if any, is that as we explore the technologies, we find that much commonality exists among them, although particular features are rather different in each.

The challenge of writing a book in such a "hot" and vivacious field is to provide a clear and concise resource to accommodate the learning process of the basic functions of the main technologies. I did try to keep this in mind throughout the course of selection and description of the topics included in this book. I hope I have succeeded, at least to a certain extent.

The book, divided into five parts, describes protocols, standards, and techniques for 2G and 3G technologies, including those specific to wireless multimedia. The first part—*Introduction*—contains three chapters and covers the basic principles of wireless communications. The second part—*2G Systems*—consists of two chapters and describes two leading technologies of the second generation. The third part—*Wireless Data*—comprises one chapter and introduces three main wireless data technologies. The fourth part—*3G Systems*—encompasses three chapters and details the general concepts of third-generation systems as well as two chief third-generation technologies. The fifth part—*Appendices*—provides a glimpse at some telecommunication issues that are relevant to the understanding of the main text and that are not covered in the introductory part of the book. A more detailed description of the book structure follows.

Part I: *Introduction*

Chapter 1—*Wireless Network*—develops the wireless network concepts within the Intelligent Network framework and describes the basic functions a telecommunication system must provide so that wireless and mobile capabilities can be implemented. General network and protocol architectures and channel structures are described that are common to the main systems. These descriptions are based on ITU Recommendations, which generalize those concepts that have been used for the various cellular networks. Specific solutions are then detailed in the other chapters.

Chapter 2—*Cellular Principles*—introduces the cellular technology fundamentals, providing a unified approach of these concepts for narrowband and wideband solutions. Topics explored in this chapter include universal frequency reuse, sectorization, power control, handoff, voice activity, interference, and others. Besides the traditional hexagonal tessellation for macrocellular networks, the chapter examines the subject of reuse pattern for microcellular systems. In addition, hierarchical cell structure, overall mean capacity for multirate systems, and the main features of narrowband and wideband networks are also addressed.

Chapter 3—*Multiple Access*—analyzes a considerable number of multiple access control techniques. Several conventional and more advanced duplexing and multiple access protocols are detailed that comply with the various classes of traffic and multirate transmission utilized in broadband services. The access and duplexing methods are explored in the frequency domain, time domain, code domain, and space domain. The performance of the techniques is investigated in terms of channel capacity, throughput, and delay.

Part II: *2G Systems*

Chapter 4—*GSM*—describes the Global System for Mobile Communication cellular network in terms of its features and services, architecture, physical channels, logical channels, signaling messages, call management, and particular features.

Chapter 5—*cdmaOne*—details the features and services, architecture, physical channels, logical channels, signaling messages, call management, and particular features for TIA/EIA/IS-95-A as well as for its evolved version TIA/EIA/IS-95-B.

Part III: *Wireless Data*

Chapter 6—*Wireless Data Technology*—depicts three data technologies applied to wireless networks, namely, General Packet Radio Service (GPRS), TIA/EIA/IS-95B, and High Data Rate (HDR). These technologies are described in terms of their basic architectures and achievable data transmission rates.

Part IV: *3G Systems*

Chapter 7—*IMT-2000*—introduces the topic on third-generation wireless networks based on the International Mobile Telecommunications-2000 (IMT-2000) concept. It describes the functional subsystems, the IMT-2000 family concept, and the capability set concept. It also develops the network functional model for IMT-2000.

Chapter 8—*UTRA*—details the IMT-2000 radio interface for direct sequence code division multiple access, the so-called Universal Terrestrial Radio Access (UTRA) or Wideband CDMA (WCDMA) 3G radio transmission technology. Descriptions include its FDD as well as its TDD options.

Chapter 9—*cdma2000*—details the IMT-2000 CDMA multicarrier radio interface, the so-called cdma2000 3G radio transmission technology. Descriptions include its various radio configurations, the 1xEV-DO radio configuration option being one of them.

Part V: *Appendices*

These *Appendices* provide tutorial information on topics such as OSI Reference Model, Signaling System Number 7, Spread Spectrum, and Positioning of Interferers in a Microcellular Grid.

The book is suitable as text as well as a reference. As a textbook, it fits into a semester course for both undergraduate and graduate levels in electrical engineering, wireless communications, and more generally in information technology. As a reference, it serves systems engineers and analysts, hardware and software developers, researchers, and engineers responsible for the operation, maintenance, and management of wireless communication systems.

Acknowledgments

I am grateful to a number of people who have generously helped with the completion of this book. Some provided me with updated material and original results, others revised parts of the manuscript, and still others stimulated discussions and lent me their ideas, suggestions, incentive, encouragement, motivation, and so many distinct forms of assistance. In the endeavor to cite their names, I may inadvertently leave some out, for they are many and my memory will certainly deceive me. To these who are not quoted here, my forgiveness.

I thank Professor Kenneth W. Cattermole, Professor Attílio J. Giarola, Professor Helio Waldman, Professor Dalton S. Arantes, Professor Rui F. de Souza, Professor Ivan L. M. Ricarte, Dr. Antônio F. de Toledo, Dr. Ailton A. Shinoda, Dr. Omar C. Branquinho, Dr. César K. d'Ávila, Dr. Ernesto L. A. Neto, Dr. Paula R. C. Gomez, Alexandre R. Esper, Alexandre R. Romero, Antônio V. Rodrigues, Cláudio R. C. M. da Silva, Edigar Alves, Fabbryccio A. C. M. Cardoso, and Gustavo Fraidenraich. I am also indebted to my colleagues of the Department of Communications (DECOM), School of Electrical Engineering (FEEC), at The State University of Campinas (UNICAMP).

I would like to express my gratitude to FEEC, CPqD, CelTec/CelPlan, Ericsson, IBM, Telesp Celular, Motorola, Instituto Eldorado, Lucent Technologies, CNPq, CAPES, and FAPESP for supporting my research efforts in wireless communications over the years.

Finally, I am very thankful to the CRC Press staff for their support and incentive and for the opportunity to publish this book.

Contents

Part II 2G Systems

Part III Wireless Data

Part IV 3G Systems

Part V Appendices

Part I
Introduction

1

Wireless Network

1.1 Introduction

First-generation (1G) wireless networks were established in the late 1970s with the primary aim of providing voice telephony services to mobile subscribers. 1G systems are basically characterized by the use of analog frequency modulation (FM) for voice transmission and frequency division multiple access (FDMA) as its multiple access architecture. Several 1G networks were independently developed in various regions of the world, with the main systems represented by the following major technologies: Advanced Mobile Phone Service (AMPS) in North America; Total Access Communication System (TACS), European TACS (ETACS), and Nordic Mobile Telephone system (NMT) in Europe; and Japan TACS (JTACS) and Nippon TACS (NTACS) in Japan.

Second-generation (2G) wireless networks emerged in the early 1990s and were totally based on digital transmission techniques. 2G systems aimed at providing a better spectral efficiency, a more robust communication, voice and low-speed data services, voice privacy, and authentication capabilities. Three major technologies are based on the 2G principles: Global System for Mobile communications (GSM), TIA/EIA/IS-136 (IS-136) or Digital AMPS (D-AMPS), and TIA/EIA/IS-95A (IS-95A). GSM and IS-136 use time division multiple access (TDMA), whereas IS-95A uses code division multiple access (CDMA) as multiple access architectures.

Although 2G systems are entirely based on digital technology, their data transmission capability is rather modest. The advent of the Internet and the increase of the demand for mobile access to Internet applications boosted the development of wireless technologies that, as an evolution of the existing 2G systems, support data transmission. Within this framework, General Packet Radio Service (GPRS), IS-95B, and High Data Rate (HDR) emerged as wireless data technologies. GPRS coupled with GSM or with IS-136, IS-95B—an evolution of IS-95A—, and HDR fulfill the aspirations of the incorporation of data transmission capabilities into wireless systems. HDR, in particular, provides for data rates as required by third-generation (3G) systems.

3G wireless network conception is embodied by the International Mobile Telecommunications–2000 (IMT-2000). IMT-2000 standards and specifications have been developed by various standards organizations worldwide under the auspices of the International Telecommunications Union (ITU). A wide and ambitious range of user sectors, radio technology, radio coverage, and user equipment is covered by IMT-2000. In essence, a 3G system must provide for multimedia services, in circuit-mode and packet-mode operations, for user sectors such as private, public, business, residential, local loop, and others, for terrestrial-based and satellite-based networks, for personal pocket, vehicle-mounted, or any other special terminal. Two major radio transmission technologies may be cited that fulfill the 3G requirements: Universal Terrestrial Radio Access (UTRA) and CDMA Multi-Carrier radio interface (cdma2000).

A wireless network is defined in terms of standards and specifications that are developed by different standardization organizations or industry associations. Hence, standards and specifications vary for different technologies. On the other hand, a common framework exists that characterizes the wireless systems. This chapter describes the wireless networks in terms of their common features. The main concepts developed here are based on an ITU recommendation for IMT-2000,[1] which generalizes those concepts that have been used for conventional cellular networks. Specific solutions are then detailed in the following chapters.

1.2 Intelligent Network

Wireless technology has gained universal acceptance, with the number of wireless subscriptions already exceeding the number of fixed lines in many countries. With the wireless market becoming increasingly competitive, rapid deployment of innovative solutions arise as the kernel of any successful wireless strategy. This can be achieved by means of the intelligent network (IN) concept.

"In an IN, the logic for controlling telecommunications services migrates from the traditional switching points to computer-based, service-independent platforms."[2] This greatly contrasts to the traditional public-switched telephone network, where the hierarchy of switching equipment and software must be upgraded in the event a new service is added. In the IN concept, services are separated from switching equipment. A centralized system is then organized so that major modifications need not be performed on multiple switches with the introduction of new services.

The implementation of the IN concept is based on the following steps:

- Creation of separate service data in a centralized database outside the switching node
- Separation of the service programs, or service logic, and definition of a protocol that allows interaction between switching systems and intelligent nodes containing the service logic and data

In such a case, open platforms are created that encompass generic service components, with these components able to interoperate with elements from different vendors. Based on the IN concept, rapid creation and deployment of enhanced services and new features are substantially eased. Note that in IN, services are detached from switching equipment, which opens markets for telecommunication-service creators and switching-equipment providers.

1.2.1 IN Protocol Architecture

The IN architecture is based on the Signaling System 7 (SS7) and its protocol architecture. The IN protocol, from the bottom of the stack upward, contains the following elements (see Appendix A and Appendix B for the OSI/ISO Reference Model and SS7, respectively):

- *Message Transfer Part (MTP).* The MTP is a common signaling transport capability that handles the physical layer, data link layer, and network layer.
- *Signaling Connection Control Part (SCCP).* The SCCP provides both connectionless-oriented and connection-oriented message transport and enables addressing capabilities for message routing.
- *Transaction Capabilities Application Part (TCAP).* The TCAP is responsible for providing procedures for real-time transaction control.
- *Intelligent Network Application Protocol (INAP).* The INAP defines the necessary operations between the various IN elements.

1.2.2 IN Elements

In an IN, several physical entities (PEs), comprising functional entities (FEs), are identified. These PEs, represented by rectangles, and their corresponding FEs, represented by ellipses, are depicted in Figure 1.1 and briefly described as follows.

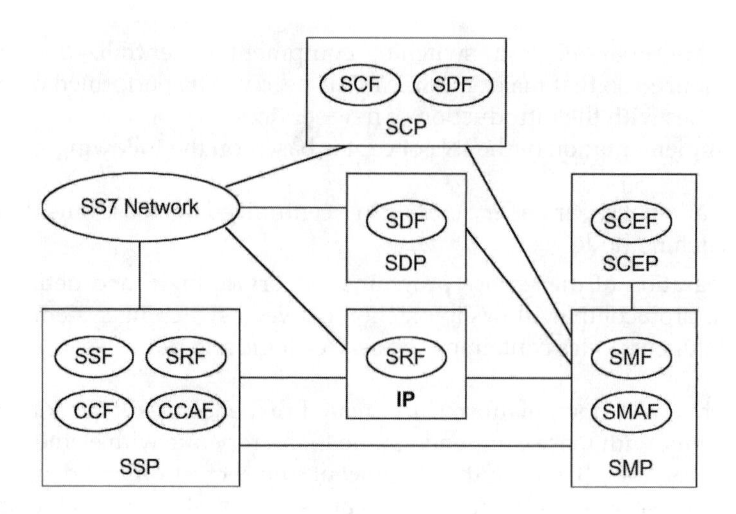

FIGURE 1.1
Physical entities and functional entities in an IN.

- *Service Switching Point (SSP).* The SSP is a stored program control switch (the switching element) that interfaces IN with SS7 signaling network. The following FEs are encompassed by SSP:

  Call Control Function (CCF). The CCF is responsible for controlling the call processing and for providing network connection services.

  Service Switching Function (SSF). The SSF is responsible for supporting IN triggering during call processing and access to IN functionalities.

  Specialized Resource Function (SRF). The SRF is responsible for supporting interaction between SSP and SCP (to be defined next).

  Call Control Agent Function (CCAF). The CCAF is responsible for providing user access to the network.

- *Service Control Point (SCP).* The SCP constitutes the intelligent node. It is a transaction-processing entity responsible for providing call-handling information to SSP. The following FEs are encompassed by SCP:

  Service Control Function (SCF). The SCF is responsible for executing IN service logic.

  Service Data Function (SDF). The SDF is responsible for managing data for real-time access by SCF in the execution of an IN service.

- *Intelligent Peripheral (IP).* The IP is described by the following FE:

Specialized Resource Function (SRF). The SRF is responsible for supporting specialized network resources. These resources are usually associated with caller interaction.

- *Service Management Point (SMP).* The SMP is described by the following FEs:

  Service Management Function (SMF). The SMF is responsible for supporting ongoing operation. It also supports deployment and provision of IN services.

  Service Management Access Function (SMAF). The SMAF is responsible for interfacing services managers and SMF.
- *Service Creation Environment Point (SCEP).* The SCEP is described by the following FE:

  Service Creation Environment (SCE). The SCE is responsible for supporting the definition, development, and test of IN services to be input into SMF.
- *Service Data Point (SDP).* The SDP is described by the following FE:

  Service Data Function (SDF). The SDF is responsible for managing data for real-time access by SCF in the execution of an IN service.

The communication between the several PEs relies on out-of-band signaling or on SS7 protocols. The SS7 protocols provide means to:

- Place service logic and service data into network elements responsible for handling control and connection remotely
- Enable the communication between intelligent applications and other applications
- Access databases located in various parts of the network

1.2.3 Wireless Service Requirements

Some service requirements that are unique or essential to wireless networks call for IN concepts, as exemplified below:

- *Roaming*. Mobility, a feature inherent to a wireless network, creates situations in which subscribers may roam out of their local calling area or out of their service provider's area. To enable subscribers to still make use of their wireless services while in a roaming condition, a seamless network must be conceived. This can be accomplished with the IN concept by exchanging information between the various devices involved in the communication. In particular, the use of SS7 makes it possible for message validations and billing reciprocation of wireless calls.

- *Carrier Select.* Carrier-select services allow providers to select the network to be used to handle a call. In the same way, they allow subscribers to route their calls selectively through their network of preference. The exchange of messages in these cases requires IN technology.
- *Hands-Free Operation.* Hands-free services require that special features be implemented within the network. For voice-activated dialing and feature activation, voice recognition technology must be available. In such a case, messages or voice signals must be collected, translated into data, and routed to the required device, the so-called intelligent peripheral (IP). In such a case, special routing features or intelligent networking are necessary. The IP is then activated to implement the required task.
- *Fee Structure.* The interaction among the various networks involved in a call, both wired and wireless, renders billing a difficult task. IN flags can be used to facilitate the billing. They can be included into the call record so that billing reflects the specific call handling and fees can be processed more easily.
- *Data-Service Capability.* Short Message Service (SMS) is a very popular wireless feature. Wireless phones are allowed to send and receive messages in addition to making or taking telephone calls. SMS works like a pager, and requires SS7 messages for the several tasks involved in its implementation: access to database, authentication, message encapsulation, paging, routing, etc. IN procedures are certainly required to implement SMS.

1.2.4 Wireless IN Services

The IN protocols and concepts can be used to implement enhanced wireless services rapidly and to have these services available across serving areas in an untethered wireless network. Some of these services are listed below:

- *Voice-Based User Identification.* This service employs a form of automatic speech recognition to validate the identity of the speaker. Access to services can then be restricted to the user whose voice (phrase) has been used to train the recognition device.
- *Voice-Based Feature Control.* This service allows the authorized user to specify feature operations, which can be carried out via feature-control strings by means of spoken commands.

- *Voice-Control Dialing*. This service allows the subscriber to place a call using spoken commands.
- *Voice-Controlled Services*. This feature allows the subscriber to control features and services using spoken commands.
- *Incoming Call-Restriction/Control*. This service allows users to impose restrictions to an incoming call as follows: it may terminate normally to the subscriber; it may terminate to the subscriber with normal alerting; it may terminate to the subscriber with special alerting; it may be forwarded to another number; it may be forwarded to voice mail; it may be routed to any specific announcement; or it may be blocked.
- *Calling Name Presentation*. This service provides the name identification of the calling party (personal name, company name, restricted, not available) as well as the date and the time of the call.
- *Password Call Acceptance*. This service allows the subscriber to restrict incoming calls only to those callers who can provide valid passwords.
- *Selective Call Acceptance*. This service allows the subscriber to restrict incoming calls only to those calling parties whose numbers are in the restricted list.
- *Short Message Service*. This service allows the short message entities (SMEs)—the short message users—to receive or send short messages (packet of data).
- *Speech-to-Text Conversion*. This service allows the user to create a short alphanumeric message by means of spoken phrases.
- *Prepaid Phone*. This service allows the user to pay before calling, i.e., not to be billed (postpaid). This can take a number of forms, for example, a debit card, a connection to a smart card, and others.

1.2.5 IN Standards

The merge of mobility systems with IN for 2G systems is a reality and takes a uniform approach.

In North America, the movement to develop a wireless intelligent network (WIN) was triggered by the Cellular Telecommunications Industry Association. In Europe, the same movement for GSM-based networks was carried out through the Customized Applications for Mobile Network Enhanced Logic (CAMEL).

3G systems—IMT2000—are already entirely based and described in terms of the IN architecture.

1.3 Network Architecture

Different wireless networks are specified by different standardization organizations and exhibit different architectures. However, a common framework may be identified in all wireless systems. Thus, some IN elements are used as the basis for an architecture common to all wireless networks. Figure 1.2 identifies the main components of a wireless system, which are as follows:

- *Mobile Station (MS)*. The MS terminates the radio path on the user side of the network enabling the user to gain access to services from the network. It incorporates user interface functions, radio functions, and control functions, with the most common equipment implemented in the form of a mobile telephone. It can work as a stand-alone device or it may accept other devices connected to it (e.g., fax machines, personal computers, etc.). For such a purpose it may be split into the terminal equipment (TE), the terminal adapter (TA), and the mobile termination (MT). A TE may be a fax, a computer, etc.; an MT is the

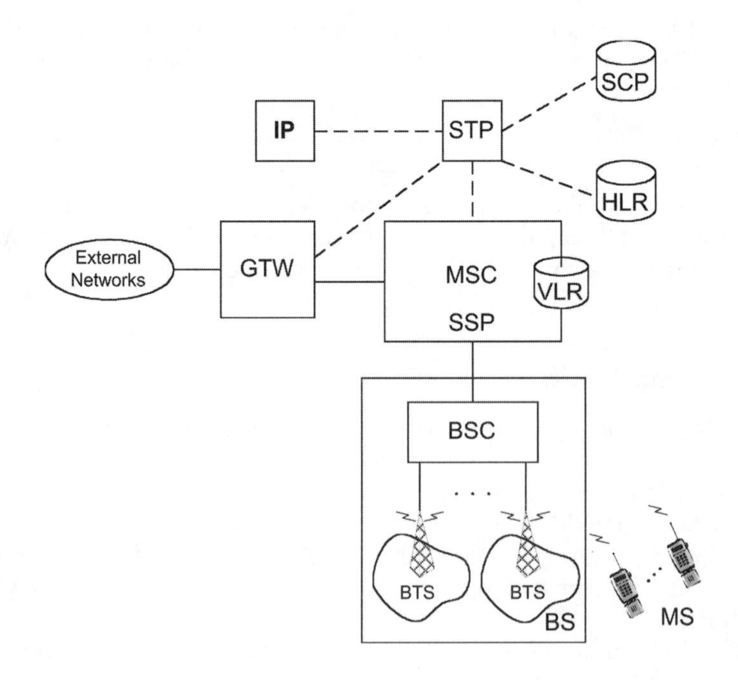

FIGURE 1.2
The main components of a wireless system.

equipment that realizes the wireless functions; and a TA works as an interface between the TE and the MT.

- *Base Station (BS).* The BS terminates the radio path on the network side and provides connection to the network. It is composed of two elements:

  Base Transceiver Station (BTS). The BTS consists of a radio equipment (transmitter and receiver–transceiver) and provides the radio coverage for a given cell or sector.

  Base Station Controller (BSC). The BSC incorporates a control capability to manage one or more BTSs, executing the interfacing functions between BTSs and the network. The BSC may be co-located with a BTS or else independently located.

- *Mobile Switching Center (MSC).* The MSC provides an automatic switching between users within the same network or other public switched networks, coordinating calls and routing procedures. In general, an MSC controls several BSCs, but it may also serve in different capacities. The MSC provides the SSP function in a wireless IN.
- *Visitor Location Register (VLR).* The VLR is a database containing temporary records associated with subscribers under the status of a visitor. A subscriber is considered a visitor if such a subscriber is being served by another system within the same home service area or by another system away from the respective home service area (in a roaming condition). The information within the VLR is retrieved from the HLR. An VLR is usually co-located with an MSC.
- *Home Location Register (HLR).* The HLR is the primary database for the home subscriber. It maintains information records on subscriber current location, subscriber identifications (electronic serial number, international mobile station identification, etc.), user profile (services and features), and so forth. An HLR may be co-located with an MSC or it may be located independently of the MSC. It may even be distributed over various locations and it may serve several MSCs. An HLR usually operates on a centralized basis and serves many MSCs.
- *Gateway (GTW).* The GTW serves as an interface between the wireless network and the external network.
- *Service Control Point (SCP).* The SCP provides a centralized element to control service delivery to subscribers. It is responsible for higher-level services that are usually carried out by the MSC in wireless networks not using IN facilities.
- *Service Transfer Point (STP).* The STP is a packet switch device that handles the distribution of control signals between different elements in the network.

- *Intelligent Peripheral (IP).* The IP processes the information of the subscribers (credit card information, personal identification number, voice-activated information, etc.) in support of IN services within a wireless network.
- *External Network.* The external network constitutes the ISDN (Integrated Services Digital Networks), CSPDN (Circuit-Switched Public Data Network), PSPDN (Packed-Switched Public Data Network), and, of course, PSTN (Public-Switched Telephone Network).

Note that a wireless network can be grossly split into a radio access network (RAN) and a core network (CN). The RAN implements functions related to the radio access to the network, whereas the CN implements functions related to routing and switching. The RAN comprises the BSC, BTS, MT, and control functionalities of the MS. The CN comprises the MSC, HLR, VRL, GTW, and other devices implementing the switching and routing functions. This book is primarily concerned with the radio aspects—the radio interface—of a wireless network.

1.4 Protocol Architecture

A radio interface implements the wireless electromagnetic interconnection between a mobile station and a base station.[1] A general radio protocol contains the three lowest layers of the OSI/ISO Reference Model, as follows:

- *Physical Layer*. The physical layer is responsible for providing a radio link over the radio interface. Such a radio link is characterized by its throughput and data quality. It is defined for the BTS and for the MT.
- *Data Link Layer*. The data link layer comprises two sublayers, as follows:

  Medium Access Control (MAC) sublayer. The MAC sublayer is responsible for controlling the physical layer. It performs link quality control and mapping of data flow onto this radio link. It is defined for the BTS and for the MT. It may or may not exist in the BSC and in the control functionalities of an MS.

  Link Access Control (LAC) sublayer. The LAC sublayer is responsible for performing functions essential to the logical link connection such as setup, maintenance, and release of a link. It is defined for BSC, BTS, MT, and control functionalities of the MS.

- *Network Layer*. The network layer contains functions dealing with call control, mobility management, and radio resource management. It is mostly independent of radio transmission technology. Such a layer can be transparent for user data in certain user services. It is defined for BSC, BTS, MT, and control functionalities of the MS.

1.5 Channel Structure

A channel provides means of conveying information between two network elements. Within the radio interface, three types of channels are specified: radio frequency (RF) channel, physical channel, and logical channel.[1] These channels are defined in the forward direction (downlink)—from BS to MS—or in the reverse direction (uplink)—from MS to BS.

1.5.1 RF Channel

An RF channel is defined in terms of a carrier frequency centered within a specified bandwidth, representing a portion of the RF spectrum. The RF channel constitutes the means of carrying information over the radio interface. It can be shared in the frequency domain, time domain, code domain, or space domain.

1.5.2 Physical Channel

A physical channel corresponds to a portion of one or more RF channels used to convey any given information. Such a portion is defined in terms of frequency, time, code, space, or a combination of these. A physical channel may be partitioned into a frame structure, with the specific timing defined in accordance with the control and management functions to be performed. Fixed or variable frame structures may be used.

1.5.3 Logical Channel

A logical channel is defined by the type of information it conveys. The logical channels are mapped onto one or more physical channels. Logical channels are usually grouped into control channels and traffic channels. Further specifications concerning these channels vary according to the wireless network. Logic channels may be combined by means of a multiplexing process, using a frame structure. The following division and definitions are based on Reference 1, and such a division, as depicted in Figure 1.3, reflects the basic structure used in most wireless networks.

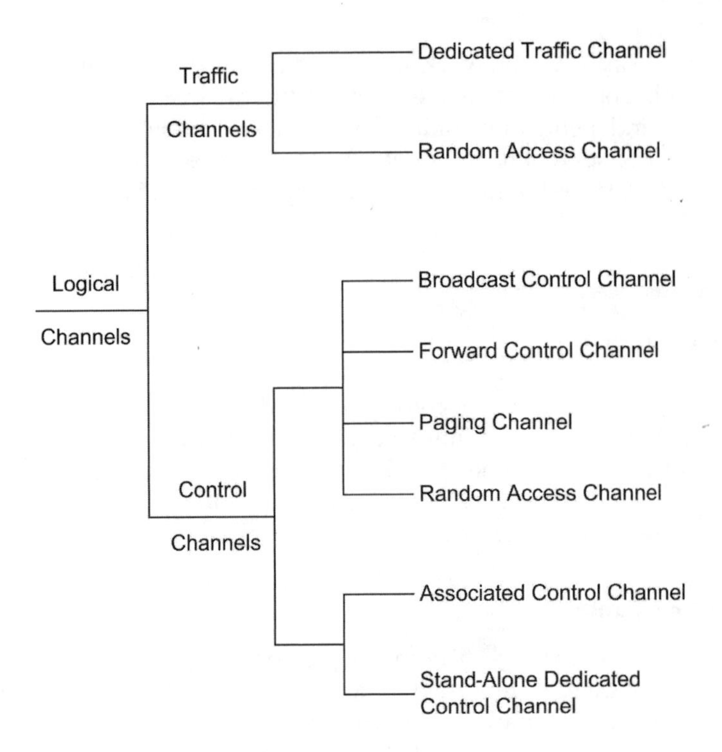

FIGURE 1.3
Logical channels.

Traffic Channels

Traffic channels convey user information streams including data and voice. Two types of traffic channels are specified:

- *Dedicated Traffic Channel (DTCH).* The DTCH conveys user information. It may be defined in one or both directions.
- *Random Traffic Channel (RTCH).* The RTCH conveys packet-type data user information. It is usually defined in one direction.

Control Channels

Control channels convey signaling information related to call management, mobility management, and radio resource management. Two groups of control channels are defined—dedicated control channels and common control channels:

- *Dedicated Control Channels (DCCH).* A DCCH is a point-to-point channel defined in both directions. Two DCCHs are specified:

  Associated Control Channel (ACCH). An ACCH is always allocated with a traffic channel or with an SDCCH.

  Stand-Alone Dedicated Control Channel (SDCCH). An SDCCH is allocated independently of the allocation of a traffic channel.

- *Common Control Channels (CCCH).* A CCCH is a point-to-multipoint or multipoint-to-point channel used to convey signaling information (connectionless messages) for access management purposes. Four types of CCCHs are specified:

  Broadcast Control Channel (BCCH). The BCCH is a downlink channel used to broadcast system information. It is a point-to-multipoint channel listened to by all MSs, from which information is obtained before any access attempt is made.

  Forward Access Channel (FACH). The FACH is a downlink channel conveying a number of system management messages, including enquiries to the MS and radio-related and mobility-related resource assignment. It may also convey packet-type user data.

  Paging Channel (PCH). The PCH is a downlink channel used for paging MSs. A page is defined as the process of seeking an MS in the event that an incoming call is addressed to that MS.

  Random Access Channel (RACH). The RACH is an uplink channel used to convey messages related to call establishment requests and responses to network-originated inquiries.

1.6 Narrowband and Wideband Systems

Wireless systems can be classified according to whether they have a narrowband or wideband architecture. Narrowband systems support low-bit-rate transmission, whereas wideband systems support high-bit-rate transmission. A system is defined as narrowband or wideband depending on the bandwidth of the transmission physical channels with which it operates. The system channel bandwidth is assessed with respect to the coherence bandwidth. The coherence bandwidth is defined as the frequency band within which all frequency components are equally affected by fading due to multipath propagation phenomena. Systems operating with channels substantially narrower than the coherence bandwidth are known as narrowband systems. Wideband systems operate with channels substantially wider than the coherence bandwidth. In narrowband systems, all the components of the signal are equally influenced by multipath propagation. Accordingly, although with different amplitudes, the received narrowband signal is essentially the same as the transmitted narrowband signal. In wideband systems, the various frequency components of the signal may be differently affected by fading. Narrowband systems, therefore, are affected by nonselective fading, whereas wideband systems are affected by selective fading.

The coherence bandwidth, B_c, depends on the environment. It is approximately given by

$$B_c = (2\pi T)^{-1}$$

in hertz, where T, in seconds, is the delay spread, as defined next. In a fading environment, a propagated signal arrives at the receiver through multiple paths. The time span between the arrival of the first and the last multipath signals that can be sensed by the receiver is known as delay spread. The delay spread varies from tenths of microseconds, in rural areas, to tens of microseconds, in urban areas. As an example, consider an urban area where the delay spread is $T = 5\mu s$. In such an environment, the coherence bandwidth is calculated as $B_c = 32$ kHz. Therefore, a system is considered to be narrowband if it operates with channels narrower than 32 kHz. It is considered to be wideband if it operates with channels several times wider than 32 kHz.

Another important definition within this context concerns coherence time. The coherence time, T_c, is defined as the time interval during which the fading characteristics of the channel remain approximately unchanged (slow change). This is approximately given as

$$T_c = (2f_m)^{-1}$$

where f_m is the maximum Doppler shift. The Doppler shift, in hertz, is given as v/λ, where v, in m/s, is the speed of the mobile terminal and λ, in m, is the wavelength of the signal.

1.7 Multiple Access

Wireless networks are multiuser systems in which information is conveyed by means of radio waves. In a multiuser environment, access coordination can be accomplished via several mechanisms: by insulating the various signals sharing the same access medium, by allowing the signals to contend for the access, or by combining these two approaches. The choice for the appropriate scheme must take into account a number of factors, such as type of traffic under consideration, available technology, cost, complexity. Signal insulation is easily attainable by means of a scheduling procedure in which signals are allowed to access the medium according to a predefined plan. Signal contention occurs exactly because no signal insulation mechanism is used. Access coordination may be carried out in different domains: the frequency domain, time domain, code domain, and space domain. Signal insulation in each domain is attained by splitting the resource available into nonoverlapping slots (frequency slot, time slot, code slot, and space slot) and assigning each signal a slot. Four main multiple access technologies are used by the wireless networks: frequency division multiple access (FDMA), time division multiple access (TDMA), code division multiple access (CDMA), and space division multiple access (SDMA).

1.7.1 Frequency Division Multiple Access

FDMA is certainly the most conventional method of multiple access and was the first technique to be employed in modern wireless applications. In FDMA, the available bandwidth is split into a number of equal subbands, each of which constitutes a physical channel. The channel bandwidth is a function of the services to be provided and of the available technology and is identified by its center frequency, known as a carrier. In single channel per carrier FDMA technology, the channels, once assigned, are used on a non-time-sharing basis. Thus, a channel allocated to a given user remains allocated until the end of the task for which that specific assignment was made.

1.7.2 Time Division Multiple Access

TDMA is another widely known multiple-access technique and succeeded FDMA in modern wireless applications. In TDMA, the entire bandwidth is made available to all signals but on a time-sharing basis. In such a case, the communication is carried out on a buffer-and-burst scheme so that the source information is first stored and then transmitted. Prior to transmission, the information remains stored during a period of time referred to as a frame. Transmission then occurs within a time interval known as a (time) slot. The time slot constitutes the physical channel.

1.7.3 Code Division Multiple Access

CDMA is a nonconventional multiple-access technique that immediately found wide application in modern wireless systems. In CDMA, the entire bandwidth is made available simultaneously to all signals. In theory, very little dynamic coordination is required, as opposed to FDMA and TDMA in which frequency and time management have a direct impact on performance. To accomplish CDMA systems, spread-spectrum techniques are used. (Appendix C introduces the concept of spread spectrum.)

In CDMA, signals are discriminated by means of code sequences or signature sequences, which correspond to the physical channels. Each pair of transmitter–receivers is allotted one code sequence with which a communication is established. At the reception side, detection is carried out by means of a correlation operation. Ideally, the best performance is attained with zero cross-correlation codes, i.e., with orthogonal codes. In theory, for a synchronous system and for equal rate users, the number of users within a given bandwidth is dictated by the number of possible orthogonal code sequences. In general, CDMA systems operate synchronously in the forward direction and asynchronously in the reverse direction. The point-to-multipoint characteristic of the downlink facilitates the synchronous approach, because one reference channel, broadcast by the base station, can be used by all mobile stations within its service area for synchronization purposes. On the other hand, the implementation of a similar feature on the reverse link is not as simple because of its multipoint-to-point transmission characteristic. In theory, the use of orthogonal codes eliminates the multiple-access interference. Therefore, in an ideal situation, the forward link would not present multiple-access interference. The reverse link, in turn, is characterized by multiple-access interference. In practice, however, interference still occurs in synchronous systems, because of the multipath propagation and because of the other-cell signals. The multipath phenomenon produces delayed and attenuated replicas of the signals, with these signals then losing the synchronism and, therefore, the orthogonality. The other-cell signals, in turn, are not time-aligned with the desired signal. Therefore, they are not orthogonal with the desired signal and may cause interference.

Channels in the forward link are identified by orthogonal sequences, i.e., channelization in the forward link is achieved by the use of orthogonal codes. Base stations are identified by pseudonoise (PN) sequences. Therefore, in the forward link, each channel uses a specific orthogonal code and employs a PN sequence modulation, with a PN code sequence specific to each base station. Hence, multiple access in the forward link is accomplished by the use of spreading orthogonal sequences. The purpose of the PN sequence in the forward link is to identify the base station and to reduce the interference. In general, the use of orthogonal codes in the reverse link finds no direct application, because the reverse link is intrinsically asynchronous. Channelization in the reverse link is achieved with the use of long PN sequences combined with some private identification, such as the electronic serial number of the mobile station. Some systems, on the other hand, implement some sort of synchronous transmission on the reverse link, as shall be detailed in the chapters that follow. In such a case, orthogonal codes may also be used with channelization purposes in the reverse link.

Several PN sequences are used in the various systems, and they will be detailed for the several technologies described in the following chapters. Two main orthogonal sequences are used in all CDMA systems: Walsh codes and orthogonal variable spreading functions (OVSF) (see Appendix C).

1.7.4 Space Division Multiple Access

SDMA is a nonconventional multiple-access technique that finds application in modern wireless systems mainly in combination with other multiple-access techniques. The spatial dimension has been extensively explored by wireless communications systems in the form of frequency reuse. The deployment of advanced techniques to take further advantage of the spatial dimension is embedded in the SDMA philosophy. In SDMA, the entire bandwidth is made available simultaneously to all signals. Signals are discriminated spatially, and the communication trajectory constitutes the physical channels. The implementation of an SDMA architecture is based strongly on antennas technology coupled with advanced digital signal processing. As opposed to the conventional applications in which the locations are constantly illuminated by rigid-beam antennas, in SDMA the antennas should provide for the ability to illuminate the locations in a dynamic fashion. The antenna beams must be electronically and adaptively directed to the user so that, in an idealized situation, the location alone is enough to discriminate the user.

FDMA and TDMA systems are usually considered to be narrowband, whereas CDMA systems are usually designed to be wideband. SDMA systems are deployed together with the other multiple-access technologies.

1.8 Summary

Wireless networks are multiuser systems in which information is conveyed by radio waves. Modern wireless networks have evolved through different generations: 1G systems, based on analog technology, aimed at providing voice telephony services; 2G systems, based on digital technology, aimed at providing a better spectral efficiency, a more robust communication, voice privacy, and authentication capabilities; 2.5G systems, based on 2G systems, aimed at providing the 2G systems with a better data rate capability; and 3G systems that aim at providing for multimedia services in their entirety.

References

1. Framework for the radio interface(s) and radio sub-system functionality for International Mobile Telecommunications-2000 (IMT-2000), Recommendation ITU-R M.1035.
2. The international intelligent network (IN), The International Engineering Consortium, available at http://www.iec.org.

2

Cellular Principles

2.1 Introduction

The electromagnetic spectrum is a limited but renewable resource that, if adequately managed, can be reused to expand wireless network capacity. *Frequency reuse,* in fact, constitutes the basic idea behind the cellular concept. In a cellular system, the service area is divided into *cells* and portions of the available spectrum are conveniently allocated to each cell. The main purpose of defining cells in a wireless network is to delimit areas within which channels or base stations are used at least preferentially. A cell, therefore, is defined as the geographic area where a mobile station is preferentially served by its base station. A mobile station moving out of its serving cell and into a neighboring cell must be provided with sufficient resources from these cells so that the already established communication will not be discontinued. Such a process is known as *handoff* or *handover*. A group of cells among which the whole spectrum is shared and within which no frequency reuse exists constitutes a *cluster*. The number of cells per cluster defines the *reuse pattern,* and this is a function of the cellular geometry. In an ideal situation, for omnidirectional transmission with antennas mounted high above the rooftops, mobile stations at the same distance from the base station receive the same mean signal power in all directions. In such a case, the cell shape can be defined as a *circle*. Its radius is determined so as to have a circular area within which base station and mobile stations receive a signal power exceeding a given threshold. Circles, on the other hand, cannot fill a plane without leaving gaps (holes) or exhibiting overlapped areas. The use of a circular geometry may impose difficulties in the design of a cellular network. Regular polygons, such as equilateral *triangles, squares,* and regular *hexagons* do not exhibit these constraints. The choice for one or another cellular format depends on the application. In practice, the coverage area differs substantially from the idealized geometric figures and "amoeboid" cellular shapes are more likely to occur.

This chapter defines the issues related to the cellular concepts. The main definitions that follow are based on an ITU Recommendation for IMT-2000.[1] The concepts developed in Reference 1 generalize those that have been used for conventional cellular networks.

2.2 Cellular Hierarchy

To maximize spectral efficiency as well as to minimize the number of handovers, it is beneficial for the cells to be designed with different sizes and formats. The design of different cells depends on several parameters, such as mobility characteristics, output power, and types of services utilized. Cellular layers are then defined with each layer containing cells of the same type in a given service area. The layering of cells does not imply that all mobile stations must be able to connect to all base stations serving the geographic area where the mobile station is positioned. For example, the mobile station may not have sufficient output power to access a given layer or may not be entitled to the service provided by the cells of a given layer. The cellular hierarchy makes use of four categories of cells: mega cells, macro cells, micro cells, and pico cells.[1]

Mega cells provide coverage to large areas and are characterized by cells presenting radii in the range 100 to 500 km. They are particularly useful for remote areas with low traffic density or for areas without access to terrestrial telecommunications networks. Mega cells are provided by low-orbit satellites and the cell radius is a function of the satellite altitude, power, and antenna aperture. Note that in a mega cell the distances between mobile stations and the base station are very large. Because of their sizes, these cells must be both flexible and robust to accommodate a wide range of user scenarios. They must be able to support low-mobility as well as very high-mobility users. Note that for nongeostationary orbits, the cells move because the satellites move with respect to the Earth. Therefore, handovers may be necessary even for stationary mobile stations.

Macro cells provide coverage to large areas and are characterized by cells presenting radii of up to 35 km. Larger cell radii may be provided with the use of directional antennas. The macro cells are outdoor cells that are illuminated by high-power sites with the antennas mounted above the rooftops—on towers or on the tops of buildings. They serve low to medium traffic density and support mobile speeds of up to 500 km/h.

Micro cells provide coverage to small areas and are characterized by cells presenting radii of up to 1 km. They are outdoor cells that are illuminated by low-power antennas with the antennas mounted below the rooftops—on lampposts or on building walls. They support medium to high traffic density and mobile speeds of up to 100 km/h.

Pico cells provide coverage to small areas and are characterized by cells presenting radii of up to 50 m. They are indoor cells supporting medium to high traffic density and mobile speeds of up to 10 km/h.

In the real world, mega cells, macro cells, micro cells, and pico cells coexist in the same environment. Digital technology has made it possible for wireless systems to take full advantage of such a coexistence so that coverage is improved, capacity is increased, load is balanced, and users are provided with different services according to the mobility characteristics. More generally, pico, micro, macro, and mega cells are displaced in a hierarchy, the so-called hierarchical cellular structure (HCS). In HCS wireless systems, very low to very high mobility and in-building to satellite coverage provide for the multimedia–anywhere–anytime wireless services. In HCS, several layers of cells may coexist with the smallest cells occupying the lowest layer in the hierarchy. The mobility and the class of service of the user determine the layer within which the required service is to be provided. In a multilayered cellular environment, the selection of which cell to serve a given call should be based on criteria such as speed of the mobile station relative to the base station, cell availability, and required transmission power to and from the mobile station.

2.3 System Management

The phases of a communication between mobile station and base station encompass the establishment, maintenance, and release of the connection. The management of these phases is carried out by several functions. These functions include link quality measurement, cell selection, channel selection/assignment, handover, and mobility support.[1]

2.3.1 Link Quality Measurement

During any given connection, forward and reverse links are continually monitored to assess the radio link quality. The assessment is based on parameters such as the received signal quality and the bit error rates.

2.3.2 Cell Selection

In advanced wireless networks, cell selection is a feature that can be provided. Cell selection may be based on several criteria, including mobility and class of service to be provided. It starts with the choice of the operator, a phase that occurs as the mobile station is powered up. The selection of the operator may be based on user preferences, available networks, mobile station capabilities, network capabilities, mobile station mobility, and service requirements. Once a system has been selected by the mobile station, a base station is then searched and its broadcast control channel monitored. Cell reselection may also occur, and the following circumstances may trigger the reselection: unsuitability of current cell due to interference or output power requirements, radio link failure, network request, traffic load considerations, and user request.

2.3.3 Channel Selection/Assignment

Channel assignment algorithms are used to ascertain conveniently the available channels and to assign one or more of these channels to a call. The algorithms vary in accordance with specific allocation policies, but they usually take into account the following: system load, traffic patterns, service types, service priorities, and interference situations. Channel assignment algorithms are added-on features that differ for different system providers.

2.3.4 Handover

Handover is defined as "the change of Physical Channel(s) involved in a call whilst maintaining the call."[1] Handover constitutes a diversity technique used to prevent mobile calls from being released when the mobile stations experience a degraded radio condition. Many factors affect the received signal quality, one of which is the distance between mobile and base stations. A signal degradation may occur, particularly when the mobile stations cross the cell boundaries. Handovers may take place in several conditions: within the cell (intracell handover), between cells in the same cell layer (intercell handover), between cells of different layers (interlayer handover), or between cells of different networks (internetwork handover).

In FDMA and TDMA wireless networks, handovers are "hard" (hard handover). In hard handover, the communication with the old base station through a given channel is discontinued, and a new communication with a new base station, and necessarily through another channel, is established. Internetwork handovers and handovers between systems of different technologies are always hard handover.

In CDMA wireless networks, handovers are "soft," and three kinds of soft-type handovers are identified: soft handover, softer handover, and soft-softer handover. In soft handover, the mobile station maintains communication simultaneously with the old base station and with one or more new base stations, provided that all base stations have CDMA channels with an identical frequency assignment. Note that the soft handover constitutes a means of diversity for both the forward and reverse paths, and is supposed to occur in the vicinities of the boundary of the cell, where the signal is presumably weaker. In softer handover, the mobile station maintains communication simultaneously with two or more sectors of the same base station and certainly within the same CDMA channel of that base station. Similar to the soft handover, the softer handover also constitutes a means of diversity and is supposed to occur in the vicinities of the boundary of the coverage area of the sectors. In soft-softer handover, the mobile station maintains communication simultaneously with two or more sectors of the same base station and with one or more base stations (or their sectors).

Note that, whereas in the hard-type handover the number of resources utilized remains unchanged, in the soft-type handover this is not true. In the hard-type handover, the new resource is utilized only after the old resource has been released. In the soft-type handover, a communication with more than one resource is maintained throughout the duration of the process. However, in the uplink direction only one channel is used by the mobile station. In such a case, the involved base station receives this very channel and the selection of the best communication is performed by the mobile switching center. In the downlink direction, on the other hand, each base station supporting the handover transmits to the same mobile station, and the combination of the received signals is carried out by the mobile station. Therefore, the involved base stations must provide for additional channels for soft handover purposes, with these additional channels used in the forward direction only.

From the transmission point of view, the handover process comprises two phases: the evaluation phase and the execution phase. In the evaluation phase, the rationale for performing a handover is continually assessed. Handovers may be initiated for such reasons as operation and maintenance, radio channel capacity optimization, poor radio transmission conditions, signal level variability, significant amount of interference, etc. The following criteria may be used to initiate a handover for radio transmission reasons: signal strength measurements, signal-to-interference ratio, bit error rates, distance between mobile station and base station, mobile station speed, mobile station mobility trends, and others. The handover is actually performed in the execution phase. A handover may be initiated by the base station or by the mobile station.

Three handover strategies are possible: mobile-controlled handover (MCHO); mobile-assisted handover (MAHO); and base-controlled handover (BCHO). In MCHO, the mobile station controls the handover evaluation phase as well as the handover execution phase. In MAHO, the base station controls the handover process with the support of the mobile station (e.g., measurements carried out by the mobile station). In BCHO, the base station controls the handover evaluation phase as well as the handover execution phase.

2.3.5 Mobility Support

User mobility is supported by the following processes: logon–logoff and location updating. In logon–logoff, messages are transmitted from the mobile station to the network to notify the network of the terminal status. Such a procedure may be initiated by the mobile station as well as by the network. In location updating, messages are exchanged between the mobile station and the network to identify the area within which the mobile station is located. A location area is defined as the geographic area, containing a group of cells, in which the mobile station is to be sought by the network. The location update is performed whenever a mobile station moves into a new location area. Note that the smaller the location area, the higher the chance of locating the mobile station within the network. On the other hand, the smaller the location area, the higher the frequency with which the exchange of updating messages must be carried out.

2.4 System Performance

Several aspects that affect the performance of the system must be addressed: interference control, diversity strategies, variable data rate control, capacity improvement techniques, and battery-saving techniques.

2.4.1 Interference Control

Synchronization is certainly one of the issues that must be examined for interference control purposes. Some technologies require that base stations belonging to the same system must be synchronized. This is particularly true for TDMA systems and some CDMA systems. In the same way, time synchronization between base stations in different but geographically co-located systems and time synchronization between user terminals and base stations are elements that have a great impact on interference. Power control is another important issue in interference control. Power control must be exercised because of the near–far phenomenon, a feature inherent to mobile communications. Because of the mobility feature, the powers of desired and interfering signals may vary according to the location of the mobile stations. Power control must be performed so that intra- and intersystem interference is minimized. The near–far phenomenon is more relevant in the multipoint-to-point transmission (mobile stations to base station) than in the point-to-multipoint one (base station to mobile stations). In the first case (multipoint-to-point), because the mobile stations may be at different distances from the base station, the various signals arriving at the base will have different strengths. In the second case (point-to-multipoint), the various signals transmitted by the base reach a given mobile station with approximately the same power loss, thus maintaining power proportionality. In practice, however, both reverse link and forward link require power control: the reverse link for the reasons already outlined, and the forward link to compensate for poor reception conditions encountered by the mobile station.

2.4.2 Diversity Strategies

Diversity strategies are used to combat fading. The diversity methods take advantage of the fact that fades occurring on independent channels (known as branches) constitute independent events. Therefore, if certain information is redundantly available on two or more branches, simultaneous fades are less likely to occur. By appropriately combining the various branches, the quality of the received signal is improved. Diversity may be achieved in several ways, such as in space (spaced antennas), frequency, and time.

2.4.3 Variable Data Rate Control

Variable data rates may be accomplished by several means: direct support of variable data rates over the air interface, variation of the number of bearer channels so that multiple bearer channels are combined to deliver the desired user rate, or packet access. Different wireless networks use different variable data rate technologies.

2.4.4 Capacity Improvement Techniques

Network capacity may be improved by means of such techniques as slow frequency hopping; dynamic power control; dynamic channel allocation; discontinuous transmission for voice, including voice activity detection, and nonvoice services; and others. The applicability of one or another technique is dependent on the multiple-access technology chosen.

2.4.5 Battery-Saving Techniques

Digital technologies facilitate the use of battery-saving techniques. These techniques include output power control, discontinuous reception, and discontinuous transmission.

2.5 Cellular Reuse Pattern

For quite a while, since the inception of modern wireless networks, the cellular grid has been dominated by macro cells. The macrocellular network makes use of a hexagonal cell array with the reuse pattern established with the supposition that reuse distances are isotropic and that a cluster comprises a group of contiguous cells. In theory, high-power sites, combined with base station antennas positioned well above the rooftops, provide for propagation symmetry, in which case, for system planning purposes, the hexagonal coverage grid has proved appropriate. Further, macro cells are adequate for low-capacity systems.

The expansion and the evolution of wireless networks can only be supported by an ample microcellular structure, not only to satisfy the high traffic demand in dense urban regions but also to provide for services requiring low mobility. The microcellular network concept differs from that of the macrocellular concept widely employed in wireless systems. In microcellular systems, with low-power sites and antennas mounted at street level (below the rooftops), the supposed propagation symmetry of the macrocellular network no longer applies and the hexagonal cell pattern does not make sense. The "microscopic" structure of the environment (e.g., street orientation, width of the streets, layout of the buildings, among others) constitutes a decisive element influencing system performance. With the antennas mounted at street level, the buildings lining each side of the street work as "waveguides," in the radial direction, and as obstructors, in the perpendicular direction. Therefore, the propagation direction of the radio waves is greatly influenced by the environment. Assuming the base stations to be positioned at the intersection of the streets, a cell in such an environment is more likely to have a diamond shape with the radial streets as the diagonals. In fact, a number of field measurements and investigations[2–8] show that an urban micro cell service area can be reasonably well approximated by a square diamond.

The ubiquitous coverage of a service area based on a microcellular network requires a much greater number of base stations as compared with the number of base stations required by macrocellular systems. Therefore, among the important factors to be accounted for in microcellular system planning (cost of the site, size of the service area, etc.), the per-subscriber cost is determinant. This cost is intimately related to how efficiently the radio resources are reutilized in a given service area. Reuse efficiency depends on the interfering environment of the network and on how the involved technology can cope with the interfering sources.

The study of interference in macrocellular systems is greatly eased by the intrinsic symmetry of the problem. In the microcellular case, the inherent asymmetry due to the microscopic structures introduces an additional complication. In such a case, the interference is dependent not only on the distance between transmitter and receiver but also, and mainly, on the line-of-sight (LOS) condition of the radio path. Assume, for example, that base stations are located at street intersections. Mobiles on streets running radially from the base station may experience an interference pattern changing along the street as they depart from the vicinity of their serving base station and approach new street intersections. Near the serving base station, the desired signal is strong and the relevant interfering signals are obstructed by buildings (non-LOS, or NLOS). Away from its serving base station and near new street intersections, mobile stations may have an LOS condition not only to their serving base station but also to the interfering stations. The interfering situation will then follow a completely distinct pattern on the perpendicular streets. Again, the asymmetry of the problem is stressed by the traffic distribution, which is more likely to comply with an uneven configuration with the main streets accommodating more mobile users than the secondary streets.

2.6 Macrocellular Reuse Pattern

A macrocellular structure makes use of a hexagonal cellular grid. For the hexagonal array, it is convenient to choose the set of coordinates as shown in Figure 2.1. In Figure 2.1, the positive portions of the u and v axes form a 60° angle and the unit distance is $\sqrt{3}R$, where R is the cell radius. The distance d between the centers of two cells, whose coordinates are, respectively, (u_1, v_1)

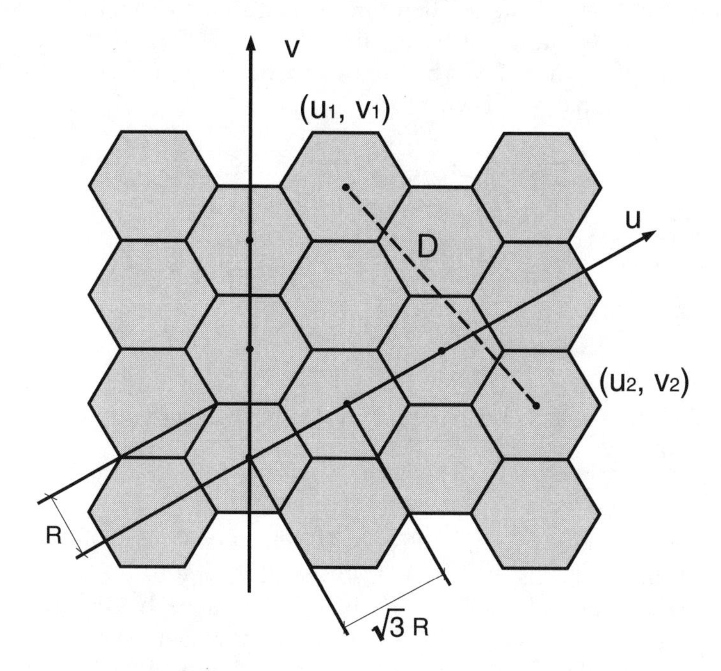

FIGURE 2.1
Hexagonal cellular geometry.

and (u_2, v_2), is

$$d^2 = 3R^2[(u_1 - u_2)^2 + (v_2 - v_1)^2 - 2\cos(60^\circ)(u_1 - u_2)(v_2 - v_1)]$$

Defining $i = u_2 - u_1$, $j = v_2 - v_1$, and $\sqrt{3}R = 1$, then

$$d^2 = i^2 + ij + j^2 \tag{2.1}$$

where i and j range over the integers.

2.6.1 Reuse Factor (Number of Cells per Cluster)

Let D be the reuse distance, that is, the distance between two co-cells. Therefore, for any given co-cells

$$D^2 = i^2 + ij + j^2 \tag{2.2}$$

Considering that reuse distances are isotropic and that a cluster is a group of contiguous cells, the format of the clusters must be hexagonal. For two adjacent clusters, D represents the distance between the centers of any two co-cells within these clusters. Therefore, D is also the distance between the centers of these two adjacent hexagonal clusters, $D/2$ their apothems, and $(D/2)/\cos 30^\circ$ their radii. Let A be the area of the hexagonal cluster and a the area of the hexagonal cell. Then,

$$A = \frac{3\sqrt{3}}{2} \times \left(\frac{D/2}{\cos 30^\circ}\right)^2 = \frac{\sqrt{3}D^2}{2}$$

and

$$a = \frac{3\sqrt{3}}{2} \times R^2 = \frac{\sqrt{3}}{2}$$

The number N of cells per cluster is $N = A/a = D^2$. Therefore,

$$N = i^2 + ij + j^2 \tag{2.3}$$

Because i and j range over the integers, the clusters will accommodate only a certain number of cells, such as 1, 3, 4, 7, 9, 12, 13, 16, 19, etc.

The layout of the cells within the cluster is attained having as principal targets symmetry and compactness. Figure 2.2 shows some hexagonal repeat patterns. The tessellation over the entire plane is then achieved by replicating the cluster in an isotropic manner. In other words, if the chosen reuse pattern

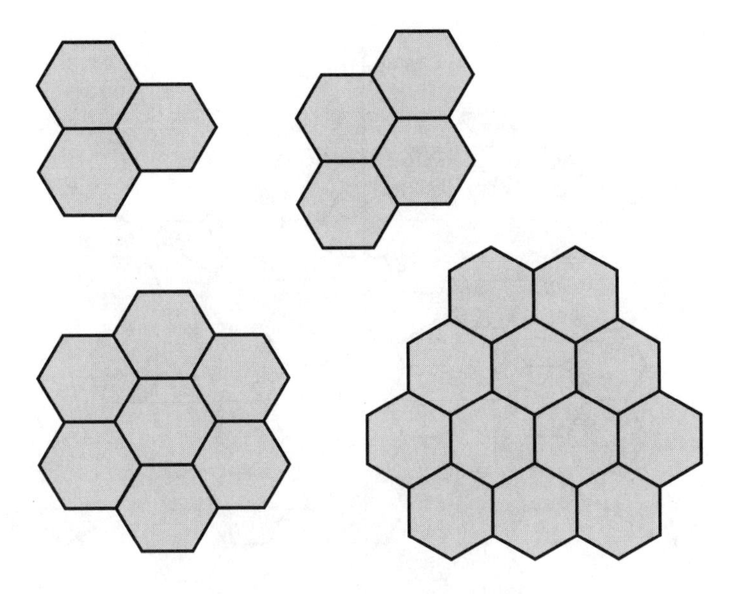

FIGURE 2.2
Some hexagonal repeat patterns ($N = 3, 4, 7, 12$ macro cells per cluster).

is such that $i = p$ and $j = q$, then for the reference cell located at the coordinates (0, 0) the six corresponding equidistant co-cells are positioned at (p, q), $(p + q, -p)$, $(-q, p + q)$, $(-p, -q)$, $(-p - q, p)$, and $(q, -p - q)$. The six surrounding adjacent clusters are then replicated from these reference co-cells. Figure 2.3 illustrates this for the case $(i, j) = (1, 2)$, and therefore reuse of $N = 7$. An equivalent procedure is to use a system of six positive axes i, j, k, l, m, n with each consecutive pair of axes i and j, j and k, k and l, l and m, m and n, and n and i at 60°, as shown in Figure 2.3. The coordinates (p, q) are then positioned using the six possible pairs of consecutive axes, i.e., $(i, j) = (p, q)$, $(j, k) = (p, q)$, $(k, l) = (p, q)$, $(l, m) = (p, q)$, $(m, n) = (p, q)$, and $(n, i) = (p, q)$, as illustrated in Figure 2.3 for the reuse of 7.

2.6.2 Reuse Ratio

An important parameter in the cellular systems is the co-channel reuse ratio. This is defined as D/R, the ratio between the co-channel reuse distance and the cell radius. From the above equations

$$\frac{D}{R} = \sqrt{3N} \tag{2.4}$$

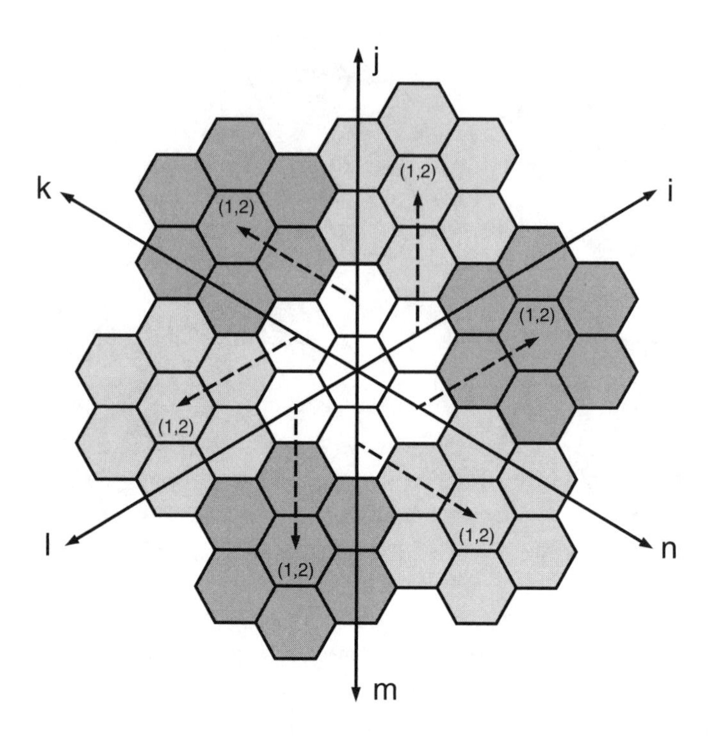

FIGURE 2.3
Partial hexagonal tessellation for reuse of 7.

The reuse ratio gives a qualitative measure of the signal quality (carrier-to-inter ference ratio) as a function of the cluster size. The larger the reuse ratio, the better the signal quality. On the other hand, the larger the reuse distance, the smaller the spectrum efficiency (fewer channels per cell).

2.6.3 Positioning of the Co-Cells

Given a reference cell, concentric rings, or tiers, of co-cells are formed around the reference cell. There are $6n$ co-cells on the nth tier, among which 6 are exactly at a distance of nD and the remaining $6(n-1)$ are at distances slightly smaller than nD. For a reference cell centered at (0, 0) of an orthogonal system of x and y axes, the $(x_l(n), y_l(n))$ coordinates of those 6 co-cells are

$$x_l(n) = nD\cos(\theta_l) \tag{2.5}$$

$$y_l(n) = nD\sin(\theta_l) \tag{2.6}$$

where $l = 1, 2, \ldots, 6$, $\theta_l = \phi(N) + (l - 1)\pi/3$, and $\phi(N)$ is a function of the cluster size. It is given by $\phi(N) = 0, \pi/6, \cot^{-1}(\sqrt{3}), \pi/6,$ and 0 for $N = 3, 4, 7, 9,$ and 12, respectively.

2.7 Microcellular Reuse Pattern

A microcellular structure makes use of a square cellular grid.[8] For the square array, it is convenient to choose the set of orthogonal coordinates as shown in Figure 2.4. In Figure 2.4, the orthogonal axes are u and v and the unit distance is $\sqrt{2}R$, where R is the cell radius. The distance d between the centers of two cells, whose coordinates are, respectively, (u_1, v_1) and (u_2, v_2), is

$$d^2 = 2R^2[(u_2 - u_1)^2 + (v_2 - v_1)^2]$$

Defining $i = u_2 - u_1$, $j = v_2 - v_1$, and $\sqrt{2}R = 1$, then

$$d^2 = i^2 + j^2 \tag{2.7}$$

where i and j range over the integers.

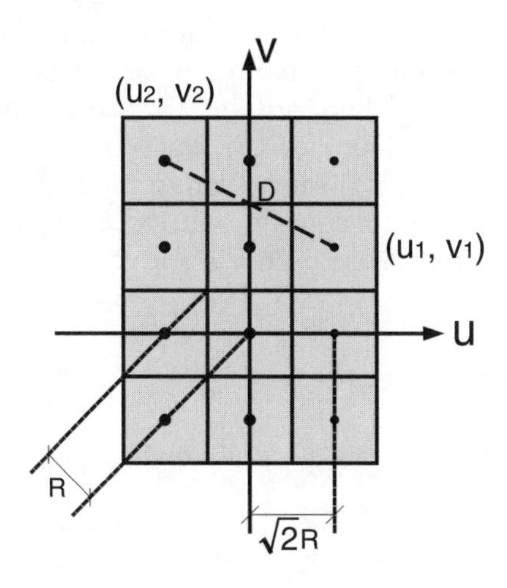

FIGURE 2.4
Square cellular geometry.

2.7.1 Reuse Factor (Number of Cells per Cluster)

Let D be the reuse distance, that is, the distance between two co-cells. Therefore, for any given co-cells

$$D^2 = i^2 + j^2 \tag{2.8}$$

Considering that reuse distances are isotropic and that a cluster is a group of contiguous cells, the format of the clusters must be a square. For two adjacent clusters D represents the distance between the centers of any two co-cells within these clusters. Therefore, D is also the distance between the centers of these two adjacent square clusters and D is also their sides. Let A be the area of the square cluster and a the area of the square cell. Then,

$$A = D^2$$

and

$$a = (2R\cos 45^\circ)^2 = 2R^2 = 1$$

The number N of cells per cluster is $N = A/a = D^2$. Therefore,

$$N = i^2 + j^2 \tag{2.9}$$

Because i and j range over the integers, the clusters will accommodate only a certain number of cells, such as 1, 2, 4, 5, 8, 9, 10, 13, 16, etc.

The layout of the cells within the cluster is attained having as principal targets symmetry and compactness. Figure 2.5 shows some square repeat patterns. The tessellation over the entire plane is then achieved by replicating the cluster in an isotropic manner. In other words, if the chosen reuse pattern is such that $i = p$ and $j = q$, then for the reference cell located at the coordinates (0, 0) the four corresponding equidistant co-cells are positioned at (p, q), $(-q, p)$, $(-p, -q)$, and $(q, -p)$. The four surrounding adjacent clusters are then replicated from these reference co-cells. Figure 2.6 illustrates this for the case $(i, j) = (1, 2)$, and therefore reuse of $N = 5$; and for the case $(i, j) = (2, 3)$, and therefore reuse of $N = 13$.

2.7.2 Reuse Ratio

The reuse ratio for the square grid is obtained using the above equations. Thus,

$$\frac{D}{R} = \sqrt{2N} \tag{2.10}$$

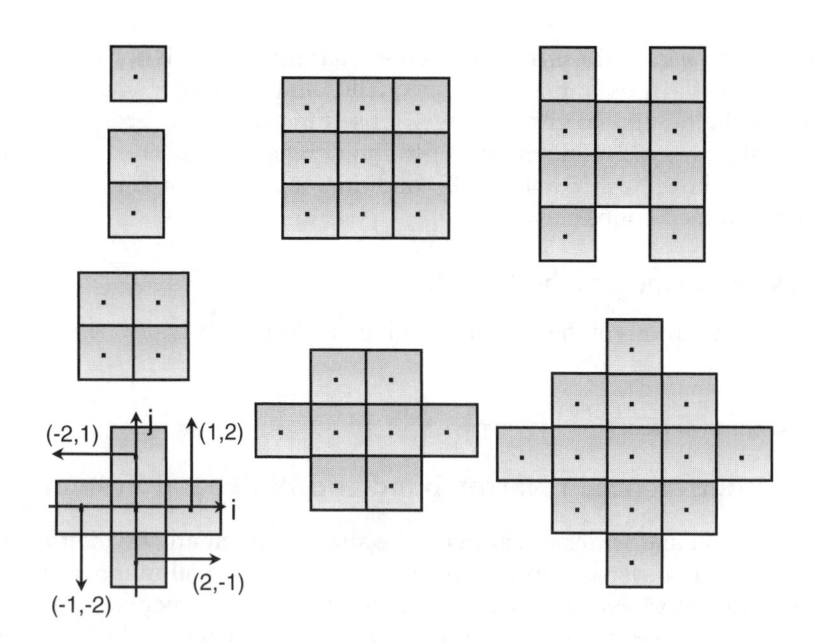

FIGURE 2.5
Square repeat patterns (N = 1, 2, 4, 5, 8, 9, 10, 13 micro cells per cluster).

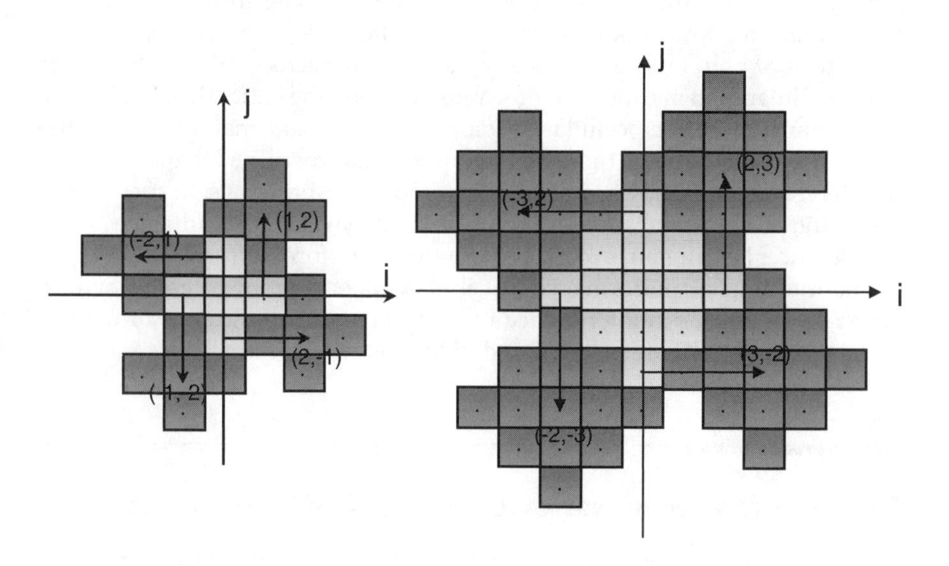

FIGURE 2.6
Partial square tessellation for reuse of 5 and reuse of 13.

Because of the inherent asymmetry of the square grid, there is not a direct relation among the reuse ratio, the signal quality, and the cluster size, as was the case for the hexagonal grid. A larger reuse ratio does not necessarily yield better signal quality (better carrier-to-interference ratio). This will greatly depend on how the LOS and NLOS conditions are experienced by the mobile stations in the various clusters.

2.7.3 Positioning of the Co-Cells

The exact positions of the co-cells are given in Appendix D.

2.8 Interference in Narrowband and Wideband Systems

Narrowband and wideband systems are affected differently by interference. In narrowband systems, interference is caused by a small number of high-power signals. Moreover, macrocellular and microcellular networks undergo different interference patterns. In addition, whereas in macrocellular systems uplink and downlink present approximately the same interference performance, in microcellular systems the interference performance of uplink and downlink is rather dissimilar. In both cases, the uplink performance is always worse than the downlink performance, but the difference between the performances of both links is drastically different in microcellular systems. For macrocellular systems, the larger the reuse pattern, the better the interference performance. For microcellular systems, it can be said that, *in general*, the larger the reuse pattern, the better the performance. In wideband systems, interference is caused by a large number of low-power signals. In such a case, the traffic profile as well as the channel activity have a great influence on the interference. Here again, uplink and downlink perform differently.

The interference performance of cellular systems is investigated here in terms of the carrier-to-interference ratio (C/I) and the efficiency of the frequency reuse (f). These are explored in the following sections.

2.9 Interference in Narrowband Macrocellular Systems

Propagation in a macrocellular environment is characterized by an NLOS condition. In this case, the mean power P received at a distance d from the transmitter is given as

$$P = Kd^{-\alpha} \tag{2.11}$$

where K is a proportionality constant and α is the propagation path loss coefficient, usually in the range $2 \leq \alpha \leq 6$. The constant K is a function of several parameters including the frequency, the base station antenna height, the mobile station antenna height, the base station antenna gain, the mobile station antenna gain, the propagation environment, and others. For the purposes of the calculations that follow it is assumed that all these parameters remain constant.

The interference performance of narrowband macrocellular systems is investigated here in terms of the C/I parameter and for the mobile station positioned for the worst-case condition, i.e., at the border of the serving cell (distance R from the base station). In the downlink direction, C/I is calculated at the mobile station. In such a case, of interest is investigation of the ratio between the signal power C received from the serving base station and the sum I of the signal powers received from the interfering base stations (co-cells). In the uplink direction, C/I is calculated at the base station. In this case, of interest is investigation of the ratio between the signal power C received from the wanted mobile station and the sum I of the signal powers received from the interfering mobile stations from the various co-cells.

In a macrocellular network, it is convenient to investigate the effects of interference with the use of omnidirectional antennas as well as directional antennas. As already mentioned in a previous subsection, there are $6n$ co-cells on the nth tier of a hexagonal cellular grid. With omnidirectional antennas, therefore, the number of interferers from each tier is given by $6n$ (all possible interferers), where n is the number of the interfering tier (layer). The use of directional antennas reduces the number of interferers by approximately s, the number of sectors used in the cell. With directional antennas, therefore, the number of interferers from the nth tier is reduced to approximately $6n/s$.

2.9.1 Downlink Interference—Omnidirectional Antenna

For the worst-case condition, the mobile station is positioned at a distance R from the base station. In addition, we assume that the $6n$ interfering base stations in the nth ring are approximately at a distance of nD. Therefore, C/I can be estimated as

$$\frac{C}{I} = \frac{R^{-\alpha}}{\sum_{n=1}^{\infty} 6n\,(nD)^{-\alpha}} \tag{2.12}$$

By using the relation $D/R = \sqrt{3N}$,

$$\frac{C}{I} = \frac{(\sqrt{3N})^{\alpha}}{6\Re(\alpha - 1)} \tag{2.13}$$

where $\Re(x) = \Sigma_{n=1}^{\infty} n^{-x}$ is the Riemann function. In particular, $\Re(x) = \infty, \pi^2/6$, 1.2021, and $\pi^4/90$, for $x = 1, 2, 3$, and 4, respectively. A good approximation for C/I is obtained by considering only the first tier ($n = 1$). Then,

$$\frac{C}{I} = \frac{(\sqrt{3N})^{\alpha}}{6} \tag{2.14}$$

For example, the exact C/I calculation for $\alpha = 4$ and $N = 7$ leads to $61.14 = 17.9$ dB, whereas the approximate C/I calculation yields $73.5 = 18.7$ dB.

2.9.2 Uplink Interference—Omnidirectional Antenna

For the worst-case condition, the mobile station is positioned at a distance R from the base station. In addition, assume that the $6n$ interfering mobile stations in the nth ring are approximately at a distance of $nD - R$. (Note that this is the closest distance the mobile station in the nth ring can be with respect to the interfered base station.) Therefore, C/I can be estimated as approximately

$$\frac{C}{I} = \frac{R^{-\alpha}}{\sum_{n=1}^{\infty} 6n\,(nD - R)^{-\alpha}} \tag{2.15}$$

By using the relation $D/R = \sqrt{3N}$,

$$\frac{C}{I} = \left[\sum_{n=1}^{\infty} 6n(n\sqrt{3N} - 1)^{-\alpha}\right]^{-1} \tag{2.16}$$

A good approximation for C/I is obtained by considering only the first tier ($n = 1$). In such a case

$$\frac{C}{I} = \frac{(\sqrt{3N} - 1)^{\alpha}}{6} \tag{2.17}$$

For example, a more exact C/I calculation for $\alpha = 4$ and $N = 7$ leads to $25.27 = 14.0$ dB, whereas the approximate calculation yields $27.45 = 14.38$ dB.

2.9.3 Downlink Interference—Directional Antenna

Following the same procedure as before,

$$\frac{C}{I} = \frac{(\sqrt{3N})^{\alpha} s}{6\Re\,(\alpha - 1)} \tag{2.18}$$

The approximation using the first tier ($n = 1$) yields

$$\frac{C}{I} = \frac{(\sqrt{3N})^{\alpha} s}{6} \tag{2.19}$$

For the same conditions as before ($\alpha = 4$, $N = 7$) and for a three-sector cell system ($s = 3$), the more exact solution yields $C/I = 183.42 = 22.6$ dB, whereas the approximate one gives $C/I = 220.5 = 23.4$ dB.

2.9.4 Uplink Interference—Directional Antenna

Following the same procedure as before,

$$\frac{C}{I} = \left[\sum_{n=1}^{\infty} \frac{6n}{s} (n\sqrt{3N} - 1)^{-\alpha}\right]^{-1} \tag{2.20}$$

A good approximation for C/I is obtained by considering only the first tier. Then,

$$\frac{C}{I} = \frac{(\sqrt{3N} - 1)^{\alpha} s}{6} \tag{2.21}$$

For the same conditions as before ($\alpha = 4$, $N = 7$), the more exact solution yields $C/I = 75.81 = 18.8$ dB, whereas the approximate one gives $C/I = 82.35 = 19.16$ dB.

2.9.5 Examples

Table 2.1 gives some examples of C/I figures for $\alpha = 4$ and for several re-use patterns, with omnidirectional and directional (120° antennas, or three-sectored cells). Note how the use of directional antennas substantially

TABLE 2.1

Examples of C/I for the Various Cluster Sizes in a Macrocellular Environment

	Uplink (dB)		Downlink (dB)	
N	Omni	Directional	Omni	Directional
3	4.0	8.7	10.5	15.3
4	7.5	12.3	13.0	17.7
7	14.0	18.7	17.9	22.7
9	16.7	21.5	20.0	24.7
12	19.8	24.5	22.5	27.3

improves the C/I performance. The choice of one or another pattern depends on how tolerant the technology is of interference. A widely deployed reuse pattern is $N = 7$ with three-sectored cells. This pattern is usually referred to as 7×21. Another widely deployed reuse pattern is $N = 4$ with three-sectored cells. This pattern is usually referred to as 4×12.

2.10 Interference in Narrowband Microcellular Systems

In the performance analysis of the various microcellular reuse patterns, a parameter of interest is the distance between the interferers positioned at the co-cell of the Lth co-cell layer and at the target cell, with the target cell taken as the reference cell.[8] We define such a parameter as n_L and, for ease of manipulation, normalize it with respect to the cell radius, i.e., n_L is given in number of cell radii. We observe that this parameter is greatly dependent on the reuse pattern. It can be obtained by a simple visual inspection, but certainly for a very limited number of cell layers. For the overall case, a more general formulation is required and this is shown in Appendix D.

The performance analysis to be carried out here considers a square cellular pattern with base stations positioned at every other intersection of the streets. This means that base stations are collinear and that each microcell covers a square area comprising four 90° sectors, each sector corresponding to half a block, with the streets running on the diagonals of this square. Figure 2.7 shows

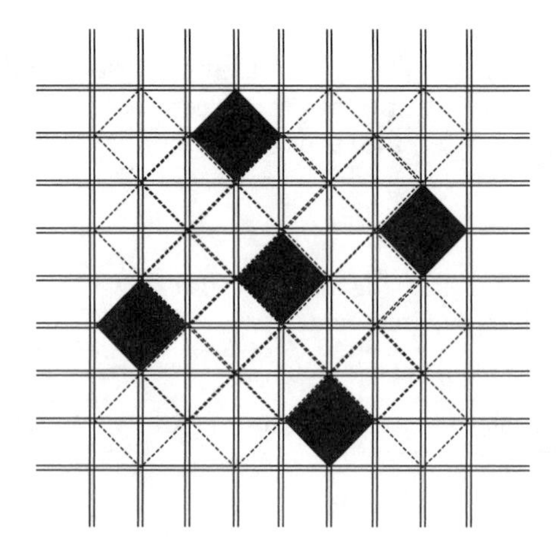

FIGURE 2.7
Microcellular layout in an urban area.

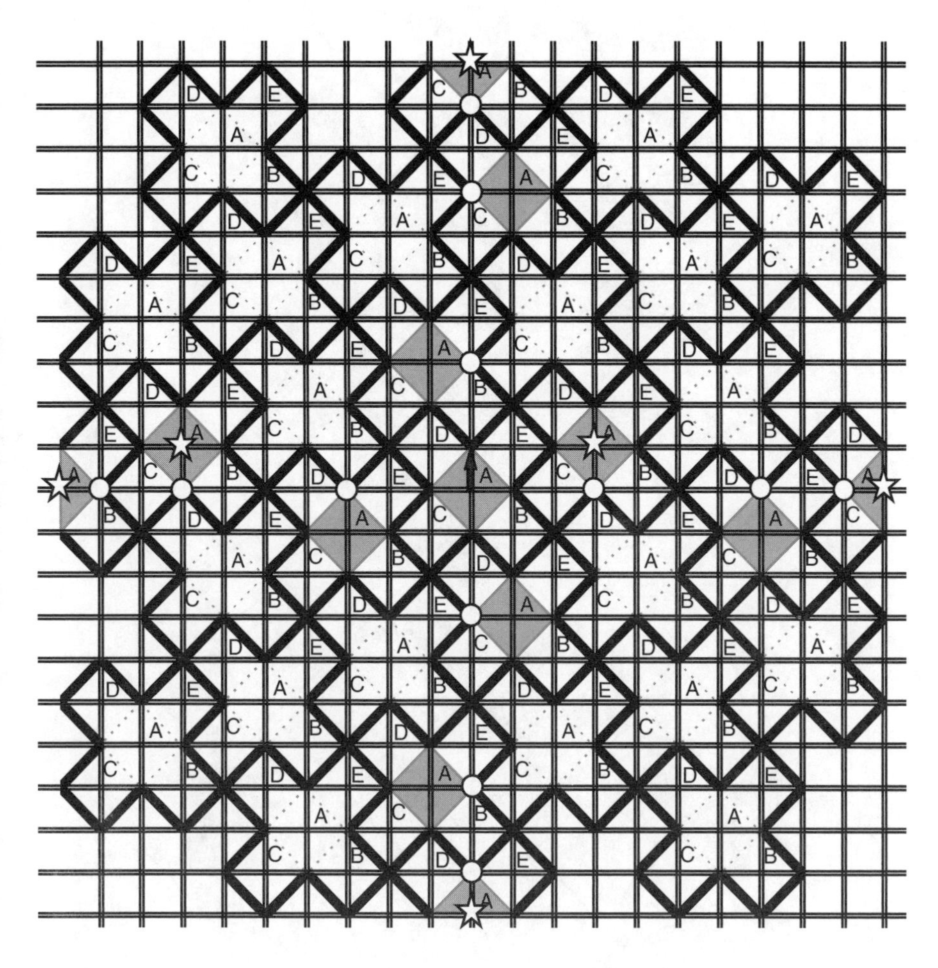

FIGURE 2.8
Five-micro-cell cluster tessellation—prime non-collinear group (see Appendix D).

the microcellular layout with respect to the streets. In Figure 2.7, the horizontal and vertical lines represent the streets and the diagonal lines represent the borders of the micro cells. The central micro cell is highlighted in Figure 2.7.

To provide insight into how the performance calculations are carried out, Figures 2.8 and 2.9 illustrate the complete tessellation for clusters containing 5 (Figure 2.8), 8, 9, 10, and 13 (Figure 2.9) micro cells, in which the highlighted cluster accommodates the target cell, and the other dark cells correspond to the co-micro-cells that at a certain time may interfere with the mobile or base station of interest. Within a microcellular structure, distinct situations are found that affect in a different manner the performance of the downlink

FIGURE 2.9
(a) Eight-micro-cell cluster tessellation—collinear $i = j$ group; (b) nine-micro-cell cluster tessellation—collinear group.

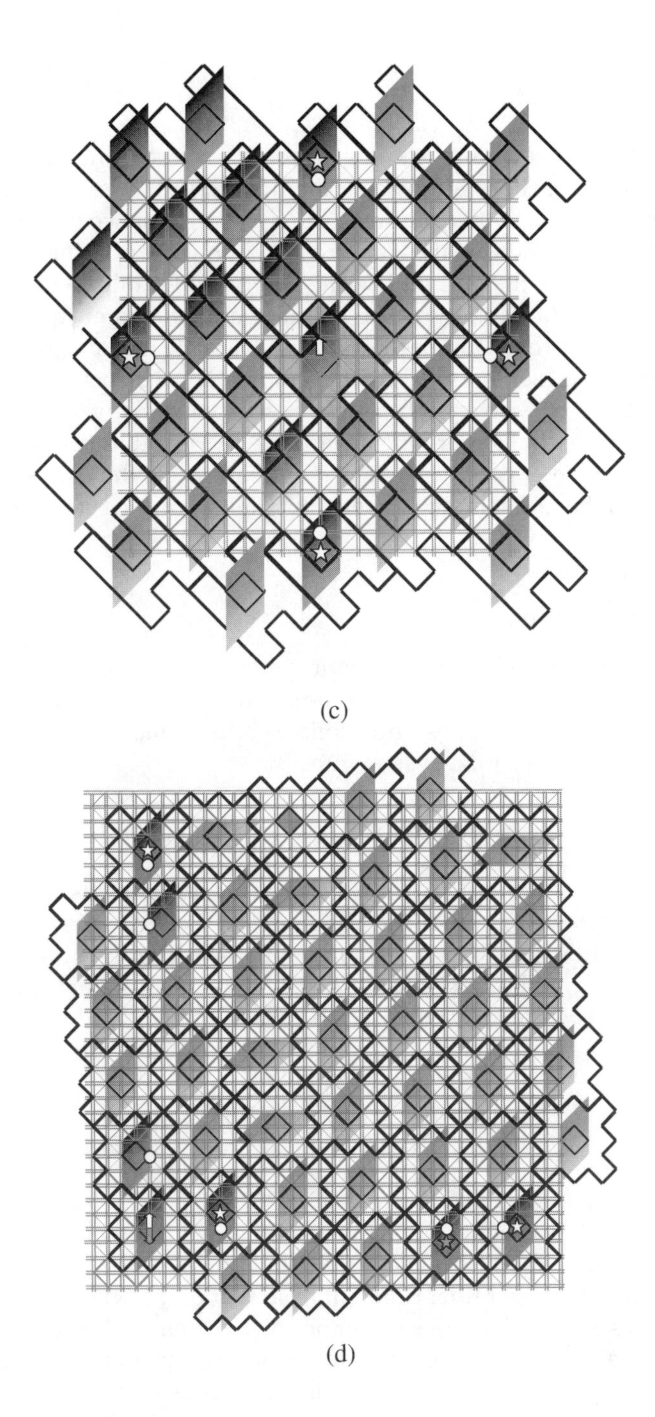

FIGURE 2.9 (continued)
(c) ten-micro-cell cluster tessellation—even noncollinear; (d) 13-micro-cell cluster tessellation—prime noncollinear group (see Appendix D).

and the uplink. In general, the set of micro cells affecting the downlink constitutes a subset of those influencing the uplink. In Figures 2.8 and 2.9, the stars indicate the sites contributing to the C/I performance of the downlink, whereas the circles indicate the worst-case location of the mobile affecting the C/I performance of the uplink. The cluster attribute (collinear, noncollinear, etc.) indicated in the captions of these figures are defined in Appendix D.

It is noteworthy that some of the patterns tessellate into staggered configurations with the closer interferers either completely obstructed or obstructed for most of the time with an LOS interferer appearing many blocks away. It is also worth emphasizing that for clusters with a prime number of constituent cells, as is the case of the five-cell cluster of Figure 2.8, the base stations that interfere with the target mobile in the downlink change as the mobile moves along the street.

2.10.1 Propagation

The propagation in a microcellular environment is characterized by both LOS and NLOS modes. In the NLOS mode, the mean signal strength P_{NLOS} received at a distance d from the transmitter follows approximately the same power law as for the macrocellular systems, i.e.,

$$P_{\text{NLOS}} = K_{\text{NLOS}} d^{-\alpha} \tag{2.22}$$

where K_{NLOS} is a proportionality constant, which depends on a series of propagation parameters (frequency, antenna heights, environment, etc.). For the LOS condition and for a transmitting antenna height h_t, a receiving antenna height h_r, and a wavelength λ, the received mean signal strength P_{LOS} at a distance d is approximately given by

$$P_{\text{LOS}} = \frac{K_{\text{LOS}}}{d^2}\left[1 + \left(\frac{d}{d_B}\right)^2\right]^{-1} \tag{2.23}$$

where K_{LOS} is a proportionality constant, which depends on a series of propagation parameters (frequency, antenna heights, environment, etc.), and $d_B = 4h_t h_r/\lambda$ is the breakpoint distance. Note that the LOS propagation mode in microcellular system is rather different from that of the NLOS. In NLOS, the mean signal strength decreases monotonically with the distance. In LOS, for distances smaller than the breakpoint distance, the mean signal strength decreases with a power law close to that of the free space condition ($\alpha \simeq 2$); for distances greater than the breakpoint distance, the power law closely follows that of the plane earth propagation ($\alpha \simeq 4$).

The C/I calculations that follow analyze the performance of a microcellular network system for the worst-case condition. In such a case, the system is assumed to operate at full load and all interfering mobiles are positioned for the highest interference situation. Because the contribution of the obstructed interferers to the overall performance is negligible if compared with that of the LOS interferers, only the LOS condition of the interferers is used for the calculations. Therefore, the results presented here are very close to the lower-bound performance of the system. A more realistic approach considers the mobiles to be randomly positioned within the network, with the channel activity of each call connection varying in accordance with a given traffic intensity. In this case, the performance of the system is found to be substantially better than the worst-case condition.[9,10]

In the C/I calculations that follow, we define $r = d/R$ as the distance of the serving base station to the mobile station normalized with respect to the cell radius ($0 < r \leq 1$) and $k = R/d_B$ as the ratio between the cell radius and the breakpoint distance ($k \geq 0$). As opposed to the macrocellular network, where the interference pattern is approximately maintained throughout the cell, in a microcellular environment the interference pattern changes along the path as the mobile station leaves the center of the cell and approaches its border. Therefore, for a microcellular network it is interesting to investigate the C/I performance as the mobile moves away from the serving base station along the radial street.

2.10.2 Uplink Interference

By using Equation 2.23 for both wanted signal and interfering signals, along with the above definitions for the normalized distances, the C/I equation can be obtained as

$$\frac{C}{I} = \frac{[1 + (rk)^2]^{-1}}{4r^2 \sum_{L=1}^{\infty} n_L^{-2}[1 + (n_L k)^2]^{-1}} \tag{2.24}$$

The parameter n_L is dependent on the reuse pattern as shown in Appendix D. A good approximation for Equation 2.24 is to consider only the first layer of interferers ($L = 1$). Then,

$$\frac{C}{I} = \frac{n_1^2[1 + (n_1 k)^2]}{4r^2[1 + (rk)^2]} \tag{2.25}$$

2.10.3 Downlink Interference

In the same way, the parameter C/I can be found for the downlink. However, this ratio greatly depends on the position of the target mobile within the micro cell. Three different interfering conditions may be identified as the mobile station moves along the street: (1) at the vicinity of the serving base station, (2) away from the vicinity of the serving base station and away from the cell border, and (3) near the cell border.

At the vicinity of the serving base station, more specifically at the intersection of the streets ($r \leq$ normalized distance from the cell site to the beginning of the block), the mobile station experiences the following propagation condition: it has a good radio path to its serving base station, but it also has radio paths to the interfering base stations on both crossing streets. Then,

$$\frac{C}{I} = \frac{r^{-2}[1+(rk)^2]^{-1}}{\sum_{L=1}^{\infty}\left\{\begin{array}{c}(n_L+r)^{-2}[1+(n_L+r)^2k^2]^{-1}+(n_L-r)^{-2}[1+(n_L-r)^2k^2]^{-1}\\ +2\left(n_L^2+r^2\right)^{-1}\left[1+\left(n_L^2+r^2\right)k^2\right]^{-1}\end{array}\right\}} \tag{2.26}$$

Away from the vicinity of the serving base station and away from the cell border, which corresponds to most of the path, the mobile station enters the block and loses LOS to those base stations located on the perpendicular street. Then,

$$\frac{C}{I} = \frac{r^{-2}[1+(rk)^2]^{-1}}{\sum_{L=1}^{\infty}\{(n_L+r)^{-2}[1+(n_L+r)^2k^2]^{-1}+(n_L-r)^{-2}[1+(n_L-r)^2k^2]^{-1}\}} \tag{2.27}$$

At the border of the cell, new interferers appear in an LOS condition. However, this is not the case for all reuse patterns. This phenomenon only happens for clusters with a prime number of cells as, for example, in the case of a five-cell cluster as illustrated in Figure 2.8. Hence, for clusters with a prime number of cells and for the mobile away from its serving base station ($1 - r \leq$ normalized distance from the site to the beginning of the block), it is found that

$$\frac{C}{I} = \frac{r^{-2}[1+(rk)^2]^{-1}}{\sum_{L=1}^{\infty}\left\{\begin{array}{c}(n_L+r)^{-2}[1+(n_L+r)^2k^2]^{-1}+(n_L-r)^{-2}[1+(n_L-r)^2k^2]^{-1}\\ +\left(\bar{n}_L^2+\bar{r}^2\right)^{-1}\left[1+\left(\bar{n}_L^2+\bar{r}^2\right)k^2\right]^{-1}\end{array}\right\}} \tag{2.28}$$

where $\bar{r} = 1 - r$ and $\bar{n}_L$ is defined in Appendix D.

A good approximation for the downlink C/I can be obtained by considering the first layer of interferers only ($L = 1$).

2.10.4 Examples

We now illustrate the C/I performance for clusters with 5, 8, 9, 10, and 13 micro cells. The performance has been evaluated with the central micro cell as the target cell and with the mobile user departing from the cell center toward its edge. This is indicated by the arrow in the respective cell in Figure 2.8. Figure 2.8 also shows, in gray, the co-micro-cells that at a certain time may interfere with the wanted mobile in an LOS condition.

For the numerical results, the calculations consider the radius of the micro cell as $R = 100$ m, a street width of 15 m, the transmitter and receiver antennas heights, respectively, equal to $h_t = 4$ m and $h_r = 1.5$ m, an operation frequency of 890 MHz ($\lambda = 3/8.9$), these leading to $k = 1.405$, and a network consisting of an infinite number of cells (in practice, 600 layers of interfering cells). Note that $k = 1.405$ indicates that the cell radius is 40.5% greater than the breakpoint distance.

Figures 2.10 and 2.11, respectively, show the uplink and downlink performances for the cases of 5-, 8-, 9-, 10-, and 13-micro-cell clusters as a function of the normalized distance. In general, the larger the cluster, the better the carrier-to-interference ratio, as expected. However, the five-micro-cell cluster

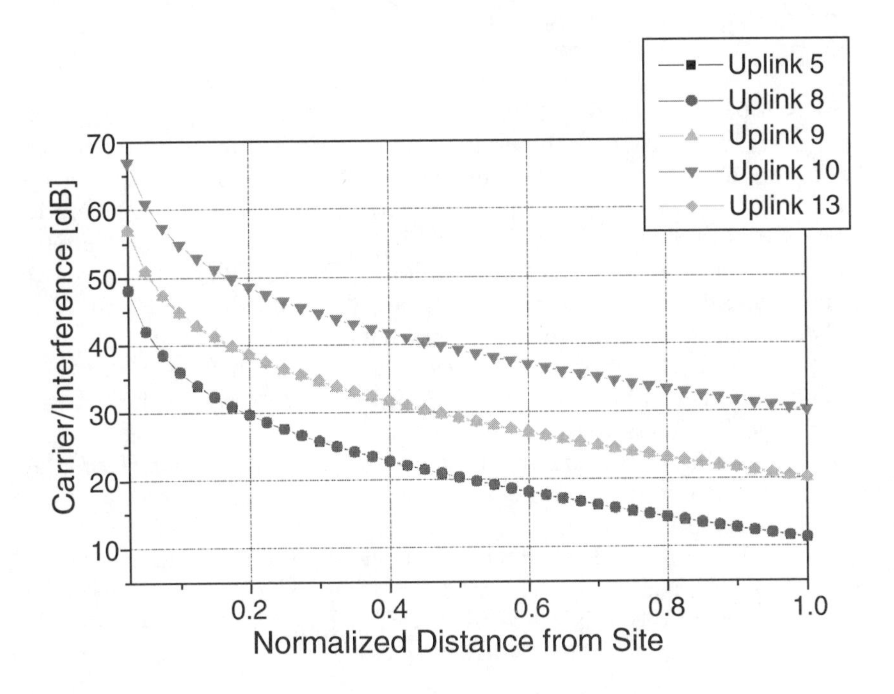

FIGURE 2.10
C/I ratio as a function of normalized distance: uplink.

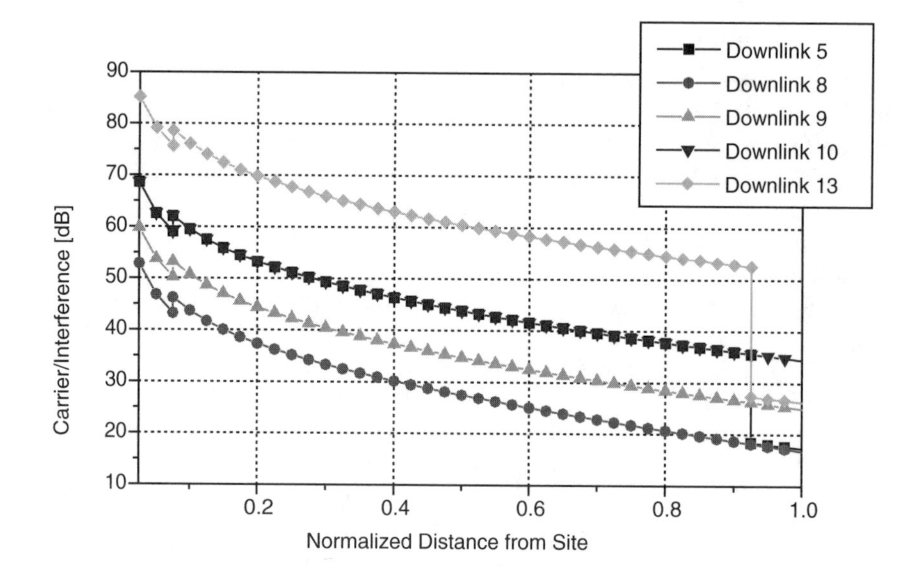

FIGURE 2.11
C/I ratio as a function of normalized distance: downlink.

exhibits notably outstanding behavior, with its C/I coinciding with that of the eight-micro-cell cluster for the uplink (lower curve in Figure 2.10) and with that of the ten-micro-cell cluster for the downlink for most of the extension of the path (curve below the upper curve in Figure 2.11), with the separation of the curves in the latter occurring at the edge of the micro cell, where two new interferers appear in an LOS condition. In Figure 2.10, the C/I curves of the nine- and thirteen-micro-cell clusters are also coincident.

There is a significant difference in performance for the uplink and downlink; this difference becomes progressively smaller with an increase in the size of the cluster. This can be better observed in Figure 2.12, where the five- and ten-micro-cell clusters are compared.

It is interesting to examine the influence of the number of interfering layers on the performance. For this purpose we analyze the performance of a one-layer network ($L = 1$ in Equations 2.24 through 2.28). Figures 2.13 and 2.14 show the performances for the uplink and downlink as a function of the normalized distance to the base station using both the exact (infinite number of layers) and the simplified (one-layer) methods for clusters of five and eight cells, respectively. The dotted lines correspond to the results for the case of an infinite number of interferers, and the solid lines represent the results for the simplified calculations. Analyzing the graphs and the numerical results, we observe that the difference between the C/I ratio for an infinite-cell network and for a one-interfering-layer network is negligible. This conclusion

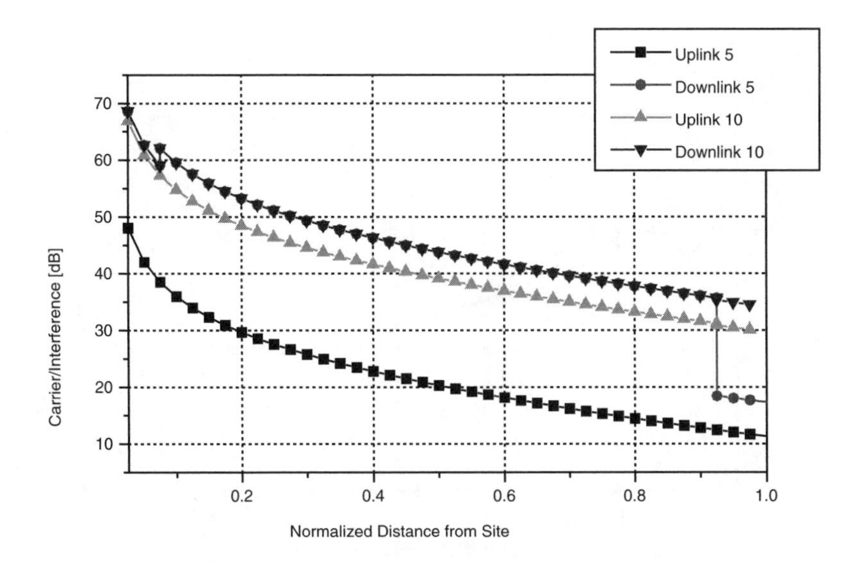

FIGURE 2.12
C/I ratio as a function of normalized distance for uplink and downlink compared.

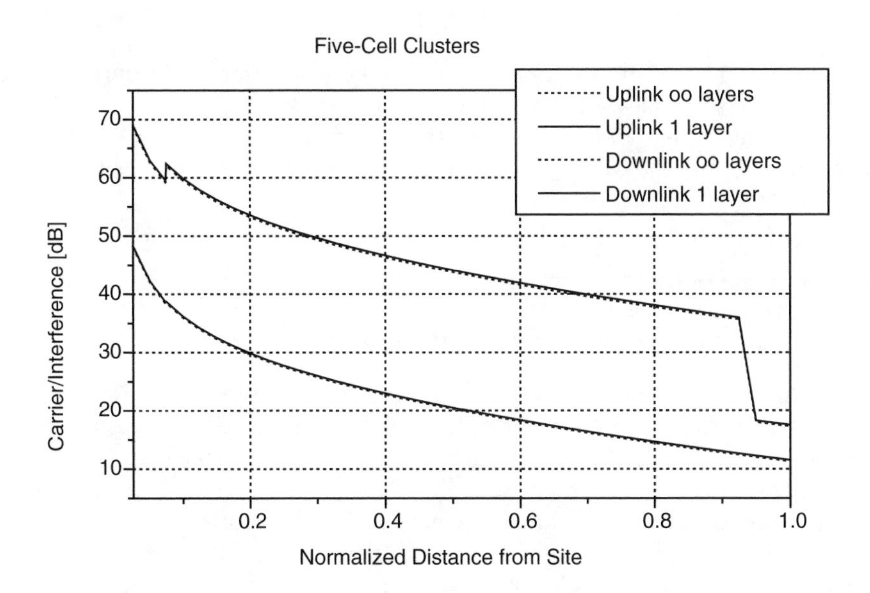

FIGURE 2.13
Performance considering an infinite number of interferers and a single layer of interferers for five-cell clusters.

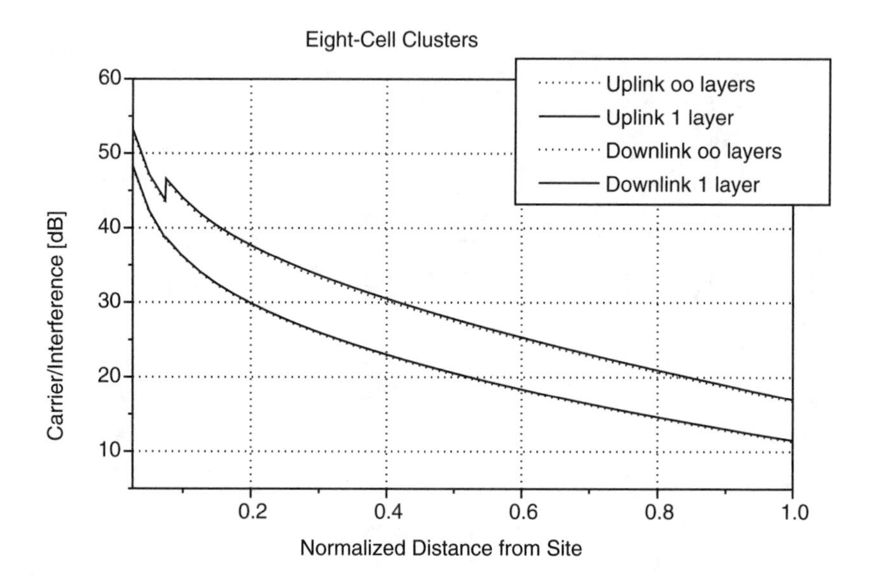

FIGURE 2.14
Performance considering an infinite number of interferers and a single layer of interferers for eight-cell clusters.

also applies to the other patterns, with the largest difference found in both methods for all reuse patterns that are less than 0.35 dB.

Therefore, for the worst-case condition, the C/I ratio can be estimated by considering only the interfering layer that is closest to the target cell.

2.11 Interference in Wideband Systems

Wideband systems operate with a unity frequency reuse factor. This means that a carrier frequency used in a given cell is reused in other cells, including the neighboring cells. As already introduced in Chapter 1, the channelization in this case is carried out by means of code sequences. In an ideal situation, with the use of orthogonal code sequences and with the orthogonality kept in all circumstances, no interference occurs. In such a case, the efficiency of the frequency reuse is 100%. We note, however, that such an ideal situation does not hold and the systems are led to operate in an interference environment. The efficiency of the reuse factor in this case is less than 100%.

Let I_S be the total power of the signals within the target cell (same cell) and I_O the interference power due to the signals of all the other cells. The frequency reuse efficiency f is defined as

$$f = \frac{I_S}{I_S + I_O} \tag{2.29}$$

Let $I = I_O/I_S$ be the (other-cell to same-cell) interference ratio. Thus,

$$f = \frac{1}{1 + I} \tag{2.30}$$

or, equivalently, $I = (1 - f)/f$. Because within a system the traffic may vary from cell to cell, the frequency reuse efficiency can be defined per cell. Assume an N-cell system. Let j be the target cell and i the interfering cell. Therefore, for cell j, $I_O = \Sigma_{i=1}^{N} I_i$, $i \neq j$, and $I_S = I_j$. The frequency reuse efficiency f_j for cell j can now be written as

$$f_j = \frac{I_j}{I_j + \sum_{i=1, i \neq j}^{N} I_i}$$

or, equivalently,

$$f_j = \left(\sum_{i=1}^{N} I_{i,j} \right)^{-1} \tag{2.31}$$

where $I_{i,j} = I_i/I_j$.

In wideband systems, the interference conditions for the uplink and for the downlink are rather dissimilar.

The multipoint-to-point communication (reverse link) operates asynchronously. In such a case, the orthogonality of codes used to separate the users is lost and all the users are potentially interferers. Efficient power control algorithms must be applied in a way that optimizes the reverse link capacity with all users contributing with the same interference power.

The point-to-multipoint communication (forward link) operates synchronously, and, ideally, because the downlink uses orthogonal codes to separate users, for any given user the interference from other users within the same cell is nil. However, because of the multipath propagation, and if there is sufficient delay spread in the radio channel, orthogonality is partially lost and the target mobile receives interference from other users within the same cell.

2.11.1 Uplink Interference

The interference condition in the reverse link is illustrated in Figure 2.15. Because of power control, the signals of all active mobile users within a given cell arrive at the serving base station with a constant and identical power. Let

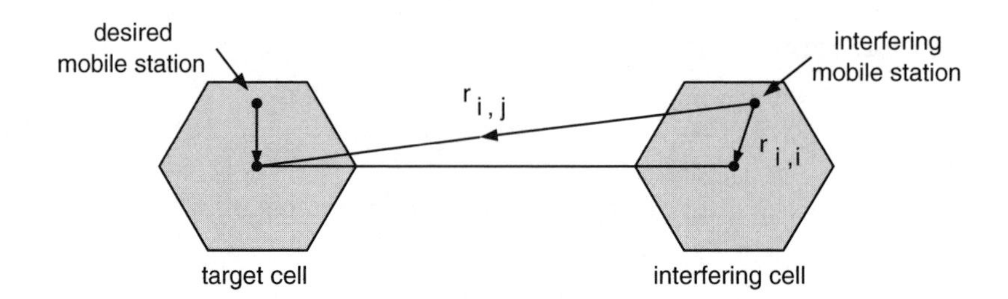

FIGURE 2.15
Interference in the reverse link.

κ be such a power. The total power from the active users within the cell is κ times the number of active users within the cell. Therefore, for cell j

$$I_j = \kappa \int \Upsilon\left(A_j\right) dA_j \tag{2.32}$$

where $\Upsilon\left(A_j\right)$ is the traffic density (users per area) of cell j whose area is A_j. Given that, for any active user i, κ is the power at its serving base station i, then the power transmitted from the mobile station distant r_{ii} from its serving base station is κr_{ii}^{α}. The power received at the base station j (interfered base station), distant r_{ij} from mobile station i, is proportionally attenuated by the corresponding distance. Therefore, the interfering power is $\kappa r_{ii}^{\alpha} r_{ij}^{-\alpha}$. Note that each interfering user i contributes with a power equal to $\kappa r_{ii}^{\alpha} r_{ij}^{-\alpha}$. For all users in cell i the total interfering power at base station j is

$$I_i = \kappa \int \Upsilon\left(A_i\right) r_{ii}^{\alpha} r_{ij}^{-\alpha} dA_i \tag{2.33}$$

where A_i is the area of cell i. Hence,

$$f_j = \frac{\int \Upsilon\left(A_j\right) dA_j}{\sum_{i=1}^{N} \int \Upsilon\left(A_i\right) r_{ii}^{\alpha} r_{ij}^{-\alpha} dA_i} \tag{2.34}$$

Note that $\int \Upsilon\left(A_n\right) dA_n = M_n$, where M_n is the number of active users within cell n. For uniform traffic distribution, $\Upsilon\left(A_n\right) = M_n/A_n$. Note further that the frequency reuse efficiency depends on both the traffic distribution as well as on the propagation conditions (path loss and fading). For uniform traffic distribution and for an infinite number of cells, all cells present the same frequency reuse efficiency. Therefore, it suffices to determine such a parameter for one cell only. The calculations in this case can be performed using only the geometry of the cellular grid. Some values for frequency reuse efficiency are

TABLE 2.2

Examples of Frequency Reuse Efficiency for the Reverse Link: Uniform Traffic Distribution

α	σ (dB)	f
3	0	0.5578
3	7	0.4340
3	8	0.3392
3	9	0.2415
4	0	0.6993
4	7	0.6253
4	8	0.5278
4	9	0.4093
5	0	0.7739
5	7	0.7301
5	8	0.6443
5	9	0.5291

presented in Table 2.2 for different path loss coefficient α, lognormal standard deviation σ, and for uniform traffic distribution.

A common practice in cellular design is to use $f = 0.6$. A simple methodology to calculate the exact frequency reuse efficiency for nonuniform traffic distributions and for realistic conditions can be found in References 11 through 13.

2.11.2 Downlink Interference

The interference condition in the forward link is illustrated in Figure 2.16. The constant-power situation, as experienced in the reverse link, no longer applies. The interference now is a function of the distance of the mobile station to the interferers. The frequency reuse efficiency $f_j(x, y)$, therefore, is a function of the mobile position variables (x, y). We may define a mean frequency reuse

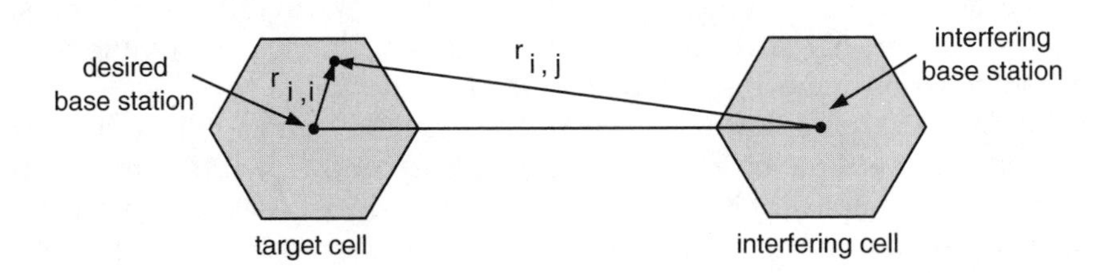

FIGURE 2.16
Interference in the forward link.

TABLE 2.3

Examples of Frequency Reuse Efficiency for the Forward Link: Uniform Traffic Distribution

α	σ (dB)	f
2	8	0.4621
3	8	0.6584
4	8	0.7687
5	8	0.8283

efficiency as

$$f_j(x, y) = \frac{1}{A_j} \int\int f_j(x, y)\, dx dy \tag{2.35}$$

Note that, as already mentioned, the own-cell interference at the mobile station depends on the degree of orthogonality of the codes. For an ideal condition, i.e., orthogonality is fully maintained, no own-cell interference occurs. The frequency reuse efficiency is 1. For a complete loss of orthogonality, the own-cell interference reaches its maximum and the reuse efficiency its minimum. Some values for frequency reuse efficiency are presented in Table 2.3 for different path loss coefficient α, lognormal standard deviation $\sigma = 8$ dB, and for uniform traffic distribution.

Here, again, a common practice in cellular design is to use $f = 0.6$.

2.12 Network Capacity

A measure of network capacity can be provided by the spectrum efficiency. The spectrum efficiency (η), as used here, is defined as the *number of simultaneous conversations per cell* (M) *per assigned bandwidth* (W). In cellular networks, efficiency is directly affected by two families of technologies: compression technology (CT) and access technology (AT).

CTs increase the spectrum efficiency by packing signals into narrower-frequency bands. Low-bit-rate source coding and bandwidth-efficient modulations are examples of CTs. ATs may be used to increase the spectrum efficiency by providing the signals with a better tolerance for interference. Within the AT family are included the reuse factor and the several digital signal processing (DSP) techniques that provide for higher signal robustness.

Narrowband systems as well as wideband systems make use of CTs and DSP solutions to improve system capacity and provide for signal robustness.

As for the reuse factor, because narrowband systems are less immune to interference as compared to wideband systems, a reuse factor greater than 1 is necessarily used. Wideband systems, on the other hand, are characterized by the use of a reuse factor equal to 1. The utilization of a reuse factor of 1 does not necessarily indicate that the wideband system will provide for a higher capacity as compared with narrowband systems. It must be emphasized that, because in wideband systems the frequency reuse efficiency is usually substantially smaller than 1, a loss in capacity occurs. This and other factors contribute to the reduction of capacity in wideband systems.

Narrowband systems are usually based on FDMA or TDMA access technologies. Wideband systems, in general, make use of CDMA access technology. This section determines the mean capacity of narrowband as well as wideband systems. Although the formulation developed here gives an estimate of the capacity, in the real world things may be substantially different, because a number of other factors, which are difficult to quantify, influence system performance.

2.12.1 Narrowband Systems

In narrowband systems, the assigned bandwidth is split into a number of subbands. The total time of each subband channel may be further split into a number of slots. Let C be the total number of resources of the system, i.e., number of slots per subband times number of subbands. The spectrum efficiency of a narrowband system is then obtained as

$$\eta = \frac{M}{W} = \frac{C}{NW} \tag{2.36}$$

given in number of simultaneous conversations per cell per assigned bandwidth. The ratio C/W is a direct result of the CTs used. The reuse factor N is chosen such that it achieves the signal-to-interference ratio required to meet transmission quality specifications. Modulation, coding, and several DSP techniques have a direct impact on this.

2.12.2 Wideband Systems

Wideband systems are typically interference limited, with the interference given by the number of active users within the system. The total interference power I_t is defined as

$$I_t = I_S + I_O + I_N \tag{2.37}$$

where I_N is the thermal noise power, I_S is the power of the signals within the target cell (same cell), and I_O the interference power due to the signals of all the other cells, as already defined. The number of active users, their geographic distribution, and their channel activity affect the interference conditions of the system. Therefore, the frequency reuse efficiency as well as the interference ratio are all affected by these same factors.

Define P_N as the signal power required for an adequate operation of the receiver in the absence of interference. Let P_I be the signal power required for an adequate operation of the receiver in the presence of interference. The ratio N_R between these two powers given as

$$N_R = \frac{P_t}{P_N} \tag{2.38}$$

is known as *noise rise*. Clearly,

$$\frac{P_t}{P_N} = \frac{I_t}{I_N} \tag{2.39}$$

Therefore, we may define the noise rise as the ratio between the total wideband power and the thermal noise power, i.e.,

$$N_R = \frac{I_t}{I_N} \tag{2.40}$$

In the absence of interference, $I_t = I_N$, $N_R = 1$, and $P_t = P_N$, i.e., the power required for an adequate operation of the receiver is the power required in the presence of thermal noise. Using Equation 2.37 in Equation 2.40, we find that

$$N_R = \frac{1}{1 - \rho} \tag{2.41}$$

where

$$\rho = \frac{I_S + I_O}{I_S + I_O + I_N} \tag{2.42}$$

is defined as the *load factor*. Note that $0 \leq \rho < 1$. The condition $\rho = 0$ signifies no active users within the system. Note that ρ increases with the increase of the number of users. Note also that as ρ approaches unity the noise rise tends to infinity, and the system reaches its pole capacity. A system is usually designed to operate with a loading factor smaller than 1 (typically $\rho \simeq 0.5$, or equivalently 3 dB of noise rise). Figure 2.17 illustrates the noise rise as a function of the load factor.

The load factor is calculated differently for the uplink and for the downlink.

2.12.3 Uplink Load Factor

Let $\gamma_i = E_i/N_i$ be the ratio between the energy per bit and the noise spectral density for user i. Define $G_i = W/R_i$ as the processing gain for user i, given

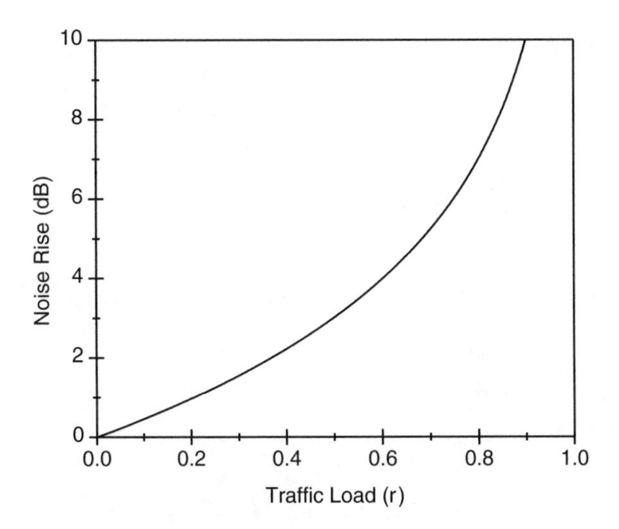

FIGURE 2.17
Noise rise as a function of the load factor.

as the ratio between the chip rate of the system (system bandwidth) and the bit rate for user i. The energy per bit is obtained as $E_i = P_i T_i = P_i / R_i$, where P_i, T_i and $R_i = 1/T_i$ are, respectively, the signal power received from user i, the bit period of user i, and the bit rate of user i. The noise spectral density is calculated as $N_i = I_N / W = (I_t - P_i) / W$. Note that these parameters assume a 100% channel activity. For a channel activity equal to a_i, $0 \leq a_i \leq 1$, and using the above definitions

$$\frac{E_i}{N_i} = \frac{W P_i}{a_i R_i (I_t - P_i)}$$

or, equivalently,

$$\gamma_i = \frac{G_i P_i}{a_i (I_t - P_i)}$$

Solving for P_i,

$$P_i = \rho_i I_t \tag{2.43}$$

where

$$\rho_i = \left(1 + \frac{G_i}{a_i \gamma_i}\right)^{-1} \tag{2.44}$$

Manipulating Equation 2.42, we obtain

$$\rho = (1 + I)\,\frac{I_S}{I_t} \tag{2.45}$$

The power I_S can be calculated as

$$I_S = \sum_{i=1}^{M} P_i \tag{2.46}$$

where M is the number of users within the cell. From Equations 2.46, 2.43, and 2.45 we obtain

$$\rho = \sum_{i=1}^{M} \rho_i = (1 + I) \sum_{i=1}^{M} \left(1 + \frac{G_i}{a_i \gamma_i}\right)^{-1} \tag{2.47}$$

which is the uplink load factor for a multirate wideband system. For a given load factor, Equation 2.47 yields the uplink capacity M.[14] Note that such a capacity is dependent on the required energy per bit and the noise spectral density γ_i on the activity factor a_i, and on the type of service that is reflected on the processing gain G_i. A load factor $\rho = 1$ gives the pole capacity of the system.

Typically,[14] a_i assumes the value 0.67 for speech and 1.0 for data; the value of γ_i depends on the service, bit rate, channel fading conditions, receive antenna diversity, mobile speed, etc.; W depends on the channel bandwidth; R_i depends on the service; and I can be taken as 0.55.

Of course, other factors, such as power control efficiency p_i, $0 \leq p_i \leq 1$, and gain s due to the use of s-sector directional antennas (s sectors per cell), can be included in the capacity Equation 2.47. The power control efficiency p_i diminishes the capacity by a factor of p_i, whereas the use of sectored antennas increases the capacity by a factor approximately equal to the number s of sectors per cell.

For a classical all-voice network, such as the 2G CDMA system, all M users share the same type of constant-bit-rate service. In this case, Equation 2.47 reduces to

$$M = \frac{\rho \times p \times s \times G}{(1 + I) \times a \times \gamma} \tag{2.48}$$

where the parameters are those already defined, but with the index dropped, and where we have assumed the condition $\frac{psG}{a\gamma} \gg 1$. The spectrum efficiency is

$$\eta = \frac{M}{W} = \frac{\rho \times p \times s \times G}{(1 + I) \times a \times \gamma W} \tag{2.49}$$

2.12.4 Downlink Load Factor

The downlink load factor can be obtained in a way similar to that used to obtain the uplink load factor. Ideally, because the downlink uses orthogonal codes to separate users, for any given user the interference from other users within the same cell is nil. However, because of the multipath propagation, and if there is sufficient delay spread in the radio channel, orthogonality is partially lost and the target mobile receives interference from other users within the same cell. An orthogonality factor t_i, $0 \leq t_i \leq 1$, can be added to account for the loss of orthogonality: $t_i = 0$ signifies that full orthogonality is kept; $t_i = 1$ signifies that orthogonality is completely lost. Another peculiarity of the downlink is that the interference ratio depends on the user location because the power received from the base stations is sensed differently at the mobile station according to its location. In this case, we define the interference ratio as I_i. Following the same procedure as for the uplink case the downlink location-dependent load factor $\rho(x, y)$ is found to be[14]

$$\rho(x, y) = \sum_{i=1}^{M} \frac{a_i \gamma_i (t_i + I_i)}{G_i} \tag{2.50}$$

where (x, y) is the mobile user coordinates. For an average position within the cell, the location-dependent parameters can be estimated as t and I and the average downlink load factor is given as

$$\rho = (t + I) \sum_{i=1}^{M} \frac{a_i \gamma_i}{G_i} \tag{2.51}$$

The same typical parameters can be used in Equation 2.51 to estimate the uplink load factor. As for the orthogonality factor, this is typically 0.4 for vehicular communication and 0.1 for pedestrian communication.[14] For a classical all-voice network, such as the 2G CDMA system, all M users share the same type of constant-bit-rate service. In such a case, Equation 2.51 reduces to

$$M = \frac{\rho \times p \times s \times G}{(t + I) \times a \times \gamma} \tag{2.52}$$

where power control efficiency as well as sectorization efficiency parameters have been included. The spectrum efficiency is

$$\eta = \frac{M}{W} = \frac{\rho \times p \times s \times G}{(t + I) \times a \times \gamma W} \tag{2.53}$$

2.13 Summary

Cellular systems are built upon the frequency-reuse principle. In a cellular system, the service area is divided into cells and portions of the available spectrum are conveniently allocated to each cell. The number of cells per cluster defines the reuse pattern, and this is a function of the cellular geometry. For a long time, since the inception of modern wireless networks, the cellular grid has been dominated by macro cells. The macrocellular network makes use of high-power sites with antennas mounted high above the rooftops. In such a case, the hexagonal cell grid has proved adequate. Further, the macrocellular structure serves low-capacity systems. However, expansion and evolution of wireless networks can only be supported by an ample microcellular structure. The microcellular network concept is rather different from that of the macrocellular one. In microcellular systems, with low power sites and antennas mounted at street level (below the rooftops), the assumed propagation symmetry of the macrocellular network no longer applies and the hexagonal cell pattern does not make sense. The "microscopic" structure of the environment constitutes a decisive element influencing system performance. With the antennas mounted at street level, the buildings lining each side of the street work as waveguides, in the radial direction, and as obstructors, in the perpendicular direction. A cell in such an environment is more likely to comply with a diamond shape with the radial streets the diagonals of this diamond. In practice, the coverage area differs substantially from the idealized geometric figures and amoeboid cellular shapes are more likely to occur.

A cellular hierarchy is structured that contains several layers, each layer encompassing the same type of cell in the hierarchy. The design of different cells depends on several parameters such as mobility characteristics, output power, and types of services utilized. The layering of cells does not imply that all mobile stations must be able to connect to all base stations serving the geographic area where the mobile station is positioned.

In a cellular design, several aspects must be addressed that affect the performance of the system: interference control, diversity strategies, variable data rate control, capacity improvement techniques, and battery-saving techniques. Interference is certainly of paramount importance. Narrowband and wideband systems are affected differently by interference.

In narrowband systems, interference is caused by a small number of high-power signals. Moreover, macrocellular and microcellular networks undergo different interference patterns. In addition, whereas in macrocellular systems uplink and downlink present approximately the same interference performance, in microcellular systems the interference performance of uplink and downlink is dissimilar. In both cases, the uplink performance is always worse than the downlink performance, but the difference between the performances of both links is drastically different in microcellular systems. For macrocellular systems, the larger the reuse pattern, the better the interference performance. For microcellular systems, it can be said that, in general, the larger the reuse pattern, the better the performance.

In wideband systems, interference is caused by a large number of low-power signals. In such a case, the traffic profile as well as the channel activity has a great influence on the interference. Here again, uplink and downlink perform differently.

Capacity is another issue that varies substantially for narrowband and wideband systems. In the first case, capacity is established given the total amount of resources and the reuse pattern. In the second case, a number of additional parameters, such as the traffic profile, channel activity, and others, may influence system capacity.

References

1. Framework for the radio interface(s) and radio sub-system functionality for International Mobile Telecommunications-2000 (IMT-2000), Recommendation ITU-R M.1035.
2. Whitteker, J. H., Measurements of path loss at 910 MHz for proposed microcell urban mobile systems, *IEEE Trans. Veh. Technol.*, 37, 125–129, Aug. 1988.
3. Rustako, A. J., Erceg, V., Roman, R. S., et al. Measurements of microcellular propagation loss at 6 GHz and 2 GHz over nonline-of-sight paths in the city of Boston, in *GLOBECOM'95 Conf.*, Singapore, 1995, 758–762.
4. Erceg, V., Rustako, A. J., Jr., and Roman, R. S., Diffraction around corners and its effects of the microcell coverage area in urban and suburban environments at 900 MHz, 2 GHz, and 6 GHz, *IEEE Trans. Veh. Technol.*, 43, 762–766, Aug. 1994.
5. Clark, M. V., Erceg, V., and Greenstein, L. J., Reuse efficiency in urban microcellular networks, *IEEE Trans. Veh. Technol.*, 46, 279–288, May 1997.
6. Goldsmith, A. and Greenstein, L. J., A measurement-based model for predicting coverage areas of urban microcells, *IEEE J. Select. Areas Commun.*, 11, 1013–1022, Sept. 1993.
7. Erceg V. et al., Urban/suburban out-of-sight propagation modeling, *IEE Commun. Mag.*, 30, 56–61, June 1992.

8. Yacoub, M. D., Toledo, A. F., Gomez, P. R. C., et al., Reuse pattern for microcellular networks, *Int. J. Wireless Inf. Networks*, 6 (1), 1–7, Jan. 1999.
9. Gomez, P. R. C., Yacoub, M. D., and Toledo, A. F., Performance of a microcellular network in a more realistic condition, Presented as 11th IEEE International Symposium on Personal Indoor and Mobile Radio Communications, 2000.
10. Gomez, P. R. C., Yacoub, M. D., and Toledo, A. F., Interference in wireless networks with a new reuse pattern [in Portuguese], Presented as Brazilian Telecommunications Symposium (Simpósio Brasileiro de Telecomunicações), SBT99, Vitória, Brazil, 1999.
11. d'Ávila, C. K. and Yacoub, M. D., Reuse efficiency for non-uniform traffic distributions in CDMA systems, *IEE Electron. Lett.*, 34 (13), 1293–1294, June 1998.
12. d'Ávila, C. K. and Yacoub, M. D., The linear method to evaluate the frequency reuse efficiency of cellular CDMA systems, Presented as IEEE Global Telecommunications Conference, Globecom General Conference, Globecom'99, Rio de Janeiro, Dec. 1999.
13. d'Ávila, C. K. and Yacoub, M. D., Reuse efficiency in a quasi-real CDMA network, Presented as 11th International Symposium on Personal, Indoor and Mobile Radio Communications, PIMRC'2000, London, Sept. 2000.
14. Holma, H. and Toskalla, A., *WDMA for UMTS: Radio Access for Third Generation Mobile Communications*, John Wiley, Chichester, U.K., 2000.

3

Multiple Access

3.1 Introduction

Wireless communication systems are multiuser systems in which information is conveyed by means of radio waves. The radio channel is essentially a broadcast medium. Therefore, the transmitted information is potentially available to all receivers. Although useful in some applications, such as in broadcast networks, nonetheless in some others, such as in wireless systems, this inherent high connectivity implies a natural source of interference, and some sort of access control must be exercised for successful communication. Access coordination can be accomplished by several mechanisms: by insulating the various signals sharing the same access medium, by allowing the signals to contend for the access, or by combining these two approaches. The choice of the appropriate scheme will have to take into account a number of factors such as type of traffic under consideration, available technology, cost, complexity, etc.

Signal insulation is easily attainable by using a scheduling procedure in which signals are allowed to access the medium according to a predefined plan. Signal contention occurs if a signal insulation mechanism is not used. Certainly, access coordination must be exercised in each direction of the communication: from the central point, hereinafter referred to as *access point*, down to the *terminals* (users)—downlink or forward link—and from the terminals (users) up to the access point—uplink or reverse link. Similarly, it must be performed in both directions for a two-way communication. Access control in the downlink direction can be implemented with techniques simpler than those used to control access in the uplink direction. The reason is obvious: because, in the first case, the communication is characterized by a point-to-multipoint transmission mechanism, the access point may exercise complete control over the access; in the second case, the communication is multipoint-to-point, therefore rendering the coordination among independent terminals more complex.

Access coordination may be carried out in different domains:[1] the frequency domain, time domain, code domain, and space domain. Signal insulation in each domain is attained by splitting the resource available into nonoverlapping slots (frequency slot, time slot, code slot, and space slot) and assigning each signal a slot. The nonoverlapping (disjoint, orthogonal) slots constitute the physical channels, which are essentially the physical medium through which information is actually conveyed. In this context, it is instructive to recall the concept of another type of channel—the logical channel. The logical channels make use of the physical channels to perform logical functions within the network. One logical channel can use one or several physical channels and, conversely, one or several logical channels can use one physical channel. Orthogonality among physical channels is an imperious requirement in order for the information conveyed on these channels to be univocally separable (distinguishable, resolvable) at the receiver. For maximum insulation, paired channels, i.e., a forward-link channel and a reverse-link channel in a two-way communication, are kept apart from each other by the farthest distance possible in each domain. This is achieved by organizing all forward channels in a contiguous group of channels and all reverse channels in another contiguous group of channels and in both groups numbering all the channels in the same sequence. For maximum packing, adjacent channels are arranged one immediately after the other. Implementation of such a philosophy, however, is impaired by practical problems. Available resources may not appear in a contiguous manner and, to account for the imperfection of the medium as well as of the equipment, a guard distance between adjacent channels must be maintained. Figure 3.1 illustrates the concepts developed here for

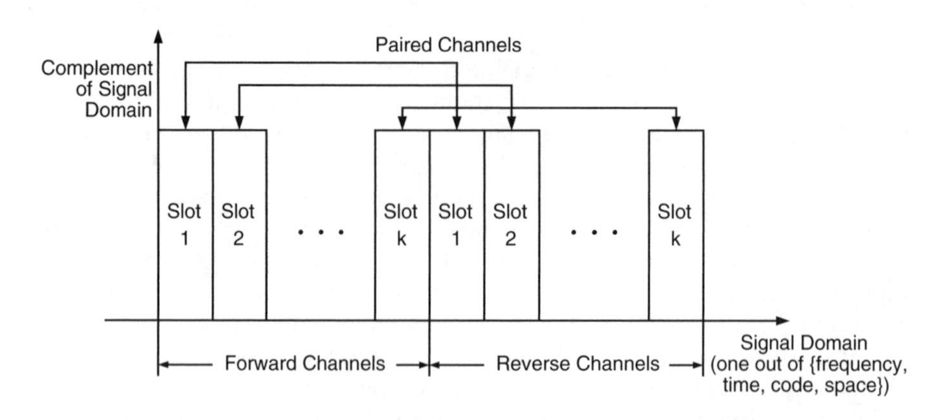

FIGURE 3.1
Signal insulation in the various signal domains.

an ideal situation in which the separation guard between adjacent channels is nil. Note from Figure 3.1 that only one dimension of these domains is shown. Practical networks may take advantage of the different domains to implement a further level of signal insulation. In such a case, a multidimensional representation may be necessary to illustrate the possible combinations.

In contrast to conventional wired networks, in a wireless communication environment signals from different terminals may arrive at the access point with different power levels due to two basic phenomena: *path loss* and *fading*. The path loss phenomenon establishes that the signal power decreases with distance d as $d^{-\alpha}$, with α usually within the range $2 \leq \alpha \leq 6$. Therefore, signals sent by terminals placed at different distances from the access point will arrive with different power levels, characterizing what is known as the *near–far effect*. The fading phenomenon establishes that the signal power fluctuates randomly due to the different environments (shadowing) and also due to the scatterers (multipath propagation) encountered along the radio paths. Contrary to what is usually expected, in some situations these phenomena may have a favorable effect in the overall multiple access performance because they constitute a natural means of splitting the terminals into different classes of access power. In other situations, however, the effects may be disastrous.

3.2 Signal Domains

The signal may be described in several domains, and an appropriate manipulation of the signals within these domains may lead to effective use of the transmission medium. The signal domains most commonly used for access purposes are frequency, time, code, and space.

3.2.1 Frequency Domain

The radio spectrum is a precious resource, the quintessence of wireless communication systems. Because radio propagation does not recognize geopolitical boundaries and because political, economic, and social aspirations may vary from country to country, international cooperation leading to an intelligent and efficient use of the frequency spectrum is mandatory. Because of the explosion of the demand for and the diversification of wireless services, special attention have been given to spectrum allocation issues. Wireless communications systems have been driven to use high frequencies due to the congestion at the lower portion of the frequency spectrum, where the available bandwidth is insufficient to satisfy the great demand for mobile services.

However, dealing with high frequencies usually leads to intricate problems that are severely aggravated by the mobility of the users. Services are assigned fixed bandwidth, not necessarily in a contiguous fashion. For competition purposes, for a given service the frequency band is split into subbands, each of which is allotted to different service operators. Each one of these bands is then further split into two halves, one for the forward link and the other for the reverse link. Subsequent divisions are carried out to form the nonoverlapping frequency slots (channels). The channel bandwidth (channel spacing) is determined according to criteria such as the services to be provided and the available technology. Each channel—the physical channel—is identified by a carrier placed in the middle of the channel band.

3.2.2 Time Domain

Signal insulation in the time domain is accomplished by allowing the information to use the frequency band during a specific period of time (time slot)—the physical channel. Nonoverlapping time slots constitute the orthogonal channels. For any given piece of information, the aim is to transmit the information in as short a period of time as possible, so that more information can be conveyed in the same frequency band; this is achieved by including more time slots per carrier. As before, this certainly depends on the services to be provided and on the available technology. Access in the time domain characterizes the transmission occurring in bursts because for the same source the information will occupy the carrier only in specific periods of time.

3.2.3 Code Domain

Signal insulation can also be accomplished by assigning each signal a different code (a key, a password)—the physical channel. A code is built as a sequence of symbols belonging to an alphabet. In an ideal situation, these codes must present zero cross-correlation so that they can be univocally discriminated (e.g., for different sources, different passwords are assigned). For a finite alphabet, the number of codes is obviously finite. Therefore, the larger the alphabet, the larger the number of orthogonal codes in the alphabet and the longer each code. For a given transmission rate, the longer the code, the longer the time to transmit the code and the longer the time to detect the code. Should it be transmitted, and detected, in a shorter period of time, the transmission rate must be increased as well as the required bandwidth. Therefore, for a limited bandwidth the number of orthogonal codes (code slots) is also limited.

3.2.4 Space Domain

Signal insulation in the space domain can be achieved in two possible dimensions: distance and angle—the physical channels. The distance dimension exploits the fact that the propagation loss increases with the increase of the distance between transmitter and receiver. Thus, signals using the same frequency but transmitted by sources sufficiently apart from each other may not strongly interfere with each other. In the same way, a given signal may reach the receiver through different paths (due to multiple reflections, for example). Each multipath signal suffers different attenuations and different delays, according to the length of the path traveled. Therefore, both attenuation and delay, jointly or independently, can be used to detect each multipath signal. The angle dimension exploits the fact that, by illuminating wedges of a circular area, signals simultaneously using the same frequency may be discriminated by these very wedges within which they are located. Smart antennas may be used to keep track of these signals.

3.2.5 Brief Remarks on Signal Domains

The most commonly used and most straightforward way of accomplishing radio signal insulation is by assigning different frequency carriers to different signals. This technique is widely employed by both analog and digital wireless systems. Insulation in the time domain has been boosted by the digital technology and is widely used in wireless communications. Insulation in the code domain is a well-known technique that has long been used for military as well as satellite communications applications. The move toward high-capacity wireless systems has found great support in this technique. Insulation in the space domain is widely used in wireless communications. More specifically, the cellular concept with its frequency reuse philosophy constitutes an example of such an application. Other examples are presented in this chapter.

3.3 Duplexing

Wireless communication systems have evolved through several stages of multiple-access control. The foremost controllable resource has always been the frequency spectrum. Other resources such as time, code, and space were initially manipulated in a very precarious and, therefore, ineffective manner. The early systems operated in the simplex mode in the forward link. Half-duplex systems soon appeared, in which forward link and reverse link shared the same channel. Access control was performed on a push-to-talk basis with the access point still competing with the terminals for access. Double half-duplex systems, in which forward link uses one channel and reverse link another channel, granted the access point the privilege of not having to contend for access. The push-to-talk procedure was the access control mechanism used on the reverse channel. The full-duplex mode, or simply duplex mode, was then the last stage in this evolutionary cycle in which push-to-talk access control was no longer necessary.

Duplex communication can be implemented by means of frequency division, time division, code division, and space division methods.

3.3.1 Frequency Division Duplexing

In frequency division duplexing (FDD), forward and reverse channels use separate frequencies. Therefore, a duplex channel is in fact a set of two distinct carriers, which constitute the physical channels. Because forward and reverse channels are continuously on and share the same antenna, the use of a duplexer (a filter between transmitter and receiver) is necessary so that reverse and forward channels do not interfere with each other. In the same way, sharp filters with strong out-of-band rejection must be used to reduce adjacent-channel interference. Adjacent-channel interference is also minimized by allowing for a guard band within each channel. For a given continuous spectrum, insulation between forward and reverse channels is maximized if paired channels are separated by half of the spectrum. FDD is a well-known technology, widely used in wireless systems.

3.3.2 Time Division Duplexing

In time division duplexing (TDD), forward and reverse channels share the same frequency band but occupy this band in nonoverlapping periods of time (slots), also known as windows. Therefore, a duplex channel is in fact a set of two nonoverlapping windows within a given carrier, which constitutes the physical channels. Because transmission and reception alternate in time, this scheme does not require the use of a duplexer. The number of possible access points (windows) within the same frequency band is a function of the technology available. As can be inferred, TDD makes more efficient use of the spectrum as compared with FDD and is more flexible. Usually, but not necessarily, an equal number of windows is dedicated to the forward channels and to the reverse channels, with paired windows symmetrically placed in time for maximum insulation. On the other hand, asymmetrical window assignment is also possible and necessary in asymmetrical traffic operation conditions, and in this case special attention must be paid regarding interference issues.

3.3.3 Code Division Duplexing

In code division duplexing (CDD), forward and reverse channels simultaneously share the same frequency band but are discriminated by means of orthogonal codes. Therefore, a duplex channel is in fact a set of two orthogonal codes within a given carrier, which constitutes the physical channels. Practical implementation of such a scheme may render the circuitry very complex. Because transmission and reception occur simultaneously and continuously within the same band and because the transmitted signal is at a much greater power than the received signal, the level of interference may impair such a communication scheme. Some sort of interference cancellation mechanism is necessary to realize this scheme.

3.3.4 Space Division Duplexing

In space division duplexing (SDD), forward and reverse channels share the same frequency band but are discriminated in space. Therefore, a duplex channel is in fact a set of two distinct locations where signals share the same frequency band. In a line-of-sight condition, directional antennas provide for the required insulation of the signals and can be used in SDD communication. In a non-line-of-sight condition, smart antennas are necessary.

3.3.5 Brief Remarks on Duplexing Techniques

FDD is certainly the duplexing technique most commonly used in wireless networks; it has been employed in all the first-generation wireless systems, in most of the second-generation systems, and its deployment in higher generations is without question. TDD is used in some second-generation systems, as well as in higher generations. CDD alone does not seem to lend itself to easy implementation for sophisticated interference cancellation mechanisms may be required. SDD alone can be used in diverse system applications mainly to increase capacity. Combination of some of these techniques is a common practice.

3.4 Multiple-Access Categories

Depending on how much coordination is required to access the shared resources, the access methods may fall into the following categories:

- *Scheduled Multiple-Access Methods*. The access methods in this category make use of the signal insulation principle in which the information of different sources is transmitted on nonoverlapping channels. In this case, collision does not occur. The methods providing for scheduled multiple-access capabilities in the frequency domain, time domain, code domain, and space domain are, respectively, frequency division (FDMA), time division (TDMA), code division (CDMA), and space division multiple access (SDMA).

- *Random Multiple-Access Methods*. The access methods in this category make use of a contention scheme in which no or little coordination is provided. The information of different sources is cast onto a common channel, and collision may occur. Collision is usually detected by an acknowledgment mechanism in which the terminal sending the packet waits for a confirmation of the successful reception of it. The widely used methods providing for random multiple-access capabilities are Pure ALOHA, Slotted ALOHA, Tree Algorithm, FCFS Algorithm, CSMA (nonslotted, slotted, 1-persistent. p-persistent), CSMA/BT, CSMA/DT, CSMA/CD, and CSMA/CA.
- *Controlled Multiple-Access Methods*. The access methods in this category make use of a deterministic or contentionless strategy in which, by means of a control signal, permission to send is granted to terminals individually, so that only one terminal is allowed to access the medium at a time. The main access schemes here are polling, token ring, and token bus.
- *Hybrid Multiple-Access Methods*. The access methods in this category make use of a combination of scheduled, contention, and deterministic schemes in which some degree of coordination is included into the random-access mechanisms or some other more sophisticated deterministic methods are employed. Examples of such schemes include R-ALOHA, Original PRMA, PRMA^{++}, Aggressive PRMA, Fast PRMA, Multirate PRMA, PRMA/DA, PRMA/ATDD, DQRUMA, DSA^{++}, DTDMA/PR, MASCARA, DTDMA/TDD, Pure RAMA, T-RAMA, F-RAMA, and D-RAMA.

For stream, steady-flow traffic, the available resources are more efficiently utilized if allotted on a scheduled access basis. This is because contention in this case may lead to a significant amount of packet loss. For intermittent, burst (random) traffic, better efficiency can be achieved if the available resources are seized on a random-access basis. This is because scheduled allocation in this case may lead to an inefficient use of the resources. For intermittent, random traffic if a high incidence of packet collision occurs, the use of controlled-access methods is recommended. For stream traffic and bursty traffic simultaneously sharing the same resource, hybrid access methods may be applied.

In a multiuser system, capacity is certainly a sensitive issue. The efficiency with which the available resources are used determines the number of users the system can support. Higher efficiency is achieved if the resources are made available to all users and they are assigned on a demand basis. This characterizes what is known as *demand-assigned multiple-access* (DAMA) or, simply, multiple access. The assignment of the resources on a demand basis suggests that some initial protocol be established before the resource is actually assigned. Handshaking protocols usually typify bursty traffic, whereas payload information can be characterized by either bursty or steady-flow traffic. Certainly, the ultimate aim of any communication network is provision of resources to convey payload information. Overhead traffic, although necessary, should be kept to a minimum to increase efficiency. The initial access protocols may use the same physical resources as those used to convey payload information. However, splitting the resources into two distinct groups—one for initial access purposes and another for payload information purposes—is a practice common to networks where a great amount of signaling is required. And this is true of wireless systems. The proportion of channels dedicated to each function varies according to the needs.

3.5 Scheduled Multiple Access

The scheduled multiple-access techniques are more efficiently utilized if applied to steady-flow traffic, in which case the resources, although seized on a demand basis, are allotted for the duration of the communication. This section describes the various techniques providing for scheduled multiple-access capabilities in the frequency, time, code, and space domains, namely, frequency division multiple access, time division multiple access, code division multiple access, and space division multiple access.

One interesting topic to investigate is the capacity of each technique for a given bandwidth. The capacity issue can be approached in terms of number of users (system capacity) or in terms of information rate (channel capacity). The first approach—number of users per bandwidth—reveals a fertile field if explored together with the wireless architecture. In this chapter, however, we shall explore the second approach—information rate per bandwidth. The first approach was explored in Chapter 2.

We assume that each access method supports k users and that the physical channel can be modeled as an ideal band-limited additive white Gaussian noise (AWGN) channel. From the Shannon capacity formula and for a single user,[2]

$$C_b = B \log_2 \left(1 + \frac{S}{N}\right) \tag{3.1}$$

where C_b is the channel capacity in bits per second, B is the channel bandwidth in hertz, S is the signal power, in watts, and N is noise power, in watts. It is useful to write this formula in terms of $\gamma = E_b/N_0$, the ratio between the energy per bit E_b and the power spectral density N_0 of the additive noise. This is carried out by recognizing that power is equal to energy divided by time. Therefore, $S = C_b E_b$ and $N = B N_0$. Therefore, the normalized capacity (dimensionless) $C = C_b/B$ is given by

$$C = \log_2 (1 + \gamma C) \tag{3.2}$$

In the descriptions and analyses that follow, the techniques are considered in only one dimension, i.e., in one domain only. Combinations of access and duplexing techniques in various ways are possible, but these will not be investigated here. Therefore, only communication in one direction is considered.

3.5.1 Frequency Division Multiple Access

FDMA is certainly the most conventional method of multiple access and was the first technique to be employed in modern wireless applications. In FDMA, the available bandwidth is split into a number of equal subbands, each of which constitutes a physical channel. The channel bandwidth is a function of the services to be provided and of the available technology. It is identified by its center frequency, known as a carrier. In the single channel per carrier FDMA technology, the channels, once assigned, are used on a non-time-sharing basis. A channel allocated to a given user remains so allocated until the end of the task for which that specific assignment was made.

Nonlinear Effects

At the access point, where broadcast occurs, the various channels share a common antenna. This can be implemented in two possible ways: either each channel has its own power amplifier and all the amplifiers are connected to a power combiner, which feeds the antenna, or all the channels share a common power amplifier, which feeds the antenna. Both the power combiner and the power amplifier present nonlinear properties, usually in the form of a saturation at high power levels. The nonlinearities may cause spectral spreading, modulation transfer, signal suppression, and intermodulation. A signal affected by the spectral spreading phenomenon exceeds its own bandwidth, therefore causing adjacent channel interference. A nonlinear change in the envelope of a multicarrier system gives rise to a change in phase of each signal component, and this is known as modulation transfer. In the same way, an amplifier operating in its nonlinear region yields an output gain that is smaller than the gain achieved if the amplifier operates in its linear region. In this case, there can be no signal enhancement, no matter the nonlinearity.

This phenomenon is known as signal suppression. A nonlinear device whose input is a sum of narrowband band-pass signals yields as outputs these very input signals, amplified by a certain gain, in addition to intermodulation products, which are unwanted signal-dependent spectral components within the bandwidth of the wanted signals. This is known as intermodulation. The effects of nonlinearities can be minimized by means of a power back-off so that the equipment operates away from its saturation point.

Channel Capacity

In FDMA, the bandwidth allocated to each user is B/k. The noise power within the allocated bandwidth is $N = (B/k)N_0$. By carrying out some simple algebraic manipulations in Equation 3.1, the channel capacity C_k per user normalized with respect to the bandwidth B is obtained as

$$C_k = \frac{1}{k}\log_2(1 + \gamma k C_k) \tag{3.3}$$

The total capacity per bandwidth of the system is defined as $C_{\text{FDMA}} = kC_k$. Therefore,

$$C_{\text{FDMA}} = \log_2(1 + \gamma C_{\text{FDMA}}) \tag{3.4}$$

which has the same form as Equation 3.2. Note that the total capacity kC_k of the FDMA system is equal to the capacity of a single channel whose average power is equal to k times its individual power. Note, however, that the individual capacity C_k decreases as k increases, because each user is allocated a smaller bandwidth.

3.5.2 Time Division Multiple Access

TDMA is another widely known multiple-access technique, which succeeded FDMA in modern wireless applications. In TDMA, the entire bandwidth is made available to all signals but on a time-sharing basis. In such a case, the communication is carried out on a buffer-and-burst scheme so that the source information is first stored and then transmitted. Prior to transmission, the information remains stored during a period of time referred to as frame. Transmission then occurs within a time interval known as a (time) slot. In order not to lose information, if k signals are to be transmitted within a frame duration of T units of time, the transmission bandwidth must exceed k/T units of frequency, the inverse of which constitutes the time slot. Certainly, the available bandwidth B must be large enough to accommodate such a transmission bandwidth. The time slot constitutes the physical channel.

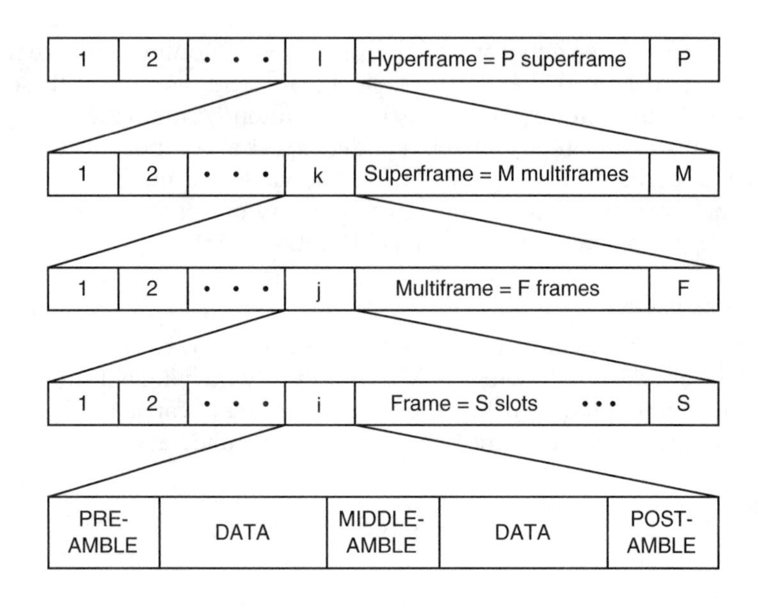

FIGURE 3.2
TDMA timing hierarchy.

Timing Hierarchy

Within TDMA, information flow must comply with a given timing hierarchy, which is specified according to the particular needs. A TDMA timing hierarchy may comprise elements such as hyperframe, superframe, multiframe, frame, and slot, with a number of sequential elements composing the immediate higher-order element in the hierarchy, as sketched in Figure 3.2. Hyperframes, superframes, and multiframes are so organized in order to distribute system and network control or signaling. A frame is a time interval over which the signal format is established within a slot and then repeated indefinitely. A slot conveys the information and is formatted into the following fields: preamble, postamble, middle amble, and data. Preamble, postamble, and middle amble carry overhead information for control purposes, such as carrier recovery, burst synchronization, signaling, training, error-monitoring, and others. A guard time is also included to better separate one slot from its adjacent slot.

Channel Capacity

In TDMA, each signal occupies the whole bandwidth B for $1/k$ of the time, but with average power equal to kS. Therefore, the capacity C_k per user normalized with respect to the bandwidth B is

$$C_k = \frac{1}{k} \log_2 (1 + \gamma k C_k) \tag{3.5}$$

The total capacity per bandwidth of the system is defined as $C_{\text{TDMA}} = kC_k$. Hence,

$$C_{\text{TDMA}} = \log_2 (1 + \gamma C_{\text{TDMA}}) \tag{3.6}$$

which is the same as that for FDMA. Note, however, that this is achieved if the transmitted power is sustained at the level of kS, which may become impracticable as k increases.

3.5.3 Code Division Multiple Access

CDMA is a nonconventional multiple-access technique that immediately found wide application in modern wireless systems. In CDMA, the entire bandwidth is made available simultaneously to all signals. In theory, very little dynamic coordination is required, as opposed to FDMA and TDMA in which frequency and time management has a direct impact on performance. Signals are discriminated by means of code sequences or signature sequences, which correspond to the physical channels. Each pair of transmitter–receivers is allotted one code sequence with which a communication is established. At the reception side, detection is carried out by means of a correlation operation. Ideally, the best performance is attained with zero-cross correlation codes, i.e., with orthogonal codes. In theory, for a synchronous system and for equal rate users, the number of users within a given bandwidth is dictated by the number of possible orthogonal code sequences. In practice, orthogonality is very difficult to achieve. Even though the codes may be designed to present zero cross-correlation, the impairments imposed by the communication channel may introduce a certain level of correlation among the codes. Hence, after the correlation procedure has taken place, the unwanted signals appear as noise, which is the noise due to multiple access. The multiple-access noise would be nil for an ideal orthogonal access. Therefore, the noise due to multiple access that affect the wanted signal is linearly proportional to the number of users within the system. Note that in such a case the degradation in the communication is graceful and increases with the successive admission of users within the system. Such a reasoning presupposes all signals to arrive at the receiver with identical power. In fact, because the multiple-access noise is proportional to the signal strength, an unwanted signal whose power exceeds the average signal power by a factor of p is equivalent to p signals, i.e., a user in such a condition is seen by the system as p users. Therefore, in CDMA, power control is fundamental for satisfactory operation. To accomplish CDMA, spread spectrum techniques are used.

Spread Spectrum

Spread spectrum is defined as a communication technique in which the signals are spread over a bandwidth in excess of the minimum bandwidth required for their transmission. It has its origin in the military scenario where robustness in communications is vital. In particular, two principal requirements must be satisfied: (1) detectability of the signal by the enemy must be made difficult and (2) enemy-introduced interference must be made easy to combat (strange expressions for a peaceful technical material!). Systems presenting these characteristics are referred to, respectively, as low-probability of intercept (LPI) and antijam communications (AJM) systems. Spread spectrum is accomplished by the use of wideband encoding signals at the transmitter, which is required to operate in synchronism with the receiver where the encoding signals, at least for the intended information, are also known. Coding, therefore, is an important element in the design of spread-spectrum signals. By allowing the signals to occupy a bandwidth in excess of the required bandwidth for their transmission, spread-spectrum signals can be given pseudorandom characteristics, which cause the signals to have a noiselike appearance.

The LPI characteristic of spread-spectrum communications can be attained by several means, such as by transmitting at a very low power level (the lowest possible), by limiting the transmission time to very small time intervals (the smallest possible), and by giving the signal a noiselike appearance. It is possible to reduce the signal power to a level below the ambient thermal noise power so that an unintended listener may find it difficult to determine that a transmission is taking place. Besides reducing the transmission time to very short time intervals, the position of these intervals within the total time can be made pseudorandom so that an unintended listener has difficulty learning the hopping pattern. Noiselike appearance is imprinted by the use of pseudorandom sequences in the modulation process.

The AJM characteristic of spread-spectrum communications is attained as follows. Spreading the signal at the transmitter and despreading the received signal at the receiver would not lend any better performance if the communication channel had not been affected by some sort of narrowband interference. Assuming that spreading and despreading operations are identical inverses of each other, the detected signal after the despreading would be identical to the transmitted signal prior to spreading. However, considerable gain is obtained in the presence of a narrowband interference, considering the interference being introduced after the spreading operation. In this case, the composite signal arriving at the receiver constitutes a wideband signal (wanted signal) and a narrowband interference (unwanted signal). The despreading operation at the receiver restores the wanted signal to its original bandwidth, whereas the interference is spread by the same amount with which the wanted signal had been shrunk. Therefore, the power density of the interference is diminished by the same amount, thus reducing its effect.

Spread-spectrum principles are introduced in Appendix C.

Channel Capacity

In CDMA, each of the k signals occupies the whole bandwidth B during the entire time of the channel with average power equal to S. Channel capacity depends on the level of coordination implemented. In a totally coordinated system, transmission is synchronous in time, i.e., the symbol transition times of all users are aligned. In such a case, the advantage of using orthogonal code sequences as spreading sequences is apparent, although this may be questionable because the initial alignment is bound to be broken by the channel. In a noncoordinated system, transmission is asynchronous, i.e., the symbol transition times of the users are not aligned. In such a case, the use of orthogonal code sequences as spreading sequences certainly renders no advantage, and the utilization of code sequences presenting small cross-correlation seems more appropriate. Synchronous architectures are simpler to implement in the forward link direction, because of the inherent facility the system exhibits in distributing synchronism from the access point. Asynchronous architectures are more applicable to the reverse link, in which case time alignment among users is more difficult to achieve.

In both cases, the achievable rate R_i of user i is

$$R_i \leq B \log_2 \left(1 + \frac{S}{N}\right) \tag{3.7}$$

And for k users the achievable rate is

$$\sum_{i=1}^{k} R_i \leq B \log \left(1 + \frac{kS}{N}\right) \tag{3.8}$$

Coordinated System. In a totally coordinated system, where perfect alignment among users exists and where accesses are discriminated by means of orthogonal code sequences, the despreading operation at the receiver restores the intended signal in its entirety with the interference of all other users completely eliminated. If all users operate with the same rate within the given bandwidth $R_i/B = C_k, 1 \leq i \leq k$, then

$$C_k = \frac{1}{k}\log_2(1 + \gamma k C_k) \tag{3.9}$$

The total capacity per bandwidth of the system is defined as $C_{\text{CDMA}} = kC_k$. Therefore,

$$C_{\text{CDMA}} = \log_2(1 + \gamma C_{\text{CDMA}}) \tag{3.10}$$

which is identical to the capacity of both FDMA and TDMA. Therefore, as far as channel capacity is concerned, CDMA, in its synchronous version, yields no improvement over the other multiple-access techniques. On the other hand, if some sort of cooperation among users is possible, then an enhancement in capacity is achieved. For example, if the multiuser receiver bears the spreading waveforms of all users and can jointly demodulate and detect all the received signals, the effect of the unintended signals can be eliminated from the intended signal by a simple subtraction operation.

Noncoordinated System. In a noncoordinated system, because some correlation among spreading sequences exists, the despreading operation at the receiver, besides restoring the intended signal, cannot eliminate the influence of the unintended signals. In such a case, the $k-1$ unwanted signals appear as interference. Assuming a perfect power control, the power of the unintended signals is given by $(k-1)S$. Therefore,

$$C_b = B\log_2\left(1 + \frac{S}{N + (k-1)S}\right) \tag{3.11}$$

or, equivalently,

$$C_k = \log_2\left[1 + \frac{\gamma C_k}{1 + (k-1)\gamma C_k}\right] \tag{3.12}$$

The total capacity per bandwidth of the system is defined as $C_{\text{CDMA}} = kC_k$. Therefore,

$$C_{\text{CDMA}} = k\log_2\left[1 + \frac{\gamma C_{\text{CDMA}}}{k + (k-1)\gamma C_{\text{CDMA}}}\right] \tag{3.13}$$

Here, again, if some sort of cooperation among users is possible, then an enhancement in capacity is achieved.

3.5.4 Space Division Multiple Access

SDMA is a nonconventional multiple-access technique that finds application in modern wireless systems mainly if combined with other multiple-access techniques. The spatial dimension has been extensively explored by wireless communications systems in the form of frequency reuse. The deployment of advanced techniques in order to take further advantage of the spatial dimension is embedded in the SDMA philosophy. In SDMA, the entire bandwidth is made available simultaneously to all signals. Signals are discriminated spatially and the communication trajectory constitutes the physical channels. The implementation of an SDMA architecture is strongly based on antenna technology coupled with advanced digital signal processing techniques.

As opposed to the conventional applications in which locations are constantly illuminated by rigid-beam antennas, in SDMA the antennas should provide for the ability to illuminate the locations in a dynamic fashion. The antenna beams must be electronically and adaptively directed to the user so that, in an idealized situation, the location alone is enough to discriminate the user. Figure 3.3 illustrates the SDMA principle. Several references treat

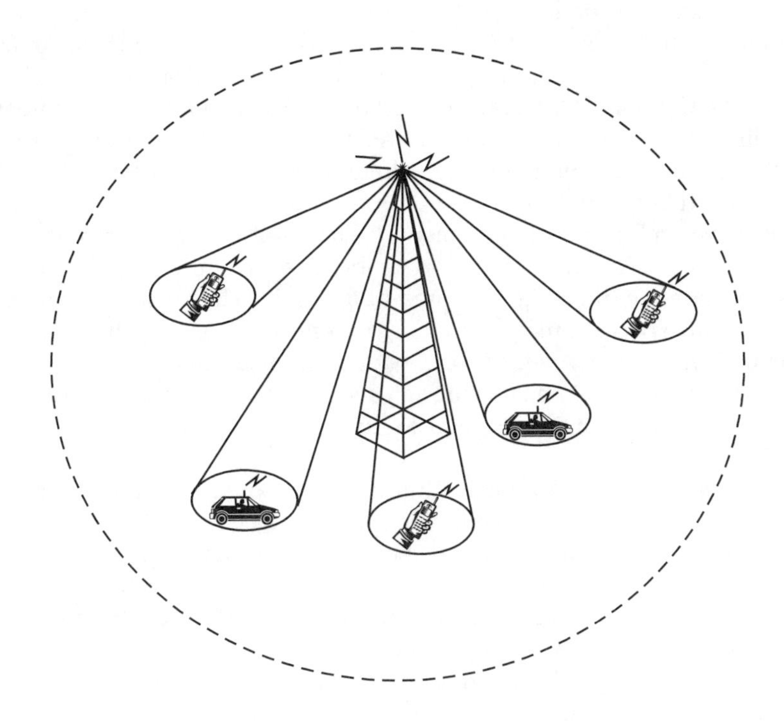

FIGURE 3.3
SDMA principle.

this subject thoroughly (e.g., References 3 to 5). What follows is a very brief introduction to the SDMA subject.

Antennas

Antenna technology is one of the keys to overall wireless communications performance. Through the appropriate use of the antennas, radio coverage can be enhanced, interference minimized, and capacity increased. Antennas are presented in several forms according to specific features.

Omnidirectional Antennas. An omnidirectional antenna presents a circular radiation pattern in the horizontal plane (azimuth), therefore radiating and detecting signals in all directions in the corresponding plane.

Directional Antennas. A directional antenna presents a radiation pattern with directional properties, radiating and detecting signals in the corresponding direction. A set of k $360°/k$-sectored antennas at an access point divides a circular area into k sectors and covers the entire circular area. In an idealized situation, such a configuration can be thought of as a primitive application of SDMA for k users.

Smart Antennas. A smart antenna is essentially a directional antenna whose beams or radiation pattern can be controlled in a dynamic fashion so that interference power is minimized and wanted signal power is maximized. Such a feature renders smart antennas suitable for application in SDMA architectures. Optimal SDMA can be implemented by means of smart antennas providing an infinitesimal beam width and infinitely fast tracking capability. In such an ideal condition, an infinite number of users will experience communication through interference-free spatial channels. Smart antennas are composed of an array of elements distributed in a given arrangement. Frequently, the elements in the array are identical or similar, copolarized, oriented in the same direction, and present low gain. They can be arranged as linear equally spaced, uniform circular, or uniformly spaced. These elements are joined into a combiner network where signal processing occurs. Smart antennas can be implemented through *switched beam* technology or *adaptive* technology. Both technologies are heavily dependent on the beam-forming techniques utilized.

Beam-forming. Beam-forming is a function performed at the antenna system through which energy is concentrated within a specific spatial direction for transmission or reception purposes. Beam-forming can be accomplished directly by constructing the antenna with the appropriate structure and shape or by including special networks to perform the task. Note that beam-forming can be thought of as spatial filtering because the signal of interest is selected according to its geographic distribution. Spatial filtering can be implemented by antenna arrays with each element of the array spatially sampling the wave front of the source at the location of the element. By appropriately processing the phases and the amplitudes of the outputs of these elements,

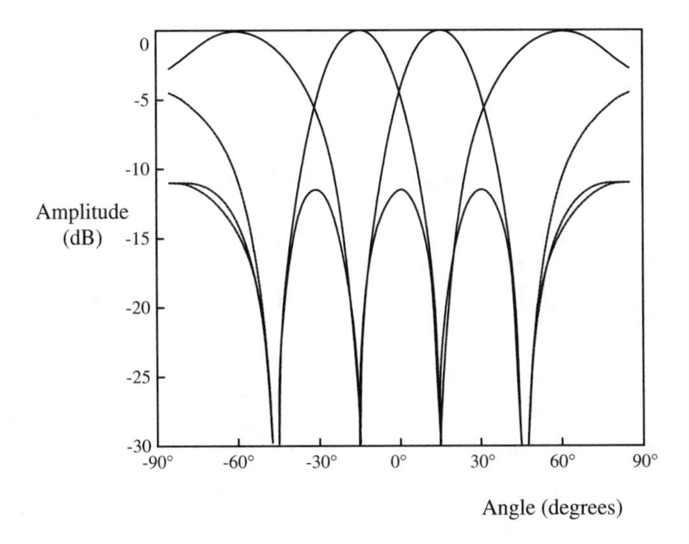

FIGURE 3.4
Four mutually orthogonal beams of the Butler matrix.

the desired beam-forming is achieved. The beam-forming network may yield a single beam or multiple beams. For better performance, these beams should present orthogonal properties, so that the maximum of each beam pattern corresponds to a null in the direction of each one of the other beam maxima. Figure 3.4 illustrates this concept for four beams. In Figure 3.4, the main lobes are shown as solid lines and the secondary lobes are shown as dashed lines. Note that, although these beams overlap, they are mutually orthogonal. An example of a multiple-beam beam-forming network, also known as a beam-forming matrix, is given by the Butler matrix.[6] Figure 3.5a shows a 90° hybrid coupler and Figure 3.5b depicts a Butler beam-forming matrix for a four-element array. Such a matrix is bidirectional; therefore, transmission and reception make use of the same beam pattern, with the ports able to both transmit and receive. Considering the signals to have the same phase at each of the elements, then the phases of the signals arriving at each one of the ports are those given in Table 3.1.

In addition, assuming the elements of the Butler configuration to be spaced by $\lambda/2$, where λ is the wavelength, the beams generated by such a network are those illustrated in Figure 3.4. More generally, a k-beam beam-forming network connected to an e-element antenna performs the following operation:

$$\mathbf{y}(t) = \mathbf{w}^H \mathbf{u}(t) \tag{3.14}$$

where $\mathbf{y}(t)$ is a k-dimensional column vector (k output signals), $\mathbf{u}(t)$ is an

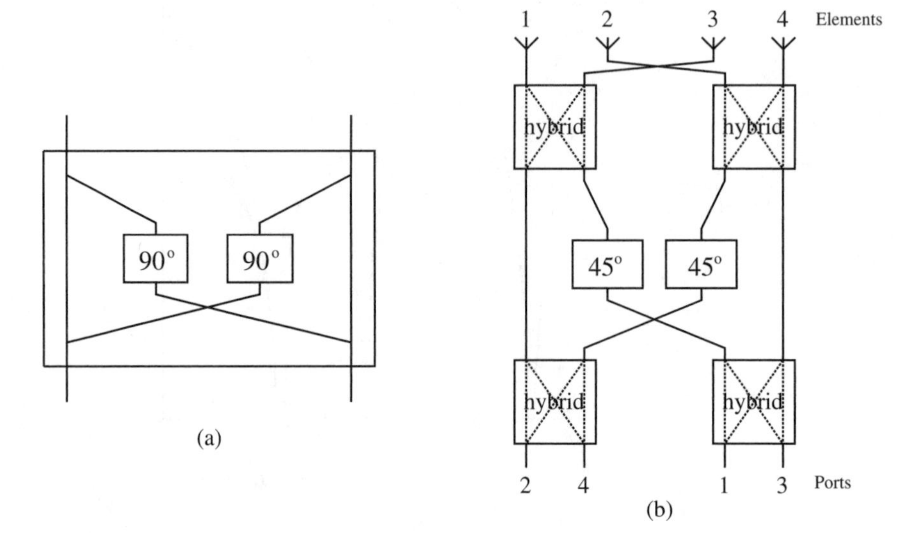

FIGURE 3.5
(a) A 90° hybrid coupler. (b) A Butler beam-forming matrix for a four-element array.

e-dimensional row vector of signals (e input signals), **w** is an $e \times k$-dimensional weight matrix, and H represents Hermitian transpose (transpose of the complex conjugate). With a suitable manipulation of the weights **w**, the beam is formed within the desired direction. At the transmission side, this is translated into transmitting the signal in the desired direction. At the reception side, this is translated into coherently adding the signals arriving from the desired direction. Orthogonal beams are produced by means of mutually orthogonal weight vectors, i.e., the desired direction is selected and the corresponding signals detected. From Table 3.1 it can be seen that the weight vectors corresponding to each port are mutually orthogonal. (The dot product for any pair of rows is always nil, e.g., for the first and the second rows: $e^{-j45^\circ}e^{j0^\circ} + e^{-j180^\circ}e^{-j45^\circ} + e^{j45^\circ}e^{-j90^\circ} + e^{-j90^\circ}e^{-j135^\circ} = 0$.)

TABLE 3.1

Element Phasing for the Butler Matrix

	Element 1	Element 2	Element 3	Element 4
Port 1	−45°	−180°	45°	−90°
Port 2	0°	−45°	−90°	−135°
Port 3	−135°	−90°	−45°	0°
Port 4	−90°	−45°	−180°	−45°

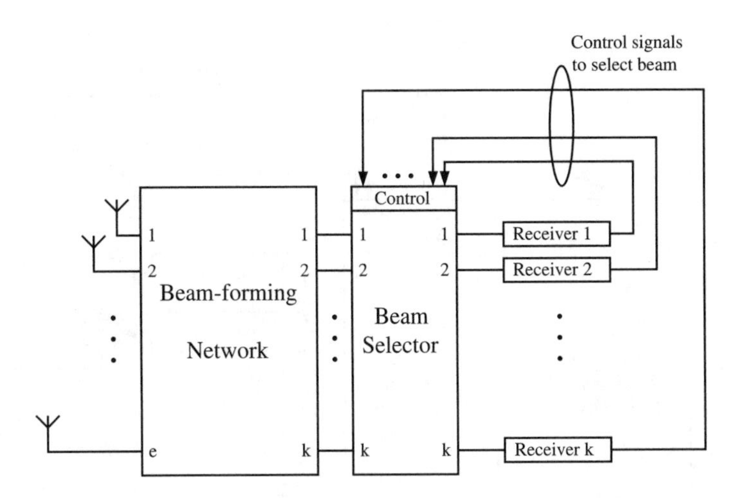

FIGURE 3.6
Switched beam system.

Switched Beam Antennas. A switched beam antenna provides a set of fixed beams that can be selected according to a given criterion. For SDMA purposes, the particular beam providing the greatest signal enhancement and interference reduction for any given signal is certainly part of the general criteria for beam selection. A switched beam system is implemented by means of a beam-forming network, an RF switch (beam selector), and a control logic as illustrated in Figure 3.6. A beam-forming network generates k beams from the e elements. By using a control logic, the RF switch selects the beam providing the largest signal enhancement and interference reduction. Switched beam systems are simple to implement and can easily fit into wireless networks already in operation. With a lower complexity and at a lower cost, as compared with the other type of smart antennas, switched beam systems empower the smart antennas' capability and increase system capacity. Their performance, on the other hand, is less satisfactory than that of systems using adaptive antennas because of the fixed characteristics of the direction of the beams. For example, by moving in an arc around the access point a user will experience a signal fluctuating in accordance with the roll-off of the antenna pattern. Note from Figure 3.4 that the beams cross at approximately −4 dB points, a figure typical of beam-forming networks.

Adaptive Antennas or *Adaptive Beam-forming*. An adaptive antenna adjusts its radiation pattern dynamically to enhance the desired signal and minimize the interfering signals. A block diagram of an adaptive antenna system is illustrated in Figure 3.7. Individual receivers, down-converters, and analog-to-digital (A/D) converters are used to produce the inputs for k e-input adaptive

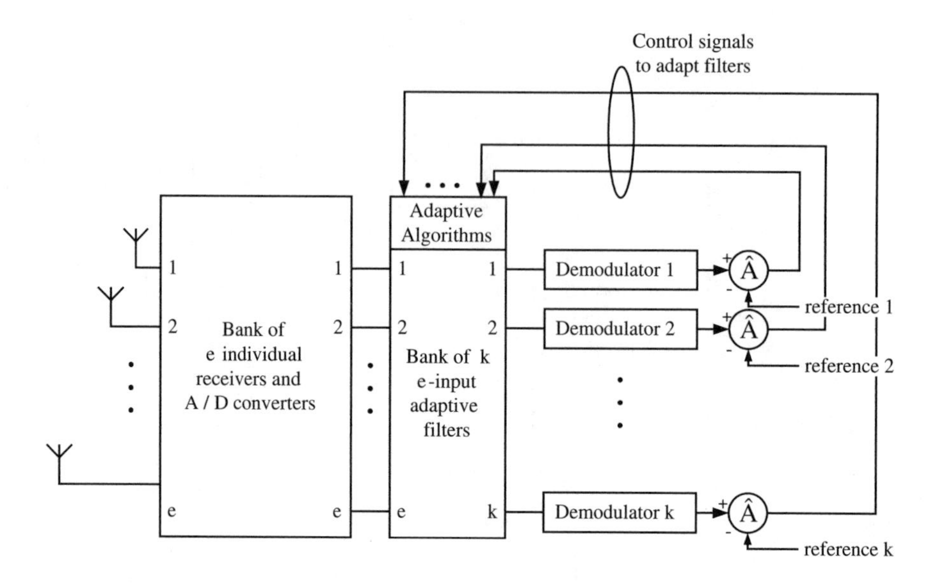

FIGURE 3.7
Adaptive antenna system.

filters. These filters have their weights adjusted by adaptive algorithms whose inputs are the errors between the detected signal and an estimate or a replica of the signal, which Figure 3.7 denotes as *reference*. The beam-forming technique determines parameters (weight vectors) that minimize a cost function so that the use of such parameters maximizes the signal quality. A typical cost function is written in terms of the square of the difference between the array output and a locally generated estimate or replica of the desired signal for a given subscriber. This essentially characterizes a digital filtering problem. The computation of the weights is carried out continuously (once each time slot for a TDMA system, for example).

Signal Processing

The signal processing techniques empowering SDMA are spatial filtering and beam-forming. Spatial filtering algorithms aim at separating the several signals arriving at the antenna array during reception. Beam-forming algorithms master the radiation patterns during transmission. Both signal processing techniques are heavily dependent on the accurate estimation of the spatial correlation among signals. Given that the signals can be discriminated according to their propagation paths, several parameters can be estimated to be used in signal processing techniques. These include number of dominant propagation paths of each user; the directions of arrival of all dominant propagation paths; and the attenuation of each dominant path. By computing the weights for beam-forming purposes, the phases at the antenna elements are adjusted to steer the main beam pointing direction toward the estimated direction of arrival of the desired user. Because these weights are computed from the parameters of the signals arriving at the antenna, a good reciprocity between forward and reverse channels must exist for better performance. Good reciprocity can be achieved in TDD systems but not in FDD systems, in which forward and reverse channels are usually kept apart from each other for half of the total bandwidth. In this case, the algorithms must compensate for different frequencies.

Channel Capacity

The implementation of SDMA requires sophisticated techniques and technology. Channel capacity becomes a very sensitive issue and cannot be evaluated in a straightforward manner. The brief calculations that follow are shown for didactic purposes only. Assume an ideal SDMA, implemented in such a way that coverage requirements are perfectly accomplished in k perfectly disjoint and contiguous wedges illuminated by directional antennas with gain equal to k. Each signal occupies the whole bandwidth B for $1/k$ of the space, but with average power equal to kS. Therefore, the capacity C_k per user normalized with respect to the bandwidth B is

$$C_k = \frac{1}{k} \log_2 (1 + \gamma k C_k) \tag{3.15}$$

The total capacity per bandwidth of the system is defined as $C_{\text{SDMA}} = kC_k$. Therefore,

$$C_{\text{SDMA}} = \log_2 (1 + \gamma C_{\text{SDMA}}) \tag{3.16}$$

which is the same as that for FDMA.

3.5.5 Brief Remarks on Scheduled Multiple-Access Techniques

FDMA is a very popular technique used in all 1G and 2G wireless systems; its deployment in higher generations is unequivocal. TDMA, CDMA, and a simple form of SDMA have always been used in conjunction with FDMA. In fact, although not widely known as such, frequency reuse can be thought of as a primitive form of SDMA. The same comment applies for the use of directional antennas. TDMA and CDMA appeared in second-generation wireless networks. A more sophisticated form of SDMA can be used in conjunction with any one of the other multiple-access techniques with the aim at increasing capacity. Figure 3.8 plots the capacity per user normalized with respect to the bandwidth for FDMA, TDMA, CDMA, and SDMA. The upper curve

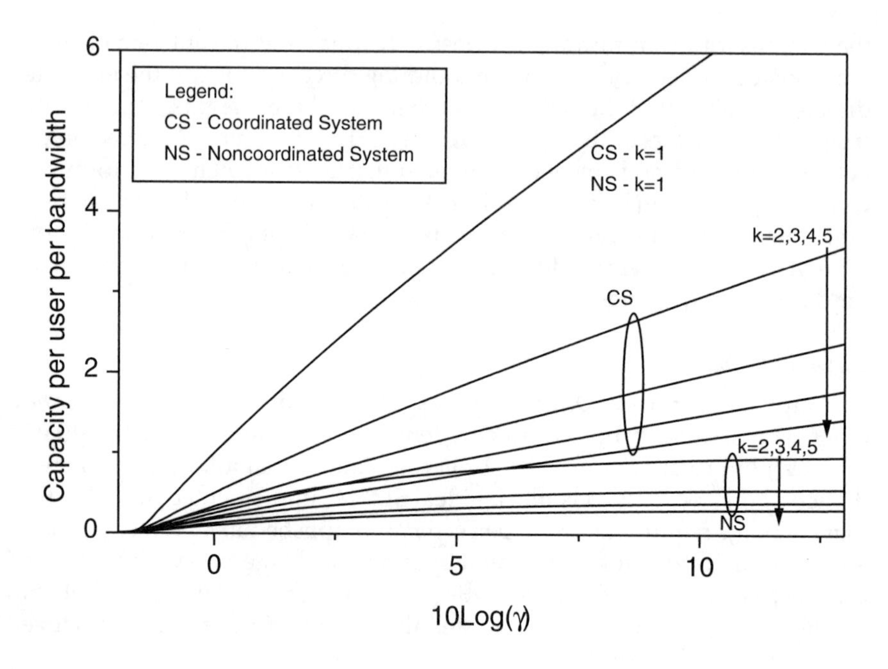

FIGURE 3.8
Capacity per user normalized with respect to the bandwidth for FDMA, TDMA, CDMA, and SDMA.

corresponds to the capacity of a single user for all scheduled multiple-access schemes. This is the same as the *total capacity* of each of the schemes with the exception of noncoordinated CDMA. Channel capacity, as analyzed in this section, applies for Gaussian channels and is determined in terms of bits per second. In a radio communication environment, where signal transmission encounters harsher conditions, what has been found here may be viewed as an upper bound performance.

3.6 Random Multiple Access

The random multiple-access techniques are more efficiently utilized if applied to bursty (random) traffic. In this case, the resources are seized as demand arises, therefore on a random basis, and are allotted for the duration of the access only. Because the protocols in this category sustain no rigid discipline in accessing the medium, collisions may occur. Terminals can find they are involved in a collision in several ways. They can listen to the channel to verify if the information sent is distorted or not, or they can wait for an acknowledgment from the access point.

Because these protocols are primarily used for initial access purposes their performance must be assessed in terms of parameters related to message (packet) transmission. In this case, *throughput* and *delay* are the applicable performance measures. The throughput S is defined as the number of successfully transmitted messages per time unit. The delay D corresponds to the time taken for the message to be successfully transmitted. For many random-access protocols the throughput can be obtained in a closed-form expression. The delay, defined as the sum of queueing delay, propagation delay, and transmission time, constitutes a more complex issue. For very simple algorithms it can be obtained in a closed-form formula. More generally, simulation mechanisms are the common means used to achieve the measures of throughput and delay.

The performance of the protocols is affected by many factors. In addition to those factors commonly influencing the performance of any wireless network such as near–far effect and multipath propagation, the *vulnerable period*, a factor specific to access methods for which the degree of coordination is not tight, is a decisive element. The vulnerable period T of a protocol is the time interval susceptible to collision. It depends on the duration of the packet and also on the propagation delay time a across the bus length.

For fixed-length messages (packets) the parameters can be normalized with respect to the packet time. Throughput then is given as the number of successfully transmitted packets per packet time ($0 \leq S \leq 1$). Delay and vulnerable period are given in units of packet times. In the same way, the offered traffic G, which represents all the traffic in the network, is given in packets per packet time and includes both newly generated packets and retransmitted packets. Therefore, for a given probability p of successful packet transmission:

$$S = pG \tag{3.17}$$

For the generation of packets following a Poisson distribution, the probability p_k that k packets are generated in T packet time is

$$p_k = \frac{(TG)^k}{k!} \exp(-TG) \tag{3.18}$$

This section describes some of the various random multiple-access methods, those more closely related to the practical wireless networks, such as pure ALOHA, slotted ALOHA, tree, FCFS, and the various forms of CSMA access protocols. It then determines the throughput and the delay of some of the protocols.

3.6.1 ALOHA

The ALOHA comprises a set of very simple protocols, which are, in fact, the basic and best-known protocols in the random-access category.

Pure ALOHA

In pure ALOHA, there is no discipline for initial transmission. As soon as the packet is ready for transmission, it may be cast into the medium. Upon collision, a random time is waited and retransmission is attempted. Assume, initially, a propagation delay time $a = 0$. Because of the lack of discipline, the vulnerable period of this protocol spans two packet times, as explained next. Refer to Figure 3.9. In the extreme condition, the initial portion of a test packet may collide with the end portion of a previously sent packet. In the same way, the end portion of the test packet may be stricken by the initial portion of a packet just accessing the medium. Therefore, for a propagation delay time equal to zero, the vulnerable period in this case equals $T = 2$. More generally, for a propagation delay time equal to a, then $T = 2(1 + a)$. By replacing this in Equation 3.18 we find the probability that an arbitrary packet is overlapped

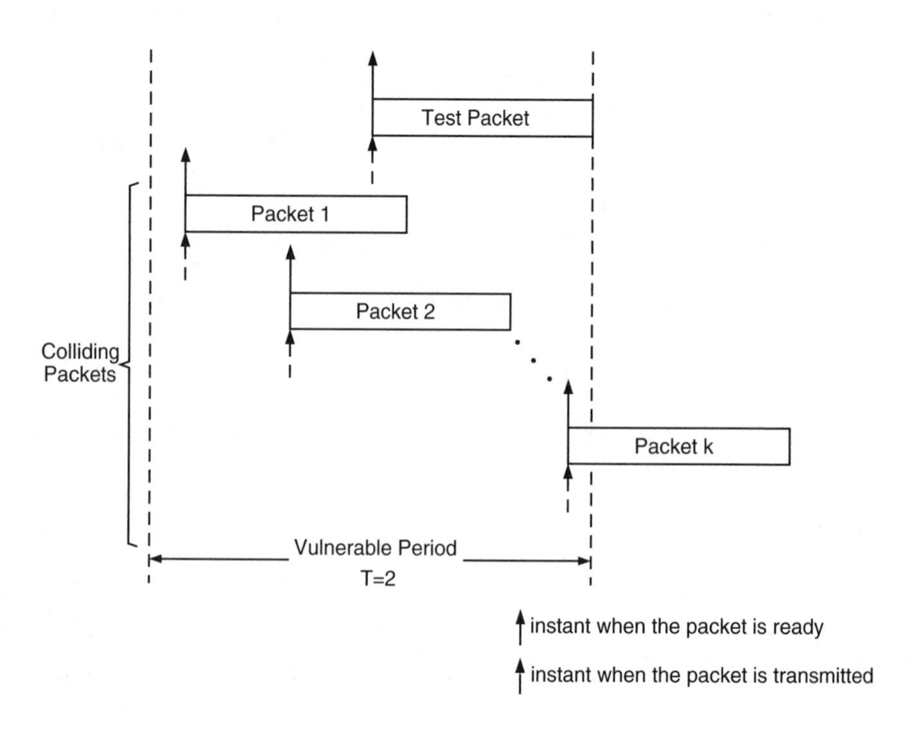

FIGURE 3.9
Vulnerable period for ALOHA.

by k packets, such that

$$p_k = \frac{[2(1+a)G]^k}{k!} \exp[-2(1+a)G] \tag{3.19}$$

The probability of a successful transmission in T packet times is the probability that no message is generated within T, i.e., $p = p_0 = \exp[-2(1+a)G]$. Therefore, from Equation 3.17, the throughput of the ALOHA protocol is

$$S = G\exp[-2(1+a)G] \tag{3.20}$$

The maximum throughput, obtained by performing $dS/dG = 0$ in Equation 3.20, is found to be

$$G = \frac{1}{2(1+a)}$$

packet per packet time, for which

$$S = \frac{1}{2(1+a)e}$$

And $S \approx 0.184$ for propagation delay time equal to zero.

As mentioned previously, the delay is determined by the addition of three terms: queueing delay, propagation delay, and transmission time. This can be estimated in an approximated fashion as follows. We consider the queueing delay time as the total time spent in unsuccessful transmissions (total time before a successful transmission), which is obtained by multiplying the expected number of transmissions times the average delay δ for one transmission. The propagation delay is simply a and the transmission time is equal to 1. Given that the probability of a successful transmission is $p = p_0 = \exp[-2(1+a)G]$, then the probability of a collision is $1 - p = 1 - \exp[-2(1+a)G]$. The probability $p_{k\text{-attempts}}$ of a transmission requiring exactly k attempts (i.e., $k-1$ collisions followed by one success) is

$$p_{k\text{-attempts}} = \exp[-2(1+a)G]\{1 - \exp[-2(1+a)G]\}^{k-1}$$

The expected number of transmissions is

$$E[k] = \sum_{k=1}^{\infty} k p_{k\text{-attempts}} = \exp[2(1+a)G]$$

And the expected number of retransmission is the expected number of transmission minus one, i.e., $\exp[2(1+a)G] - 1$. Therefore, the delay is given by

$$D = \{\exp[2(1+a)G] - 1\}\delta + a + 1$$

The parameter δ (delay for one transmission), by its turn, equals four terms: the time $1 + a$ taken for the packet to reach its destination; the time w the receiver takes to generate the acknowledgment; the propagation time a for the acknowledgment to reach the terminal; and the time r taken for the terminal to decide for the retransmission. Therefore, $\delta = 1 + 2a + w + r$. The time r depends on the retransmission policy (known as back-off algorithm) whose aim is to spread retransmissions out over an interval of time to ensure that an increase in traffic load does not trigger a decrease in throughput. If r is selected from a uniform distribution ranging from 0 to f packet transmission times, then $r = f/2$. Note that the minimum value of δ is 1 obtainable for $a = w = r = 0$. In this case $D = \exp(2G)$. If the time taken to generate the acknowledgment and the time taken for the terminal to decide for retransmission are nil, then $D = (1 + 2a) \exp[2(1 + a)G] - a$.

Slotted ALOHA

The slotted ALOHA introduces some sort of discipline to reduce the vulnerable period. The time is divided into fixed-length time slots, with the time slot chosen to be equal to the packet transmission time and with the packet allowed to be sent only at the beginning of a time slot. Assume, initially, a propagation delay time $a = 0$. In this case, as can be visualized in Figure 3.10, the vulnerable period is equal to one packet time, i.e., $T = 1$. More generally, for a propagation delay time equal to a, then $T = 1 + a$. By replacing this in Equation 3.18 we find the probability that an arbitrary packet is overlapped by k packets, such that

$$p_k = \frac{[(1 + a)G]^k}{k!} \exp[-(1 + a)G] \tag{3.21}$$

The probability of a successful transmission in T packet times is the probability that no message is generated within T, i.e., $p = p_0 = \exp[-(1 + a)G]$. From Equation 3.17, the throughput of the slotted ALOHA is

$$S = G \exp[-(1 + a)G] \tag{3.22}$$

In this case, the maximum throughput is obtained for

$$G = \frac{1}{1 + a}$$

packet per packet time, for which

$$S = \frac{1}{(1 + a)e}$$

and $S \approx 0.368$ for a propagation delay time equal to zero.

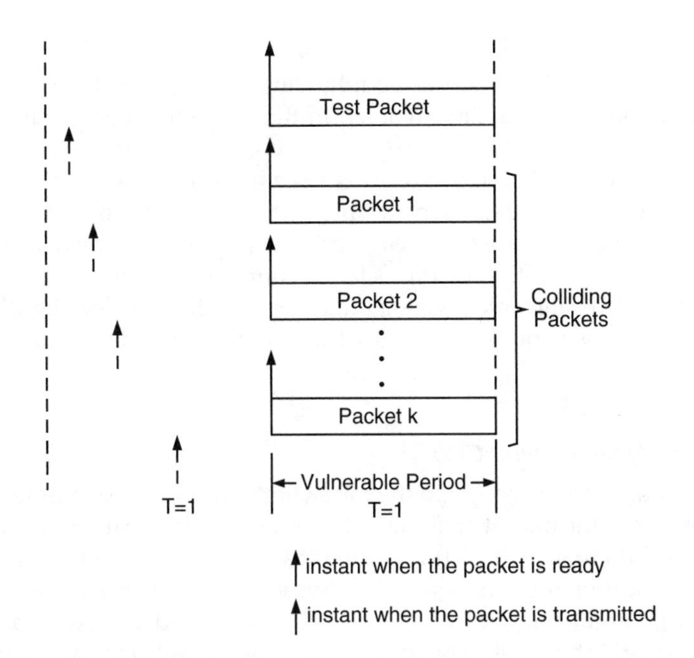

FIGURE 3.10
Vulnerable period for slotted ALOHA.

A reasoning similar to that used for the pure ALOHA can be applied to the slotted ALOHA to estimate the delay. In such a case,

$$D = \{\exp[(1+a)G] - 1\}\delta + a + 1.5$$

where $\delta = 1.5 + 2a + w + r$. In this case, we have assumed that, on average, the packet is ready for transmission, or retransmission, in the middle of the time slot. Note that the minimum value of δ is 1.5 (obtained for $a = w = r = 0$), for which $D = 1.5\ \exp(G)$. If the time taken to generate the acknowledgment and the time taken for the terminal to decide for retransmission are nil then $D = (1.5 + 2a)\exp[2(1+a)G] - a$.

3.6.2 Splitting Algorithms

The splitting algorithms comprise a set of protocols whose common feature is the application of some sort of segregation to resolve the conflicts.

Tree

The tree algorithm is based on the following strategy. At the occurrence of a collision, the terminals not involved in the collision enter a waiting state. Those involved are split into two groups according to a given criterion (e.g., by flipping a coin). The first group is permitted to use one time slot, and the second group can use the next time slot, if the first group is successful. However, if another collision occurs, a further splitting is carried out until, eventually, a group with only one active terminal will be allowed to transmit successfully. This algorithm yields a maximum performance slightly better than the previous protocol ($S \approx 0.47$ for a propagation delay time equal to zero[7]).

First-Come First-Served (FCFS)

The FCFS algorithm is based on the following strategy. At each time slot, say, time slot k, only the packet arriving within a specified allocation time interval, (say, from $T(k)$ to $T(k) + t(k)$) is entitled to be transmitted. If a collision occurs, the allocation interval is split into two equal subintervals—from $T(k)$ to $T(k) + t(k)/2$ and from $T(k) + t(k)/2$ to $T(k) + t(k)$—and a packet that arrived in the first subinterval is sent. In case of another collision, a further splitting ($t(k)/4$) is required, and so on, until the transmission is successful. This algorithm yields a maximum performance slightly better than the previous protocol ($S \approx 0.487$ for a propagation delay time equal to zero[8]).

3.6.3 Carrier Sense Multiple Access

Carrier sense multiple access (CSMA) comprises a class of protocols whose common feature is the sensing of the status (busy or idle) of the transmission medium before any transmission decision policy is exercised. This does not necessarily require the use of a carrier but simply the ability to detect idle or busy periods. Note, however, that a terminal may sense the medium idle although, in fact, a packet may be traveling through it and, because of the bus length, the bus occupancy is not detected at the time of the sensing. These protocols may appear in nonslotted and slotted versions. In the nonslotted variant, there is no rigid time to initiate the transmission of the packet. In the slotted mode, the time axis is divided into slots, whose length is chosen to be equal to a submultiple of the packet time, and the packet transmission will always occur at the beginning of a slot. In such a case, the status of the medium is sensed at the beginning of the slot next to the time of a packet arrival at the terminal. The performance analyses derived for these protocols assume the sensing time to be nil. For details of the derivations, the reader is referred to References 8 and 9.

Nonslotted Nonpersistent CSMA

In this protocol, the *listen-before-talk* strategy is used. If the terminal senses the medium to be idle, the packet is transmitted immediately. Otherwise, a random time is waited and the process is reinitiated. The throughput in this case is given by

$$S = \frac{G \exp(-aG)}{(1+2a)G + \exp(-aG)} \tag{3.23}$$

Slotted Nonpersistent CSMA

In this protocol, the listen-before-talk strategy is again used. If the terminal finds the medium idle the packet is transmitted immediately. Otherwise, the process is reinitiated (the medium is sensed again in the next slot). Recall that the medium is sensed at the beginning of the time slot and transmission occurs synchronously with the slot. The throughput in this case is given by

$$S = \frac{aG \exp(-aG)}{1 + a - \exp(-aG)} \tag{3.24}$$

Nonslotted 1-Persistent CSMA

In this protocol, the listen-before-talk strategy is again used. If the terminal finds the medium idle, the packet is transmitted immediately, i.e., transmission occurs with probability 1. Otherwise, the process is reinitiated. Note here that the medium is constantly being sensed until it is found idle for the transmission. Note also that if one or more terminals sense the medium to be busy, collision will certainly occur. This is because these terminals will sense the medium to be idle at the same time and will then transmit their packets. The throughput in this case is given by

$$S = \frac{G[1 + G + aG(1 + G + aG/2)]\exp[-(1+2a)G]}{(1+2a)G - [1 - \exp(-aG)] + (1+aG)\exp[-(1+a)G]} \tag{3.25}$$

Slotted 1-Persistent CSMA

In this protocol, the listen-before-talk strategy is again used. If the terminal finds the medium idle the packet is transmitted immediately. Otherwise, the process is reinitiated. Note here that the medium is constantly sensed until it is found to be idle for transmission. But note also that these events occur synchronously with the time slot. The throughput in this case is given by

$$S = \frac{G[1 + a - \exp(-aG)]\exp[-(1+a)G]}{(1+a)[1 - \exp(-aG)] + a\exp[-(1+a)G]} \tag{3.26}$$

p-Persistent CSMA

In this protocol, which is applicable to slotted channels, the listen-before-talk strategy is again used. If the terminal finds the medium to be idle the packet is transmitted with probability p. With probability $1 - p$ transmission is deferred to the next slot. This process is repeated until either the packet is transmitted or the channel is sensed to be busy. In such a case, a random time is waited and the process is reinitiated, which, in the same way, corresponds to the action taken when a collision occurs. At the first transmission, when the medium is sensed to be busy the access procedure is deferred to the next slot. Note that this protocol reduces to the 1-persistent if p is chosen to be equal to 1.

Busy Tone CSMA (CSMA/BT) or Busy Tone Multiple Access (BTMA)

In this protocol, the listen-before-talk strategy is again used. A busy tone is transmitted by the access point on the forward channel to indicate that the reverse channel is in use. Note that this protocol assumes a network with a centralized topology. The aim of the busy tone approach is to solve the *hidden terminal problem*, as explained next. In packet radio networks, some terminals are within range and line-of-sight of each other, but some are not. Those within range may successfully sense the medium to be busy and therefore may use the access protocol as prescribed. On the other hand, those out of range cannot detect whether or not the medium is in use. Therefore, for an access point whose position is conveniently chosen to be within range of all terminals, the CSMA protocol can be applied as defined. The basic CSMA protocols (nonpersistent, 1-persistent, and p-persistent, with their slotted and nonslotted versions) can be used here.

Digital Busy Tone CSMA (CSMA/DT) or Digital (or Data) Sense Multiple Access (DSMA)

In this protocol, the listen-before-talk strategy is again used. A busy/idle bit is transmitted by the access point on the forward channel to indicate that the reverse channel is in use. Note that this is just a slight variation on the busy tone CSMA protocol; the difference is the use of a busy/idle bit instead of a tone. It applies for digital networks where a busy/idle bit is included within the forward channel frame structure for the purposes of indicating the occupancy status of the reverse channel. The basic CSMA protocols (nonpersistent, 1-persistent, and p-persistent, with their slotted and nonslotted versions) can be used here.

CSMA with Collision Detection (CSMA/CD)

In this protocol, in addition to the listen-before-talk scheme, the *listen-while-talk* strategy is also used. This means that, besides sensing the channel prior to transmission, the medium continues to be monitored by the terminals while their own transmissions are on course. As soon as a collision is detected, transmission is ceased and a jamming signal is sent to force collision consensus among users. A retransmission back-off procedure is initiated. Collision detection can be easily performed on a wired network by simply sensing voltage levels. In this case, if the sensed voltage level is different from the voltage level of the initial transmission, then collision has occurred. For a wireless network, this simple procedure is not applicable. Detecting the received signal and comparing it with its own transmitted signal is not effective since the signal of the terminal overrules all other signals received in its vicinity. Any better solution would certainly lead to complex signal processing techniques to compensate for this discrepancy. Typically, an acknowledgment for collision detection, as used in all the other protocols, is the simplest solution to this problem. The basic CSMA protocols (nonpersistent, 1-persistent, and p-persistent, with their slotted and nonslotted versions) can be used here.

CSMA with Collision Avoidance (CSMA/CA), Multiple Access with Collision Avoidance (MACA), or Multiple Access with Collision Avoidance and Acknowledgment (MACAW)

In this protocol, the listen-before-talk strategy is used. Unlike the other protocols in which an access attempt is carried out by casting the information packet itself onto the medium, this protocol provides for access in two steps: first by outputting a short frame—a request to send (RTS) message—and second, after receiving an acknowledgment—a clear to send (CTS) message—by transmitting the information packet. Both messages, RTS and CTS, encompass the data length to be sent after CTS has been successfully received. This protocol is basically designed for wireless applications. Suppose terminal A sends an RTS to B to initiate a communication. Terminals within reach of A can hear this RTS and therefore be able to follow the progress of the communication without interfering. In similar manner, terminals within reach of B can hear the CTS and be able to follow the progress of the communication so as not to interfere. Certainly, a collision may occur when more than one terminal attempts to send an RTS. In such a case, if no CTS is received within a given time, a random time is waited and access is retried. Performance is further improved by including an acknowledgment after the information message is successfully received (the MACA protocol is then renamed MACAW).

3.6.4 Brief Remarks on Random Multiple-Access Techniques

Several aspects can be looked into when dealing with contention-based protocols. In particular, stability considerations and capture effects are two relevant issues that can be explored to choose the appropriate protocol for any given application.

Stability Analysis

The performance analysis of the contention-based protocols, as examined here, assumes an infinite number of users and a statistical equilibrium, with the offered traffic modeled as a Poisson process with a fixed average arrival rate. This certainly simplifies the derivation of the expressions.

Refer to Figure 3.11 where the plots show the throughput vs. offered traffic for the various random multiple-access techniques. For any of the contention-based protocols, assume, initially, that the system operates at some steady value G of traffic, $0 < G \ll G_{max}$, where G_{max} is the traffic for which the throughput is maximum. Consider that a sudden increase of the traffic occurs and that the increase is such that the throughput is still kept below the maximum throughput. An increase in the traffic results in an increase of the throughput which, in turn, produces a decrease in the backlog of messages to be retransmitted. This leads to a decrease in the offered traffic, thus taking the system to operate around the initial steady-state value of the offered traffic. The region of the curve for traffic G within the range $0 < G \ll G_{max}$ is recognized as that within which the operation of the system is stable.

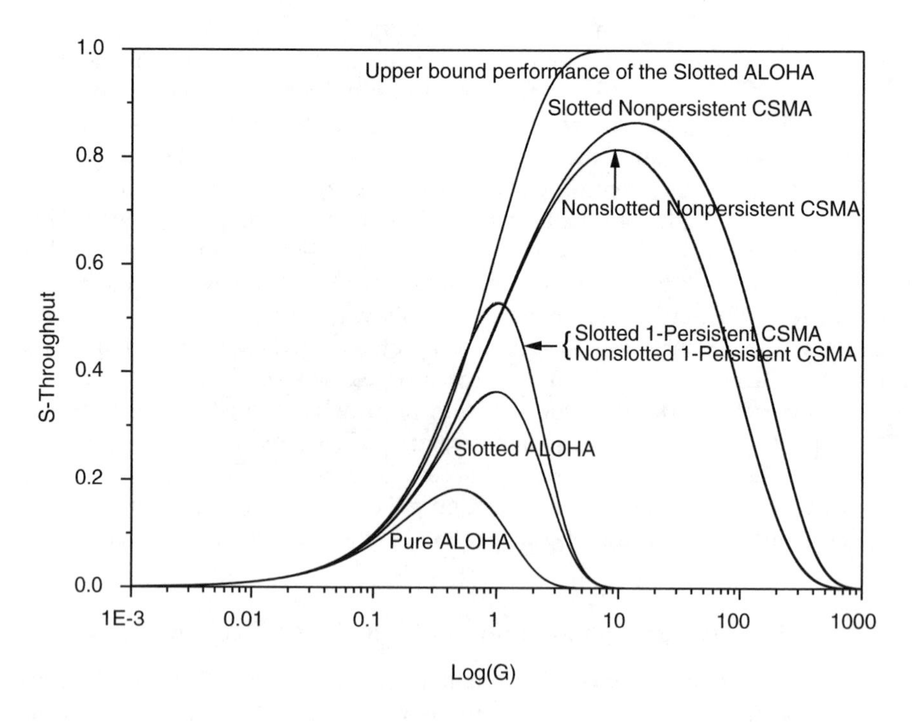

FIGURE 3.11
Throughput vs. offered traffic for the various random multiple-access techniques ($a = 0.01$).

Consider now that the system is led to operate at a value of traffic G, $G_{\max} < G < \infty$, and that an increase in the offered traffic occurs. In such a case, a decrease in the throughput also occurs, i.e., fewer packets are successfully transmitted with more collisions taking place. This leads to an increase in the number of retransmissions, thus in the offered traffic, which, as already observed, leads to a smaller throughput, and so on, until eventually the traffic escalates to infinity and the throughput goes to zero. The region of the curve for traffic G within the range $G_{\max} < G < \infty$ is recognized as that within which the operation of the system is unstable.

The considerations on stability carried out here apply for all contention-based protocols operating with an infinite number of users. The infinite number of users condition implies that the number of messages to be retransmitted has no influence in the number of new messages being generated, i.e., the number of messages of both types can increase without limit. For a finite number of terminals, on the other hand, if the number of collisions increases, some terminals may choose to leave the game, thence reducing the offered traffic. Back-off algorithms may be adequately used so that the retransmissions occur in a time conveniently chosen to restore stability.

Delay

Figure 3.12 shows the delay vs. offered traffic. The increase in the delay with the increase of the traffic is related to the fact that, as the traffic increases, more collisions occur and more retransmissions are attempted. As can be observed in the formulation presented here, the performance of the protocols is dependent on the propagation delay. For the signal assumed to propagate at the speed of the light 1 km is traveled within 3.33 μs, i.e., the propagation delay is approximately 3.33 μs/km. In wireless applications, the duration of a time slot is usually on the order of several hundreds of microseconds. Therefore, the propagation delay is indeed a very small proportion of the duration of a time slot. Figure 3.13 plots the throughput vs. the propagation delay for maximum throughput traffic. Note that, apart from ALOHA and slotted ALOHA, the throughput of all the other protocols is very sensitive to the propagation delay.

Capture Effect

The expressions for throughput as shown previously are derived on the assumption that if collisions occur, the packets involved in the collision are rendered unusable and retransmission takes place. Note that such an assumption is plausible if a perfect power control mechanism is used, in which case the packets competing for access are given equal chances. In a wireless communications environment, as mentioned before, signals from different terminals may arrive at the access point with different power levels as a result of path loss and fading. Path loss and fading act independently on the users, naturally

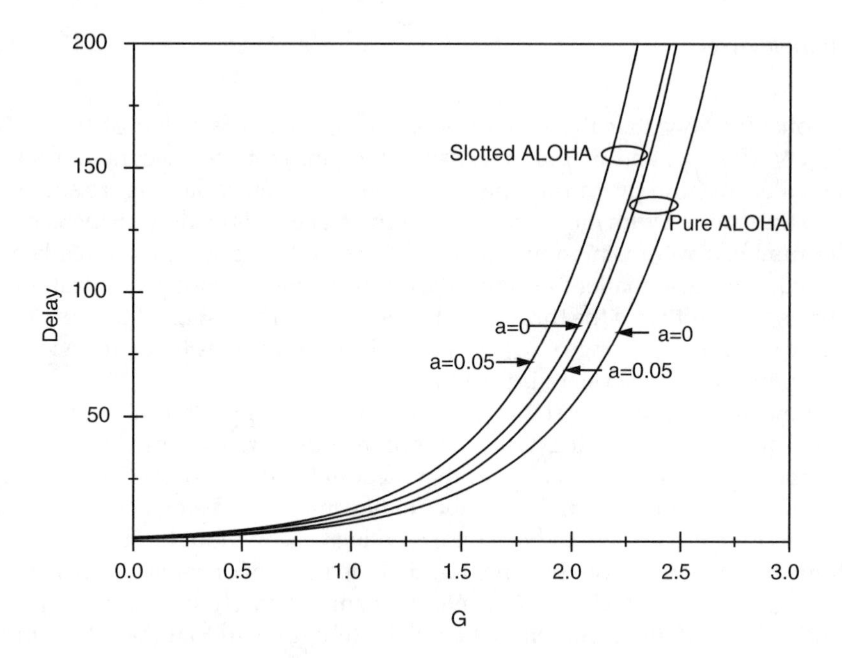

FIGURE 3.12
Delay vs. offered traffic ($r = w = 0$).

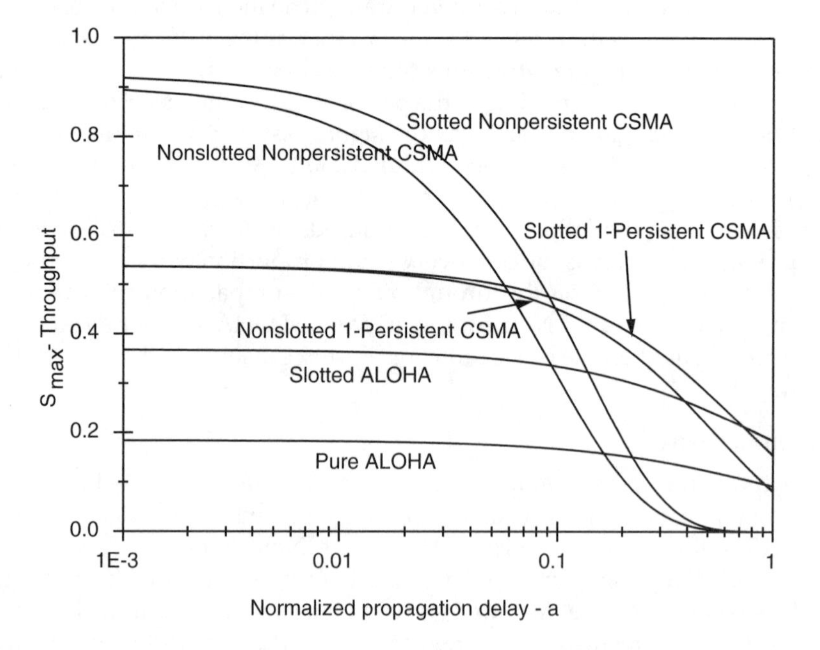

FIGURE 3.13
Throughput vs. propagation delay for maximum throughput traffic ($G = G_{max}$).

splitting them into classes of access power. Therefore, the wireless channel imposes an intrinsic priority in the accesses with this priority changing dynamically.

Given that the terminals are naturally split into classes of access, it may be possible that in case $k+1$ users are competing for access one of the packets is successfully received, i.e., *capture* occurs. The k unsuccessful packets are said to have been *captured* by the successful packet. The capture phenomenon depends on a series of factors such as modulation technique, coding scheme, average signal-to-noise ratio, the minimum power-to-interference ratio (capture ratio or capture parameter), and channel characteristics.

This subsection estimates the upper- and lower-bound performances of the slotted ALOHA in the presence of the capture effect. The approach used here is certainly a very simple one.[10] It is shown with an aim at illustrating the phenomenon and can be easily extended to other types of protocols. A more rigorous derivation may be found, for example, in Reference 11. For protocols using fixed-length packets, for which the parameters can be normalized with respect to the packet time, the throughput represents the probability of a successful packet reception. For example, for both ALOHA and slotted ALOHA, the throughput can be obtained directly from Equations 3.19 and 3.21, respectively, by fixing $k = 1-$ the probability of having one packet in the system. Now assume that $k+1$ packets are competing for access, with one of them the wanted packet and the remaining k the interfering packets. Let $p_{k-\text{capture}}$ be the probability that the wanted packet captures the k interfering packets. The probability of a successful reception given that k packets are captured is $p_{k+1|k-\text{capture}}$. The unconditional probability of a successful reception is the throughput, such that

$$S = \sum_{k=0}^{\infty} p_{k+1|k-\text{capture}} \times p_{k-\text{capture}} \tag{3.27}$$

The conditional probability $p_{k+1|k-\text{capture}}$ depends on the access protocol. Specifically, for the ALOHA protocol it is given by Equation 3.19 and for the slotted ALOHA protocol it is given by Equation 3.21. The probability $p_{k-\text{capture}}$, as already mentioned, depends on several factors including the modulation technique, coding scheme, channel conditions, and others and equals the unity for $k = 0$, always ($p_{0-\text{capture}} = 1$).

At one extreme, upon collision all colliding packets are destroyed. In this case, successful reception is achieved if no collision occurs. Therefore, $p_{k-\text{capture}} = 0 \;\forall\, k \neq 0$. Hence, $S = p_{1|0-\text{capture}}$. For the slotted ALOHA, Equation 3.21 is used to yield:

$$S = G \exp\left[-(1+a)G\right] \tag{3.28}$$

as in Equation 3.22, representing the lower-bound performance (no capture) of the slotted ALOHA in a wireless medium. At the other extreme, upon collision one packet survives the collisions with the k interfering packets. Therefore, $p_{k-\text{capture}} = 1 \ \forall \ k$. Hence, $S = \sum_{k=0}^{\infty} p_{k+1|k-\text{capture}}$. For slotted ALOHA, Equation 3.21 is used to yield:

$$S = 1 - \exp\left[-(1 + a)G\right] \tag{3.29}$$

representing the upper-bound performance (perfect capture) of the slotted ALOHA in a wireless medium. Equations 3.28 and 3.29 are plotted in Figure 3.11. As already mentioned, Equation 3.28, which represents the lower-bound performance of slotted ALOHA, coincides with Equation 3.22, which yields the throughput if no capture is considered. Equation 3.29, which gives the upper-bound performance of slotted ALOHA, shows that the throughput tends to unity as the traffic load goes to infinity. This is because for a perfect capture any small difference in signal power is enough to resolve the contention.

3.7 Controlled Multiple Access

Intermittent traffic bursts require a high degree of flexibility for resource assignment purposes in low traffic conditions, and such a flexibility is provided by random multiple-access methods. As the traffic increases, because of the lack of a proper access coordination, a high incidence of packet collision occurs and delay becomes a critical factor leading to performance degradation. In this case, controlled multiple-access methods are recommended. The access methods in this category make use of deterministic or noncontention strategies in which, by means of a control signal, permission to send is granted to terminals individually, so that only one terminal is permitted to access the medium at a time. The access control can be performed on a centralized basis or on a distributed basis. The methods in the first case are referred to as polling controlled whereas those in the second case are referred to as token controlled.

3.7.1 Polling Controlled

Polling-controlled access comprises a set of centralized control access techniques through which terminals take turns to access the medium. A master station—a controller—periodically polls all the terminals in some prescribed order to determine whether or not they require access. In the positive case, the respective terminal sends its packet; otherwise the next terminal in the list is polled. The sequence with which the terminals are queried is determined in a such way that a high efficiency may be achieved. For example, a given terminal may be polled more than once before all the terminals are polled. Note that polling is highly dependent on the exchange of overhead messages between controller and terminals.

3.7.2 Token Controlled

Token-controlled access comprises a set of distributed control access techniques through which a token circulating among nodes is used as an access control to the medium. Tokens are special packets composed of bits arranged in a predefined pattern, identifiable by all terminals. In the absence of traffic, the token circulates from node to node so that whenever a terminal requires access to the medium it removes the token from circulation and holds it. With the token under its custody, exclusive access to the network is granted. The terminal then may transmit its packets immediately.

Token Ring

In a token ring, terminals are logically and physically ordered into a ring, with transmission occurring from one terminal to the next in a sequential manner. Therefore, a ring is not a broadcast medium, but an aggregate of individual point-to-point links laid in a circle. A bit stream received from the preceding terminal whose destination is not the present terminal is relayed to the subsequent terminal with at least one bit delay. This allows for the information to be read and regenerated, if necessary. A packet addressed to a given terminal will reach its destination after being received by and relayed to as many terminals in the path as required. Therefore, all nodes must provide for store-and-forward operation. The responsibility of withdrawing a packet from circulation is left to the originating station and this occurs when the packet reaches the originating station back after the round-trip is completed.

Token Bus

In a token bus, terminals are logically organized into a ring but physically attached to a broadcast medium, a bus, which can be of the bidirectional or unidirectional type. In the bidirectional configuration, the token-passing procedure requires that an address be appended to the token message so that the broadcast message may be captured by the appropriate terminal. A terminal is permitted to transmit if it receives a free token addressed to it. After transmitting, the terminal directs the idle token to the next terminal. In the unidirectional configuration, an implicit token-passing procedure is used. The bus in this case comprises an inbound channel and an outbound channel, the latter used for transmitting data and the former for reading the transmitted data. Both channels are connected in such a way that the information transmitted on the outbound channel is repeated on the inbound channel, thereby accomplishing broadcast communication among terminals. A natural ordering among terminals is provided because of the asymmetry created by the signal propagating unidirectionally. A round-robin access protocol can be easily implemented.

3.7.3 Brief Remarks on Controlled Multiple-Access Techniques

The polling techniques perform well if the following conditions are fulfilled: the round-trip propagation delay is not critical, the overhead due to polling messages is small, and the number of terminals is not excessive. It is not difficult to infer why in case any one of the mentioned conditions is not fulfilled performance is degraded. The token ring system has a known worst-case time to send the packets. If there are k terminals and a packet takes a time t to be sent, then the maximum time a packet will have to wait to be sent is kt. Note, however, that the ring itself does not provide for robustness, because a break in the chain brings the entire network down. A bus ring implementation in such a case is more appropriate.

3.8 Hybrid Multiple Access

The scheduled multiple-access techniques, with their rigid resource assignment, efficiently support stream, steady-flow traffic given a number of simultaneous users with a specific traffic profile sharing the resources. If the number of active users falls below that number or if the traffic changes into a different profile, those resources become inadequately used. The random multiple-access techniques, with their flexible resource assignment, efficiently support bursty traffic and perform adequately in low-traffic conditions with low average data rate and potentially high peaks, and operate with little or no centralized control. On the other hand, they can become very inefficient as the traffic load increases with the throughput degrading and the delay augmenting. The controlled multiple-access techniques, although overcoming the limitations of the random-access methods, serve very specific applications.

As communications networks evolve to support a variety of services in an integrated fashion, the choice of an appropriate access method among those supporting the conventional services becomes a challenge. Audio, video, and data and their various combinations may be characterized by steady flow traffic at one instant and by random traffic at another instant. Both constant bit rate and variable bit rate and the provision for different qualities of service are requirements of these networks. Therefore, choosing one access method, either scheduled or random-access, will lead to unsatisfactory system performance. More advanced access schemes, especially designed to accommodate traffic of dissimilar sources, must be considered.

Although in the long run the individual streams of traffic composing multimedia traffic exhibit a diverse behavior, in a given short term they can be made similar. For example, by eliminating the speechless (quiet) periods of a conversation the steady flow, continuous characteristic of the voice signal is destroyed. The voice signal in such a case is composed of *talkspurts* and can be transmitted in packets and, therefore, is able to share both a transmission medium and a switching network common to the traffic of different sources. Although data packets can experience delays in the case of system congestion, conversational speech requires packet delivery within a maximum permissible delay. The various multiple-access methods in this hybrid category attempt to control the transmission delay as well as access to the medium. Because these protocols comply with traffic of different sources, in addition to throughput a performance parameter of interest is the number of simultaneous conversations that can take place with satisfactory performance.

The extension of broadband services to the wireless environment has led to the introduction of the wireless asynchronous transfer mode (W-ATM) into the network. Besides its ability to handle both asynchronous and synchronous services, ATM provides for an end-to-end multimedia capability with guaranteed quality of service (QoS). Certainly, there may be quantitative differences in the achievable service characteristics because of the fundamental limitations of the radio medium, but the great flexibility of the ATM philosophy must be kept within the W-ATM. In particular, a subset of available bit rate (ABR), constant bit rate (CBR), unspecified bit rate (UBR), and variable bit rate (VBR) services should be provided by this technology. The access control methods, in this case, although they use different algorithms, heavily rely on conventional protocols. This section describes some of the various hybrid multiple-access methods, namely, R-ALOHA, Original PRMA, PRMA^{++}, Aggressive PRMA, Fast PRMA, Multirate PRMA, PRMA/DA, PRMA/ATDD, DQRUMA, DSA^{++}, DTDMA/PR, MASCARA, DTDMA/TDD, Pure RAMA, T-RAMA, F-RAMA, and D-RAMA. Unless otherwise specified, the methods described are essentially developed for an FDD configuration.

3.8.1 Reservation-ALOHA (R-ALOHA)

The R-ALOHA[12] can be thought of as a flexible TDMA in which slotted ALOHA is combined with time division multiplexing. Several schemes exist that implement R-ALOHA.

In one scheme, the time axis is divided into fixed-length frames, with the frame arranged into equal-duration slots, with the slot duration chosen to be longer than the longest propagation delay in the network. The protocol may operate in two modes: *unreserved* and *reserved*. In the unreserved mode, the slots are further divided into reservation subslots where the slotted ALOHA protocol can be applied. The slots in this case are known as reservation slots. A terminal with packets to be sent requests slot allocation by sending a reservation request on one of these subslots. It then waits for an acknowledgment and the corresponding assignment (one or more slots, as requested). The system thus switches to the reserved mode. In the reserved mode, in addition to a reservation slot, message slots to convey the information packets compose the frame. The proportion of reservation subslots to message slots is a design parameter that depends on the traffic profile. Such a proportion should be kept to a minimum so that resources are not wasted with transmission overhead.

In another scheme, the time axis is also divided into equal-length frames, which are organized into equal-length slots, with the slot duration chosen to be equal to the packet transmission time. A slot reservation is granted to a terminal in all subsequent frames if it succeeds in transmitting in that slot in any given frame. The transmission attempt is carried out by means of slotted ALOHA. Note, therefore, that resource reservation is achieved in an implicit manner by a successful transmission in the respective resource. A slot becomes available for a new access when it is no longer required, in which case it is released (goes empty).

3.8.2 Packet Reservation Multiple Access (PRMA)

The PRMA, like R-ALOHA, can be thought of as a flexible TDMA in which slotted ALOHA is combined with time division multiplexing. It encompasses a family of protocols with the basic aim to increase the radio interface capacity. Originally, PRMA was designed to accommodate voice and data services only. The latest versions, however, aim at multimedia applications.

Original PRMA

In the original PRMA,[13] the channel bit stream is organized in slots and frames, as in TDMA. Each slot within a frame can either be idle (available for contention) or busy (reserved). Terminals with new information to transmit contend for access to the idle slots, as in slotted ALOHA. Once a terminal is successful in the contention for a time slot in one frame, this slot is reserved for its exclusive use in subsequent frames, until it has no more packets to send. Terminals are informed about the status (available or reserved) of each time slot by a continuous signal stream broadcast by the access point. Note that such a protocol applies for a centralized network.

More specifically PRMA works as follows. At the beginning of a talkspurt, the terminal randomly chooses an idle slot and, with a given probability—permission probability, a design parameter in PRMA—sends its first packet. In case of an unsuccessful transmission, either due to collision or due to excessive error rate, this procedure is repeated until the packet is successfully transmitted. An acknowledgment is sent by the access point and the corresponding slot is rendered reserved in the next frames. The terminal then sends its packets in this slot with probability 1 until it has no more packets to send, at which point the slot is released. Two procedures can be used in the release process. In one of them—late release—an empty slot indicates that it is free. In another, a bit in the last talkspurt packet indicates that the slot is available. In both cases, the access point broadcasts a message that the slot has been released.

In PRMA, each terminal provides for a first-in/first-out buffer where the packets are stored prior to transmission. The size of the buffer depends on the application. Voice applications require that voice packets remaining in a queue must be discarded after some time so that a given grade of service can be achieved. Because PRMA guarantees slot reservation after a successful transmission is accomplished, old voice packets may be eliminated so that only the most recent packets are transmitted in a continuous mode. Voice quality is controlled by a packet-dropping probability parameter that is related to the maximum time a voice packet can be held before being discarded.

Assume that the channel rate is c bits per second and that the source rate is s bits per second. In a conventional TDMA system, the capacity, defined as the number of simultaneous users k_{TDMA} within the system, is given by $k_{\text{TDMA}} = c/s$. The gain over TDMA obtained with the application of PRMA is therefore $Gain = k_{\text{PRMA}}/k_{\text{TDMA}}$, where k_{PRMA} is the number of simultaneous users within the system with the application of PRMA. Assuming that the frame duration is f and that the frame comprises information of the source overhead message of h bits, then the number of slots per frame is $t = \lfloor cf/(sf + h) \rfloor$ where $\lfloor x \rfloor$ indicates the largest integer less than or equal to x. For a buffer able to hold up to b time units of information the corresponding size of the buffer given in number of slots is $\lfloor tb/f \rfloor$.

An important element in PRMA is the voice activity detector (VAD). By properly identifying and adequately using the various activity states of a voice signal, a statistical multiplexing gain can be achieved. In its simplest version—the slow version—a VAD identifies two states only: speech or silence. A more sophisticated version of a VAD—fast version—identifies other states such as mini-speech (syllabic activity) or mini-silence (intersyllabic activity). Medium access contention occurs during the active period of the conversation.

The performance of PRMA depends on a series of design parameters such as permission probability, frame duration, voice activity detection complexity, and others. If the permission probability is chosen to be small, an increase in the packet loss may occur because terminals will be likely to wait a long time before a new transmission is attempted. A long delay in packet transmission may imply that the packet must be discarded. A large permission probability imposes less restriction on the transmission attempts and more packets will be cast onto the medium, increasing the chances of packet collision and, thus, decreasing the capacity. Depending on the traffic profile an optimum permission probability may be encountered, but as the traffic changes dynamically, in order to have the protocol operating adequately, such a parameter would also have to change dynamically. By varying the frame duration, and assuming a constant overhead, the following effect in the performance may be expected. If the frame duration is made small the number of slots per frame is correspondingly small ($1/h$, in the limit), thus increasing the probability of packet collision. A large frame duration (c/f, in the limit) yields a correspondingly large number of slots per frame and, for a fixed buffer size in units of time, a small buffer size in number of slots. With a small buffer size, the probability of a terminal being granted a slot before the packet is discarded increases, therefore increasing the packet loss. As for the influence of the VAD, it is observed that a fast VAD provides a larger gain as compared with that given by a slow VAD in case the frame duration is kept small. In the same way, the gain is smaller otherwise. For a channel rate $c = 720$ kbit/s, a source rate $s = 32$ kbit/s, a frame duration $f = 16$ ms, and an overhead $h = 64$ bits, the number of slots per frame equals $t = 20$; the time slot is 0.8 ms; and the buffer size is 40 slots. For a permission probability of 0.3 and a packet-dropping probability of 0.01, the number of simultaneous conversation is found to be $k_{\text{PRMA}} = 37$. The corresponding TDMA capacity is $k_{\text{TDMA}} = 720/32 = 22.5$. In such a case $Gain = 37/22.5 = 1.644$.[13]

Note that in the original PRMA the same slot is used for contention and for information flow purposes. Therefore, in high-traffic conditions the reservation bandwidth may reduce to zero.

PRMA^{++}, Aggressive PRMA, Fast PRMA, Multirate PRMA

In PRMA^{++},[14] a slot is split into minislots, which are used for reservation purposes. Terminals contend for access on these minislots and the successful terminal will have a slot reserved for its packet transmission. Note that in this case the reservation bandwidth is always kept to a minimum, improving the stability problem of original PRMA. In aggressive PRMA,[17] a preemptive algorithm is used, in which voice services are given priority over data services. In fast PRMA,[15,16] the terminal has its slot allotted immediately after it has successfully accessed the medium. In multirate PRMA, the access point grants more or less bandwidth according to the traffic demand of the terminals.

PRMA with Dynamic Allocation (PRMA/DA)

In PRMA/DA,[18] data traffic, CBR, and VBR services are supported. Frames on the uplink are designed to be of equal duration and are divided into four variable-duration subframes each of which encompasses different quantities of equal-duration slots. The four subframes contain, respectively, data reservation slots, available reservation slots, CBR reservation slots, and VBR reservation slots. The downlink frames comply with the TDM format and operate on a contention-free basis. Access is carried out as follows. All terminals are initialized in the inactive state. Upon generation of a packet that is to be sent, the terminal then moves into the contending state and utilizes the slotted ALOHA procedure for this purpose, randomly choosing one of the available slots. It remains in this state until the terminal is successful or else the maximum setup time is exceeded. In the latter, unsuccessful, case, the packet is discarded and the terminal returns to the initial state. In the former, successful, case, the terminal then moves into the reserving state, where it will be assigned a certain number of slots in a given subframe. The type of service and the required bandwidth (number of slots) are specified in the initial access. The slots remain assigned to the terminal until the end of its active session. The dynamic allocation mode of operation of this method is based on four parameters: number of available slots, number of slots where contention occurred, number of slots where successful access occurred, number of unused slots. Generally speaking, if the number of collisions increases, then the number of available slots also increases; conversely, if the number of successful access attempts increases, then the number of available slots decreases.

One interesting feature of this method is certainly the provision for dynamic allocation of bandwidth with the aim to adapt the resource allocation to the traffic demand. On the contrary, because access attempts are carried out with packets already conveying payload information, i.e., no slots or minislots specially dedicated to contention purposes exist, an access attempt by itself demands a considerable bandwidth. As already observed in the performance of the protocols with this same characteristic, whereas this solution is satisfactory for low traffic load, it is certainly inadequate for heavy traffic.

PRMA with Adaptive TDD (PRMA/ATDD)

In PRMA/ATDD,[19] ABR, CBR, and VBR services are supported. As the name implies, the TDD scheme is used, where a frame is composed of a downlink subframe and an uplink subframe. The TDD frame is designed to support asymmetric subframes although the frame itself has a constant duration (fixed number of time slots). A slot, here referred to as an extended cell, comprises one ATM cell plus overhead for wireless transmission purposes. The first slot of the frame is used for synchronization and the second slot conveys broadcast information such as number of slots in one of the subframes, terminals vs. slot mapping, PRMA parameters, and others. The remaining slots carry information packets.

Four network elements are directly related to the multiple-access control layer: the static list handler (SLH), dynamic list handler (DLH), broadcast packet generator (BPG), and PRMA parameter computer (PPC). The SLH deals with the parameters to be stored in the static list. These parameters include call identifier, maximum allowable delay in the downlink, maximum allowable delay in the uplink, maximum bit rate, and others. These parameters are updated on a call-by-call basis for each call. The DLH deals with the dynamic parameters to be stored in the dynamic list. These parameters contain information about a specific ATM packet waiting in the access point buffer or in the terminal buffer. An example of such a parameter includes the maximum allowable delay to be complied with before the packet is discarded. The BPG generates the broadcast packets with the information available from the various sources (DLH, PPC, access point, and others). The PPC, based on the instantaneous traffic levels and on the QoS requirements, computes the permission probabilities related to the various transport services and to the number of available slots per frame.

One interesting feature of this method is the provision for two different list handlers, one with parameters used for the duration of the call, and another for the current packet within the buffer. By means of these lists, priorities to ATM packets can be appropriately handled so that the smallest number of packets is discarded. Like PRMA/DA, this protocol does not provide for slots, or minislots, specially dedicated to contention.

3.8.3 Distributed Queuing Request Update Multiple Access (DQRUMA)

In DQRUMA,[20] both ABR and VBR services are treated as bursty traffic for which no priority mechanism is provided. DQRUMA makes use of the time-slotted approach but no frame reference is specified. The uplink stream is divided into request access channels (RAC) and packet transmission channels (PTC), whereas the downlink stream is split into acknowledgment channels (AKC) and packet transmission channels. Each PTC is provided by one slot. The other two channels (RAC and AKC) are subslots within a given slot. In fact, any slot can be converted into a number of slots for these respective purposes, as required. A successful access, which is the result of the application of a contention protocol, is immediately followed by an acknowledgment in the appropriate AKC. Permission to transmit, however, is granted by the access point based on the current traffic load and on a round-robin policy. Each ATM packet transmission then may include a piggyback message in case the terminal has more packets to be transmitted. An absence of the piggyback message determines the end of the slot reservation.

Note that in this algorithm the reception of the acknowledgment is carried out on a slot-by-slot basis, which speeds the process. The inclusion of a piggyback reservation field saves bandwidth by avoiding further requests. The use of the minislot scheme for access contention increases the chances of successful access by decreasing the probability of collision (the smaller the contention packet, the smaller the chances of collision). In the same way, the loss of a contention packet has a small impact on channel utilization. The disadvantage of this protocol is that it treats ABR and VBR as bursty traffic.

3.8.4 Dynamic Slot Assignment (DSA^{++})

In DSA^{++},[21] terminals are allotted resources according to the QoS and instantaneous capacity requirements. ABR, CBR, VBR, and UBR services are supported. DSA^{++} makes use of variable-duration frames with downlink and uplink frames having the same duration and with each slot being of an ATM packet size plus overhead. The uplink frame contains message slots and random-access slots. The random-access slots are divided into minislots, which are used for access request purposes, which are carried out on a contention basis. The downlink frame conveys information such as a reservation message for each uplink slot, an announcement message for each downlink slot, signaling messages (collision resolution, paging, etc.), and feedback messages (empty, success, collision) for each random-access slot of the previous signaling period. The order with which backlogged terminals waiting for a feedback message are advised of the status of their requests is decided by means of a splitting algorithm. Resource assignment is determined on a priority basis and takes into account a set of dynamic parameters transmitted by the terminal along with each packet; one of them is the number of ATM packets and their due date. The dynamic parameters can be updated on request, in response to which the terminal will either carry out a random-access procedure or attend to a polling, as prescribed beforehand. The number of minislots for contention purposes to be provided in the next frame is determined according to a series of parameters, such as probability of a new packet arrival at each terminal in the contention mode since the last transmission of its dynamic parameters; number of terminals in contention mode; and throughput of the random-access procedure. CBR, VBR, ABR, and UBR ATM classes of services are assigned priorities in this very order.

Note that in this algorithm the broadcast of the information defining the next signaling period renders the protocol very flexible. On the other hand, a loss of a broadcast packet compromises a whole signaling period.

3.8.5 Dynamic TDMA with Piggyback Reservation (DTDMA/PR)

In DTDMA/PR,[22] ABR, CBR, and VBR services are supported. In addition, fixed-duration frames containing fixed-duration slots are used. The uplink is divided into three subframes, namely, reservation minislots subframe, long-term reservable slots subframe, and short-term reservable slots subframe with the boundary between the long-term and the short-term reservable subframes adjustable according to the traffic load. A voice activity detector is used to determine the active period of voice traffic. During any active period, CBR and VBR packets are generated periodically, the latter in groups of different sizes with the number of packets in each group chosen according to a given probability distribution function. In contrast, ABR packets are generated in bursts. Terminals with packets to be sent randomly choose a minislot in the next frame and transmit a reservation packet. At the end of the reservation period, the access point broadcasts a message containing the identification of the successful terminals, the number of slots assigned to each terminal, and the assigned slot positions. Reservations for CBR and VBR services can only occur in the long-term subframe and remain for the duration of the active period. Reservations for ABR services occur in the short-term subframe with the resource released immediately after the transmission of the packet. VBR traffic is guaranteed service by means of the piggyback reservation message. This message is sent along with the VBR packet whenever the number of assigned slots is smaller than the number of packets generated in the current slot. Because the traffic is delay sensitive, packets making use of CBR and VBR facilities are discarded after a waiting time limit is exceeded. Conversely, ABR packets can be buffered until a slot is assigned for transmission. A rearranging mechanism gathers all unused slots at the end of the frame.

Note that a high degree of flexibility is provided by this algorithm because of the movable-boundary subframes and because each subframe is dedicated to a different type of traffic. On the other hand, it does not account for the different QoS involved in each virtual channel. Hence, traffic with different characteristics shares the same priority when competing for resources.

3.8.6 Mobile Access Scheme Based on Contention and Reservation for ATM (MASCARA)

In MASCARA,[23] ABR, CBR, UBR, real-time VBR, and non-real-time VBR services are supported. In addition, a hierarchical mode of operation is used in which the master–slave scheduling functions, respectively, are performed by the access point and the terminals. MASCARA is developed for TDD applications with its variable-duration frames containing variable-duration uplink and downlink subframes. The subframes are subdivided into a variable number of slots. A slot is defined so that it comprises one ATM packet. The uplink subframe contains the contention period, for access purposes, and the up period, for information packet transportation purposes, both with variable durations. In the same way, the downlink subframe, transmitted in a TDM mode, contains the frame header period, for overhead information purposes, and the down period, for information packet transportation purposes, both with variable durations. More specifically, the contention period is used by the terminals to transmit reservation requests for subsequent frames or for some control information (e.g., registration). The frame header, transmitted at the beginning of each frame, is used by the access point to broadcast the length of each period, the results of the contention accesses in the previous frame, and the slot allocation for each active terminal. The number of slots within each period is determined as a function of the instantaneous traffic. In such a case, the period operating in the reservation mode may be reduced to zero slots, whereas that for contention purposes will always maintain a minimum number of slots to allow for registration. Payload information is transmitted by means of a cell train, which is a sequence of ATM packets belonging to a terminal with a common header. The payload information itself constitutes the MASCARA protocol data unit (MPDU) and the terminal determines if it will receive or transmit MPDUs in the current frame by means of the frame header. The type and the volume of traffic to be transmitted in the next frame by the terminal is decided by taking into account the service class of the current ATM virtual channels, the negotiated QoS, the amount of traffic, and the number of reservation requests. Transmissions over the radio are scheduled by means of the priority regulated allocation delay–oriented scheduling (PRADOS) algorithm, which considers the priority class, the agreed characteristics, and the delay constraints of each active connection. Priorities are given in a decreasing order to CBR, real-time VBR, non-real-time VBR, ABR, and UBR services in that order.

Note that the introduction of the cell train concept renders the algorithm flexible with the provision of variable capacity to the terminals. Moreover, the allocation of slots on a frame-by-frame basis facilitates the fulfillment of the negotiated QoS parameters for each connection. On the other hand, each access packet is relatively large (equivalent to two ATM packets—one for synchronism and overhead and one for control information), which increases the probability of collision, thence reducing the throughput. Indeed, the use of variable-duration frames introduces an extra difficulty in assigning capacity to terminals with CBR services.

3.8.7 Dynamic TDMA with Time Division Duplex (DTDMA/TDD)

In DTDMA/TDD, ABR, CBR, UBR, and VBR services are supported.[24] DTDMA/TDD, as the name implies, has been developed for TDD applications with its fixed-duration frames containing variable-duration uplink and downlink subframes, with the subframes subdivided into a variable number of slots. The uplink subframe is divided into four slot groups: request group (RQG), dynamic allocation group (DAG), fixed and shared allocation group (SAG), and fixed allocation group (FAG). The RQG contains minislots for access request purposes. The DAG conveys ABR and UBR traffic, the SAG carries VBR traffic, and the FAC transports CBR traffic. The downlink subframe, transmitted in a TDM mode, consists of two parts: one containing control and feedback signals, and another used to transmit information packets. Like the boundary between the uplink and downlink, the boundaries between the groups within the subframes are also movable and they are adjusted dynamically in accordance with the traffic demand. Terminals with packets to be transmitted access the request slots through the slotted ALOHA algorithm. The result of the access, along with the slots to be assigned and other control information, is transmitted by the downlink in the next subframe. Virtual channels for CBR traffic are allotted on a permanent basis for the duration of the active period. Virtual channels for ABR and UBR are allotted on a burst-by-burst basis with the slots chosen dynamically for the ABR/UBR group and from the unused slot of the CBR or VBR groups. Virtual channels for VBR traffic are assigned on a fixed shared basis, where some of the slots are allotted for the duration of an active period. In addition, some extra slots can be assigned according to a usage parameter control. The DTDMA/TDD protocol can be viewed as having two components, namely, the supervisory component (SC) and the core component (CC). The SC manages the call admission control, performs the channel scheduling, and builds a schedule table based on the relevant QoS parameters. The CC interfaces the data link control with the physical layer. In addition, for each virtual channel it (de)multiplexes the packets for transmission into the wireless medium, with this carried out in accordance with the schedule table supplied by the SC.

Note that the introduction of the data link control layer facilitates the inclusion of control tasks related to the transmission link itself. For example, a buffer is included to guarantee that ATM packet jitter is kept below acceptable limits. In the same way, for erroneous CBR packets retransmission is carried out through ABR channels, therefore keeping the CBR channels reserved for the transmission of new CBR packets. An upgrade of the protocol allows for retransmission of reservation packets to give priority to terminals requesting service for CBR or real-time VBR traffic.

3.8.8 Resource Auction Multiple Access (RAMA)

The RAMA comprises a set of deterministic protocols in which contention is avoided and the resources are assigned to the terminals by means of an auction mechanism. Terminals transmit their request through orthogonal, nonoverlapping signals within the same access medium so that these signals are univocally separable at the access point. Originally, RAMA was designed to accommodate voice and data services only. The latest versions, however, aim at multimedia applications. The best-known protocols within this family are Pure RAMA, Tree-Search RAMA, Fair RAMA, and Dynamic Priority RAMA.

Pure RAMA

In Pure RAMA,[25] the auction mechanism declares as the winner the terminal with the highest identification number, but the process occurs on a digit-by-digit basis as follows. Terminals requiring access transmit the most significant digit of their identification. The access point then selects the largest digit among those received and loops it back in a broadcast mode. Terminals whose identification does not contain the broadcast digit are withdrawn from the process, whereas those whose identification contains the digit remain in the process. These terminals then transmit their second most significant digit and this process is repeated until one and only one terminal is selected. The system may choose to carry out scheduled or nonscheduled auctions. In the first case, terminals with packets to be transmitted will have to wait for the next auction to initiate the process. In the second case, an auction may start as soon as the terminals require transmission. The access point may or may not be equipped with a buffer. In the first case, in the occurrence of an auction and if no resource is available the winner may be placed on a waiting list and assigned a resource as soon as one is made available. In the second case, the auction must be performed only if resources are available.

Note that in RAMA the information about the terminals that have been turned down is lost and these terminals will have to reenter the next auction process from the beginning. Note also that there is no equity in the process and terminals with low identity number will certainly have to take part in more than one auction to be assigned a resource.

Tree-Search RAMA (T-RAMA)

In T-RAMA,[26] terminals contending for access remain in the process until they are all served. In particular, terminals requiring access transmit the most significant digit of their identification. The access point then selects the largest digit among those received and loops it back in a broadcast mode. At the same time, the access point stores the remaining digits so that they can be used later in the process. Terminals whose identification contains the broadcast digit transmit their second most significant digit and the same procedure as previously described follows. This process is repeated until a terminal is declared the winner and a resource is assigned to it. The access point then resumes the auction process by broadcasting the least significant digit selected among the remaining digits. This is carried out until all terminals initially involved in the contention course are served. At this point, a new contention process is initiated with new terminals now allowed to participate.

Note that in T-RAMA the auction cycles may vary considerably because they depend on the number of terminals participating in the process. In such a case, periodicity makes no sense. (In RAMA, independently of the number of terminals participating in the process, only one terminal is served per cycle and the others are discarded). On the other hand, as in RAMA, buffers can be used so that an auction may be performed even though no resource is available. In this case, the identity of the winner remains in the buffer until a resource is released. It can be easily inferred that the time required to serve a given number of terminals is substantially shorter in T-RAMA than it is in

RAMA. This is because, whereas in RAMA the information of the losers is lost and they have to start the access process from the beginning, in T-RAMA the losers remain in the system and the access process starts from the point they were discarded. T-RAMA also facilitates the partition of users into classes, so that users subscribing to special services may be grouped into classes of services. Within the same class, however, equity becomes a more-complicated issue.

Fair RAMA (F-RAMA)

In F-RAMA,[27] the auction mechanism is replaced by a lottery process that makes the assignment procedure more equitable. Terminals requiring access transmit the most significant digit of their identification number. The access point draws a number from a uniform distribution and selects the digit among those received that is closest to the drawn number and loops it back in a broadcast mode. Terminals whose identification does not contain the broadcast digit are withdrawn from the process, whereas those whose identification contains the digit remain in the process. These terminals then transmit their second most significant digit and this process is repeated until one and only one terminal is selected. The system may choose to carry out scheduled or non-scheduled auctions. In the first case, terminals with packets to be transmitted will have to wait for the next auction to initiate the process. In the second case, an auction may start as soon as the terminals require transmission. The access point may or may not be equipped with a buffer. In the first case, in the occurrence of an auction and if no resource is available the winner may be placed on a waiting list and assigned a resource as soon as one is made available. In the second case, the auction must be performed only if resources are available.

Note that F-RAMA is very similar to RAMA but its assignment procedure is more equitable.

Dynamic Priority RAMA (D-RAMA)

In D-RAMA,[28] a more complex auction mechanism is introduced. In particular, besides the identification number of the terminal a priority parameter is included to meet the requirements of multimedia services. Terminals requesting access transmit the most significant digit of their priority parameter. The access point then selects the largest digit among those received and loops it back in a broadcast mode. Terminals whose priority parameter does not contain the broadcast digit are withdrawn from the process, whereas those whose priority parameter contains the digit remain in the process. These terminals then transmit their second most significant digit and this process is repeated until only the terminals sharing the same class of priority are kept. These terminals then send the most significant digit of their identification. The access point draws a number from a uniform distribution and selects the digit among those received that is closest to the drawn number and loops it back in a broadcast mode. Terminals whose identification does not contain the broadcast digit are withdrawn from the process, whereas those whose identification contains the digit remain in the process. These terminals then transmit their second most significant digit and this process is repeated until one and only one terminal is selected. After the terminal is selected, it indicates the number of slots required. Note that the first part of the auction process follows the RAMA procedure, whereas the second part uses the F-RAMA scheme. A sensitive issue in this protocol is the priority parameter P. This is defined as

$$P = P_{\min} + \lceil \max \{\text{contention}, \text{buffer}\} \rceil \quad (3.30)$$

where $P_{\min}$ defines the minimum priority for that particular access, contention gives the number of times the access is being tried without success, buffer informs the number of packets still in the buffer to be transmitted, and $\lceil x \rceil$ signifies the smallest integer less than or equal to x. The parameter $P_{\min}$ is given a nonzero value for voice transmission and a zero value for other types of traffic. The protocol chooses to use integer values for the priority parameter to facilitate its encoding and to save bandwidth. The contention parameter is introduced to control the maximum number of slots a packet is allowed to wait before it is discarded. The buffer parameter is used to control the loss of packets occurring for the number of packets exceeding the buffer size.

Note that D-RAMA is more complex than the other RAMA-type protocols, but it addresses the priority issue, which is certainly useful for multimedia applications.

3.8.9 Brief Remarks on Hybrid Multiple-Access Techniques

The hybrid multiple-access techniques attempt to solve the question of integration of traffic with dissimilar characteristics into a common stream in order to use the wireless medium efficiently. With such a purpose the protocols in this category provide for flexibility and may use a variety of parameters. Because these protocols are usually evaluated in different environments, comparing their performances becomes a difficult task. Basically, it can be said that the protocols in this category are of either the contention resolution type or the deterministic type. PRMA-based protocols belong to the first type, whereas RAMA-based protocols belong to the second type.

3.9 Summary

The inherent high connectivity provided by the radio channel, which is a broadcast medium, requires that adequate access control be provided so that the wireless network may operate adequately. Access control can be accomplished by several mechanisms: by insulating the various signals sharing the same access medium, by allowing the signals to contend for access, or by combining these two approaches. It can carried out in the different domains, such as frequency, time, code, and space. Certainly, access coordination must be carried out in the forward-link direction, in the reverse-link direction, and in both directions simultaneously. The techniques providing for two-way communication are called duplexing techniques. In theory, these techniques can be built over all the four domains, although, in general, only FDD and TDD find application in practical networks. Depending on how much coordination is required to access the shared resources, the access methods may fall into the following categories: *scheduled multiple-access methods*, *random multiple-access methods*, *controlled multiple-access methods*, and *hybrid multiple-access methods*. The scheduled multiple-access techniques are more efficiently utilized if applied to steady flow traffic, in which case the resources, although seized on a demand basis, are allotted for the duration of the communication. The access methods in this category make use of the signal insulation principle in which information of different sources is transmitted on nonoverlapping channels. In this case, collision does not occur. The methods providing for scheduled multiple-access capabilities in each domain are named FDMA, TDMA, CDMA, and SDMA.

The random multiple-access techniques are more efficiently utilized if applied to bursty (random) traffic, in which case the resources are seized on a random access basis and are allotted for that specific access only. Because the protocols in this category sustain no rigid discipline in accessing the medium, collisions may occur. The widely used methods providing for random multiple-access capabilities are pure ALOHA, slotted ALOHA, tree algorithm, FCFS algorithm, CSMA (nonslotted, slotted, 1-persistent, p-persistent), CSMA/BT, CSMA/DT, CSMA/CD, and CSMA/CA.

The controlled multiple-access methods make use of deterministic or noncontention strategies. With a control signal, permission to send is granted to terminals individually, so that only one terminal is allowed to access the medium at a time. They improve the delay performance of a burst-in-nature traffic network operating at high traffic levels. Access control can be performed on a centralized basis or on a distributed basis. The methods in the first case are referred to as polling controlled, whereas those in the second case are referred to as token controlled, with the token ring and the token bus the best-known strategies of the token controlled type.

The hybrid multiple-access methods make use of a combination of scheduled, contention, and deterministic schemes in which some degree of coordination is included in the random-access mechanisms or some other more sophisticated deterministic methods are employed. They aim at combining streams of traffic of different nature such as voice, video, and data.

The various multiple-access methods in this category attempt to control transmission delay as well as access to the medium. Because these protocols support traffic of different sources, besides throughput a performance parameter of interest is the number of simultaneous conversations that can take place with satisfactory performance. The extension of broadband services to the wireless environment has led to the introduction of the W-ATM into the network. In addition to its ability to handle both asynchronous and synchronous services, ATM provides for an end-to-end multimedia capability with guaranteed QoS. In particular, a subset of ABR, CBR, UBR, and VBR services should be provided by this technology. Examples of such schemes include R-ALOHA, original PRMA, $PRMA^{++}$, aggressive PRMA, fast PRMA, multirate PRMA, PRMA/DA, PRMA/ATDD, DQRUMA, DSA^{++}, DTDMA/PR, MASCARA, DTDMA/TDD, pure RAMA, T-RAMA, F-RAMA, and D-RAMA.

References

1. Paris, P.P., Access methods, in *The Mobile Communications Handbook*, 2nd ed., Gibson, J.D., Ed., CRC Press, Boca Raton, FL, 1998.
2. Proakis, J.G., *Digital Communications*, McGraw-Hill International, New York, 1995.
3. Liberti, J.C. and Rappaport, T.S., *Smart Antennas for Wireless Communications: IS-95 and Third Generation Cdma Applications*, Prentice-Hall, Upper Saddle River, NJ, 1999.
4. Van Rooyen, P., Lotter, M.P., and Van Wyk, D., *Space-Time Processing for CDMA Mobile Communications*, Kluwer Academic Publishers, Dordrecht, the Netherlands, 2000.
5. Litva, J., *Digital Beamforming in Wireless Communications*, Artech House, Norwood, MA, 1997.
6. Butler, J.L., Digital, matrix, and intermediate frequency scanning, in Hansen, R.C., Ed., *Microwave Scanning Arrays*, Academic Press, New York, 1966.
7. Capetanakis, J.I., Tree algorithms for packet broadcasting channels, *IEEE Trans. Inform. Theory*, IT-25, 505–515, September 1979.
8. Bertsekas, D. and Gallager, R., *Data Networks*, 2nd ed., Prentice-Hall, Englewood Cliffs, NJ, 1992.
9. Kleinrock, L. and Tobagi, F.A., Packet switching in radio channels: I. Carrier sense multiple access models and their throughput delay characteristics, *IEEE Trans. Commun.*, COM-23(12), 1400–1416, December 1975.
10. Pahlavan, K. and Levesque, A.H., *Wireless Information Networks*, John Wiley & Sons, New York, 1995.
11. Arnbak, J.C. and Van Blitterwijk, W., Capacity of slotted ALOHA in Rayleigh fading channels, *IEEE J. Selected Areas Commun.*, SAC-5(2), 261–269, February 1987.

12. Tasaka, S. and Ishibashi, Y., A reservation protocol for satellite packet communication—a performance analysis and stability considerations, *IEEE Trans. Commun.*, COM-32 (8), 920–927, 1984.
13. Goodman, D.J. and Wei, S.X., Efficiency of packet reservation multiple access, *IEEE Trans. Veh. Technol.*, VT-40(1), 170–176, February 1991.
14. Dunlop, J. et al., Performance of a statistically multiplexed access mechanism for a TDMA radio interface, *IEEE Personal Commun.*, 2(3), 56–64, June 1995.
15. Jang, J.S. and Shin, B.C., Performance evaluation of fast PRMA protocol, *Electron. Lett.*, 31(5), 347–349, March 1995.
16. Jang, J.S. and Shin, B.C., Performance evaluation of a voice/data integrated fast PRMA protocol, *IEICE T. Commun.*, E80B(7), 1074–1089, July 1997.
17. Khan, F. et al., Analysis of aggressive reservation multiple access schemes for wireless PCS, in *Proceedings of the IEEE International Conference on Communications*, Dallas, TX, June 1996, 1750–1755.
18. Kim, J. and Widjaja, I., PRMA/DA: a new media access protocol for wireless ATM, in *Proceedings of the IEEE International Conference on Communications*, Dallas, TX, June 1996, 1–19.
19. Priscoli, F.D., Medium access control for MEDIAN system, in *Proceedings of ACTS Mobile Summit'96*, Granada, Spain, November 1996, 1–8.
20. Karol, M.J., Liu, Z., and Eng, K.Y., Distributed-queuing request update multiple access (DQRUMA) for wireless packet (ATM) networks, in *Proceedings IEEE INFOCOM'95*, 1995, 1224–1231.
21. Petras, D. and Krämling, A., MAC protocol with polling and fast collision resolution for an ATM air interface, presented as IEEE ATM Workshop, San Francisco, CA, August 1996.
22. Qiu, X., Li, V.O.K., and Ju, J.-H., A multiple access scheme for multimedia traffic in wireless ATM, *J. Spec. Top. Mobile Networks Appl.* (MONET), 1(3), 259–272, December 1996.
23. Bauchot, F. et al., MASCARA, a MAC protocol for wireless ATM, in *Proceedings of ACTS Mobile Summit'96*, Granada, Spain, November 1996, 1–8.
24. Raychaudhuri, D. et al., WATMnet: a prototype wireless ATM system for multimedia personal communication, *J. Sel. Area Commun.*, 15(1), 83–95, January 1997.
25. Amitay, N., Resource auction multiple access (RAMA)—efficient method for fast resource assignment in decentralized wireless PCS, *Electron. Lett.* 28(8), 799–801, April 1992.
26. Karol, M.J. and Lin C., A protocol for fast resource assignment in wireless PCS, *IEEE Trans. Veh. Technol.*, 43(3), 727–732, August 1994.
27. Pinheiro, A.L.A. and deMarca, J.R.B., Fair deterministic packet access protocol: F-RAMA (fair resource assignment multiple access), *Electron. Lett.* 32(25), 2310–2311, December 1996.
28. Santivañez, J. and deMarca, J.R.B., D-RAMA: a new media access protocol for wireless communications, in *Proceedings of IEEE ICUPC'97*, Helsinki, 1997, 1043–1048.

Part II

2G Systems

4

GSM

4.1 Introduction

The existence of incompatible analog air interfaces for the cellular networks already in operation in various countries was certainly a deterrent that had to be removed if a truly economic integration were to be achieved in Europe. Full roaming in all countries was then a target to be pursued, but this was impossible in light of the different standards adopted by each country. Table 4.1 shows these standards and some of their features, such as operating bands, channel widths, and maximum number of channels.

A system that would serve this purpose would also have to provide for different service plans to accommodate different needs and different policies. In the early 1980s, stimulated by the authorities, the Conference of European Postal and Telecommunications (CEPT) administrations created the Group Special Mobile with the aim of developing a Pan-European standard for digital cellular communications. The project was named GSM and the system implementing the corresponding standard, Global System for Mobile Communications, was also referred to as GSM. The responsibility for the development of the complete standard was transferred to the European Telecommunications Standard Institute (ETSI) and a memorandum of understanding (MoU) was signed by the various network operators. The MoU details the operational matters concerning the procedures to bring GSM systems into commercial service and to govern the business arrangements between different operators of GSM networks. Field trials were carried out in the mid-1980s and the system was finally put into operation in the beginning of the 1990s.

From that time, GSM networks, either in the original GSM conception or as an evolution of it, spread worldwide and are unanimously considered a very successful project. Its success can be assessed not only by the number

TABLE 4.1

European Analog Cellular Systems

System	Reverse (MHz)	Forward (MHz)	Channel Width (kHz)	No. of Channels	Countries
TACS	890–915	935–960	25	1000	Austria, Spain
ETACS	872–905	917–950	25	1240	United Kingdom, Italy
NMT–450	435–457.5	463–467.5	25	180	Nordic countries, France, Germany, The Netherlands, Spain
NMT–900	890–915	935–960	12.5	1999	Nordic countries, France, Germany, The Netherlands
C–450	450–455.74	460–465.74	10	573	Austria, Germany, Portugal
RTMS	450–455	460–465	25	200	Italy
Radiocom 2000	192.5–199.5	200.5–207.5	12.5	560	France
	215.5–233.5	207.5–215.5		640	
	165.2–168.4	169.8–173		256	
	414.8–418	424.8–428		256	

of GSM networks in operation but also by the number of books dedicated to describing it.[1–15] GSM systems are found operating in frequency bands around 900 MHz (GSM-900), 1.8 GHz (GSM-1800), or 1.9 GHz (GSM-1900). GSM-900 is the original GSM cellular network initially conceived to serve large areas (macro cells) and to operate with high power terminals. GSM-1800 and GSM-1900 incorporate the personal communication service concepts. GSM-1800 is designed to operate in Europe and GSM-1900 is designed to operate in America, and both comprise low power terminals and serve small areas (micro cells). A new revision of the GSM specifications define an E-GSM system. In E-GSM, the original GSM-900 operating band is extended and lower power terminals and smaller serving areas (micro cells) are specified. GSM-900, E-GSM, GSM-1800, and GSM-1900 comprise the GSM family and their respective operating bands are shown in Table 4.2. The number of channels shown in Table 4.2 indicates the maximum possible number of GSM channels within the available band. In fact, because of interference problems, a guard band is recommended and fewer channels are actually used.

TABLE 4.2

GSM Bands

System	Reverse (MHz)	Forward (MHz)	Channel Width (kHz)	Max. No. of Channels
GSM-900	890–915	935–960	200	125
E-GSM	880–915	925–960	200	175
GSM-1800	1710–1785	1805–1880	200	375
GSM-1900	1850–1910	1930–1990	200	300

4.2 Features and Services

The GSM project embraces an ambitious set of targets:

- International roaming
- Open architecture
- High degree of flexibility
- Easy installation
- Interoperation with ISDN (Integrated Services Digital Networks), CSPDN (Circuit-Switched Public Data Network), PSPDN (Packed-Switched Public Data Network), and PSTN (Public-Switched Telephone Network)
- High-quality signal and link integrity
- High spectral efficiency
- Low-cost infrastructure
- Low-cost, small terminals
- Security features

These objectives have been gradually achieved and a broad collection of services are provided. The GSM services are grouped into three categories:

1. Teleservices (TS)
2. Bearer services (BS)
3. Supplementary services (SS)

TS cover, in essence, telephony, BS encompass basically data transmission, and SS are the value-added features.

4.2.1 Teleservices

Regular telephony, emergency calls, and voice messaging are within TS. Telephony, the old bidirectional speech calls, is certainly the most popular of all services. An emergency call is a feature that allows the mobile subscriber to contact a nearby emergency service, such as police, by dialing a unique number. Voice messaging permits a message to be stored within the voice mailbox of the called party either because the called party is not reachable or because the calling party chooses to do so.

4.2.2 Bearer Services

Data services, short message service (SMS), cell broadcast, and local features are within BS. Rates up to 9.6 kbit/s are supported. With a suitable data terminal or computer connected directly to the mobile apparatus, data may be sent through circuit-switched or packet-switched networks. Short messages containing as many as 160 alphanumeric characters can be transmitted to or from a mobile phone. In this case, a message center is necessary. The broadcast mode (to all subscribers) in a given geographic area may also be used for short messages of up to 93 alphanumeric characters. Some local features of the mobile terminal may be used. These may include, for example, abbreviated dialing, edition of short messages, repetition of failed calls, and others.

4.2.3 Supplementary Services

Some of the SS are as follows:

- Advice of charge. This SS details the cost of a call in progress.
- Barring of all outgoing calls. This SS blocks outgoing calls.
- Barring of international calls. This SS blocks incoming or outgoing international calls as a whole or only those associated with a specific basic service, as desired.
- Barring of roaming calls. This SS blocks all the incoming roaming calls or only those associated with a specific service.
- Call forwarding. This SS forwards all incoming calls, or only those associated with a specific basic service, to another directory number. The forwarding may be unconditional or may be performed when the mobile subscriber is busy, when there is no reply, when the mobile subscriber is not reachable, or when there is radio congestion.
- Call hold. This SS allows interruption of a communication on an existing call. Subsequent reestablishment of the call is permitted.

- Call waiting. This SS permits the notification of an incoming call when the mobile subscriber is busy.
- Call transfer. This SS permits the transference of an established incoming or outgoing call to a third party.
- Completion of calls to busy subscribers. This SS allows notification of when a busy called subscriber becomes free. At this time, if desired, the call is reinitiated.
- Closed user group. This SS allows a group of subscribers to communicate only among themselves.
- Calling number identification presentation/restriction. This SS permits the presentation or restricts the presentation of the calling party's identification number (or additional address information).
- Connected number identification presentation. This SS indicates the phone number that has been reached.
- Freephone service. This SS allocates a number to a mobile subscriber, and all calls to that number are free of charge for the calling party.
- Malicious call identification. This SS permits the registration of malicious, nuisance, and obscene incoming calls.
- Three-party service. This SS permits the establishment of conference calls.

4.3 Architecture

The Public Land Mobile Network System (PLMN) with GSM architecture is illustrated in Figure 4.1. It consists of four major blocks:

1. Mobile Station Subsystem (MSS)
2. Base Station Subsystem (BSS)
3. Network and Switching Subsystem (NSS)
4. Operation and Support Subsystem (OSS)

The MSS consists of the apparatus supporting the interface between the user and the PLMN. The BSS provides and manages radio access between the MSS and the rest of the GSM network. The NSS is responsible for communication within the same PLMN or with other networks, such as another PLMN, PSTN, ISDN, CSPDN, and PSPDN. The OSS provides the means for operation and maintenance of the GSM network, which include monitoring, diagnoses, and troubleshooting.

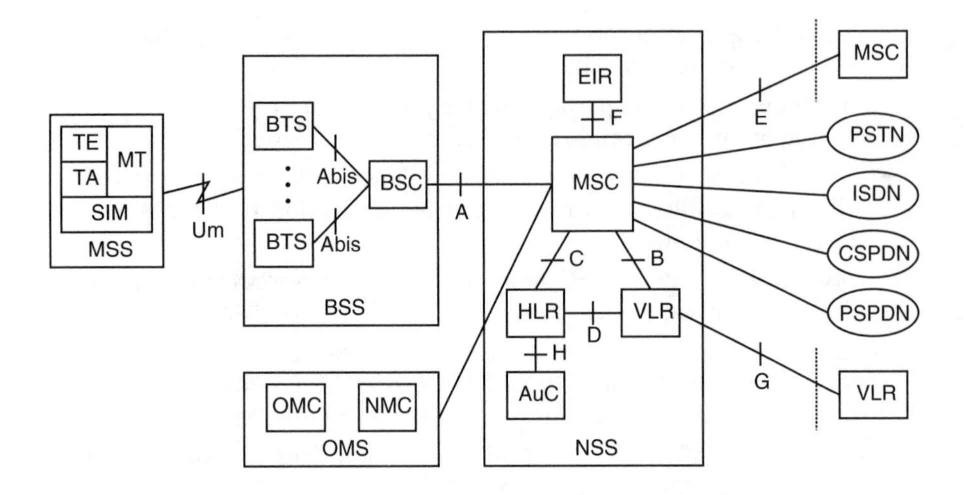

FIGURE 4.1
GSM architecture.

4.3.1 Mobile Station Subsystem

The MSS is basically a human–machine interface performing functions to connect the user and the PLMN. These functions include voice and data transmission, synchronization, monitoring of signal quality, equalization, display of short messages, location updates, and others. To carry out all its functions, an MSS includes the terminal equipment (TE), the terminal adapter (TA), the mobile termination (MT), and a subscriber identity module (SIM). TE, TA, and MT compose the mobile equipment, and SIM enables the use of the mobile equipment.

Mobile Equipment

A TE may be a fax, a computer, or another nonspecific GSM device. An MT is the equipment that realizes the standard GSM mobile terminal functions. A TA works as an interface between the TE and the MT. These devices may all be integrated into one piece or just partially integrated. In addition to its international equipment identification, the mobile equipment is identified through its classmark, which includes the following information:

- Revision level. This indicates the GSM specification version implemented within the terminal.
- Encryption capability. This indicates the type of cryptography supported by the equipment.
- Frequency capability. This indicates the frequency band in which the equipment can operate.

TABLE 4.3

Equipment Class and Their Maximum Power

Class	GSM-900 (watt)	GSM-1800 (watt)	GSM-1900 (watt)
I	20	1	1
II	8	0.25	0.25
II	5	4	2
IV	2	—	—
V	0.8	—	—

- Short message capability. This indicates whether or not the equipment is able to receive/send short messages.
- RF power capability. This indicates the maximum RF power of the equipment.

Five different categories of MSSs are specified according to their maximum RF power. They are identified as Class I, II, III, IV, and V and their maximum power levels are shown in Table 4.3.

Subscriber Identity Module

A GSM MSS is not associated with the subscriber, i.e., subscriber identity and equipment identity are independent elements. This is achieved through the SIM card, which is a removable card containing all subscriber specific data, such as identification numbers, contracted services, and personal features, as chosen by the subscriber. The user may never own an equipment set. It suffices to have an SIM card that can be inserted into any type of GSM MSS. A validation process is utilized with the insertion of SIM. This is accomplished when the subscriber provides the four-digit personal identification number (PIN), which should match that number stored within the SIM card. An SIM card can be found in credit card–sized or plug-in-sized formats. Note that this is in fact an incorporation of the personal communications services philosophy within the cellular network. The SIM card makes it possible for the subscriber to be charged according to the subscriber's identification. Except for emergency calls, any other call may only be performed with the use of the SIM card.

A number of identification codes are used for authentication purposes. These include:

- International Mobile Subscriber Identity (IMSI). An IMSI is a 15-digit subscriber's directory number that includes the mobile country code (3 digits), the mobile network code (3 digits), and the PLMN mobile identification number (9 digits). The IMSI is used only at the initialization procedure (to be detailed later).

- Temporary Mobile Subscriber Identity (TMSI). A TMSI is an identification number assigned to a subscriber on a temporary basis (per-call basis). It is mainly used for security reasons to avoid sending the IMSI over the air interface, thereby deterring fraudulent access.
- Mobile Station ISDN Number (MS-ISDN). An MS-ISDN is the subscriber ISDN number.
- Mobile Station Roaming Number (MSRN). An MSRN is a temporary identity assigned to a mobile subscriber in a roaming condition.
- International Mobile Equipment Identity (IMEI). An IMEI identifies the equipment itself. The pair IMEI–IMSI ensures that only authorized users are granted access to the system.
- Location Area Identity (LAI). An LAI identifies a group of cells. The SIM stores the LAI of the group of cells that has most recently been visited by the terminal.
- Subscriber Authentication Key (Ki). A Ki is a secret assigned by the operating company to a subscriber.

4.3.2 Base Station Subsystem

The BSS is responsible for the radio coverage of a given geographic region and for appropriate signal processing. It is functionally and physically separated into two components: a base transceiver station (BTS) and base station controller (BSC). A third element within the BSS (not shown in Figure 4.1) is the transcoder (XCDR), also named transcoder/rate adapter unit (TRAU).

Base Station Transceiver

A BTS is responsible for the radio coverage itself and basically consists of the radio equipment. The BSC is responsible for functions concerning network operations and signal processing.

Base Station Controller

A BSC controls one or more (typically several) BTSs. Its functions include radio resource management, signaling transmission, power control, handover control, frequency hopping control, and others.

Transcoder/Rate Adapter Unit

A TRAU is a device placed between two GSM elements—BTS, BSC, or mobile switching center (MSC)—and is used to conserve transmission resources. It takes four 13-kbit/s speech channels (to be defined later), inserts overhead information to bring each channel to 16 kbit/s, and combines the four channels into one data stream of 64 kbit/s. Channels with lower data rates are also raised to 16 kbit/s and processed in a similar manner. The 30 64-kbit/s channels can be multiplexed to form a 2.048-Mbit/s data stream. A TRAU can be physically located within a BTS, a BSC, or an MSC; the second option is the most common configuration.

4.3.3 Network and Switching Subsystem

The network and switching subsystem (NSS) carries out the GSM switching procedures and the manipulation of the databases for mobility management of the subscribers. Its functions include coordination of call setup, paging, resource allocation, location registration, encryption, interfacing with other networks, handover control, billing, synchronization, and others. The NSS consists of the MSC, the home location register (HLR), the visitor location register (VLR), the authentication center (AuC), and the equipment identity register (EIR).

Mobile Switching Center

The MSC performs the switching functions and coordinates the calls and routing procedures within GSM. In general, an MSC controls several BSSs. Thus, it is responsible for traffic management and the radio coverage of a given geographic area, the MSC area. A GSM network may have one or more MSCs, depending on the traffic to be controlled. An MSC is responsible for several functions such as paging, coordination of call setup, allocation of resources, interworking with other networks, handover management, billing, encryption, echo canceling, control, synchronization with BSSs, and others. It interfaces the GSM PLMN with the external networks such as PSTN, ISDN, CSPDN, and PSPDN. Such interfacing may be carried out through a gateway MSC connected to a serving MSC.

Home Location Register

The HLR is a database containing a list of those subscribers belonging to one or more MSC areas within which they have originally been registered. An HLR, therefore, defines the subscription area. In the HLR, these subscribers are associated with information records relevant to call management. Both permanent and temporary data are held within the HLR. The permanent data constitute data that are modified only for administrative reasons and are kept for every call. The temporary data comprise data that are modified to accommodate the transient status of the subscribers' parameters and can be changed from call to call. The permanent data include IMSI, MS-ISDN, information on roaming restriction, permitted supplementary services, and authentication key for security procedures. The temporary data consist of

MSRN, data related to ciphering, VLR address, MSC address, and information on roaming restriction. An HLR is usually centralized, but it can also be distributed within the network, with the configuration chosen in accordance with the operator's needs.

Visitor Location Register

The VLR is a database containing a list of those subscribers belonging to another subscription area, but who are now in a roaming condition to this MSC area. In the VLR, these roaming subscribers are associated with information records relevant to the call management. In essence, the VLR contains the same information associated with the HLR. In fact, when a subscriber roams into another MSC area, the relevant data belonging to this subscriber stored in the HLR are transferred to the corresponding VLR. The data of the roaming subscriber, retrieved from the HLR, remain in the respective VLR as long as the subscriber is found in a roaming condition. A VLR is usually co-located with the MSC.

Authentication Center

The AuC is a functional entity that manipulates the authentication functions and encryption keys for each subscriber within the system. An authentication key, kept in the SIM card and in the AuC, is provided for each subscriber in the system and is never transmitted over the air. Instead, a random challenge and the response to this challenge are transmitted (to be detailed later). The random challenge and the respective response are based on authentication keys and on some ciphering algorithms. Note that the contents of the information exchanged in such a procedure may change for each call. This, in conjunction with the TMSI, constitutes an interesting procedure that renders GSM robust with respect to unauthorized accesses. On the other hand, vulnerability is present when the authentication key must be transmitted from an HLR to a VLR in a roaming situation.

Equipment Identity Register

The EIR is a database containing the IMEIs of all subscribers. The IMEIs are grouped in three categories as follows:

1. White List, containing the IMEIs known to belong to equipment with no problems
2. Black List, containing the IMEIs of equipment reported stolen
3. Gray List, containing the IMEIs of equipment with some problems not substantial enough to warrant barring

4.3.4 Operation and Support Subsystem

The operation and support subsystem (OSS) performs the operation and maintenance functions through two entities, namely, the operation and maintenance center (OMC) and the network management center (NMC). The GSM standards do not fully specify these elements. Therefore, different manufacturers may have different implementations, which may be a problem for interoperability between different GSM systems. The NSS uses lines leased on the PSTN or other fixed networks to communicate with the GSM entities. Protocols used for data transfer are SS7 and X.25. In general, the operation and management functions performed by the OSS include alarm handling, fault management, performance management, configuration control, traffic data acquisition, etc. In many circumstances, the actions may be taken remotely and automatically and, upon detection of an abnormal operation, tests, diagnoses, and fault removal can be carried out to place the system back in service.

Operation and Maintenance Center

The network resources may be activated or deactivated via the OMC functions. There may be one or several OMCs depending on the size of the network. An OMC is a regional entity used for daily maintenance activities.

Network Management Center

The NMC serves the entire network. It performs a centralized network management and is used for long-term planning.

4.3.5 Open Interfaces

A large number of open interfaces are specified within the GSM architecture. Open interfaces favor market competition with operators able to choose equipment from different vendors. The various interfaces are identified as follows.

- A-Interface: The interface between BSC and MSC. It supports signaling and traffic (voice and data) information transmitted by means of one or more 2.048-Mbit/s transmission systems.
- Abis-Interface: The interface between BTS and BSC. It handles common control functions within a BTS. It is physically supported by a digital link using the link access data protocol (LAPD).
- B-Interface: The interface between MSC and VLR.
- C-Interface: The interface between MSC and HLR.
- D-Interface: The interface between HLR and VLR.

- E-Interface: The interface between MSCs.
- F-Interface: The interface between MSC and EIR.
- G-Interface: The interface between VLRs.
- H-Interface: The interface between HLR and AuC.
- Um-Interface: The interface between MSS and BSS.

The great majority of the GSM interfaces make use of those well-consolidated protocols such as SS7, X.25, and LAPD (the ISDN data link layer). The 2.048-Mbit/s E1 digital links are used to interconnect the several GSM elements and external elements. For example, E1 is used to interconnect MSC and PSTN, MSC and MSC, and BSC and MSC. Each E1 channel is then used to convey payload information or control information. The control information may use protocols such as SS7, LAPD, and X.25.

4.4 Multiple Access

GSM is a fully digital system with a multiple-access architecture based on the narrowband FDMA/TDMA/FDD technology. GSM-900 uses a total 50-MHz band, divided into two 25-MHz bands, with uplink and downlink separated by 45 MHz. With a carrier spacing of 200 kHz the maximum number of carriers per direction is $25,000 \div 200 = 125$. E-GSM adds 10 MHz, and correspondingly 50 carriers, to each direction of transmission; the separation between uplink and downlink (45 MHz) is maintained. GSM-1800 uses of a total 150-MHz band, divided into two 75-MHz bands. The same 200-kHz carrier spacing is used, this leading to a maximum of $75,000 \div 200 = 375$ carriers per direction, with the separation between uplink and downlink 95 MHz. GSM-1900 makes use of a total 120-MHz band, divided into two 30-MHz bands. The number of carriers per direction is $60,000 \div 200 = 300$, with a separation between uplink and downlink of 80 MHz. In all cases, each carrier is shared in time by eight accesses.

4.4.1 Signal Processing

The signal processing operations in each direction (uplink and downlink) are briefly described as follows. At the mobile station, and for the uplink direction, the voice signal is A/D converted at a sampling rate of 8 kHz with each sample using 13 bits for uniform encoding. The resulting $8 \times 13 = 104$ kbit/s rate is reduced to 13 kbit/s by means of the regular pulse excitation–long-term prediction linear prediction coding (RPE-LTP-LPC) speech coding algorithm.

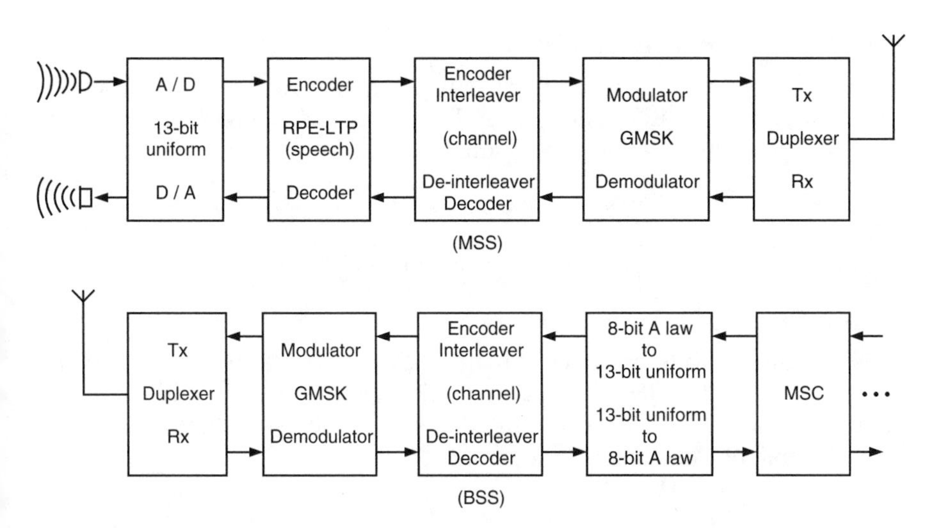

FIGURE 4.2
GSM transmission/reception chain.

Forward error correction code is then introduced and interleaving is applied so that burst errors are minimized. The resulting data stream modulates the carrier by means of the Gaussian minimum shift keying (GMSK) modulation scheme, and radio transmission occurs. At the base station, and still in the uplink direction, the received signal is demodulated, deinterleaved, error-corrected, decoded into a 13-bit/sample uniform code, and converted to the standard A-Law 64 kbit/s. This is then transmitted to the MSC. For the downlink direction, the reverse operations are carried out culminating with the D/A conversion at the mobile station. These operations are illustrated in Figure 4.2.

Speech Coding

The LPC-RPE speech coder manipulates blocks of speech data of 20 ms (160 samples at 8 kHz). This is a period of time within which the vocal tract is considered to be reasonably stationary. Each 20-ms block is represented by 260 bits divided as follows: 36 for the linear prediction coefficients, 36 bits for the long-term prediction, and 188 for excitation. These lead to a speech coding rate of $260 \div 0.02 = 13$ kbit/s. The 260 bits are split into three classes according to their significance for speech intelligibility:

1. Class 1a. Class 1a contains 50 bits identified as *essential*.
2. Class 1b. Class 1b contains 132 bits identified as *important*.
3. Class 2. Class 2 contains 78 bits identified as *least important*.

Check Redundancy Cyclic Coding

The 50 Class 1a bits pass through a parity generator block, which uses cyclic redundancy check (CRC) coding. The CRC encoder adds three parity bits and uses the following generator polynomial:

$$g(D) = 1 + D + D^3$$

The output of the CRC encoder, therefore, yields 53 bits.

Convolutional Coding

These 53 resulting bits are attached to the 132 Class 1b bits and to four all-zero tail bits. The resulting 189 bits are convolutionally encoded by a half-rate convolutional encoder whose constraint length is 5. The generator polynomials of such a convolutional encoder are

$$g_0(D) = 1 + D^3 + D^4$$
$$g_1(D) = 1 + D + D^3 + D^4$$

The output of the half-rate convolutional encoder, therefore, yields 378 bits.

4.4.2 Multiple Access

The FDMA/TDMA/FDD multiple-access scheme allows the terminal to transmit and receive simultaneously. In this particular case, transmission and reception make use of different carriers. GSM, on the other hand, adopts the solution of aligned frames on the downlink and uplink but with the time reference for the reverse link retarded by three time slots with respect to the time reference of the forward direction. This eliminates the requirement for the terminal to transmit and receive simultaneously.

GSM makes use of two types of channels: physical channels and logical channels. The physical channels constitute the physical medium through which information flows. The logical channels support the logical functions within the network.

4.4.3 Physical Channels

Although the 25-MHz spectrum per direction for GSM-900 admits 125 200-kHz carriers, in order to allow for 100 kHz of guard bands at both edges of the spectrum, only 124 carriers are effectively specified. The absolute radio frequency channel number (N) and the corresponding center frequency, in MHz, for the uplink and for the downlink are related to each other as

$$f_{\text{uplink}} = 0.2N + 890$$
$$f_{\text{downlink}} = 0.2N + 935$$

with $1 \le N \le 124$. For GSM-1800, the corresponding formulas are

$$f_{\text{uplink}} = 0.2N + 1607.8$$
$$f_{\text{downlink}} = 0.2N + 1702.8$$

with $512 \le N \le 885$. And for GSM-1900, these are

$$f_{\text{uplink}} = 0.2N + 1747.8$$
$$f_{\text{downlink}} = 0.2N + 1827.8$$

with $512 \le N \le 810$.

The multiple-access structure is defined in terms of time slot, frame, multiframe, superframe, and hyperframe, as illustrated in Figure 4.3. One time slot contains 156.25 bits and has a duration of 577 μs. Therefore, the bit duration is 3.69 μs. Eight time slots compose one 4.615-ms frame. There are two types of multiframes: traffic multiframe, with 26 frames and a duration of 120 ms, and control multiframe, with 51 frames and a duration of 235.4 ms. In

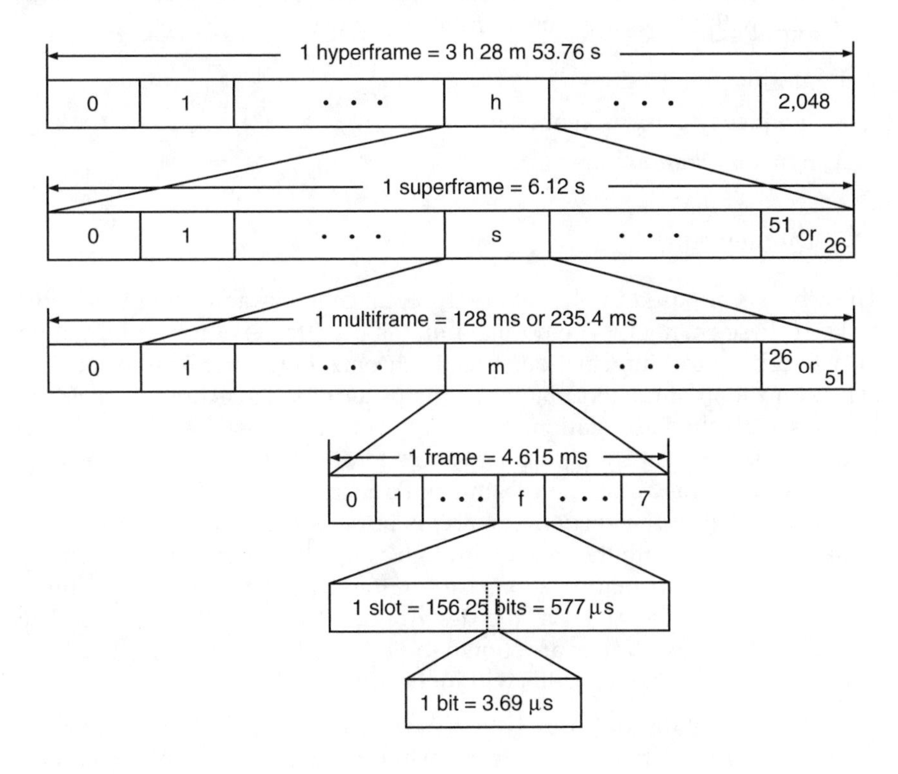

FIGURE 4.3
GSM timing structure.

the same way, there are two types of superframes, both with the same 6.12 s duration: traffic superframe, with 51 multiframes, and control superframes, with 26 multiframes. Finally, 2048 superframes compose a hyperframe (encryption hyperframe) with a total duration of 3 h 28 m 53.76 s.

The exact 120-ms traffic multiframe duration conforms with two basic time intervals: that of some digital networks (125 ms for ISDN, for example), and that of the duration of each block of speech (20 ms, as is the case of GSM speech coding). Note that the GSM timing structure is very elaborate, with definitions of time intervals ranging from one quarter of a bit (900 ns) to 3,394,560,000 bits (3 h 28 m 53.76 s). The GSM transmission rate can be obtained by dividing the number of bits per traffic multiframe ($156.25 \times 8 \times 26 = 32{,}500$) by the traffic multiframe duration (120 ms, a round number), which gives 270.83333... kbit/s. A physical channel is then specified by a carrier and a time slot within the whole multiple access structure.

4.4.4 Burst Formats

GSM specifies five different time slot formats (156.25 bits), or burst formats, to comply with the various functions to be performed:

1. Normal burst
2. Frequency correction burst
3. Synchronization burst
4. Access burst
5. Dummy burst

These bursts are used by the different logical channels according to specific tasks of these channels. The normal burst is used to carry user information or network control information. The synchronization burst is used for time synchronization of the terminal, i.e., it helps terminals synchronize their operations with the base station. The frequency correction burst is used for frequency synchronization purposes of the terminal with respect to the base station; during this burst, the BTS transmits a signal whose frequency is 67.7 kHz above that of the carrier. The access burst is used by the terminals to initiate a dialogue with the system through a signaling protocol. The dummy burst carries no information and is transmitted by the BTS. Details on burst format usage will be given in the description of the logical channels. The formats of these five bursts are shown in Figure 4.4. The following is a brief description of the various fields within the bursts.

- *Data*. Except for the normal burst, in which the data field may convey user information or network control information, in all the other burst formats this field conveys network control information only.

Tail 3	Data 57	Flag 1	Training 26	Flag 1	Data 57	Tail 3	Guard 8.25

Normal Burst

Tail 3	All zeros 142	Tail 3	Guard 8.25

Frequency Correction Burst

Tail 3	Data 39	Synchronization 64	Data 39	Tail 3	Guard 8.25

Synchronization Burst

Tail 8	Synchronization 41	Data 36	Tail 3	Guard 68.25

Access Burst

Tail 3	Mixed 58	Training 26	Mixed 58	Tail 3	Guard 8.25

Dummy Burst

FIGURE 4.4
GSM burst formats.

- *Flag*. The flag indicates the type of information that is being transmitted (user information or control network information).
- *Training*. The training field carries a training sequence used by the adaptive equalizer to estimate the channel.
- *Tail*. The tail field contains all-zero bits used to indicate the start and the end of the burst. They are also used by the adaptive equalizer as a start/stop bit pattern.
- *Guard*. The guard field serves several purposes. It uses a ramp time for the transmitter to turn off at the end of the time slot and to turn on at the beginning of the next time slot. It is also used to avoid overlapping between adjacent time slots. In the case of the access burst, it makes it possible for mobile transmissions from all parts of the cell to arrive at the base station within the duration of the burst.
- *Synchronization*. The synchronization field conveys a synchronization sequence (a known bit pattern), which is used for time synchronization purposes.

Some other specific information on these fields will be given in the description of the logical channels.

4.4.5 Logical Channels

The tasks performed in a GSM platform are supported by a number of functional channels, the logical channels. These logical channels are of two types: traffic channels, which convey payload information (speech, data), and signaling channels, which carry overhead (control) information.

The traffic channels carrying speech information are of two types: full rate traffic channel (TCH/F) and half rate traffic channel (TCH/H). The traffic channels carrying data information are named TCH/9.6, for 9.6 kbit/s; TCH/4.8, for 4.8 kbit/s; and TCH/2.4, for 2.4 kbit/s.

The signaling channels are grouped into three categories: broadcast channels, common control channels, and dedicated control channels. The following are the broadcast channels: frequency correction channel (FCCH), synchronization channel (SCH), and broadcast control channel (BCCH). The following are the common control channels: paging channel (PCH), access grant channel (AGCH), and random-access channel (RACH). And the following are the dedicated control channels: stand-alone dedicated control channel (SDCCH), slow associated control channel (SACCH), and fast associated control channel (FACCH). These channels will be detailed later in this chapter. The broadcast channels and the common control channels are one-way channels operating in the forward direction. They occupy time slot 0 of some specific carriers, those defined as standard broadcast carriers. In particular, if necessary, the common control channels may also occupy time slots 2, 4, and 6 of the same carriers. The dedicated control channels, on the other hand, are two-way channels. They may occupy any time slot, with the exception of time slot 0 of those carriers defined as standard broadcast carriers.

Figure 4.5 shows the logical channels specified by GSM.

4.4.6 Multiframes

There are two types of multiframes: traffic multiframe and control multiframe. The traffic multiframe comprises 26 frames, 24 of which accommodate traffic channels and the remaining two convey one signaling channel, the SACCH. The traffic channels occupy frames 0 to 11 and frames 13 to 24, whereas the signaling channel occupies frames 12 and 25. A TCH always has a SACCH associated with it. More specifically, a TCH/F occupies one time slot in 24 frames in every multiframe and a SACCH associated with it occupies one time slot in only one frame, either 12 or 25. In other

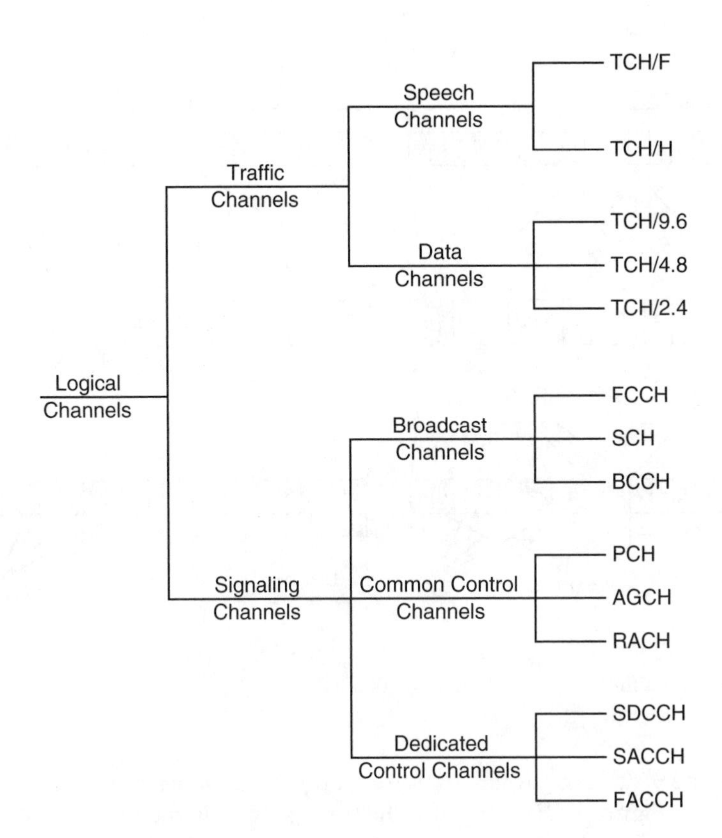

FIGURE 4.5
GSM logical channels.

words, if a SACCH, which is associated with a given TCH/F, appears in a specified time slot in a given frame (say, 12), then the corresponding time slot in the other frame (25, in this example) is kept idle. In such a case, each GSM carrier can convey eight TCH/Fs together with their corresponding SACCHs. On the other hand, a TCH/H occupies one time slot in 12 alternate consecutive frames, with the other 12 occupied by another TCH/H. Eight TCH/Hs will have their corresponding (eight) SACCHs in frame 12 and the eight remaining TCH/Hs will have theirs in frame 25. In this case, the total capacity is doubled. The traffic multiframe structure is shown in Figure 4.6.

The control multiframe comprises 51 frames and appears in two formats. In one of them (Format 1), the frames are occupied by SDCCHs and SACCHs. In the other one (Format 2), the frames comprise FCCHs, SCHs, BCCHs, PCHs, AGCHs, and RACHs. In Format 1, the SDCCHs occupy 32 consecutive

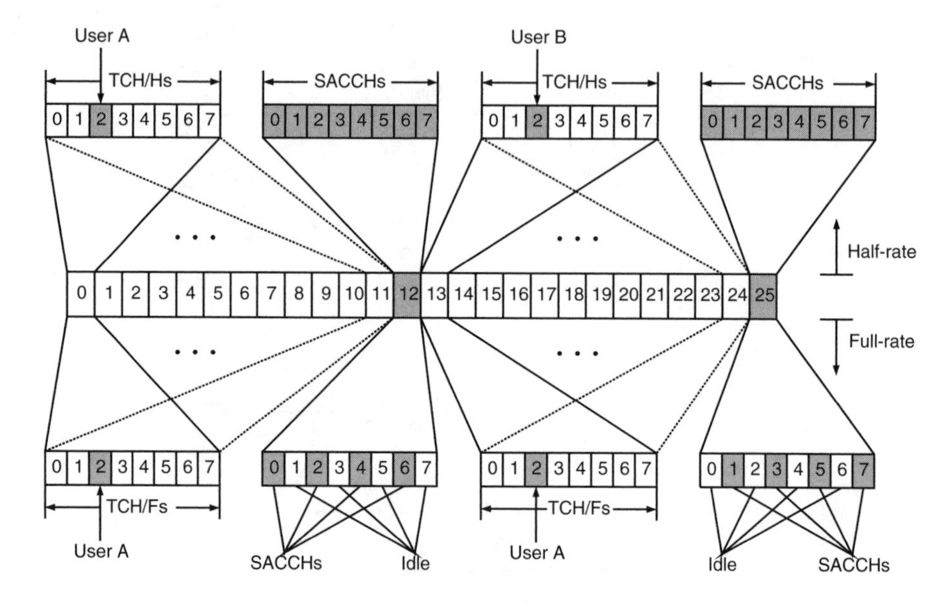

FIGURE 4.6
GSM traffic multiframe.

frames, the SACCH occupies 16 frames, and three frames remain idle. This is depicted in Figure 4.7. In Format 2, the frames differ in the forward and in the reverse directions. In the latter, the 51 frames are occupied by the RACHs only. In the forward direction, the control channels FCCH, SCH, BCH, PCH, and AGCH form their respective frames whose positions within the multiframe are as shown in Figure 4.7. In particular, the control channels PCH and AGCH share the same frames. Note that the control multiframe contains five control blocks, each block with ten frames.

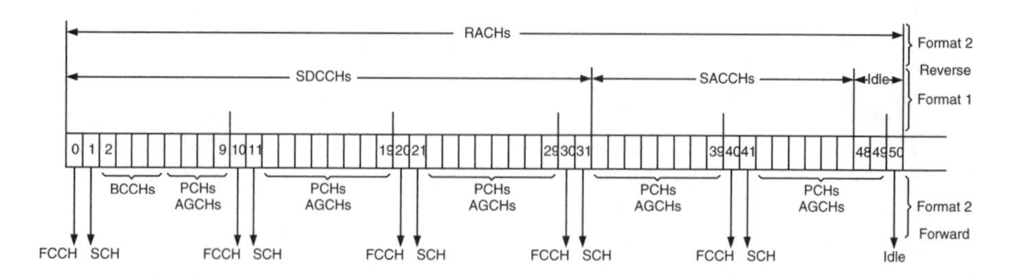

FIGURE 4.7
GSM control multiframe.

4.5 The Logical Channels

In this section, the GSM logical channels are described in more detail.

4.5.1 Traffic Channels

The traffic channels bear speech and data information. The traffic channels are two-way channels using the normal burst format. They occupy time slots 1 to 7 on each carrier and also time slot 0 on the carriers other than those specially dedicated to signaling. A traffic channel is associated with a terminal when a call is established. GSM defines two types of traffic channels: full-rate and half-rate. As described before, a TCH/F occurs 24 times within the traffic multiframe and has associated with it a SACCH to convey control information. As stipulated in the normal burst format, $2 \times 57 = 114$ bits are used for data transmission purposes (data field in the normal burst). Therefore, the TCH/F rate is $24 \times 114 \div 0.120 = 22.8$ kbit/s. In the same way, a TCH/H occurs 12 times within the traffic multiframe, which leads to a TCH/H bit rate of 11.4 kbit/s. A TCH/H also has a SACCH associated with it.

The traffic channel bit rate, as calculated before, is a consequence of the following. The 13 kbit/s speech coder delivers blocks of 260 bits within a period of 20 ms. The 50 Class 1a bits are CRC encoded and, as a consequence, three parity bits are attached to this Class 1a block. The 53 bits together with the 132 Class 1b bits and four all-zero tail bits (total of 189 bits) are half-rate convolutionally encoded, the convolutional encoder yielding, therefore, 378 bits. These 378 bits are then multiplexed with 78 Class 2 bits, resulting in 456 bits per block within the 20-ms speech block. These lead to a rate of $456 \div 0.02 = 22.8$ kbit/s, as calculated before. Two 456-bit blocks are interleaved (total of 912 bits) and transmitted over $912 \div 114 = 8$ frames. The interleaver is a 114×8 memory into which data are written by rows and from which data are read by columns. The data from the eight columns of the interleaver, each of which contains 114 bits, are then transmitted, respectively, over eight frames. Figure 4.8 shows the traffic channel structure.

4.5.2 Frequency Correction Channel

The FCCH bears information for frequency acquisition purposes. The FCCH is a one-way channel operating in the forward direction and using the frequency correction burst format. The 142 all-zero bits in this burst causes the GMSK modulator to deliver an unmodulated carrier for the entire duration of the time slot. The modulator produces a carrier with an offset of $1625 \div 24$ kHz above the nominal carrier frequency. Upon detecting this sine wave, the

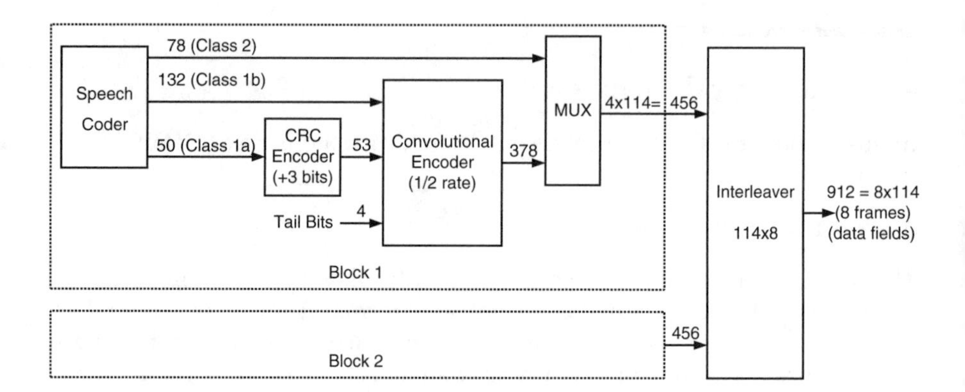

FIGURE 4.8
Traffic channel structure.

terminal can adjust its frequency reference appropriately, and is thus able to keep track of the occurrences of the time slots. As already mentioned, the FCCHs occur in time slot 0 of some specific carriers, and their relative positions within the control superframe are shown in Figure 4.7.

4.5.3 Synchronization Channel

The SCH bears data information for timing synchronization and BTS identification purposes. The SCH is a one-way channel operating in the forward direction and using the synchronization burst format. The raw data information for the SCH comprises 89 bits, which are appropriately encoded to compose the synchronization burst. Of these 89 bits, 64 bits, which are the same for all the cells, help the terminal to achieve the exact timing synchronization with respect to the GSM frame. Of the remaining 25 bits, 6 bits are used to identify the BTS and they are mapped onto the BTS identification code (BSIC). The remaining 19 bits represent the reduced frame number. These 19 bits are used as follows: 11 bits identify the superframe within the hyperframe (1 hyperframe = 2048 superframes); 5 bits specify the multiframe within the superframe (1 superframe = 26 control multiframes); and 3 bits identify the control block within the control multiframe (1 control multiframe = 5 control blocks).

The 64 synchronization bits are mapped directly onto the synchronization field of the synchronization format. The 25 data bits are CRC encoded with 10 parity bits attached to these 25 bits. The resulting 35 bits are half-rate convolutionally encoded, yielding 78 bits. These 78 bits are split in two 39-bit blocks and included within the two data fields of the synchronization burst. As already mentioned, the SCHs occur in time slot 0 of some specific carriers

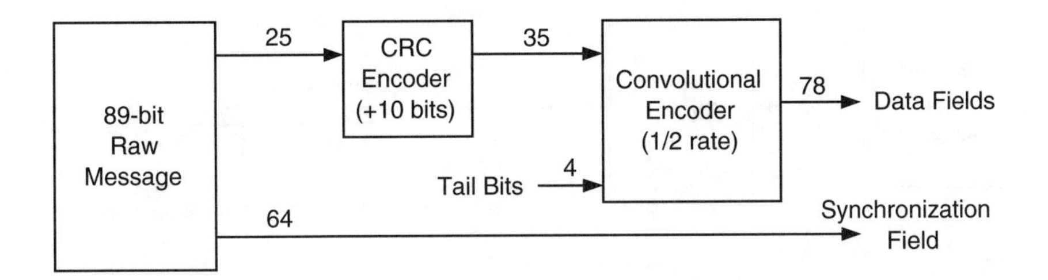

FIGURE 4.9
Synchronization channel structure.

and their relative position within the control superframe are shown in Figure 4.7. Figure 4.9 shows the synchronization channel structure.

4.5.4 Broadcast Control Channel

The BCCH bears data information for call setup purposes. The BCCH is a one-way channel operating in the forward direction and using the normal burst format. More specifically, it conveys the following information:

- Cell identity
- Network identity
- Control channel structure
- List of channels in use
- Congestion status
- Details of the access protocol

The raw message containing this information uses 184 bits. These 184 bits are protected by an error-correcting block code—a fire code—adding 40 bits to the message. A fire code is specially used to detect and correct bursty errors. The resulting 224 bits together with 4 tail bits are half-rate convolutionally encoded, yielding 456 bits. Finally, these 456 bits are split into 4 × 114 bits and sent in four time slots. As already mentioned, the BCCHs occur in time slot 0 of some specific carriers and their relative positions with the control superframe are shown in Figure 4.7. Figure 4.10 shows the BCCH structure.

4.5.5 Paging Channel

The PCH bears data information for paging purposes. The PCH is a one-way channel operating in the forward direction and using the normal burst format. The same coding scheme used for the BCCH is also used for the PCH,

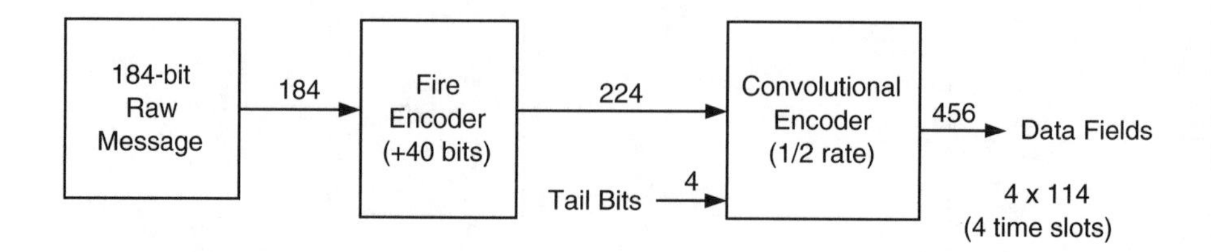

FIGURE 4.10
Broadcast control channel, paging channel, access grant channel, stand-alone dedicated control channel, slow associated control channel, and fast associated control channel structure.

as shown in Figure 4.10. Therefore, each paging message contains 4×114 bits and occupies four time slots. Note, from Figure 4.7, that there are 36 paging frames per control multiframe and, therefore, $36 \div 4 = 9$ paging messages per control multiframe (235.4 ms). Both paging channel and access grant channel share the same resources within the control multiframe. This is done by assigning blocks of four frames to either paging or access grant tasks, such an assignment being informed through the BCCH. The PCH blocks are then divided into groups. The terminal can therefore choose to monitor only the respective PCH groups to which it belongs, and this is a way of saving battery. This is the principle of the sleep-mode operation, in which the terminal is on only for a fraction of the time.

4.5.6 Access Grant Channel

The AGCH bears data information for access acknowledgment purposes. The AGCH is a one-way channel operating in the forward direction and using the normal burst format. More specifically, it carries information on confirmation of a successful access and on the particular SDCCH assigned to a terminal. As mentioned before, the AGCH shares the physical resources with the PCH. The same coding scheme used for the BCCH is also used for the AGCH, as shown in Figure 4.10.

4.5.7 Random Access Channel

The RACH bears data information for access purposes. The RACH is a one-way channel operating in the reverse direction and using the access burst format. More specifically, the terminal may use the RACH to perform the following tasks:

- Call origination
- Short message transmission
- Acknowledgment to paging messages

- Location registration
- IMSI attachment
- IMSI detachment

Access occurs on a contention basis with the slotted ALOHA protocol as the access algorithm. Basically, an access attempt occurs as follows. The terminal chooses a time slot, transmits its message, and waits for an acknowledgment for a fixed time. If no acknowledgment arrives, a random time is waited, and a new trial is carried out. This procedure is repeated up to a maximum number of times, as specified on the BCCH. An acknowledgment consists of sending the following information:

- The number of the time slot in which the access was made
- A 5-bit codeword transmitted in the access procedure (a loop back of the codeword)
- The time slot number of the SDCCH

Note, from Figure 4.4, that in the access burst 68.25 of 156.25 bits are used for guard time purposes. This is because before the initial access no information on the terminal timing is known. Therefore, the guard time ensures that in the initial access the information bits remain within a single time slot upon arrival at the base station when transmitted from any part of the cell. By determining the arrival time, the base station calculates the timing advance, information that is sent to the terminal to be used in the subsequent transmissions. The 252-μs guard time (68.25×3.69) due to the 68.25 bits corresponds to a propagation distance of approximately 75 km, which, therefore, establishes that a maximum cell radius is of 37.5 km.

The raw access message is, in fact, embodied by only eight bits. These eight bits are split into two fields, one containing three bits and the other containing five bits. The three-bit field identifies the type of access (call origination, paging acknowledgment, etc.). The five-bit field contains a randomly generated code used to distinguish the messages of two or more terminals transmitting in the same time slot (contending for the time slot). These eight bits are CRC encoded, which adds six parity bits to the eight bits. The resulting 14 bits together with 4 tail bits (total of 18 bits) are half-rate convolutionally encoded yielding data 36 bits. Figure 4.11 shows the RACH structure.

4.5.8 Stand-Alone Dedicated Control Channel

The SDCCH bears data information for signaling purposes. The SDCCH is a two-way channel using the normal burst format. More specifically, the SDCCH is used for signaling related to mobility management and call setup

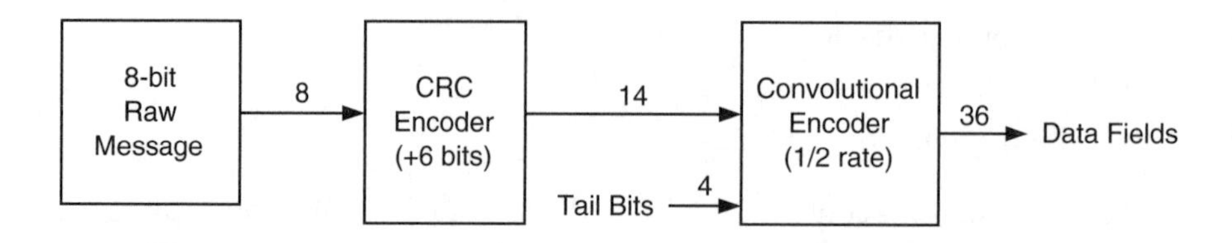

FIGURE 4.11
Random-access channel structure.

management. The following tasks require the use of the SDCCH:

- Registration
- Authentication
- Location updates

The use of an SDCCH usually follows that of the RACH for access purposes and precedes the allocation of a TCH after the call setup signaling has been completed. The SDCCH employs a set of four time slots within the 51-frame control multiframe. Knowing that each time slot within a normal burst uses 114 bits and that the duration of a superframe is 6.12 s, the SDCCH rate is $4 \times 114 \times 26 \div 6.12 = 1937.25$ bit/s. Like the TCH, the SDCCH also has associated with it an SACCH for control purposes. The same coding scheme used for the BCCH is also used for the SDCCH, as shown in Figure 4.10.

4.5.9 Slow Associated Control Channel

The SACCH bears data information for control purposes. The SACCH is a two-way channel using the normal burst format. A SACCH is always associated with a TCH or with an SDCCH. It uses the same carrier frequency of the logical channel with which it is associated. The SACCH is a continuous data channel carrying control information from the terminal to the base station, and vice versa. In the forward link, it supports power level commands and timing adjustments directives. In the reverse link, it conveys measurements reports related to the signal quality of the serving base station and of the neighboring cells.

When associated with a TCH the SACCH occurs in frames 12 or 25 of each 26-frame traffic multiframe. It then occupies one time slot (114 bits) per multiframe (0.120 s). Therefore, the SACCH rate is $114 \div 0.120 = 950$ bit/s. Each message comprises 456 bits, meaning that four traffic multiframes (480 ms) are used to transmit a message. When associated with an SDCCH the SACCH occupies two time slots per control multiframe. It follows that in this case the SACCH rate is $2 \times 114 \times 26 \div 6.12 = 969$ bit/s.

The same coding scheme used for the BCCH is also used for the SACCH, as shown in Figure 4.10.

4.5.10 Fast Associated Control Channel

The FACCH bears data information for signaling purposes. The FACCH is a two-way channel using the normal burst format. It conveys messages that must be treated in an expeditious manner and that cannot rely on the 480-ms transmission time provided by SACCH. An example of this is the message concerning a handover request. An FACCH empowers the characteristic of an in-band signaling channel both for TCH and SDCCH, operating in a stealing mode. That is, if necessary, a TCH or an SDCCH can be interrupted and replaced by an FACCH to transmit urgent messages. The time slot is recognized as operating as FACCH or as TCH or SDCCH by appropriately setting the 2-bit flag field in the message of the normal burst. The same coding scheme used for the BCCH is also used for the FACCH, as shown in Figure 4.10.

4.6 Messages

The signaling channels, with the exception of FCCH, RACH, and SCH, use the LAPDm format to transmit information. The LAPD *m* protocol in the *mobile* network is equivalent to the LAPD protocol in the fixed network. The messages are transmitted in segments of 184 bits. In general, the messages fit into a single segment and, as already mentioned, the 184 bits of raw information are processed to yield 456 bits. These 456 bits are then transmitted through four time slots.

The structure of a segment is shown in Figure 4.12. Apart from the length indicator field, which appears in every message, the presence of the other fields will depend on the message itself. For example, there may be messages

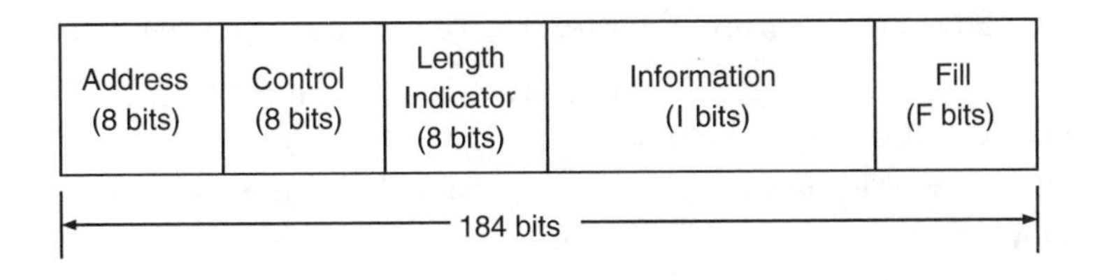

FIGURE 4.12
GSM message segment.

with zero length, in which case, with the exception of the 8-bit length indicator field, all the other fields (176 bits) are filled with 1s.

Six of the bits in the length indicator field denote the number of octets in the variable-length information field. Another bit in the length indicator field determines whether or not the current message segment is the final segment in the corresponding message. The address field contains the following fields: 1 bit indicating whether the message is a command or a response; 3 bits indicating the current version of the GSM protocol; 1 bit as an extension of the address (set to 0 in the initial version of GSM); and 3 bits indicating network management messages or short message service messages. The control field contains 3 bits to indicate the sequence number of the current message and another 3 bits to indicate the sequence number of the last message received by the entity that is sending the present message. In case the complete message encompasses fewer than 184 bits, the fill field is stuffed with 1s.

A number of network management messages are specified in GSM. According to their specific functions the messages can be of three types: supervisory (S), unnumbered (U), and information (I). The S and U messages precede or follow the I messages to control the flow of messages between terminals and base stations. The I messages perform the main tasks concerning network management. An S message may request (re)transmission or may suspend transmission of I messages. An U message may initiate or may terminate a transfer of I messages or may confirm a command. The S and U messages are Layer 2 messages and, more specifically, data link control (DLC) messages. The I messages are Layer 3 messages. More specifically, they carry out the network management operations such as radio resources management (RRM), mobility management (MM), and call management (CM). The RRM messages involve interactions between mobile station, base station, and mobile switching center. The MM and CM messages use the base station as a relay node between mobile stations and MSC where they are effectively treated.

The next subsections summarize the main GSM messages.

4.6.1 DLC Messages

The main DLC messages and their respective purposes are listed below:

- *Set Asynchronous Balanced Mode*. This is a U command message. It initiates a transfer of I messages.
- *Disconnect*. This is a U command message. It terminates a transfer of I messages.
- *Unnumbered Acknowledgment*. This is a U response message. It confirms a command.

- *Receive Ready*. This is an S command or an S response message. It requests transmission of an I message.
- *Receive Not Ready*. This is an S command or an S response message. It requests retransmission of an I message.
- *Reject*. This is an S command or an S response message. It suspends transmission of I messages.

4.6.2 RRM Messages

The main RRM messages and their respective purposes are listed below:

- *Sync Channel Information*. This is a downlink message running on the SCH. It conveys the base station identifier and the frame number, the latter allowing the terminal to achieve time synchronization.
- *System Information*. This is a downlink message running on the BCCH. It contains the location area identifier, the number of the physical channel carrying signaling information, the parameters of the random-access protocols, and the radio frequency carriers active in the neighboring cells.
- *System Information*. This is a downlink message running on the SACCH. It provides local system information to those active terminals that are moving away from the cell where the call was originated.
- *Channel Request*. This is an uplink message running on the RACH. It is used to respond to a page, to set up a call, to update the location, to attach the IMSI, to detach the IMSI.
- *Paging Request*. This is a downlink message running on the PCH. It is employed to set up a call to a terminal.
- *Immediate Assignment*. This is a downlink message running on the AGCH. It is utilized to assign an SDCCH to a terminal at the setup procedure as a result of a channel request message.
- *Immediate Assignment Extended*. This is a downlink message running on the AGCH. It is utilized to assign two terminals to two different physical channels.
- *Immediate Assignment Reject*. This is a downlink message running on the AFCH. It is utilized as a response to channel request messages from as many as five terminals when the system is not able to provide these terminals with dedicated channels.
- *Assignment Command*. This is a downlink message running on the SDCCH. It is used at the end of the setup call process to move the terminal to a TCH.

- *Additional Assignment*. This is downlink message running on the FACCH. It is used to assign another TCH to a terminal already operating on a TCH.
- *Paging Response*. This is an uplink message running on the SDCCH. It is used to respond to a page with the aim of identifying the terminal and causing the initiation of the authentication procedure.
- *Measurement Report*. This is an uplink message running on the SACCH. It is used to indicate the signal level of the terminal and the signal quality of the active physical channel and of the channels of the surrounding cells, for intracell or intercell handover purposes.
- *Handover Command*. This is a downlink message running on the FACCH. It is used to move a call from one physical channel to another physical channel. It is also used for the terminal to adjust its timing advance.
- *Handover Access*. This is an uplink message running on the TCH. It is used to provide the base station with the necessary information so that the base can instruct the terminal on the timing adjustment needed in a handover process.
- *Physical Information*. This is a downlink message running on the FACCH. It is used to transmit the timing adjustment the terminal requires in a handover process.
- *Handover Complete*. This is an uplink message running on the FACCH. It is utilized after the terminal has adjusted its transmission time within the newly assigned physical channel.
- *Ciphering Mode*. This is a downlink message running on the FACCH. It indicates whether or not user information is to be encrypted.
- *Channel Release*. This is a downlink message running on the FACCH. It informs the terminal that a given channel is to be released.
- *Frequency Redefinition*. This is a downlink message running on the SACCH as well on the FACCH. It informs the terminal about the new hopping pattern to be used.
- *Classmark Change*. This is an uplink message running on the SACCH as well as on the FACCH. It informs the network about the terminal's new class of transmission power. This message occurs, for example, when a phone is plugged in or removed from an external apparatus with high power.
- *Channel Mode Modify*. This is a downlink message running on the FACCH. It commands the terminal to change from one channel mode (speech or data) to another. The channel mode defines the specific source coder (for speech) or the data speed (for data).

- *RR Status*. This is a two-way message running on the FACCH as well as on the SACCH. It reports the error conditions of the radio resource (RR).

4.6.3 CM Messages

The main CM messages and their respective purposes are listed below:

- *Setup*. This is a two-way message running on the SDCCH. It is used to initiate a call.
- *Emergency Setup*. This is an uplink message running on the SDCCH. It is used to initiate a call.
- *Call Proceeding*. This is a downlink message running on the SDCCH. It is used as a response to a setup message.
- *Progress*. This is a downlink message running on the SDCCH. It is used to inform the calling party, through an audible tone, that the call is being transferred to a different network (from a public to a private one, for example).
- *Call Confirmed*. This is an uplink message running on the SDCCH. It is used as a response to a setup message.
- *Alerting*. This is a two-way message running on the SDCCH. It is used to indicate to the calling party that the called party is being alerted.
- *Connect*. This is a two-way message running on the SDCCH. It is used to indicate that the call is being accepted.
- *Start DTMF*. This is an uplink message running on the FACCH. It is used to indicate that a button of the phone keypad has been pressed. This causes the network to send to the terminal a dual-tone multiple frequency.
- *Stop DTMF*. This is an uplink message running on the FACCH. It is used to indicate that a button of the phone keypad has been released. This causes the network to turn off a dual-tone multiple frequency.
- *Modify*. This is a two-way message running on the FACCH. It is used to indicate that the nature of the transmission is being modified (e.g., from speech to facsimile).
- *User Information*. This is a two-way message running on the FACCH. It is used, for example, to carry user-to-user information as part of GSM supplementary services.
- *Disconnect/Release/Release Complete*. This is a two-way message running on the FACCH. It is used to end a call. For example, if the terminal is concluding a call, it sends a disconnect message to the network,

which responds with a release message, and this causes the terminal to send a release complete message to the network. The same sequence of messages flow in the opposite direction if the other party terminates the call.

- *Disconnect*. This is a two-way message running on the FACCH. It is used to indicate that a call is terminating.
- *Release*. This is a two-way message running on the FACCH. It is used as a response to a disconnect message.
- *Release Complete*. This is a two-way message running on the FACCH. It is used as a response to a release message.
- *Status*. This is a two-way message running on the FACCH. It is used as a response to a status enquiry message to describe error conditions.
- *Status Enquiry*. This is a two-way message running on the FACCH. It causes the network element (either the terminal or the base) to respond with a status message.
- *Congestion Control*. This is a two-way message running on the FACCH. It is used to initiate a flow control procedure, in which case the flow of call management messages is retarded.

The CM messages occur at different stages of a call. At the beginning of a call, the following messages run on the SDCCH:

- Setup
- Emergency setup
- Call proceeding
- Progress
- Call confirm
- Alerting
- Connect

During a call, the following messages run on the FACCH of the assigned channel:

- Start DTMF
- Stop DTMF
- Modify
- User Information

At the end of a call, the following messages run on the FACCH of the assigned channel:

- Disconnect
- Release
- Release complete

During abnormal conditions, the following messages run on the FACCH of the assigned channel:

- Status
- Status enquiry
- Congestion control

4.6.4 MM Messages

The MM messages travel on the SDCCH. The MM messages and their respective purposes are listed below.

- *Authentication Request*. This is a downlink message. It is used to send a 128-bit random number (RAND) to the terminal, which, by means of an encryption algorithm, computes a 32-bit number to be sent to and checked up at the base.
- *Authentication Response*. This is an uplink message. It is used as a response to an authentication request message, conveying the 32-bit number generated out from the encryption algorithm.
- *Authentication Reject*. This is a downlink message. It is used to abort the communication between the terminal and the network as a result of an unsuccessful authentication.
- *Identity Request*. This is a downlink message. It is used to request any of the three identifiers: IMSI (stored on the SIM), IMEI (stored in the terminal), and TMSI (assigned by the network to a visiting terminal).
- *Identity Response*. This is an uplink message. It is used as a response to the identity request message.
- *TMSI Reallocation Command*. This is a downlink message. It is used to assign a new TMSI to the terminal.
- *Location Updating Request*. This is an uplink message. It is used by the terminal to register its location.
- *Location Updating Accept*. This is a downlink message. It is used to accept a location registration.
- *Location Updating Reject*. This is a downlink message. It is used to reject a location registration. A location updating may be rejected in any of the following events: unknown subscriber, unknown location area, roaming not allowed, or system failure.

- *IMSI Detach Indication*. This is an uplink message. It is used to cancel the terminal registration when the terminal is switched off.
- *CM Service Request*. This is an uplink message. It is used to initiate an MM operation. As a consequence, one or more MM messages will follow.
- *CM Re-Establishment Request*. This is an uplink message. It is used to reinitiate an MM operation that has been interrupted.
- *MM Status*. This is a two-way message. It is used to report error conditions.

4.7 Call Management

This section outlines some call management procedures, namely, mobile initialization, location update, authentication, ciphering, mobile station termination, mobile station origination, handover, and call clearing.

4.7.1 Mobile Initialization

There are three main goals of the mobile initialization procedure:

1. Frequency synchronization
2. Timing synchronization
3. Overhead information acquisition

Frequency Synchronization. As the terminal is switched on, it scans over the available GSM RF channels and takes several readings of their RF levels to obtain an accurate estimate of the signal strengths. Starting with the channel with the highest level, the terminal searches for the frequency correction burst on the BCCH. If no frequency correction burst is detected, it then moves to the next highest level signal and repeats the process until it is successful. In this event, the terminal will then synchronize its local oscillator with the frequency reference of the base station transceiver.

Timing Synchronization. After frequency synchronization has been achieved, the terminal will search for the synchronization burst for the timing information present on the SCH. If it is not successful, it then moves to the next highest level signal and repeats the process starting from the frequency synchronization procedure until it is successful. In this event, it moves to the BCCH to acquire overhead system information.

Overhead Information Acquisition. After timing synchronization has been achieved, the terminal will search for overhead information on the BCCH. If the BCCH information does not include the current BCCH number, it will restart the mobile initialization procedure. In a successful event, the terminal will have acquired, from the BCCH and through the system information message present on the BCCH, the following main information:

- Country code
- Network code
- Location area code
- Cell identity
- Adjacent cell list
- BCCH location
- Minimum received signal strength

The terminal checks if the acquired identification codes coincide with those in the SIM card. In a successful event, it will maintain the link and monitor the PCH. Otherwise, it will start a location update procedure.

4.7.2 Location Update

A location update procedure is carried out in one of the following events:

- The terminal is switched on and verifies that the identification codes present on the current BCCH do not coincide with those in the SIM card.
- The terminal moves into a location area different from that within which it is currently registered.
- There has been no activity for a preestablished amount of time. As part of the process used to speed the paging procedure, location reports are used. These location reports are periodic reports used to update the location of the terminal so that, in the event of a page, the latest reported location is used as an initial guess to locate the terminal. The time span between location reports constitutes a system parameter whose value is indicated on the BCCH, varying in accordance with the network loading.

The location update procedure starts with the uplink channel request message on the RACH. The network answers with an immediate assignment message on the AGCH indicating the SDCCH number to be used throughout the location update procedure. The terminal moves to this SDCCH and sends a location updating request message with its identification (IMSI or, preferably, TMSI). An authentication procedure is then carried out. In case the authentication is unsuccessful, the procedure is aborted. In a successful event, the ciphering procedure is performed. The network then uses the location updating accept message to assign a new TMSI to the terminal. The terminal stores its TMSI and responds with a TMSI allocation complete message. The location update is concluded with a channel release message from the network to the terminal. The terminal then resumes its PCH monitoring procedure.

4.7.3 Authentication

An authentication procedure may be required at the location update procedure or at the request of a new service. The authentication procedure starts with the network sending an authentication request message to the terminal; the message conveys a 128-bit random number (RAND). The terminal uses the RAND, the secret key, Ki, stored at SIM, and the encryption algorithm, referred to as A3, to compute a 32-bit number, referred to as a signed response (SRES). Another 64-bit key, the ciphering key, Kc, is computed using another encryption algorithm, referred to as A8. The Kc parameter is later used in the ciphering procedure. After these computations, the terminal responds with an authentication response message, which contains the SRES. The network uses the same parameters and the same algorithm to compute another SRES. The terminal SRES and the network SRES are then compared with each other. If a match occurs, the network accepts the user as an authorized subscriber. Otherwise, the authentication is rejected.

4.7.4 Ciphering

Ciphering (or encryption) is usually required for user transactions over the RF link after authentication has been successful. The network transmits a ciphering mode message to the terminal indicating whether or not encryption is to be applied. In case ciphering is to be performed, the secret key Kc (64 bits), which was generated previously in the authentication procedure, the frame number (22 bits), and an encryption algorithm, referred to as A5, are used to compute a 114-bit encryption mask. This mask is modulo-2 added to the $2 \times 57 = 114$ bits of the data fields, in the bursts. Deciphering is obtained at the base station by performing the same procedure. The terminal answers with a ciphering mode acknowledgment message. Note that the ciphering to be used is continuously changing (on a frame-by-frame basis), because it depends on the current frame number.

4.7.5 Mobile Station Termination

After the mobile initialization procedure, the terminal camps on the PCH. It eventually detects a paging request message conveying its TMSI. This impels the terminal to access the RACH to transmit a channel request message. An immediate assignment with the SDCCH number is sent by the network on the AGCH. The terminal moves to SDCCH and the following occurs. The terminal transmits a paging response message indicating the reason for the specific message (response to a paging). An authentication procedure is carried out, as already described. In a successful event, a ciphering procedure is accomplished, as already described. The base station then sends a setup message. The terminal responds with a call confirmed message followed by an alerting message to indicate that the subscriber is being alerted. At the subscriber's call acceptance, the terminal sends a connect message and removes the alerting tone. The network responds with an assignment command message indicating the traffic channel number to be used for the conversation. The subscriber, still on the SDCCH, responds with an assignment acknowledgment message and moves to the traffic channel that has been assigned. The network confirms the acceptance of the call by the other party by means of a connect acknowledgment message on the FACCH of the assigned TCH. And the conversation proceeds on the TCH.

4.7.6 Mobile Station Origination

The terminal detects a user-originated call. It then accesses the RACH to send a channel request message. An immediate assignment with the SDCCH number is sent by the network on the AGCH. The terminal moves to this channel and the following occurs. The terminal transmits a paging response message indicating the reason for the specific message (call setup). The base station responds with an unnumbered acknowledgment message. An authentication procedure is carried out, as already described. In a successful event, a ciphering procedure is performed, as already described. The terminal then sends a setup message. The base station responds with a call confirmed message followed by an alerting message in which case the terminal applies the ring-back tone. At the called party's call acceptance, the network sends an assignment command message informing the traffic channel number to be used for the conversation. The subscriber, still on the SDCCH, responds with an assignment acknowledgment message and moves to the traffic channel that has been assigned. The network confirms the acceptance of the call by the other party by means of a connect acknowledgment message on the FACCH of the assigned TCH. And the conversation proceeds on the TCH.

4.7.7 Handover

The handover process in a GSM network has the mobile terminal as an integral part of the procedure. The whole process is named mobile-assisted handover (MAHO). While making use of the traffic channel, the mobile monitors the signal levels of its own channel, of the other channels of the same cell, and of the channels of six surrounding cells. The measurements are then reported to the base on an SACCH. Concerning the control of the process, handovers may occur:

- Within the same BTS or between BTSs controlled by the same BSC
- Between different BSCs controlled by the same MSC
- Between different BSCs controlled by different MSCs
- Between different BSCs controlled by different MSCs belonging to different PLMNs

In addition, there are two modes of handovers: synchronous or asynchronous. In the synchronous mode, the origin cell and the destination cell are synchronized. By measuring the time difference between their respective time slots, the mobile itself may compute the timing advance. This is used to adjust its transmissions on the new channel, therefore, speeding up the handover process. In the asynchronous mode, the origin cell and the destination cell are unsynchronized. The timing advance, in this case, must be acquired by means of a procedure involving the terminal and the new BTS, as follows. The mobile terminal sends a series of access bursts with a zero timing advance through several handover access messages. The BTS then computes the required timing advance using a round-trip time delay of the messages. On the average, the handover processing time in the synchronous mode (200 ms) is twice as long as that of the synchronous mode (100 ms).

Next a simple asynchronous handover procedure occurring between BTSs of the same BSC is described. While in conversation on a TCH, the terminal monitors the signal levels of several channels. These measurements are reported to the base station on a periodic basis by means of the measurement report message running on the SACCH. Whenever suitable, the base sends a handover command message on the FACCH, indicating that a handover is to take place. The number of the new TCH is included within the message. The terminal then moves to this new channel and sends a series of handover access messages so that the base may compute the timing advance to be transmitted to the terminal. This is done in the physical information message transmitted to the terminal on the FACCH. The timing adjustment is carried out and the terminal responds with a handover complete message.

4.7.8 Call Clearing

The call clearing process may be initiated either by the network or by the mobile. In either case, the channel used for the exchange of information is the BCCH. Assuming the network initiates the clearing, the base sends a disconnect message to the terminal. The terminal responds with a release message. The base replies with a release complete message. If the terminal initiates the clearing, then the same messages flow, but in the opposite direction.

4.8 Frequency Hopping

GSM cellular reuse planning is given an additional level of sophistication by the introduction of frequency hopping (FH). In FH, the signal in a given time slot moves from one frequency to another according to a preestablished hopping pattern. By changing frequencies periodically, the transmission becomes less vulnerable to fading, because the probability of encountering more than one faded frequency at the same time diminishes with the increase of the number of frequencies utilized. Therefore, the signal is affected by fading only during part of its time. Besides, FH may reduce co-channel interference if co-cells are assigned different hopping patterns.

As opposed to fast FH, in which the change of frequency (or frequencies) occurs during the symbol time, GSM uses slow FH with the hopping taking place at each TDMA frame (one hopping at each 4.615 ms). The FH facility is implemented in all GSM terminals. The application of FH, however, is decided by the network operator. The hopping algorithm is based on such parameters as the set of hopping frequencies, the hopping pattern, frame number, and others, which are transmitted over the SCH. There are two possible FH operation modes: cyclic and random. In the cyclic mode, the hopping occurs sequentially over the set of frequencies. In the random mode, the hopping is performed in a pseudorandom fashion according to one out of 63 allowable patterns. Furthermore, the FH facility may be implemented at the baseband or at the RF levels. The baseband FH implementation makes use of a number of transceivers, each of which operates on a fixed frequency. In this case, the FH occurs as the baseband information moves from one input of one transceiver to the input of another transceiver, according to the preestablished hopping sequence. The RF FH implementation provides the FH at the frequency synthesizer level for a given transceiver. In this case, the FH occurs as the synthesizer changes its frequency in accordance with the given hopping pattern.

All carriers in the GSM band, with the exception of the standard broadcast carriers, are entitled to hop. The standard broadcast carrier, also known as the base channel, contains the FCCH, the SCH, and the BCCH, and is the beacon upon which the terminals carry out their measurements and extract the necessary information. All signals within a cell and also within a group of cells hop in a coordinated manner. In other words, the hopping sequences are chosen so that frequency overlapping is avoided (the hopping sequences are orthogonal to each other). Both the uplink and the downlink operate with the same FH sequence.

When a terminal is to use FH, it is informed about the available set of hopping channels and the hopping sequence number.

4.9 Discontinuous Transmission

GSM equipment is designed for discontinuous transmission (DTx), a feature utilized to conserve battery power and to reduce interference. DTx takes advantage of the fact that in a normal conversation, on average, the voice activity factor is far less than 100% (typically less than 60%). Therefore, in theory, the transmitter must be on only during the time the voice is effectively active, and off otherwise.

To implement this feature, the terminal is equipped with a voice activity detector (VAD). Upon detecting voice (in the presence of noise) or noise, the VAD outputs a corresponding signal that controls a transmitter switch. Therefore, the design of such a VAD plays a decisive role in transmission performance. A decision in favor of a wrong detection will certainly produce annoying effects in the transmission. In such a case, a clipping effect in speech may be noticed. Note that, as the transmitter is switched off between talkspurts, at the receiving end the background acoustic noise present in the conversation abruptly disappears. It has been observed that this can be annoying to the listener. To minimize the effect of such a noise "modulation," a synthetic signal, known as *comfort noise signal* (CNS), with characteristics matching those of the background noise, is introduced at the receiver. Note that the background noise thus generated is not standard noise. It tries to conform with the characteristics of the background noise of the current transmission. To accomplish this, the noise parameters are computed at the transmit end during a time span of four frames after the VAD detects the end of a talkspurt. It is very unlikely that the speech restarts during this time, so that the noise is present for the parameter evaluation. These parameters are sent to recompose the noise at the receiver.

Besides VAD and CNS functions, another function of the DTx feature is speech frame extrapolation (SFE). The SFE aims to replace a speech frame badly corrupted by error with a preceding uncorrupted speech frame. This replacement is based on the fact that consecutive speech frames are highly correlated. Therefore, the use of SFE improves the signal quality or, equivalently, allows for a reduced carrier-to-interference ratio.

4.10 Power Control

GSM employs power control techniques to adjust the power of both mobile stations and base stations for better performance. Power control reduces co-channel interference and increases battery life. Mobile stations can have their power adjusted in steps of 2 dB with the power levels ranging over 30 dB, i.e., 16 power levels are permitted. The time span between power adjustments is 60 ms, corresponding to 13 frames. Base stations can also control their own power, this process involving the mobile station: the mobile station monitors the signal received from the base station and the base station transmitting power can be changed to improve the signal reception at the mobile station.

4.11 Spectral Efficiency

GSM uses powerful interference counteraction techniques such as adaptive equalization, powerful error-correcting codes, efficient modulation, speech frame extrapolation, and others. These render GSM system robust and capable of operating at a low carrier-to-interference ratio, in which case, reuse factors of three or four cells per cluster, depending on the environment, are admissible.

A number of spectral efficiency definitions are available. In accord with Reference 12, the spectral efficiency parameter η is defined as

$$\eta = \text{conversations/cell/spectrum}$$

The number of physical channels in the 50-MHz GSM spectrum is 124 carriers $\times$ 8 channels/carrier = 992 (GSM-900). It can also be assumed that all these 992 channels can be used for conversation. Therefore, for a reuse factor of 4:

$$\eta = \frac{992}{4 \times 50} = 4.96$$

And for a reuse factor of 3:

$$\eta = \frac{992}{3 \times 50} = 6.61$$

The same calculations can be performed for the other GSM systems. The results will be very close to these.

4.12 Summary

GSM has emerged as a digital solution to the incompatible analog air interfaces of the differing cellular networks operating in Europe. Among the set of ambitious targets to be pursued, *full roaming* was indeed a very important one. In addition, a large number of *open interfaces* have been specified within the GSM architecture. Open interfaces favor market competition with operators able to choose equipment from different vendors. GSM was the first system to stimulate the incorporation of the *personal communication services* philosophy into a cellular network. These and other innovative features rendered GSM networks, either in the original GSM conception or as an evolution of it, a very successful project with worldwide acceptance. GSM systems are found operating in frequency bands around 900 MHz (GSM-900), 1.8 GHz (GSM-1800), or 1.9 GHz (GSM-1900). A new revision of the GSM specifications define an E-GSM that extends the original GSM-900 operation band and stipulates lower power terminals and smaller serving areas.

References

1. Eberspaecher, J., Bettstetter, C., and Vhogel, H.-J., *GSM: Switching, Services and Protocols*, John Wiley & Sons, New York, 2001.
2. Tisal, J., *The GSM Network: The GPRS Evolution: One Step Towards UMTS*, John Wiley & Sons, New York, 2001.
3. Steele, R., Gould, P., and Chun, L. C., *GSM, cdmaOne and 3G Systems*, John Wiley & Sons, New York, 2000.
4. Nielsen, T. and Wigard, J., *Performance Enhancements in a Frequency Hopping GSM Network*, Kluwer Academic Publishers, Dordrecht, the Netherlands, 2000.
5. Heine, G. and Horrer, M., *GSM Networks: Protocols, Terminology, and Implementation*, Artech House, Norwood, MA, 1999.

6. Zvonar, Z., Jung, P., and Kammerlander, K., *GSM: Evolution towards 3rd Generation Systems*, Kluwer Academic Publishers, Dordrecht, the Netherlands, 1999.
7. Garg, V. K. and Wilkes, J. E., *Principles and Applications of GSM*, Prentice-Hall, Englewood Cliffs, NJ, 1998.
8. Redl, S., Weber, M., and Oliphant, M., *GSM and Personal Communications Handbook*, Artech House, Atlanta, 1998.
9. Levine, R. and Harte, L., *GSM Superphones: Technologies and Services*, McGraw-Hill Professional, New York, 1998.
10. Lamb, G., Lamb, B., and Batteau, Y., *GSM Made Simple*, Cordero Co., 1997.
11. Mehrotra, A. K., *GSM System Engineering*, Artech House, Atlanta, 1997.
12. Goodman, D. J., *Wireless Personal Communications Systems*, Addison-Wesley Longman, Reading, MA, 1997.
13. Tisal, J., *GSM Cellular Radio*, John Wiley & Sons, New York, 1996.
14. Redl, S., Weber, M., and Oliphant, M., *An Introduction to GSM*, Artech House, Atlanta, 1998.
15. Mouly, M. and Pautet, M.-B., *The GSM System for Mobile Communications*, Telecom Publishing, Palaiseau, France, 1992.

5

cdmaOne

5.1 Introduction

The need for increased capacity was the great motivation for the advent of American digital cellular technology. As demand for wireless services increased, mainly in dense urban areas, the old analog standard, known as AMPS (Advance Mobile Phone Service), proved inadequate to satisfy the demand. Time Division Multiple Access technology, based on the EIA/TIA/IS-54 specifications (later on enhanced and renamed EIA/TIA/IS-136) was the first solution to the capacity problem of the old analog system. By offering roughly a threefold increase in capacity by dividing each 30 kHz AMPS channel into three time slots, this system was the first American response to the European cellular second generation, the GSM.

This digital novelty, however, was not enough to soothe a number of service providers, who argued that such a technology would not be adequate for future growth in service. Other alternatives were then considered, and a technical committee was formed to study and generate cellular standards for wideband services. In the late 1980s and early 1990s, QUALCOMM, Inc. of San Diego proposed a Code Division Multiple Access, CDMA, system and together with Pacific Telesis demonstrated its operation. Extensive successful field trials and network refinement led the Telecommunication Industry Association (TIA) and the Electronic Industry Association (EIA) to adopt QUALCOMM system as their interim standard, the "TIA/EIA/IS-95—Mobile Station–Base Station Compatibility Standard for Dual-Mode Wideband Spread Spectrum Cellular System."[2,3]

The TIA/EIA/IS-95 specifications establish that the system operate on a dual-mode (analog and digital) basis, both modes within the same frequency band. The dual-mode capability facilitates the transition from the analog environment to a digital environment. Although compatible, analog and digital systems are rather different; details of the digital system are the subject of this chapter. TIA/EIA/IS-95 supports a direct sequence spread spectrum technology with 1.25 MHz band duplex channels. Therefore, an operating company that chooses this CDMA technology must deactivate about 42 contiguous 30-kHz channels of its analog system. Coexistence of analog and digital systems implies that dual-mode mobile stations are able to place and receive calls in any system and, conversely, all systems are able to place and receive calls from any mobile station. Handoff operations in such a scenario require some attention. A mobile station may initiate a call in the CDMA system and, while the call is still in progress, it may migrate to the analog system, if required. The search for one or another system for the initial registration is not specified by the standard and the exact action is dependent on the manufacturer. In fact, the standard leaves a number of issues to be detailed by the manufacturer. Those recommendations in the standard appearing with the verbal forms "shall" and "shall not" identify the requirements from which no deviation is permitted. Those with "should" and "should not" indicate that several possibilities are permitted. There are still others with "may" and "need not" and "can" and "cannot," which are certainly much less restrictive. Therefore, solutions may be implemented differently by different manufacturers.

A number of innovations have been introduced in the CDMA system as compared with earlier cellular systems. Soft handoff is certainly a great novelty. In soft handoff, handoff from one base station to another occurs in a smooth manner. In soft handoff, the mobile station keeps its radio link with the original base station and establishes a connection with one or more base stations. The excess connections are given up only when and if the new link has sufficient quality. Another innovation introduced in the CDMA system is the use of Global Positioning System (GPS) receivers at the base stations. GPSs are utilized so that base stations be synchronized, a feature vital to the soft handoff operation. Vocoders at variable rates are specified to accommodate different voice activities aimed at controlling interference levels, thence increasing system capacity. Sophisticated power control mechanisms are used so that the full benefit of spread spectrum technique is realized.

The first CDMA systems were employed under the TIA/EIA/IS-95A specifications. The A version of the specifications evolved to TIA/EIA/IS-95B, in which new features related to higher data rate transmission, soft handoff algorithms, and power control techniques have been introduced. The name cdmaOne is then used to identify the CDMA technology operating with either specification.

This chapter describes the cdmaOne specifications. Most of the descriptions concern TIA/EIA/IS-95A specifications. A final section describes the new features included in TIA/EIA/IS-95B.

5.2 Features and Services

TIA/EIA/IS-95 specifications establish two types of features: voice features and short message service features.

5.2.1 Voice Features

The following are the primary voice features.

- *Call Delivery (CD).* CD allows the reception of a call while in a roaming condition.
- *Call Forwarding Busy (CFB)/Call Forwarding Busy No Answer (CFNA)/ Call Forwarding Busy Unconditional (CFU).* CFB, CFNA, and CFU allow a called subscriber to have the system send incoming calls, addressed to the called subscriber's directory number, to another directory number (forward-to number), or to the called subscriber's designated voice mailbox. This happens when the subscriber is engaged in a call or service (for CFB active), or when the subscriber does not respond to paging, does not answer the call within a specified period after being alerted, or is otherwise inaccessible (CNFA active). The inaccessibility may be characterized by the following: no paging response, unknown subscriber's location, inactive subscriber, CD not active for a roaming subscriber, *Do Not Disturb* active, etc. If CFU is active, calls are forwarded regardless of the condition of the termination.
- *Call Transfer (CT).* CT enables the subscriber to transfer an in-progress established call to a third party. The call to be transferred may be an incoming or outgoing call.
- *Call Waiting (CW).* CW provides notification to a controlling subscriber of an incoming call while the subscriber's call is in the two-way state. Subsequently, the controlling subscriber can either answer or ignore the incoming call. If the controlling subscriber answers the second call, it may alternate between the two calls.
- *Calling Number Identification Presentation (CNIP)/Calling Number Identification Presentation Restriction (CNIR).* CNIP provides and CNIR restricts the number identification of the calling party to the called subscriber. The termination network receives the calling number identification (CNI) as part of the basic call setup. This CNI may include one or two calling parties numbers (CPNs), a calling party

subaddress (CPS), redirecting numbers (RNs), and a redirecting subaddress (RS).

- *Conference Calling (CC).* CC provides a subscriber with the ability to conduct a multiconnection call, i.e., a simultaneous communication between three or more parties (conferees). If any of the conferees to a conference call disconnects, the remaining parties remain connected until the controlling subscriber disconnects.
- *Do Not Disturb (DND).* DND prevents a called subscriber from receiving calls. When this feature is active, no incoming calls shall be offered to the subscriber. DND also blocks other types of alerting, such as the CFU abbreviated (or reminder) alerting and message waiting notification alerting. DND makes the subscriber inaccessible for call delivery.
- *Flexible Alerting (FA).* FA causes a call to a pilot directory number to branch the call into several legs to alert several termination addresses simultaneously. The first leg to be answered is connected to the calling party and the other call legs are abandoned.
- *Message Waiting Notification (MWN).* MWN informs enrolled subscribers when a voice message is available for retrieval. MWN may use pip tone or alert pip tone to inform a subscriber of an unretrieved voice message(s).
- *Mobile Access Hunting (MAH).* MAH causes a call to a pilot directory number to search a list of termination addresses for one that is idle and able to be alerted, in a way that only one termination address is alerted at a time.
- *Password Call Acceptance (PCA).* PCA is a call-screening feature that allows a subscriber to limit incoming calls to only those calling parties who are able to provide a valid PCA password (i.e., a series of digits).
- *Preferred Language (PL).* PL provides the subscriber with the ability to specify the language for network services.
- *Priority Access and Channel Assignment (PACA).* PACA allows a subscriber to have priority access to voice or traffic channels on call origination by queuing these subscribers' originating calls when channels are not available. The subscriber is assigned one of several priority levels and the invocation of PACA is determined to one of two options: permanent, in which the feature is always available, and demand, in which the feature is available only on request.
- *Remote Feature Control (RFC).* RFC allows a calling party to call a special RFC directory number to specify one or more feature operations.

- *Selective Call Acceptance (SCA).* SCA is a call-screening service that allows a subscriber to receive calls only from parties whose CNPs are in an SCA screening list of specified CNPs.
- *Subscriber PIN Access (SPINA).* SPINA allows subscribers to control whether their mobile station is allowed to access the network. This feature may be used by subscribers to prevent unauthorized use of their own mobile station or fraudulent use by a clone.
- *Subscriber PIN Intercept (SPINI).* SPINI enables subscribers to restrict outgoing calls originated from their mobile station. The subscriber requires a SPINI PIN authorization code to originate calls meeting specified criteria (e.g., international call type). SPINI PIN shall not be required on unrestricted call types (e.g., emergency) and may not be required for a list of frequently called numbers, regardless of their call type.
- *Three-Way Calling (3WC).* 3WC provides the subscriber with the ability to add a third party to an established two-party call, so that all three parties may communicate in a three-way call.
- *Voice Message Retrieval (VMR).* VMR permits a subscriber to retrieve messages from a voice message system (VMS).
- *Voice Privacy (VP).* VP provides a degree of privacy for the subscriber over the base station to mobile station (BS–MS) radio link.

5.2.2 Short Message Service Features

The following are the primary short message service features:

- *Short Message Delivery–Point-to-Point Bearer Service (SMD-PP).* SMD-PP provides bearer service mechanisms for delivering a short message as a packet of data between two service users, known as short message entities (SMEs). The length of the bearer data may be up to 200 octets. Implementations and service providers may further restrict this length. The SMD-PP service attempts to deliver a message to an MS-based SME whenever the MS is registered even when the MS is engaged in a voice or data call.
- *Cellular Paging Teleservice (CPT).* CPT conveys short textual messages (up to 63 characters) to an SME for display or storage.
- *Cellular Messaging Teleservice (CMT).* CMT conveys and manages short messages to an SME for display or storage. This teleservice should coordinate the use of the display and arbitrate between conflicting users or services. Each message includes attributes for management of the messages received by the SME.

5.3 Architecture

The TIA/EIA/IS-95 system uses the ANSI/TIA/EIA-41 (or ANSI-41, for short) platform, which is basically built on the TIA/EIA/IS-41-C (or IS-41-C, for short) specifications with some minor protocol changes. The main elements constituting the ANSI-41 network reference model is depicted in Figure 5.1. The model shows the functional entities and interface points between these entities. These entities or physical interfaces do not necessarily imply a physical implementation. In fact, more than one functional entity can be implemented on a single physical device. In such a case, the internal interfaces between these functional entities need not comply with the standards. The entities and interfaces shown in Figure 5.1 are described next.

5.3.1 Mobile Station

The mobile station terminates the radio path on the user side of the network enabling the user to gain access to services from the network. It incorporates user interface functions, radio functions, and control functions, with the most

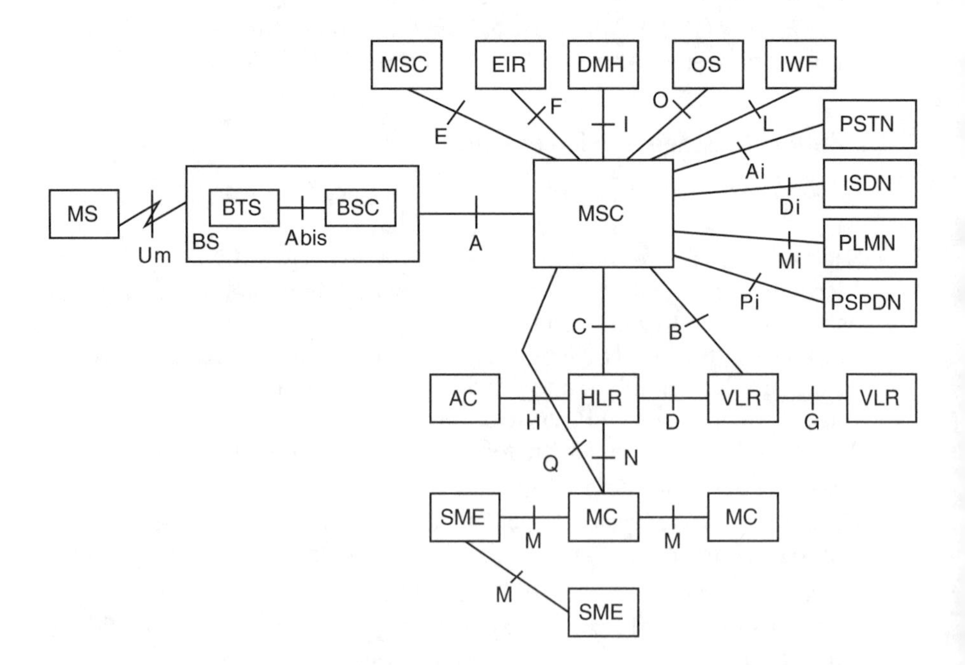

FIGURE 5.1
ANSI-41 network reference model.

common equipment implemented in the form of a mobile telephone. It can work as a stand-alone device or it may accept other devices connected to it (e.g., fax machines, personal computers, etc.).

Mobile stations operating in the analog mode are identified by the mobile identification number (MIN). The MIN is a 34-bit number composed of two fields, namely, MIN1 (24 bits) and MIN2 (10 bits). Mobile stations operating in the CDMA mode use the international mobile station identity (IMSI). The IMSI consists of up to 15 numerical characters (decimal numbers, 0 to 9), the first three digits comprising the mobile country code (MCC), and the remaining digits corresponding to the national mobile station identity (NMSI). In particular, the NMSI contains the mobile network code (MNC) and the mobile station identification number (MSIN). These identifications are programmed into the mobile apparatus by the cellular service provider. Another key identification is the electronic serial number (ESN). The ESN is a 32-bit binary number that is factory-set and not readily alterable in the field. Its alteration requires special facilities not normally available to subscribers.

Mobile stations also store the class-of-information, referred to as the station class mark. Among others, the station class mark specifies functions related to the mode of operation, such as dual mode (CDMA or CDMA and analog), slotted class (slotted or nonslotted), transmission (continuous or discontinuous), and power class (Class I, Class II, Class III). The power class is specified in terms of the effective radiated power at maximum output (ERP_{max}), which must be within the ranges as follows:

- $1.25\text{ W} \leq ERP_{max} \leq 6.3\text{ W}$, for Class I
- $0.5\text{ W} \leq ERP_{max} \leq 2.5\text{ W}$, for Class II
- $0.2\text{ W} \leq ERP_{max} \leq 1.0\text{ W}$, for Class III

5.3.2 Base Station

The base station terminates the radio path on the network side and provides connection to the network. It is composed of two elements: the base transceiver system (BTS) and the base station controller (BSC). The BTS consists of radio equipment (*trans*mitter and re*ceiver* = *transceiver*) and provides the radio coverage for a given cell or sector. The BSC incorporates control capability to manage one or more BTSs, executing the interfacing functions between BTSs and network. The BSC may be co-located with a BTS or else independently located.

All base stations transmissions are commonly referenced to the GPS timescale. The GPS timescale is synchronous with and traceable to the universal coordinated time (UTC), both differing from each other by an integer number of seconds. In particular, this difference equals the number of leap second corrections added to UTC since January 6, 1980 00:00:00 UTC, which is the start of both CDMA system time and of GPS time.

The base station transmit carrier frequency shall be maintained within $\pm 5 \times 10^{-8}$ of the CDMA frequency assignment. The maximum effective radiated power and antenna height above average terrain must be coordinated locally on an ongoing basis.

5.3.3 Mobile Switching Center

The mobile switching center (MSC) provides automatic switching between users within the same network or other public-switched networks, coordinating calls and routing procedures. In general, an MSC is found controlling several BSCs, but it may also serve in different capacities. Depending on the function performed, the MSC is identified by different names, as follows:

- *Anchor MSC*: The first MSC to serve a mobile station-originated or a mobile station-terminated call.
- *Border MSC or Neighbor MSC*: The MSC controlling BSCs responsible for BTSs adjacent to the current serving system.
- *Candidate MSC*: The MSC considered for intersystem handoff purposes.
- *Originating MSC*: The MSC initiating the mobile station-terminated call delivery procedures.
- *Receiving MSC*: The MSC receiving the request for inter-MSC trunk release.
- *Serving MSC or Visited MSC*: The MSC currently serving the mobile station.
- *Tandem MSC*: The MSC in the handoff chain providing only trunk connections for the handoff process; it is neither the anchor MSC nor the serving MSC.
- *Target MSC*: The MSC chosen by the serving MSC to carry out the serving MSC responsibilities in a handoff.

Each MSC within the network has an identification (MSC ID), which is a unique three-octet number, one octet to identify the switch number (SWNO) and two octets to identify the market (market ID). The SWNO identifies a specific MSC within the market area and is allocated by the service provider. The market ID is split into system identification (SID) and billing identification (BID). The SID is a 15-bit value assigned by the Federal Communications Commission (FCC) to each cellular geographic service area covered by a licensed operator. The BID is a 15- or 16-bit value assigned by CIBERNET, a subsidiary of CTIA, to each cellular provider to identify the subset of the entire area represented by an SID for market purposes.

5.3.4 Home Location Register

The home location register (HLR) is the primary database for the home subscriber. It maintains information records of subscriber current location, subscriber identifications (electronic serial number, IMSI, etc.), user profile (services and features), and others. An HLR may be co-located with an MSC or it may be located independently of the MSC. It may even be distributed over various locations and it may serve several MSCs.

5.3.5 Visitor Location Register

The visitor location register (VLR) is a database containing temporary records associated with subscribers under the status of a visitor. A subscriber is considered as a visitor if such a subscriber is being served by another system within the same home service area, or by another system away from the respective home service area (in a roaming condition). The information within the VLR is retrievable by the HLR. A VLR may serve one or more MSCs.

5.3.6 Authentication Center

An authentication center (AC) is a functional unit performing authentication tasks. It includes a database with encryption and authentication keys related to the mobile station. Encryption and authentication algorithms are performed with the involvement of the AC.

5.3.7 Equipment Identity Register

The equipment identity register (EIR) is a database containing mobile station equipment–related data (status of the equipment). These data are used to prevent fraudulent manipulation of the equipment (for example, stolen or cloned equipment may be denied access to the network).

5.3.8 Message Center

The message center (MC) is responsible for storing and forwarding short messages for SMS. As opposed to the voice services, which require real-time operation, the SMS is a store-and-forward service, and is, therefore, less restrictive than voice calls. The MC operates with mobile station-originated or MS-terminated messages. SMS can be provided if an SME is implemented.

5.3.9 Short Message Entity

The SME is a functional unit able to originate, terminate, or originate and terminate short messages. An SME may be associated with any one of the ANSI-41 functional entities or with any of the external entities.

5.3.10 Data Message Handler

The data message handler (DMH) is responsible for collecting billing data.

5.3.11 Operations System

The operations system (OS), responsible for the operations, administration, maintenance, and provisioning of the system, is also referred to as an OAM&P system. The OAM&P systems are responsible for monitoring the system as a whole, carrying out addition and removal of equipment, hardware and software tests, functional diagnosis, billing, and other tasks.

5.3.12 Interworking Function

The interworking function (IWF) provides means for the MSCs to interconnect with other networks.

5.3.13 External Networks

The external networks with which interconnection is possible include the Public Switched Telephone Network (PSTN), the Integrated Services Digital Network (ISDN), the Public Land Mobile Network (PLMN), and the Public Switched Packet Data Network (PSPDN).

5.3.14 Interface Reference Points

The following are the interface reference points in ANSI. Some of these points are not completely defined.

- *A-Interface*: The interface between BS and MSC. It supports signaling and traffic (voice and data). Three protocols are standardized for use: SS7-based, frame relay-based, both defined in TIA/EIA/IS-634, and ISDN-based, defined in TIA/EIA/IS-653.
- *Abis-Interface*: The interface between BTS and BSC.
- *Ai-Interface*: The interface between MSC and PSTN. It is defined in TIA/EIA/IS-93 as an analog interface that uses dual-tone multi-frequency (DTMF) signaling or multifrequency signaling (MF).

- *B-Interface*: The interface between MSC and VLR, which is defined in TIA/EIA/IS-41-C.
- *C-Interface*: The interface between MSC and HLR, which is defined in TIA/EIA/IS-41-C.
- *D-Interface*: The interface between HLR and VLR. It is based on SS7 and is defined in TIA/EIA/IS-41C.
- *Di-Interface*: A digital interface between MSC and ISDN. It uses Q-931 signaling and is defined in TIA/EIA/IS-93.
- *E-Interface*: The interface between MSCs, i.e., between wireless networks, which is defined in TIA/EIA/IS-93.
- *F-Interface*: The interface between MSC and EIR (not defined).
- *G-Interface*: The interface between VLRs (not defined).
- *H-Interface*: The interface between HLR and AC, which is defined in TIA/EIA/IS-41C.
- *I-Interface*: The interface between DMH and MSC.
- *L-Interface*: The interface between MSC and IWF, which is defined by the interworking function.
- *M-Interface*: The interface between SME and SME, SME and MC, and MC and MC, which is defined in TIA/EIA/IS-41C.
- *Mi-Interface*: The interface between MSC and PLMN.
- *N-Interface*: The interface between MC and HLR, which is defined in TIA/EIA/IS-41C.
- *O-Interface*: The interface between MSC and OS.
- *Pi-Interface*: The interface between MSC and PSPDN.
- *Q-Interface*: The interface between MC and MSC.
- *Um-Interface*: The interface between MS and BS, which, specifically for the CDMA system, is defined in TIA/EIA/IS-95. (Other standards are ANSI/TIA/EIA-553-A, IS-91, and IS-136.)

5.4 Multiple-Access Structure

TIA/EIA/IS-95 is a dual-mode system with a multiple-access architecture based on the narrowband FDMA/CDMA/FDD technology. A total 50-MHz band, divided into two 25-MHz bands, labeled A and B, constitutes the spectrum available. For competition purposes each of these bands is allotted to a different operator. And for historical reasons these bands are not contiguous, with the A-band split into three segments named A, A′, and A″, and the

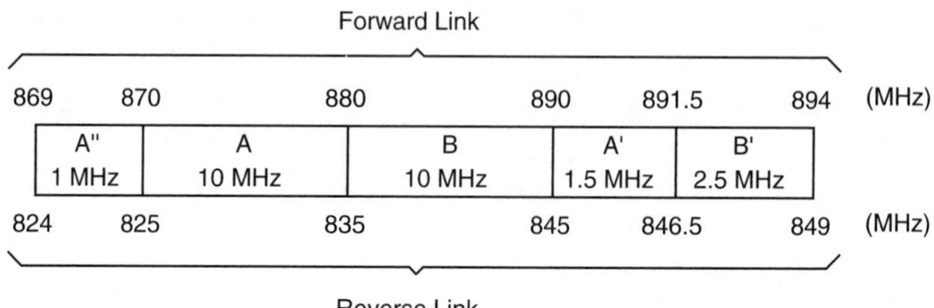

FIGURE 5.2
Frequency allocation for the AMPS and IS-95 systems.

B-band also split into three segments named B, B′, and B″. Each band is further split into equal parts, one half band for the downlink or forward link (base station to mobile station) and the other half band for the uplink or reverse link (mobile station to base station). Therefore, each link is allotted 12.5 MHz, with uplink and downlink separated by 45 MHz. The CDMA frequency spectrum is shown in Figure 5.2.

TIA/EIA/IS-95 specifies a CDMA carrier with a nominal bandwidth of 1.25 MHz. The choice of such a bandwidth was partially influenced by the fact that a carrier should be chosen to fit within the A′ band (1.5 MHz), which is the smallest contiguous segment of the allotted spectrum. Moreover, 1.25 is a submultiple of 10. Thus, in theory, if no guard band is used, it is possible to have up to ten carriers in each 12.5-MHz band, A, A′, A″ or B, B′, B″.

Forward and reverse links comprise different structures and features, details of which are given later in this chapter. The main difference, however, lies in the fact that forward link operates synchronously and reverse link asynchronously. The point-to-multipoint characteristic of the downlink facilitates the synchronous approach, because one reference channel, broadcast by the base station, can be used by all mobile stations within its service area for synchronization purposes. On the other hand, the implementation of a similar feature on the reverse link is not as simple because of its multipoint-to-point transmission characteristic.

Multiple access is supported by functions providing for signal reference, synchronism information, paging, access, and traffic carrying. In TIA/EIA/IS-95, the following channels implement such functions: pilot channel, sync channel, paging channel, access channel, and traffic channel.

5.4.1 Forward Link

The forward link is composed of the pilot channel, sync channel, paging channel, and traffic channel. Figure 5.3 depicts the forward-link main blocks where

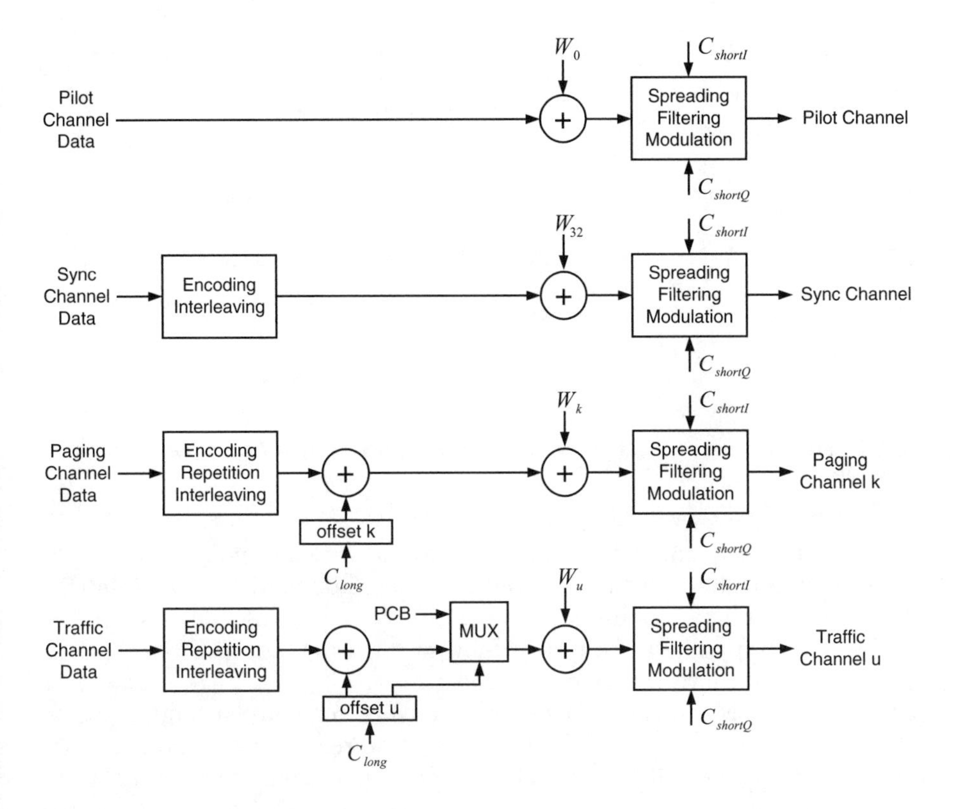

FIGURE 5.3
Forward-link main blocks.

PCB are power control bits that are multiplexed with traffic data. The binary system depicted in Figure 5.3 is of the bipolar type, in which the logical level 0 is represented by a positive voltage +1 and the logical level 1 is represented by a negative voltage −1. In this case, the modulo-2 addition operation, or exclusive OR operation, is equivalent to a multiplication. The spreading, filtering, and modulation block in Figure 5.3 is detailed in Figure 5.4.

In the forward link, channelization is based on orthogonal code division multiplexing, with the channels distinguished by 64 Walsh functions. The Walsh sequence can be generated by means of the Rademacher functions or by the Hadamard matrices. The Hadamard matrix is defined as

$$H_0 = [1]$$

$$H_n = \begin{bmatrix} H_{n-1} & H_{n-1} \\ H_{n-1} & -H_{n-1} \end{bmatrix}$$

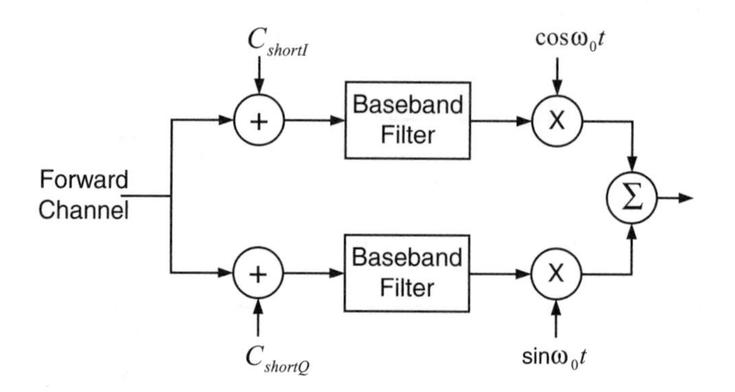

FIGURE 5.4
Quadrature spreading modulation: spreading, filtering, and modulation for the forward-link channels.

The Walsh sequences are indexed by the row of the matrix. Hence, W_i corresponds to the sequence in row i of the Hadamard matrix. Base stations are identified by shifted versions of PN sequences, called short PN codes, which modulate the forward-link waveforms to achieve the spread spectrum characteristics. Two PN sequences, $C_{\text{short}I}$ and $C_{\text{short}Q}$, at a chip rate of 1.2288 Mchip/s, are used, respectively, to modulate the in-phase (cosine) and quadrature (sine) RF carriers, leading to a *form* of quaternary phase shift keying (QPSK) modulation. The same data stream is assigned to both in-phase and quadrature branches of the modulator, which corresponds to a form of diversity. Two pulse-shaping filters of the finite impulse response (FIR) type, one at the in-phase branch and the other at the quadrature branch, are used to confine the radiated spectrum within the 1.25 bandwidth.

Except for the pilot channel, all forward channels use half-rate convolutional coding, symbol repetition, and 20-ms span interleaving. Coding and interleaving combat burst error patterns. Symbol repetition adjusts the data rate to a maximum of 19.2 kbit/s. For traffic channels, vocoders at variable data rates of 9.6, 4.8, 2.4, and 1.2 kbit/s for Rate Set 1, and 14.4, 7.2, 3.6, and 1.8 kbit/s for Rate Set 2 are specified. The various transmission rates are used to accommodate different voice activities. In addition to the specific Walsh sequences, paging channels and traffic channels are further identified by long PN code sequences, specific for the respective channels.

The TIA/EIA/IS-95 system uses 64 Walsh sequences denoted $W_i, i = 0, 1, \ldots, 63$ and they are assigned to the respective channels as follows. W_0 is assigned to the pilot channel; W_1 to W_7 are assigned to the paging channels; W_{32} is assigned to the sync channel; and the remainder W_i code sequences are assigned to traffic channels. The highest baseband data rate is

19.2 ksymb/s which, combined with the assigned 64-chip Walsh sequence, yields a 64×19.2 ksymb/s = 1.2288 Mchip/s baseband data stream. The in-phase PN code, $C_{\text{short}I}$, and quadrature PN code, $C_{\text{short}Q}$, are generated by 15-stage linear feedback shift registers and have their offset phases chosen as a multiple of 64 chips. The codes are lengthened by the insertion of one chip per period, which implies that their period is $2^{15} - 1 + 1 = 32,768$ chips. At a rate of 1.2288 Mchip/s these sequences repeat every 26.666... ms, or 75 times per 2 s. Given that the offset phases are multiple of 64 chips, there are, therefore, $32,768/64 = 512$ possible code offsets that are used to identify the base stations. The characteristic polynomials for the in-phase and quadrature branches are given by

$$C_{\text{short}I} = 1 + x^2 + x^6 + x^7 + x^8 + x^{10} + x^{15} \tag{5.1}$$

$$C_{\text{short}Q} = 1 + x^3 + x^4 + x^5 + x^9 + x^{10} + x^{11} + x^{12} + x^{15} \tag{5.2}$$

The long PN code, C_{long}, is generated by 42-stage linear feedback shift registers and its offset phases are different for the paging channels and for the traffic channels. For the paging channels the phase offset is chosen to identify the base station and the particular paging channel. For the traffic channels the phase offset is chosen as a function of the electronic serial number (ESN) of the subscriber using that particular channel. The long PN codes run at 1.2288 Mchip/s. They repeat after $2^{42} - 1$ chips. At a rate of 1.2288 Mchip/s these sequences repeat every 41.4 days. The characteristic polynomial of the long code is given by

$$\begin{aligned} C_{\text{long}} = {} & 1 + x + x^2 + x^3 + x^5 + x^6 + x^7 + x^{10} + x^{16} + x^{17} + x^{18} \\ & + x^{19} + x^{21} + x^{22} + x^{25} + x^{26} + x^{27} + x^{31} + x^{33} + x^{35} + x^{42} \end{aligned} \tag{5.3}$$

5.4.2 Reverse Link

The reverse link is composed of the access channel and traffic channel. In the reverse link, channelization is based on a conventional spread spectrum CDMA, with the channels identified by distinct phases of a long PN code C_{long}, which is generated by the same characteristic polynomial used for the forward link. Figure 5.5 depicts the reverse-link main blocks. The spreading, filtering, and modulation block shown in Figure 5.5 is detailed in Figure 5.6.

As in the forward-link case, the binary system depicted in Figure 5.5 is of the bipolar type, in which the logical 0 is represented by a positive voltage +1 and the logical 1 is represented by a negative voltage −1. In such case the modulo-2 addition operation, or exclusive OR operation, is equivalent

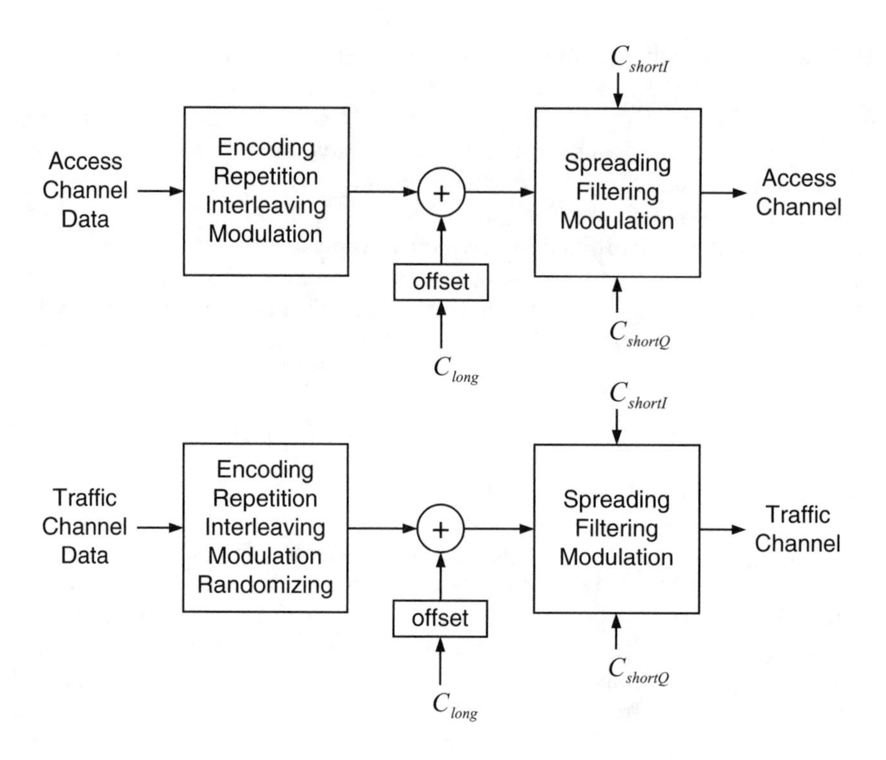

FIGURE 5.5
Reverse-link main blocks.

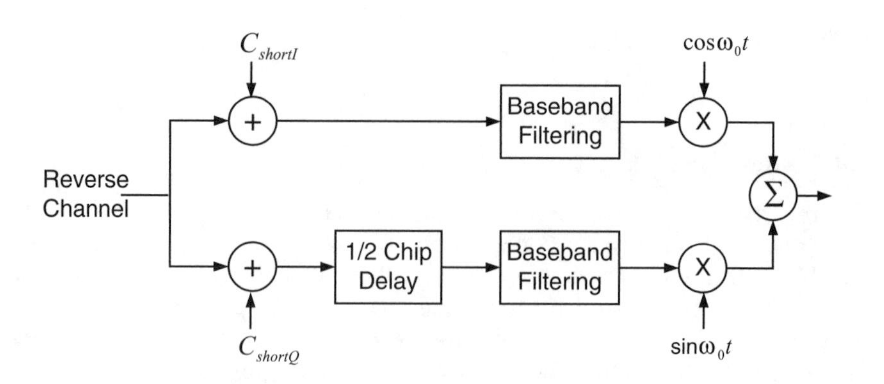

FIGURE 5.6
Quadrature spreading modulation: spreading, filtering, and modulation for the reverse-link channels.

to a multiplication. The reverse-link data stream also experiences quadrature spreading, with the in-phase and quadrature branches modulated by the same two short PN codes of the forward link, respectively, $C_{\text{short}I}$ and $C_{\text{short}Q}$. All codes run at a rate of 1.2288 Mchip/s. The quadrature branch, on the other hand, is shifted by half a PN chip, generating a form of offset quaternary phase shift keying (OQPSK) modulation. The main purpose of this chip delay is to prevent the signal envelope from instantaneously collapsing to zero, as occurs in the QPSK modulation for the symbol transitions from (0,0) to (1,1) and from (0,1) to (1,0). By avoiding these transitions, and as consequence always maintaining the signal envelope above a certain level, the amplifier is able to operate on the linear region over a small dynamic range. This saves power, as is desirable in a mobile terminal. Two pulse-shaping filters of the FIR type, one at the in-phase branch and the other at the quadrature branch, are used to confine the radiated spectrum within the 1.25 bandwidth. Unlike the forward link, the 64 Walsh sequences in the reverse link are used for orthogonal modulation purposes. In this respect, each group of six binary data symbols is represented by the corresponding row of the Hadamard matrix, i.e., by a 64-bit sequence. Therefore, a symbol $(b_0, b_1, b_2, b_3, b_4, b_5)$ is represented by the Walsh sequence W_i, where i is computed as

$$i = b_0 + 2b_1 + 4b_2 + 8b_3 + 16b_4 + 32b_5$$

Note that this corresponds to a 64-ary orthogonal modulation. For the traffic channels, vocoders at variable data rates of 9.6, 4.8, 2.4, and 1.2 kbit/s for Rate Set 1 and 14.4, 7.2, 3.6, and 1.8 kbit/s for Rate Set 2 are specified to accommodate different voice activities. One-third-rate convolutional coding and 20-ms span interleaving are used to combat burst error patterns.

For every paging channel on the forward link there must be at least one access channel, and up to a maximum of 32 access channels on the reverse link. The number of traffic channels on the reverse link equals that of the forward. The long PN code used to identify each channel is obtained by a 42-stage linear feedback shift register with its phase determined by a mask specific for each channel. For the access channel such a mask is chosen as a function of (1) the access channel number itself and of (2) the paging channel with which it is associated. For the traffic channel the mask is chosen as a function of the ESN of the subscriber using that particular channel. The orthogonal modulation implemented by mapping each group of six binary data symbols onto a 64-bit sequence of the Hadamard matrix embodies an error correction capability. Such encoding corresponds to using a code of the kind (64, 6, 32), where 64 is the number of bits of the code word, 6 is the number of bits of information, and 32 is the minimum distance (therefore, the code is able to correct up to 15 bits). The OQPSK modulation scheme presents no transitions through the origin, providing for a relatively constant envelope. The same data stream is assigned to both the in-phase and quadrature branches of the modulator, which corresponds to a form of diversity.

5.4.3 Physical Channels

The frequency band used by TIA/EIA/IS-95 was initially allotted to the AMPS system. In AMPS, the channels are 30 kHz wide and, for a 50-MHz band, 832 bidirectional channels, 416 for each 25-MHz band, can be used. The correspondence between each channel N and the center frequency, in MHz, f_{uplink} for the uplink and f_{downlink} for the downlink are given, respectively, by

$$f_{\text{uplink}} = 0.030N + 825 \qquad 1 \leq N \leq 799$$

$$f_{\text{uplink}} = 0.030\,(N - 1023) + 825 \qquad 990 \leq N \leq 1023$$

$$f_{\text{downlink}} = 0.030N + 870 \qquad 1 \leq N \leq 799$$

$$f_{\text{downlink}} = 0.030\,(N - 1023) + 870 \qquad 990 \leq N \leq 1023$$

Note that a TIA/EIA/IS-95 (1.25-MHz) carrier encompasses 41.67 AMPS carriers. Therefore, the introduction of a CDMA carrier implies that, at least, 42 analog channels must be deactivated. In addition, a guard band of approximately nine analog channels must be added whenever the CDMA carrier is adjacent to any other band used by another system.

The physical channel in a CDMA system is composed of a carrier and a binary code sequence. More specifically, for the forward link the code sequence is 1 of 64 possible rows of the 64×64 Hadamard matrix. In other words, the physical channel is a 64-bit Walsh sequence within a CDMA carrier. For the reverse link, the binary code is a 42-bit-long code mask within a carrier.

Each CDMA system, namely, System A and System B, is allotted a primary channel and a secondary channel. These channels, one of them as the first option and the other as the second option, are acquired by the terminal at the initialization process as the terminal is powered on. The specific channel acquisition order is left to the mobile station manufacturer and may be set as primary and secondary, or vice versa. TIA/EIA/IS-95 recommends that the primary channel be channel 283 for System A and 384 for System B and that the secondary channel be channel 691 for System A and 777 for System B.

5.4.4 Logical Channels

The logical channels in TIA/EIA/IS-95 are composed of the following channels: pilot channel, sync channel, paging channel, access channel, and traffic channel. Pilot channel, sync channel, paging channel, and traffic channel constitute the forward-link logical channels. Access channel and traffic

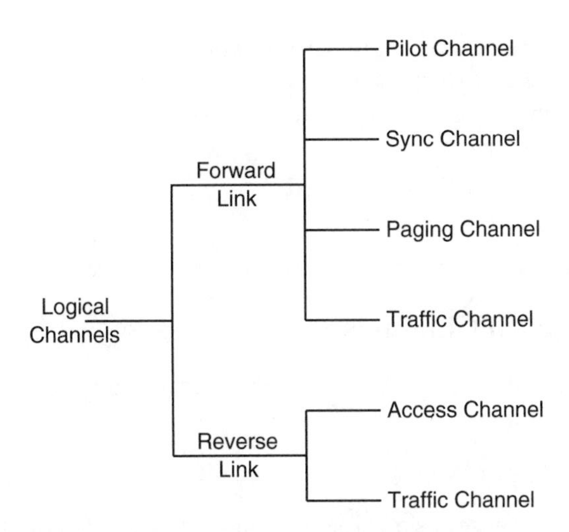

FIGURE 5.7
Forward and reverse logical channels.

channel constitute the reverse-link logical channels. Figure 5.7 illustrates the TIA/EIA/IS-95 logical channels.

The pilot channel, in number of one per carrier and transmitted by the base station, is used for coherent carrier phase and timing references at the mobile stations served by this base station. It is monitored by the terminals served by other base stations as part of the mobile-initiated handoff procedure. The channel is identified by the W_0 Walsh sequence, which is a succession of 64 zeroes.

The sync channel, in number of one per carrier and transmitted by the base station, repeatedly broadcasts one message containing the following information: system time (acquired from the GPS), time delay introduced into the PN sequences, base station and network identifiers, the protocol revision supported by the base station, and the information rate of the base station paging channels. The channel is identified by the W_{32} Walsh sequence.

The paging channel, in number of up to seven per carrier and transmitted by the base station, conveys overhead and paging messages. The overhead messages contain the system parameters and the paging messages contain a page to one terminal or to a group of terminals. The paging channels are identified by the Walsh sequences from W_1 to W_7.

The access channel, in number of at least one and up to 32 per paging channel and transmitted by the mobile station, is used by the terminal with one of the following purposes: to originate a call, to respond to a paging message, or to register its location.

The traffic channel, nominally in number of up to 55 (for seven paging channels) or up to 62 (for one paging channel) per carrier and transmitted by both the mobile and the base stations carry digital payload information (voice or data) and signaling.

5.5 The Logical Channels

This section describes the TIA/EIA/IS-95 logical channels in more detail.

5.5.1 Pilot Channel

The pilot channel serves as a coherent phase reference for demodulating the other channels. It carries no baseband information and is continuously transmitted by the base station. It is formed by an exclusive OR logical combination (multiplication) of a stream of zeroes with the W_0 Walsh sequence, which is also a succession of 64 zeroes, both running at 1.2288 Mchip/s. The resulting stream of zeroes is then fed into the quadrature spreading modulator as shown in Figure 5.8.

5.5.2 Sync Channel

The sync channel conveys information on system synchronization and system parameters, which are repeatedly broadcast via a specific message by the base station. The baseband information is transmitted at a rate of 1.2 kbit/s and is error-control-protected by means of a half-rate (constraint length 9) convolutional encoder, which results in a 2.4 ksymb/s data rate. Each bit is then repeated once, which constitutes a form of time diversity, yielding an information rate of 4.8 ksymb/s. A protection against burst errors is provided by means of a $16 \times 8 = 128$-symbol block interleaver. The resulting 4.8 ksymb/s

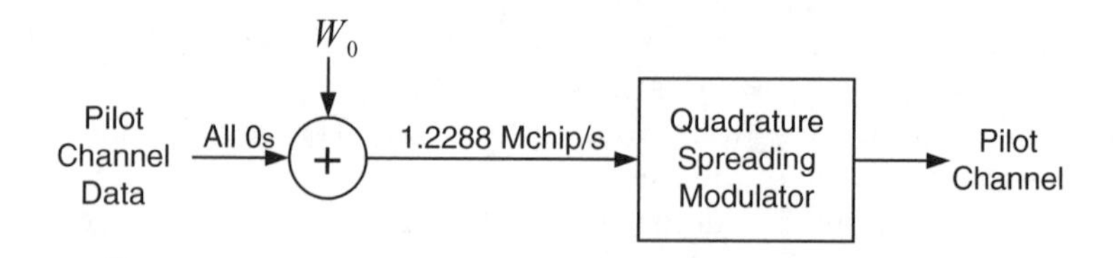

FIGURE 5.8
Pilot channel structure.

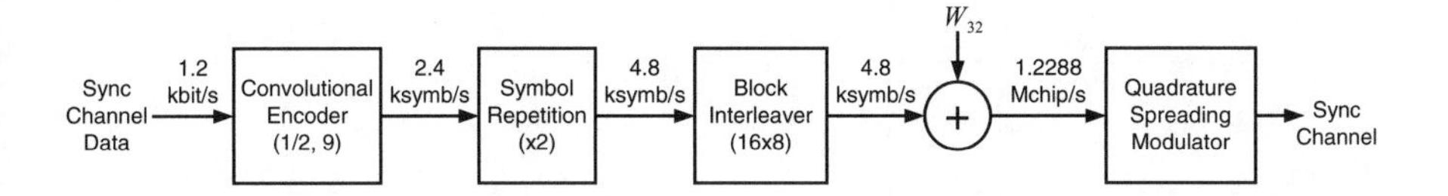

FIGURE 5.9
Sync channel structure.

rate information is combined with the W_{32} Walsh sequence, which runs at a rate of 1.2288 Mchip/s. The combined stream is then fed into the quadrature spreading modulator, as shown in Figure 5.9.

5.5.3 Paging Channel

The paging channels, as the name implies, are used for paging purposes. But more than this, the paging channels convey channel assignments and system overhead information. The baseband information is transmitted at two possible rates: 4.8 or 9.6 kbit/s. The rate at which the paging channel runs is informed by the sync channel. Although there may be as many as seven paging channels serving a base station (sector), the mobile monitors one paging channel only. As opposed to the pilot channel and to the sync channel, the paging channels make use of a long PN code, which is generated by a 42-stage linear feedback shift register with the respective characteristic polynomial given by Equation 5.3. The phase offset is chosen such that it identifies the base station and the particular paging channel. Beginning with the most significant bit, the paging channel mask is given by

$$11000110011010000ppp000000000000bbbbbbbbb$$

where ppp indicates the paging channel number and $bbbbbbbbb$ indicates the base station pilot PN sequence offset index. The long PN code runs at a rate of 1.2288 Mchip/s but is decimated to a 19.2 ksymb/s rate by sampling every 64th PN code chip (a decimation of 64:1). The resulting decimated long PN code is further combined with a channel information running at the same rate, as detailed next.

The baseband information, at 4.8 or 9.6 kbit/s, is first error-control-protected by means of a half-rate convolutional encoder (constraint length 9), which results in a 9.6 or 19.2 ksymb/s data rate, respectively. Each symbol is then repeated once, for the initial rate of 4.8 kbit/s, or not, if the initial rate is 9.6 kbit/s, yielding an information rate of 19.2 ksymb/s. A protection against burst errors is provided by means of a $24 \times 16 = 384$-symbol block interleaver. The output of the interleaver runs at 19.2 ksymb/s rate and is combined with

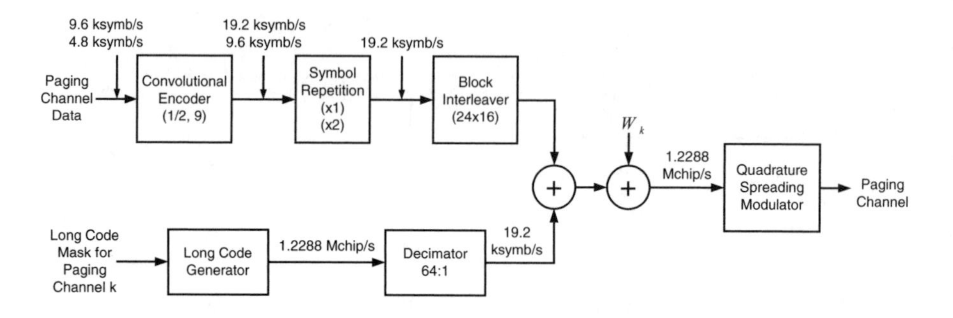

FIGURE 5.10
Paging channel structure.

the decimated version of the masked long PN sequence, running at the same rate. The resulting 19.2 ksymb/s rate information is combined with one of the seven Walsh sequences, from W_1 to W_7, which identifies the specific paging channel and runs at a rate of 1.2288 Mchip/s. The combined 1.2288 Mchip/s rate stream is then fed into the quadrature spreading modulator as shown in Figure 5.10.

5.5.4 Access Channel

The access channel is used by the mobile station to initiate a communication with the base station and to respond to a paging. The channel is identified by a long PN code, which is generated by a 42-stage linear feedback shift register with the respective characteristic polynomial given by Equation 5.3. The phase offset is chosen as a function of the access channel itself and of the paging channel with which the access channel is associated. The resulting long PN sequence runs at 1.2288 Mchip/s.

The baseband information, running at 4.8 kbit/s, is first error-control-protected by means of a 1/3-rate convolutional encoder (constraint length 9) which results in a 14.4 ksymb/s data rate. Each bit is then repeated once, yielding an information rate of 28.8 ksymb/s. A protection against burst errors is provided by means of a $32 \times 18 = 576$-symbol block interleaver. The output of the interleaver runs at 28.8 ksymb/s rate and is fed into the (64,6) Walsh encoder. In other words, each group of 6 bits of the interleaved information is encoded into 64-bit data and this results in a data rate of $28.8 \times 64/6 = 307.2$ kchip/s. This is then combined with the masked long PN sequence, which runs at 1.2288 Mchip/s. The combined 1.2288 Mchip/s rate stream is then fed into the quadrature spreading modulator as shown in Figure 5.11.

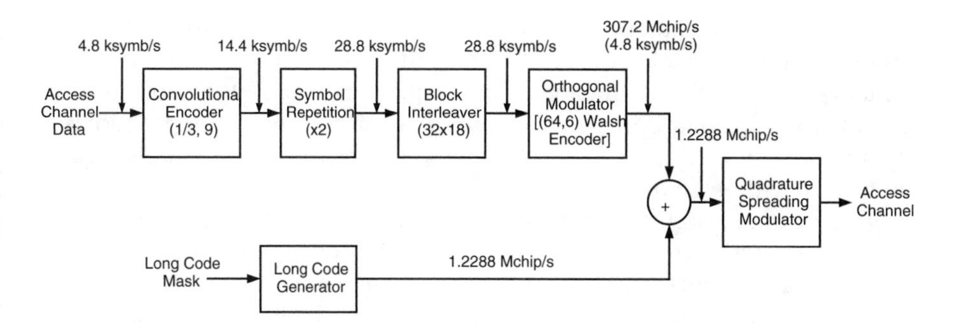

FIGURE 5.11
Access channel structure.

5.5.5 Traffic Channel—Forward and Reverse Links

Forward and reverse traffic channels operate at a variable bit rate: full rate, half rate, one-fourth rate, and one-eighth rate. These rates are 9.6, 4.8, 2.4, and 1.2 kbit/s for Rate Set 1, and 14.4, 7.2, 3.6, and 1.8 kbit/s for Rate Set 2. The variable bit rate provides a means of adjusting the output energy according to the voice activity so that power is not transmitted unnecessarily. By varying the output signal power, the interference provoked by the corresponding signal also varies and an increase in capacity may be achieved at reduced rates.

As far as traffic channels are concerned, forward and reverse links employ different approaches to have the link power varied in accordance with voice activity. In the forward link, at any vocoder output rate the symbols are repeated as many times as necessary to yield a constant rate equal to that of the full rate at all rates. However, repeated symbols are produced at a reduced energy such that at any given period the forward link operates at a reduced power. Assume, for example, that at full rate each of the s symbols per second can be successfully detected if the energy per symbol is E during the symbol period T. If, due to voice activity, the symbol rate is reduced to s/n symbols per second, n real, instead of producing symbols with energy E over the symbol period $n \times T$ (power equal to $n \times E \times T$), the symbol period is reduced to T. Each symbol with energy E/n is repeated $n - 1$ times so that each symbol is transmitted with a power of $(E/n) \times T$ and the n resultant symbols have a power equal to $E \times T$, and not $n \times E \times T$ as in the full-rate case.

In the reverse link, symbol repetition also occurs so that independently of the voice activity, and thence of the bit rate, the channel operates at a constant rate, equal to that dictated by the full rate. On the other hand, variation on the link power as a function of the voice activity is accomplished by conveniently gating off (masking out) redundant symbols, which, in effect, corresponds to reducing the symbol duty cycle. The repeated symbols are eliminated (masked out) in a pseudorandom manner by means of a data burst randomizer, which yields the symbol masking pattern. The symbol masking pattern is obtained as a function of the vocoder rate and of the control bits drawn from the long PN code sequence at a certain time. This pseudorandom masking guarantees that the elimination of the repeated symbols occurs uniformly within the frame.

Observe that, to detect a symbol of the forward link, the receiver must perform an integration during a period of $n \times T$ so that enough power is accumulated for the appropriate detection. Such a scheme is satisfactory in a link for which the speed requirement for power control is not critical. In other words, a delay in adjusting the link power upon monitoring the link quality does not compromise the overall performance. As a matter of fact, the forward-link quality measure is the frame error rate (FER). On the other hand, because power control is a very sensitive matter for the reverse link, the monitoring of the link quality must be carried out expeditiously, even when the vocoder is operating at lower rates. Therefore, in the reverse link, the scheme used for the forward-link power variation as a function of the voice activity is inappropriate. As a matter of fact, the reverse-link quality parameter is the ratio of energy per bit to noise density, E_b/N_o, a quantity that is very quickly obtainable.

There may be as many as 55 (for seven paging channels) or as many as 62 (for one paging channel) traffic channels per carrier. Besides primary data and secondary data, the full-rate (only the full-rate) traffic channels also convey signaling messages. Primary data correspond to voice information and secondary data correspond to data information. The multiplexing of voice, data, and signaling is carried out in two possible modes: blank and burst or dim and burst. In the blank-and-burst mode, the entire frame is used to carry secondary data or signaling messages. (Primary data are completely blanked out within the frame.) In the dim-and-burst mode, primary data and secondary data or signaling messages are transmitted in the same frame. (Primary data are partially overwritten by either secondary data or signaling messages within the frame. In other words, the frame is shared by voice and data or voice and signaling.) The first bit in the frame indicates the type of data being transmitted. If this bit is set to zero, then the frame conveys primary data only. If it is set to one, then the next bit will indicate if the frame contains primary data and secondary data or primary data and signaling messages. Two other bits instruct on the proportion of primary data to other types of data (secondary or signaling as applicable). Of course, the proportion zero (i.e., zero bits of voice) is also pertinent.

Traffic Channel—Forward Link

The vocoder yields variable data rate information, depending on the voice activity. For the Rate Set 1 these rates are 9.6, 4.8, 2.4, and 1.2 kbit/s. For the Rate Set 2 they are 14.4, 7.2, 3.6, and 1.8 kbit/s. These data are first error-control-protected by means of a half-rate (for Rate Set 1) or three quarter-rate (for Rate Set 2) convolutional encoder (constraint length 9). In both cases the output of the convolutional encoder yields the respective rates of 19.2, 9.6, 4.8, and 2.4 ksymb/s. Each symbol is then repeated conveniently so that the resultant rate is constant and equal to 19.2 ksymb/s. A protection against burst errors is provided by means of a $24 \times 16 = 384$-symbol block interleaver. The output of the interleaver runs at 19.2 ksymb/s rate and is combined with the decimated version of the masked long PN sequence, running at the same rate. This combination in fact corresponds to a data scrambling, which yields privacy to the encoded voice. This 19.2 ksymb/s scrambled information is then "multiplexed" with 800 bit/s power control bits (PCB). This is not really a multiplexing but a "puncturing." Multiplexing usually implies no loss of information, which is achieved by means of an increase of the output data rate. In the present case, the output data rate remains the same and what really happens is a data puncturing in which the scrambled information is overwritten by the PCBs, as detailed next.

The long PN code is generated by a 42-stage linear feedback shift register with the respective characteristic polynomial given by Equation 5.3. Its phase offset is obtained by a mask for user u given by the ESN of the subscriber using that particular channel. The long PN code runs at a rate of 1.2288 Mchip/s but is decimated to a 19.2 ksymb/s rate by sampling every 64th PN code chip (64:1 decimation). This decimated PN code is combined with the output of the interleaver, which also runs at 19.2 ksymb/s. A further decimation of the 19.2 ksymb/s decimated PN code is carried out by means of sampling every 24th PN code chip (24:1 decimation). The four most significant bits of the resulting 800 bit/s decimator output are used to control the multiplexer whose inputs are the scrambled information and the power control bits.

The resulting 19.2 kbit/s rate information (data and power control bits) is combined with one of W_8 to W_{31} or W_{33} to W_{63} Walsh sequences, which identifies the specific traffic channel (user) and runs at a rate of 1.2288 kbit/s. The Walsh sequences that are not used for the paging channels may be utilized by the traffic channels. The combined 1.2288 kbit/s rate stream is then fed into the quadrature spreading modulator as shown in Figure 5.12.

Traffic Channel—Reverse Link

The vocoder yields variable data rate information, depending on the voice activity. For Rate Set 1 these rates are 9.6, 4.8, 2.4, and 1.2 kbit/s. For Rate Set 2 they are 14.4, 7.2, 3.6, and 1.8 kbit/s. These data are first error-control-protected by means of a one third-rate (for Rate Set 1) or half-rate (for Rate Set 2) convolutional encoder (constraint 9). In both cases the output of the convolutional encoder yields the respective rates of 28.8, 14.4, 7.2, and 3.6 ksymb/s. Each symbol is then repeated conveniently so that the resultant

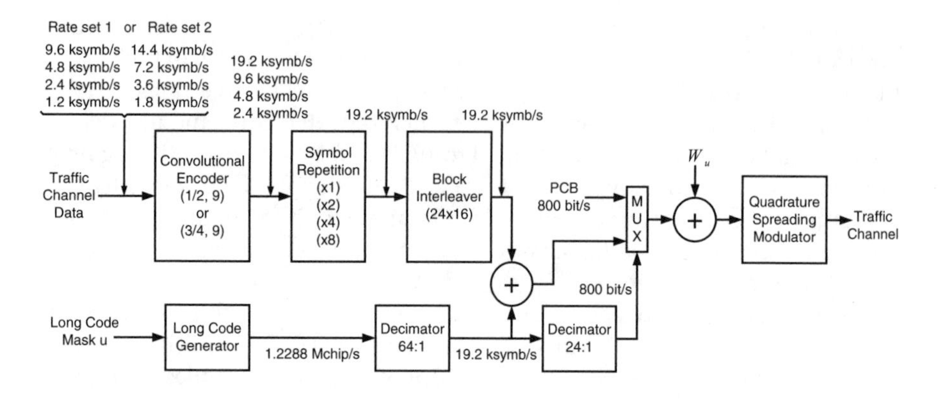

FIGURE 5.12
Forward traffic channel structure.

rate is constant and equal to 28.8 ksymb/s. Protection against burst errors is provided by means of a $32 \times 18 = 576$-symbol block interleaver. The symbols are written in by column and read out by row. This means that for the full-rate case each row contains a different set of symbols. For the half-rate case two consecutive rows contain the same set of symbols. For the one fourth-rate case four consecutive rows contain the same set of symbols. For the one eighth-rate case, eight consecutive rows contain the same set of symbols. The output of the interleaver runs at 28.8 ksymb/s rate and is fed into the (64, 6) Walsh encoder. In other words, each group of 6 bits of the interleaved information is encoded into 64-bit data and this results in a data rate of $28.8 \times 64/6 = 307.2$ kchip/s. Note, therefore, that each row of 18 symbols of the interleaver generates three Walsh-encoded orthogonal modulation symbols. And six modulation symbols compose what is called a power control group (PCG), a PCG containing symbols of two rows of repeated symbols or two rows of nonrepeated symbols. Each PCG lasts $6 \times 64/307.2 = 1.25$ ms and 16 PCG compose a frame of 20 ms. The rows of the interleaver are read out in such a way that in a frame one PCG of nonrepeated symbols alternates with $6(k - 1)$ PCGs of repeated symbols, where k is the order of repetition ($k = 1$ for full-rate–no repetition; $k = 2$ for half-rate–1 repetition; $k = 4$ for one fourth-rate–3 repetitions; $k = 8$ for one eighth-rate–7 repetitions). The redundant $16(1 - 1/k)$ PCGs are then gated off in a pseudorandom manner by the data burst randomizer to reduce the total amount of power of the reverse link in accordance with the voice activity. The symbol masking pattern is determined as a function of the vocoder rate and of the control bits obtained out of the long PN code sequence bits at a certain time. The output of the data burst randomizer is then combined with the long PN code, which is generated by a 42-stage linear feedback shift register with the respective characteristic polynomial given by Equation 5.3.

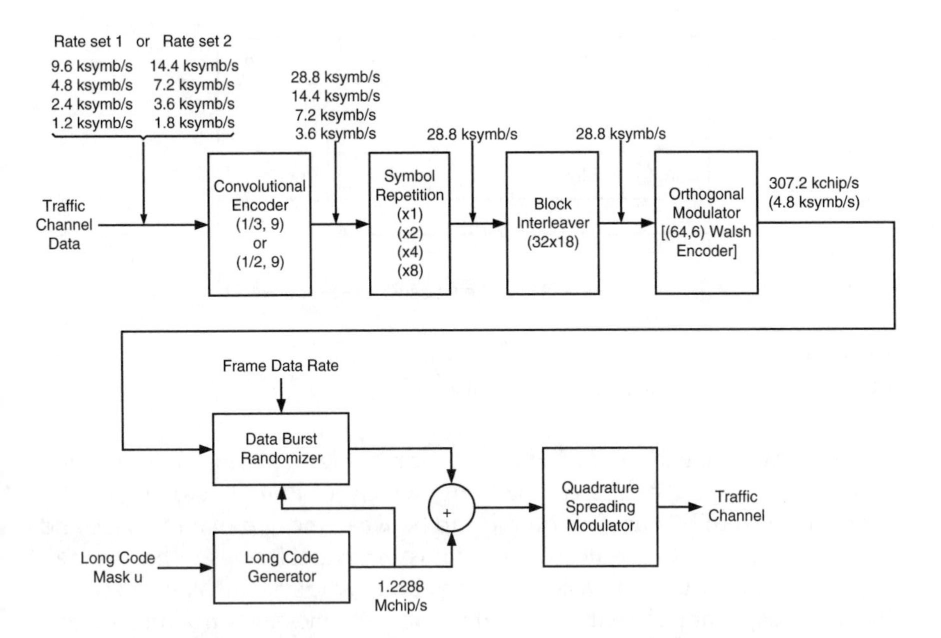

FIGURE 5.13
Reverse traffic channel structure.

Its phase offset *u* is obtained by a mask given by the ESN of the subscriber using that particular channel and it runs at a rate of 1.2288 Mchip/s. The combined 1.2288 kbit/s rate stream is then fed into the quadrature spreading modulator as shown in Figure 5.13.

5.6 Signaling Format

The TIA/EIA/IS-95 standard defines *message* as "a data structure that conveys control information or application information." It defines *order* as "a type of message that contains control codes for either the mobile station or the base station." In general, orders are simpler than other messages, many of them constituting confirmations of messages or simple directions, with only a minority of them carrying numerical parameters. All messages contain three fields as follows:

1. Length field, which conveys information on the size of the message
2. Message body, which conveys the message information
3. CRC field, which conveys the CRC part associated with the message

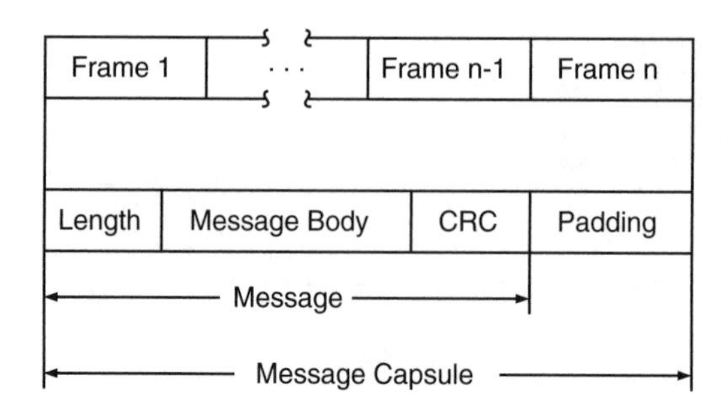

FIGURE 5.14
CDMA message structure and signaling format.

A message is always followed by a padding. Message plus padding compose the message capsule. The padding, which may be of zero length, is a sequence of zero bits used to fill the gap between the end of a message and the end of a message capsule, which comprises several frames, or half frames. The frame constitutes the basic timing interval in the system. Note, therefore, that a message may be sent in several frames. The message structure and the signaling format are shown in Figure 5.14.

5.7 Messages, Orders, and Parameters

This section lists and briefly describes some of the important identification codes and parameters that appear in messages or orders. They are shown in terms of their TIA/EIA/IS-95 identification, name, and description, in that sequence:

- ACC_CHAN—Number of Access Channels. This field conveys the number of access channels associated with the paging channel.
- ACC_MSG_SEQ—Access Parameters Message Sequence Number. This field conveys the access configuration sequence number, which is incremented whenever the access parameter message is modified.
- ACK_REQ—Acknowledgment Required. As part of acknowledgment procedures this field indicates whether or not an acknowledgment is required.
- ACK_SEQ—Acknowledgment Sequence Number. As part of acknowledgment procedures this field conveys the value of the message sequence number from the most recently received message requiring acknowledgment.

- ACTION_TIME—Action Time. This field conveys the system time at which the action specific to the message should take effect.
- ANALOG_CHAN—Analog Voice Channel Number. This field conveys the channel number of the analog voice channel.
- ASSIGN_MODE—Assignment Mode. This field informs the mode with which the channel assignment is to be performed. These modes comprise traffic channel assignment, page channel assignment, acquire analog system, analog voice channel assignment.
- AUTHBS—Challenge Response. As part of the authentication process AUTHBS is an 18-bit quantity obtained by the base station through the authentication signature procedure and sent to the mobile station.
- AUTHU—Authentication Challenge Response. As part of the authentication process AUTHU is an 18-bit quantity obtained by the mobile station through the authentication signature procedure and sent to the base station.
- BASE_CLASS—Base Station Class. This field conveys the class of service provided by the base station (public system or some other to be defined).
- BASE_ID—Base Station Identification. This field conveys the base station identification number.
- BASE_LAT—Base Station Latitude. This field conveys the base station latitude in units of 0.25 s.
- BASE_LONG—Base Station Longitude. This field conveys the base station longitude in units of 0.25 s.
- BCAST_INDEX—Broadcast Slot Cycle Index. This field conveys the broadcast slot cycle index (in the range 1 to 7) when periodic broadcast paging is enabled. The value zero in this field indicates that periodic paging is disabled.
- BURST_TYPE—Data Burst Type. This field indicates the type of data burst in the message. (This is defined in TSB58, Administration Parameter Value Assignments for TIA/EIA Wideband Spread Spectrum Standards.)
- CDMA_CHAN—CDMA Channel Number. Each occurrence of this field conveys the associated CDMA channel number.
- CDMA_FREQ—CDMA Channel Frequency Assignment. The order in which this field is included in the message indicates the CDMA channel containing a paging channel that is supported by the base station. Each occurrence of this field is set to the CDMA channel number corresponding to the CDMA frequency.

- CHARi—Dialed Digit or Character. This field conveys the dialed digit according to the DIGIT_MODE.
- CODE_CHAN—Code Channel. This field conveys the code channel index (one out of 64 Walsh sequences).
- CONFIG_MSG_SEQ—Configuration Message Sequence Number. This field conveys the configuration sequence number, which is incremented whenever the base station modifies the system parameter message, the neighbor list message, the CDMA channel list message, or, if sent, the extended system parameters message and the global service redirection message.
- DIGIT_MODE—Digit Mode Indicator. This field indicates the code used for the dialed digits (4-bit DTMF codes using unknown numbering plan or 8-bit ASCII codes using a specified numbering plan).
- DIGITi—DTMF Digit. This field conveys the DTMF digit to be generated by the base station.
- DTMF_OFF_LENGTH—DTMF Interdigit Interval Code. This field conveys the value of the requested minimum interval between DTMF pulses to be generated by the base station.
- DTMF_ON_LENGTH—DTMF Pulse Width Code. This field conveys the value of the requested width of DTMF pulses to be generated by the base station.
- ENCRYPT_MODE—Message Encryption Mode. This field indicates the encrypting mode (encryption disabled or encrypt call control messages) to be used for messages sent on the forward and reverse traffic channels.
- ENCRYPTION—Message Encryption Indicator. This field conveys the mode with which the current message is being encrypted. (This is maintained for retransmitted messages.) This is made equal to the content of the ENCRYPT_MODE field of the last received channel assignment message, handoff direction message, or message encryption mode order.
- ERRORS_DETECTED—Number of Frame Errors Detected. This field conveys the number of bad frames in PWR_MEAS_FRAMES.
- ESN—Electronic Serial Number. As part of MSID this field contains the electronic serial number assigned by the manufacturer to the mobile station.
- FEATURE—Feature Identifier. This field conveys the supplementary service or feature requested.
- FRAME_OFFSET—Frame Offset. This field conveys the offset by which the forward and reverse traffic channel frames must be delayed with respect to the system timing.

- HARD_INCLUDED—Hard Handoff Parameters Included. This field indicates whether or not the mobile station should change the parameters relative to the hard handoff (FRAME_OFFSET, PRIVATE_LCM, ENCRYPT_MODE, NOM_PWR, BAND_CLASS, CDMA_FREQ).
- IMSI—International Mobile Station Identity. As part of the MSID this field contains the mobile station identification as specified in CCITT Recommendation E.212 for mobile station in the land mobile service.
- INIT_PWR—Initial Power Offset. This field conveys the correction factor to be used by the mobile stations in the open-loop power estimate for the initial transmission on the access channel.
- LC_STATE—Long Code State. This field conveys the long code state at the time given by the SYS_TIME field of the message.
- LP_SEC—Leap Seconds. This field indicates the leap seconds that have occurred since the start of system time.
- MCC—Mobile Country Code. As part of the IMSI this field contains the MCC.
- MEM—Message Encryption Mode Indicator. This field enables or disables analog control message encryption on the analog voice channels.
- MIN—Mobile Identification Number. As part of MSID this field contains the directory number assigned by the operating company to the mobile station.
- MOB_FIRM_REV—Firmware Revision Number. This field indicates the firmware number of the mobile station assigned by the manufacturer.
- MOB_MFG_CODE—Manufacturer Code. This field indicates the manufacturer of the mobile station.
- MOB_MODEL—Model Number. This field conveys the model of the mobile station assigned by the manufacturer.
- MOB_P_REV—Protocol Revision of the Mobile Station. This field indicates the TIA/EIA/IS-95 version supported by the mobile station.
- MOB_TERM_FOR_NID—Foreign NID Roaming Registration Enable Indicator. This field indicates whether or not the mobile station is configured to receive terminated calls when it is a foreign NID roamer.
- MOB_TERM_FOR_SID—Foreign SID Roaming Registration Enable Indicator. This field indicates whether or not the mobile station is configured to receive terminated calls when it is a foreign SID roamer.

- MOB_TERM_HOME—Home (Nonroaming) Registration Enable Indicator. This field indicates whether the mobile station is configured to receive terminated calls when not roaming or when roaming.
- MORE_PAGE—More Slotted Pages to Follow Indicator. This field indicates whether or not more slotted page messages should follow after the current message.
- MSG_NUMBER—Message Number. This field conveys the number of the message within the data burst.
- MSG_SEQ—Message Sequence Number. As part of acknowledgment procedures this field conveys the value of the message sequence number for the message in which it appears.
- MSG_TYPE—Message Type. This field identifies the type of message or order within the channel. The types of messages or orders conveyed by the channel will be detailed later.
- MSID—Mobile Station Identifier. This field conveys the mobile station identification.
- NA_CHAN_TYPE—Analog Channel Type. This field indicates the type of analog channel to be assigned. The types of analog channels comprise narrow channel 10 kHz below ANALOG_CHAN; narrow channel 10 kHz above ANALOG_CHAN; narrow channel centered on ANALOG_CHAN; and wide channel on ANALOG_CHAN. This is to comply with the narrowband AMPS in which a 30-kHz channel encompasses 3 × 10 kHz channels. These are the narrow channels, whereas 30-kHz channels of the standard AMPS are identified as "wide" channels.
- NGHBR_CONFIG—Neighbor Configuration. This field conveys the configuration of the neighbor base station (if its configuration is the same as or different from this base station, if it has a paging channel on the current CDMA frequency assignment, etc.).
- NGHBR_PN—Neighbor Pilot PN Sequence Offset Index. This field conveys the pilot PN sequence offset of the neighboring base station.
- NID—Network Identification. This field uniquely identifies a network within a cellular system. (A network is a subset of a cellular system. It may comprise a set of base stations set up to manage some special requirements.)
- NOM_PWR—Nominal Transmit Power Offset. This field conveys the correction factor to be used by the mobile stations in the open-loop power estimate.
- NUM_CHANS—Number of CDMA Channels. This field conveys the number of occurrences of the CDMA_CHAN field.

- NUM_DIGITS—Number of DTMF Digits. This field conveys the number of DTMF digits included in the message.
- NUM_FIELDS—Number of Dialed Digits in the Message. This field conveys the number of dialed digits included in the message.
- NUM_MSGS—Number of Messages in the Data Burst Stream. This field conveys the number of messages in the data burst stream.
- NUM_PILOTS—Number of Pilots. This field conveys the number of pilots in the current active set.
- NUM_STEP—Number of Access Probes. This field conveys the number of access probes (to be defined later) in a single-access probe sequence (access probe sequence will be defined later).
- ORDER—Order Code. This field conveys the type of order in the message.
- PAGE_CHAN—Number of Paging Channels. This field indicates the number of paging channels on the CDMA channel.
- PARAMETER—Parameter Value. This field conveys the parameters that can be retrieved and set in the mobile station by means of special messages. Some of these parameters are defined by the TIA/EIA/IS-95 standard, whereas others are defined by the mobile station manufacturers.
- PILOT_PN—Pilot PN Sequence Offset. This field indicates the PN sequence offset, in number of 64 PN chips.
- PILOT_PN_PHASE—Pilot Measured Phase. This field conveys the phase of the pilot PN sequence relative to the zero offset pilot PN sequence of this pilot, in units of one PN chip.
- PILOT_STRENGTH—Pilot Strength. This field conveys the strength of the pilot used by the mobile station to derive its time reference.
- PM—Privacy Mode Indicator. This field indicates voice privacy requested.
- PREF_MSID_TYPE—Preferred Access Channel Mobile Station Identifier Type. This field conveys the type of MSID (IMSI, ESN, IMSI, and ESN) that the mobile station should use on the access channel.
- PRIVATE_LCM—Private Long Code Mask Indicator. This field conveys whether or not the private long code mask is to be used after a hard handoff.
- PWR_MEAS_FRAMES—Number of Forward Traffic Channel Frames in the Measurement Period. This field conveys the number of forward traffic channel frames monitored during a given period.

- PWR_PERIOD_ENABLE—Period Report Mode Indicator. This field indicates whether or not the mobile station is to generate periodic power measurement report messages.
- PWR_REP_DELAY—Power Report Delay. This field conveys the period the mobile station should wait before restarting frame counting for power control purposes.
- PWR_REP_FRAMES—Power Control Reporting Frame Count. This field conveys the number of frames over which mobile stations are to count frame errors.
- PWR_REP_THRESH—Power Control Reporting Threshold. This field conveys the number of bad frames permitted to be received by the mobile station in a measurement period before reporting bad frame reception to the base station.
- PWR_STEP—Power Increment. This field conveys the step value by which the mobile stations must increment their transmit power between successive access probes (access probes will be defined later).
- PWR_THRESH_ENABLE—Threshold Report Mode Indicator. This field indicates whether or not the mobile station is to generate measurement report messages.
- RAND—Random Challenge Memory. As part of the authentication process RAND is a 32-bit random value challenge sent by the base station (access parameters message) and held in the mobile station to be used by the mobile station for authentication purposes.
- RANDC—Random Challenge Value. As part of the authentication process RANDC corresponds to the eight most significant bits of RAND.
- RANDSSD—Random Data for the Computation of SSD. As part of the authentication process RANDSSD is a quantity stored in the HLR/AC associated with the mobile station that is used to compute SSD.
- RANDU—Random Challenge Data. As part of the authentication process RANDU is a 24-bit quantity generated by the base station.
- RECORD_TYPE—Information Record Type. This field conveys the information record type such as display (information to be displayed by the mobile station), called party number (identifies the called party's number), calling party number (identifies the calling party's number), connected number (identifies the responding party to a call), signal (conveys information to a user by means of tones and other alerting signals), message waiting (conveys the number of messages waiting).
- RECORD_TYPE—Redirection Record Type. This field conveys the redirection record type such as redirection to an analog system or to a CDMA system.

- REDIRECT_ACCOLC—Redirected Access Overload Classes. This field conveys the access overload classes of mobile stations that are to be redirected to a system to obtain service. The mobile station overload classes comprise the following: normal subscribers, test mobile stations, emergency mobile stations, and overload classes to be specified.
- REF_PN—Time Reference PN Sequence Offset. This field indicates the PN sequence offset of the pilot used by the mobile station to derive its time reference.
- REG_DIST—Registration Distance. This field conveys the distance beyond which the mobile station should perform a distance-based registration if the mobile station is to perform such a type of registration. The null value in this field indicates that this type of registration should not be performed.
- REG_TYPE—Registration Type. This field indicates the type of event generating the registration attempt.
- REG_ZONE—Registration Zone. This field conveys the zone (group of base stations within a given system or network) within which a base station is registered.
- RELEASE—Origination Completion Indicator. This field indicates whether or not the message is used to complete an origination request.
- REQUEST_MODE—Requested Mode Code. This field conveys the current configuration mode (CDMA only, wide analog only, narrow analog only, and combinations).
- RETURN_IF_FAIL—Return If Fail Indicator. This field indicates whether or not the mobile station must return to the original system from which it is being redirected upon failure to obtain service in the system to which it has been redirected.
- SCC—SAT Color Code. This field conveys the supervisory audio tone associated with the designated analog voice channel.
- SCM—Station Class Mark. This field indicates the class with which the mobile station operates (CDMA mode, dual mode, slotted class, nonslotted class, power class—class I, class II, class III, etc.).
- SERVICE_OPTION—Requested Service Option. This field indicates the type of special service requested. (This is defined in TSB58, Administration Parameter Value Assignments for TIA/EIA Wideband Spread Spectrum Standards.)
- SID—System Identification. This field uniquely identifies the cellular system.

- SLOT_CYCLE_INDEX—Slot Cycle Index. This field conveys the slot cycle to be used by the mobile station to monitor the paging channel. This is only applicable to mobile stations operating in the slotted mode. (A mobile station operating in the slotted mode monitors the paging channel for one or two slots per slot cycle.)
- SPECIAL_SERVICE—Special Service Option Indicator. This field indicates whether special service or default service is requested.
- SRCH_WIN_A—Search Window Size for Active Set and Candidate Set. This field conveys the search window size (range of PN offsets) to be used by the mobile station to search for multipath components of the pilots in the active set and candidate set.
- SRCH_WIN_N—Search Window Size for Neighbor Set. This field conveys the search window size (range of PN offsets) to be used by the mobile station to search for components of pilots in the neighbor set.
- SRCH_WIN_R—Search Window Size for Remaining Set. This field conveys the search window size (range of PN offsets) to be used by the mobile station to search for components of pilots in the remaining set.
- SSD—Shared Secret Data. As part of the authentication procedure, encryption, and voice privacy, SSD is 128-bit information stored in the semipermanent memory of the mobile station. It is composed of two equal-sized parts: SSD_A, used to support the authentication procedures, and SSD_B, used to support voice privacy and message encryption. SSD is not accessible to the user. SSD is generated according to a given cryptographic algorithm and is initialized with mobile station specific information, random data, and the A-key of the mobile station (a 64-bit secret key known only to the mobile station and to its associated HLR/AC).
- SYS_ORDERING—System Ordering. This field conveys the order in which the mobile station should attempt to obtain service from the analog system (System A only, System B only, System A first, then System B if the first option was unsuccessful, etc.)
- SYS_TIME—System Time. This field indicates the system time that is the time reference used by the system. Except for leap seconds, the system time is synchronous with the UTC (universal coordinated time) using the same time origin as GPS time. SYS_TIME is set by the base station as four sync channel superframes after the end of the last superframe containing any part of this sync channel message, minus the pilot PN sequence offset, in units of 80 ms.
- T_ADD—Pilot Detection Threshold. This field conveys the pilot strength level above which a pilot is to be transferred to the candidate

set. (The candidate set contains the pilots that are not currently in the active set but have sufficient strength to indicate that the associated forward traffic channels could be successfully demodulated.) This triggers the mobile station to send a pilot strength measurement message.

- T_COMP—Comparison Threshold. This field conveys the margin by which the strength of a pilot in the candidate set should exceed that of the active set so that the mobile station can send a pilot strength measurement message.
- T_DROP—Pilot Drop Threshold. This field conveys the pilot strength level below which the mobile station should start the handoff drop timer for pilots in the active set and in the candidate set. (The active set contains the pilots associated with the forward traffic channels assigned to the mobile station.)
- T_TDROP—Drop Timer Value. This field conveys the timer value after which one of the following actions can be taken by the mobile station for pilots belonging to the active set or to the candidate set and whose signal strength has not become greater than T_DROP: if the pilot is a member of the active set a pilot strength measurement message is issued; if the pilot is a member of the candidate set, this pilot is moved to the neighbor set. (The neighbor set contains the pilots that are not currently in the active set but are likely candidates for handoff.)
- TOTAL-ZONES—Number of Registration Zones to Be Retained. This field conveys the number of registration zones the mobile station is to retain for zone-based registration purposes.
- USE_TIME—Use Action Time Indicator. This field indicates whether an ACTION_TIME is specified in the message.
- VMAC—Voice Mobile Station Attenuation Code. This field conveys the power level associated with the analog voice channel.

5.8 Messages and Orders and Logical Channels

In this section, the CDMA channels, namely, pilot channel, sync channel, paging channel, forward traffic channel, access channel, and reverse traffic channel, are described in terms of the messages and orders and parameters they convey and in terms of the format in which these messages and orders are encapsulated. The messages are described in terms of those fields that are relevant to the task they perform. The orders, most of them constituting confirmation of messages or simple directions and having their names directly associated with the task they perform, are either very briefly explained or simply quoted. The purposes of the specific messages and orders are better understood in conjunction with the call processing descriptions. Mobile station call processing and base station processing are the subjects of the following sections.

5.8.1 Pilot Channel

The pilot channel conveys no baseband information. From the base station, it continuously transmits a stream of zeroes for coherent carrier phase and timing reference purposes. Thus, messages and therefore message structure as previously described do not apply.

5.8.2 Sync Channel

The sync channel structure is organized in frames and a superframe that contains three-equal duration frames. A superframe accommodates 96 bits. At a rate of 1.2 kbit/s these 96 bits are transmitted in 80 ms, which is the duration of the superframe. A frame contains 32 bits and, therefore, lasts 26.666... ms, exactly the same duration as the period of the short PN code. Each frame, and therefore the superframe, begins at the same time instant as the short PN sequence. Hence, when synchronization is achieved through the pilot channel, the alignment for the sync channel becomes known. Note that, after being half-rate convolutional encoded and repeated once, an initial 32-bit piece of information (a frame) turns into a $32 \times 4 = 128$-bit stream, which is the size of the block interleaver.

The sync channel conveys only one message: the sync channel message. The sync channel message may occupy more than one superframe but always an integer number of superframes. The start of the message coincides with the beginning of a superframe and is indicated by a bit one in its first position. Subsequent frames have a bit zero in its first position indicating the continuation of the message started at the beginning of the corresponding superframe. If necessary, padding of bits is used to complete the remaining positions of that superframe.

The sync channel message may include the following main fields: system identification (SID), network identification (NID), pilot PN sequence offset index (PILOT_PN), long code state (LC_STATE), system time (SYS_TIME), paging channel data rate (PRAT). After acquiring the sync channel message, the precise time to start running the long PN sequence is known. After starting running the long PN sequence, such a sequence can be accessed whenever necessary. The field PRAT indicates whether the paging channel is running at 4.8 or 9.6 kbit/s.

5.8.3 Paging Channel

The paging channel structure is organized into half-frame, frame, slot, and slot cycle. A half-frame lasts 10 ms and contains 96 bits. Two half-frames constitute one frame (20 ms, 192 bits). Four frames form one slot (80 ms, 768 bits). And 2048 slots (163,840 ms, 1,572,864 bits) comprise the maximum slot cycle. The choice of a particular paging channel and of a particular slot within that channel is carried out by the base station in a random fashion but is based on a hash function. Because the selection makes use of the mobile station identification number and of known system parameters, by computing the hash function the mobile station will be able to identify and then monitor that particular slot within the particular paging channel. The hash function aims at balancing the load on all the active paging channels and on their slots.

The messages on the paging channel are of the synchronized or unsynchronized type. They may occupy more than one half-frame and may not necessarily use integer number of half-frames. When a message ends 7 bits or less before the end of the current half-frame a padding of bits is added to complete the half-frame. In the same way, if a message ends before the end of the current half-frame and the message to be sent is of the synchronized type, then a padding of bits is also included to complete the half-frame. On the other hand, if a message ends 8 bits or more before the end of the current half-frame and the next message to be sent is of the unsynchronized type, then such a message is immediately transmitted. A bit one at the beginning of the half-frame indicates the start of a new message. A bit zero is set otherwise.

As far as information is concerned, there are basically two types of messages: overhead and paging. A paging message alerts the mobile or a group of mobiles of an incoming call. The paging messages include the general page message, the page message, and the slotted page message. An overhead message conveys information on system configuration, such as handoff parameters, forward and reverse power control parameters, access parameters, list of neighboring sectors, and their pilot PN sequences, etc. Examples of overhead messages include system parameter message, access message, and neighbor list message. All messages on the paging channel convey the mobile identification number (MIN), the ESN, and the IMSI.

Next listed and briefly described are the messages and orders on the paging channel.

Messages

- *System Parameters Message*. Among others, this message may include the following fields: PILOT_PN, SID, NID, REG_ZONE, TOTAL_ZONES, BASE_ID, BASE_CLASS, PAGE_CHAN, BASE_LAT, BASE_LONG, SRCH_WIN_A, SRCH_WIN_N, SRCH_WIN_R, T_ADD, T_DROP, T_COMP, T_TDROP.

- *Access Parameters Message*. This message defines the parameters used by the mobile station when transmitting on the access channel. Among others, this message conveys the following fields: PILOT_PN, ACC_CHAN, NOM_PWR, INIT_PWR, PWR_STEP, NUM_STEP, RAND.
- *Neighbor List Message*. Among others, this message may include the following fields: PILOT_PN, NGHBR_CONFIG, NGHBR_PN.
- *CDMA Channel List Message*. Among others, this message may include the following fields: PILOT_PN, CDMA_FREQ.
- *Slotted Page Message*. Among others, this message may include the following fields: ACC_MSG_SEQ, CONFIG_MSG_SEQ, MORE_PAGES, SPECIAL_SERVICE, SERVICE_OPTION.
- *Page Message*. Among others, this message may include the following fields: ACC_MSG_SEQ, CONFIG_MSG_SEQ, MSG_SEQ, SPECIAL_SERVICE, SERVICE_OPTION.
- *Order Message*. Among others, this message may include the following fields: ACK_SEQ, MSG_SEQ, ACK_REQ, ORDER, order-specific fields (field specific to the order message).
- *Channel Assignment Message*. Among others, this message may include the following fields: ACK_SEQ, MSG_SEQ, ACK_REQ, ASSIGN_MODE, CODE_CHAN, CDMA_FREQ, FRAME_OFFSET, PILOT_PN, ANALOG_CHAN.
- *Date Burst Message*. Among others, this message may include the following fields: ACK_SEQ, MSG_SEQ, ACK_REQ, MSG_NUMBER, BURST_TYPE.
- *Authentication Challenge Message*. Among others, this message may include the following fields: ACK_SEQ, MSG_SEQ, ACK_REQ, RANDU.
- *SSD Update Message*. Among others, this message may include the following fields: ACK_SEQ, MSG_SEQ, ACK_REQ, RANDSSD.
- *Feature Notification Message*. Among others, this message may include the following parameters: ACK_SEQ, MSG_SEQ, ACK_REQ, RELEASE, RECORD_TYPE (information record type).
- *Extended System Parameters Message*. Among others, this message may include the following fields: PILOT_PN, MCC, PREF_MSID_TYPE, BCAST_INDEX.
- *Service Redirection Message*. Among others, this message may include the following fields: ACK_SEQ, MSG_SEQ, ACK_REQ, RECORD_TYPE (redirection record type), RETURN_IF_FAIL, SYS_ORDERING, NUM_CHANS, CDMA_CHANS.
- *General Page Message*. Among others, this message may include the

following fields: CONFIG_MSG_SEQ, MSG_SEQ, ACC_MSG_SEQ, SPECIAL_SERVICE, SERVICE_OPTION, MCC, BURST_TYPE.

- *Global Service Redirection Message*. Among others, this message may include the following fields: PILOT_PN, CONFIG_MSG_SEQ, REDIRECT_ACCOLC, RECORD_TYPE (redirection record type), RETURN_IF_FAIL, SYS_ORDERING, NUM_CHANS, CDMA_CHANS.

Orders

- *Abbreviated Alert Order*. This order causes the mobile station to emit an audible tone.
- *Base Station Challenge Confirmation Order*. As part of the authentication procedure this order includes the AUTHBS field.
- *Reorder Order*. As part of the call management procedures, this order causes the mobile station to move to an idle state (to be defined later). It may be sent by the base station after receiving the origination message sent by the mobile station.
- *Audit Order*. This order is used for operations, administration, and maintenance purposes.
- *Intercept Order*. As part of the call management procedures, this order causes the mobile station to move to an idle state (to be defined later). It may be sent by the base station after receiving the origination message sent by the mobile station.
- *Base Station Acknowledgment Order*. This order is used as a response to a message requiring acknowledgment.
- *Lock until Power-Cycled Order*. This order instructs the mobile station to switch off its transmitter until an unlock order is received. This order also indicates the lock reason.
- *Maintenance Required Order*. This order instructs the mobile station to display that maintenance is required. This order also indicates the maintenance reason.
- *Unlock Order*. This order instructs the mobile station to return to normal operation.
- *Release Order*. This order is used as a simple direction (no reason given) or to indicate that the requested service option is rejected.
- *Registration Accepted Order*. This order is used to reject a registration as a response to a registration message sent by the mobile station.
- *Registration Request Order*. This order is used by the base station to command the mobile station to register in the ordered registration type.

- *Registration Rejected Order*. This order is used to accept a registration as a response to a registration message sent by the mobile station.
- *Local Control Order*. This order is used to request a specific order as designated within the message and as determined by each system.

5.8.4 Access Channel

The access channel structure is organized into frame and slot. A frame lasts 20 ms and contains 96 bits, leading to a rate of 4.8 kbit/s. Each slot contains a maximum number of frames, with this information conveyed on the paging channel through the access parameters message. Although a maximum number of frames is allowed the mobile may choose to use fewer frames according to its own needs.

As far as information is concerned, there are basically two types of messages: response and request. A response message occurs as a result of a paging to the mobile. A request message occurs whenever the mobile is initiating a communication with a base station. Some of the fields common to all the messages on the access channel include among others: ACK_SEQ, MSG_SEQ, ACK_REQ, MSID, ESN, IMSI, MCC, RANDC.

Next listed and briefly described are the messages and orders on the access channel.

Messages

- *Registration Message*. Among others, this message may include the following fields: REG_TYPE, SLOT_CYCLE_INDEX, MOB_P_REV, SCM, MOB_TERM.
- *Order Message*. Among others, this message may include the following fields: ORDER, order-specific fields (field specific to the order message).
- *Data Burst Message*. Among others, this message includes the following fields: RANDC, MSG_SEQ, BURST_TYPE, NUM_MSG, CHARi.
- *Origination Message*. Among others, this message includes the following fields: MOB_TERM, SLOT_CYCLE_INDEX, MOB_P_REV, SCM, REQUEST_MODE, SPECIAL_SERVICE, SERVICE_OPTION, PM.
- *Page Response Message*. Among others, this message includes the following fields: RANDC, MOB_TERM, SLOT_CYCLE_INDEX, MOB_P_VER, SCM, REQUEST_MODE, SERVICE_OPTION, PM.
- *Authentication Challenge Response Message*. Among others, this message includes the following fields: MSID, AUTHU.

Orders

- *Base Station Challenge Order.* As part of the authentication procedure this order includes the AUTHBS field.
- *SSD Update Confirmation Order.* As part of the authentication procedure this order may be issued as a response to the base station challenge confirmation order sent by the base station.
- *SSD Update Rejection Order.* As part of the authentication procedure this order may be issued as a response to the base station challenge confirmation order sent by the base station.
- *Mobile Station Acknowledgment Order.* This order is used as a response to a message requiring acknowledgment.
- *Local Control Response Order.* This order is used as a response to the local control order sent by the base station.
- *Mobile Station Reject Order.* This order is used to reject a message or an order. The following constitute the reasons for rejection: unspecified reason, message not accepted in the present state, message field not in valid range, message type or order code not understood, requested capability not supported by the mobile station, message not handled by the current mobile station configuration.

5.8.5 Traffic Channel: Forward and Reverse Links

Forward and reverse links operate with 20-ms frames and, because the transmission bit rate is variable, the number of bits per frame vary according to the rate. For example, for the 9.6 kbit/s rate the number of bits per frame is 192 and for the 1.2 kbit/s rate the number is 24. All frames contain an 8-bit encoder tail whose objective is to ensure that the convolutional encoder returns to the all-zero state at the end of the frame. Additionally, the full-rate and the half-rate frames are encoded with a CRC block code.

5.8.6 Forward Traffic Channel

The field common to the messages running on the forward traffic channel include among others: ACK_SEQ, MSG_SEQ, ACK_REQ, ENCRYPTION. Next listed and briefly described are the messages and orders on the forward traffic channel.

Messages

- *Order Message.* Among others, this message includes the following fields: USE_TIME, ACTION_TIME, ORDER.

- *Authentication Challenge Message*. Among others, this message includes the following field: RANDU.
- *Alert with Information Message*. Among others, this message includes the following field: RECORD_TYPE (information record type).
- *Data Burst Message*. Among others, this message includes the following fields: MSG_NUMBER, BURST_TYPE, NUM_MSGS, NUM_FIELDS, CHARi.
- *Handoff Direction Message*. Among others, this message includes the following fields: USE_TIME, ACTION_TIME, SRCH_WIN_A, T_ADD, T_DROP, T_COMP, T_TDROP, FRAME_OFFSET, PRIVATE_LCM, ENCRYPT_MODE, CDMA_FREQ, PILOT_PN, CODE_CHAN.
- *Analog Handoff Direction Message*. Among others, this message includes the following fields: USE_TIME, ACTION_TIME, SID, VMAC, ANALOG_CHAN, SCC, MEM.
- *In-Traffic System Parameters Message*. Among others, this message includes the following fields: SID, NID, SRCH_WIN_A, SRCH_WIN_N, SRCH_WIN_R, T_ADD, T_DROP, T_COMP, T_TDROP.
- *Neighbor List Update Message*. Among others, this message includes the following field: NGHBR_PN.
- *Send Burst DTMF Message*. Among others, this message includes the following fields: NUM_DIGITS, DTMF_ON_LENGTH, DTMF_OFF_LENGTH, DIGITi.
- *Power Control Parameters Message*. Among others, this message includes the following fields: PWR_REP_THRESH, PWR_REP_FRAMES, PWR_THRESH_ENABLE, PWR_PERIOD_ENABLE, PWR_REP_DELAY.
- *Retrieve Parameters Message*. Among others, this message includes the following field: PARAMETER.
- *Set Parameters Message*. Among others, this message includes the following field: PARAMETER.
- *SSD Update Message*. Among others, this message includes the following field: RANDSSD.
- *Flash with Information Message*. Among others, this message includes the following fields: RECORD_TYPE (information record type), type-specific fields (fields specific to the message).
- *Mobile Station Registered Message*. Among others, this message includes the following fields: SID, NID, REG_ZONE, TOTAL_ZONES, ZONE_TIMER, BASE_LAT, BASE_LOGN, REG_DIST.
- *Extended Handoff Direction Message*. Among others, this message includes the following fields: USE_TIME, ACTION_TIME, SRCH_WIN_A,

T_ADD, T_DROP, T_COMP, T_TDROP, HARD_INCLUDED, FRAME_OFFSET, PRIVATE_LCM, ENCRYPT_MODE, NOM_PWR, CDMA_FREQ, PILOT_PN, CODE_CHAN.

Orders

- *Base Station Challenge Confirmation Order*. As part of the authentication procedure this order includes the AUTHBS field.
- *Message Encryption Mode Order*. This order informs in its ENCRYPT_MODE field the encryption mode (encryption disabled or encrypt call control messages) to be used for the messages.
- *Parameter Update Order*. This order directs the mobile station to update (increment) its call history parameter (COUNTs-p), which is a modulo-64 count held in the mobile station. This order is usually issued after a successful channel assignment procedure.
- *Audit Order*. This order is used for operations, administration, and maintenance purposes.
- *Maintenance Order*. This order is used as a result of a malfunction detection by the base station.
- *Base Station Acknowledgment Order*. This order is used as a response to a message requiring acknowledgment.
- *Pilot Measurement Request Order*. This order is used by the base station to direct the mobile station to send a pilot strength measurement message.
- *Lock until Power-Cycled Order*. This order instructs the mobile station to switch off its transmitter until an unlock order is received. This order also indicates the lock reason.
- *Maintenance Required Order*. This order instructs the mobile station to display that maintenance is required. This order also indicates the maintenance reason.
- *Service Option Request Order*. This order is used by the base station to request a given service option. The service option code is informed in the SERVICE_OPTION field.
- *Service Option Response Order*. This order is used by the base station either to accept or to reject a service option request, whose code is informed in the SERVICE_OPTION field.
- *Release Order*. This order is used as a simple direction (no reason given) or to indicate that the requested service option is rejected.
- *Long Code Transition Request Order*. This order is used to request a transition to a private long code or to a public long code as a result of a request for voice privacy.

- *Continuous DTMF Tone Order*. This order conveys the DTMF tone.
- *Status Request Order*. This order requests status information records and causes the mobile station to respond with a status message.
- *Service Option Control Order*. This order is used to invoke specific control as determined by each service option.
- *Local Control Order*. This order is used to request a specific order as designated within the message and as determined by each system.

5.8.7 Reverse Traffic Channel

The fields common to all the messages running on the reverse traffic channel include ACK_SEQ, MSG_SEQ, ACK_REQ, ENCRYPTION. Next listed and briefly described are the messages and orders on the reverse traffic channel.

Messages

- *Order Message*. Among others, this message includes the following field: ORDER.
- *Authentication Challenge Response Message*. Among others, this message includes the following field: AUTHU.
- *Flash with Information Message*. Among others, this message includes the following field: RECORD_TYPE (information record type).
- *Data Burst Message*. Among others, this message includes the following fields: MSG_NUMBER, BURST_TYPE, CHARi.
- *Pilot Strength Measurement Message*. Among others, this message includes the following fields: REF_PN, PILOT_STRENGTH, PILOT_PN_PHASE.
- *Power Measurement Report Message*. Among others, this message includes the following fields: ERRORS_DETECTED, PWR_MEAS_FRAMES, NUM_PILOTS, PILOT_STRENGTH.
- *Send Burst DTMF Message*. Among others, this message includes the following fields: NUM_DIGITS, DTMF_ON_LENGTH, DTMF_OFF_LENGTH, DIGITi.
- *Status Message*. Among others, this message includes the following field: RECORD_TYPE (information record type).
- *Origination Continuation Message*. Among others, this message includes the following fields: DIGIT_MODE, NUM_FIELDS, CHARi.
- *Handoff Completion Message*. Among others, this message includes the following field: PILOT_PN.
- *Parameters Response Message*. Among others, this message includes the following field: PARAMETER.

Orders

- *Base Station Challenge Order*. As part of the authentication procedure, this order includes the AUTHBS field.
- *SSD Update Confirmation Order*. As part of the authentication procedure, this order may be issued as a response to the base station challenge confirmation order sent by the base station.
- *SSD Update Rejection Order*. As part of the authentication procedure, this order may be issued as a response to the base station challenge confirmation order sent by the base station.
- *Parameter Update Confirmation Order*. This order is issued as a response to the parameter update order sent by the base station.
- *Request Wide Analog Service Order*. This order is used by the mobile station when the mobile station is directed by the user to operate in wide analog mode. In such a case, the base station may respond with an analog handoff direction message.
- *Request Narrow Analog Service Order*. This order is used by the mobile station when the mobile station is directed by the user to operate in narrow analog mode. In such a case, the base station may respond with an analog handoff direction message.
- *Request Analog Service Order*. This order is used by the mobile station when the mobile station is directed by the user to operate in analog mode, allowing operation in either wide or narrow analog mode. In such a case, the base station may respond with an analog handoff direction message.
- *Mobile Station Acknowledgment Order*. This order is used as a response to a message requiring acknowledgment.
- *Service Option Request Order*. This order is used by the mobile station to request a given service option. The service option code is informed in the SERVICE_OPTION field.
- *Service Option Response*. This order is used by the base station either to accept or to reject a service option request, whose code is informed in the SERVICE_OPTION field.
- *Release Order*. This order is used as a normal release or with a power-down indication.
- *Long Code Transition Request Order*. This order is used to request a transition to a private long code or to a public long code as a result of a request for voice privacy.
- *Long Code Transition Response Order*. This order is used as a response to the long code transition request order indicating that the mobile

station accepts or not the long code transition requested in the long code transition request order sent by the base station.

- *Connect Order*. This order is used by the mobile station when directed by the user to answer a call. The transmission of this order causes the mobile station to move to a conversation state (to be detailed later).
- *Continuous DTMF Tone Order*. This order is used to convey the DTMF tone or to stop the DTMF tone.
- *Service Option Control Order*. This order is used to invoke specific control as determined by each service option.
- *Local Control Response Order*. This order is used as a response to the local control order sent by the base station.
- *Mobile Station Reject Order*. This order is used to reject a message or an order. The following constitute the possible reasons for such a rejection: unspecified reason, message not accepted in the present state, message field not in valid range, message type or order code not understood, requested capability not supported by the mobile station, message not handled by the current mobile station configuration.

5.9 Mobile Station Call Processing

This section describes the states and procedures mobile stations go through to process a call. TIA/EIA/IS-95 specifies these states and procedures separately for mobile stations and base stations. From the time the mobile station is powered up until a traffic channel is used, a call processing goes through four different states in a sequential manner as illustrated in Figure 5.15. These states are as follows:

1. Mobile Station Initialization State
2. Mobile Station Idle State
3. System Access State
4. Mobile Station Control on the Traffic Channel State

When the mobile station is powered up it enters the mobile station initialization state, where a system is selected and acquired. After system selection and system acquisition, the mobile station moves to the mobile station idle state, where the messages on the paging channel are monitored. If the mobile station is unable to receive a paging channel message, it then returns to the

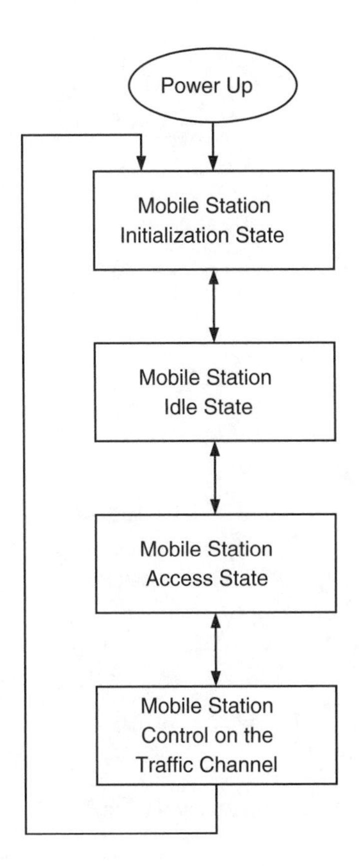

FIGURE 5.15
Mobile station call processing states.

mobile station initialization state. The system access state is reached from the mobile station idle state when the mobile station sends a message to the base station through the access channel. Such a message may be due to an acknowledgment, a response, an origination of a call, or a registration. The mobile station moves into the mobile station control on the traffic channel state when it is directed to a traffic channel. It remains in this state until the traffic channel is released, in which case it returns to the mobile station initialization state.

5.9.1 Mobile Station Initialization State

The mobile station enters the mobile station initialization state immediately after it is powered up. This state contains four substates as follows:

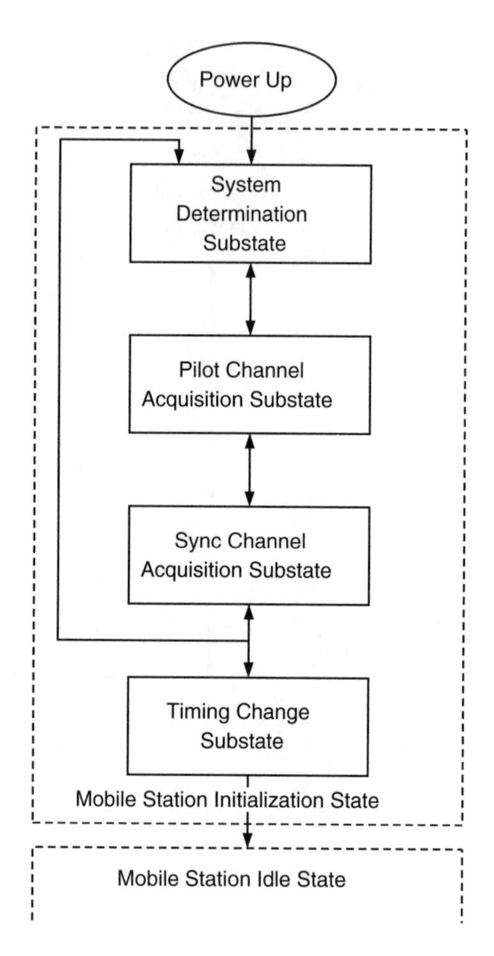

FIGURE 5.16
Mobile station initialization state.

1. System Determination Substate
2. Pilot Channel Acquisition Substate
3. Sync Channel Acquisition Substate
4. Timing Change Substate

These states are visited in a sequential manner, as shown in Figure 5.16.

System Determination Substate

In this substate, the mobile station selects which cellular system to use in accordance with the custom system selection process. The custom system selection is left to the mobile station manufacturer and may be set as

(1) system A (or B) only; (2) system A (or B) preferred; (3) CDMA (or analog) system only; (4) CDMA (or analog) system preferred. If the selected system is an analog system, the mobile shall enter an initialization task. If the selected system is a CDMA system, it sets the carrier number to the primary or secondary CDMA carrier number and enters the pilot channel acquisition substate.

Pilot Channel Acquisition Substate

In this substate, the mobile acquires the pilot channel of the selected CDMA system. With this purpose, it tunes to the CDMA carrier number, sets its code channel for the pilot channel (which is always 0—Walsh sequence W_0—for the pilot channel), and searches for the pilot channel. If it fails to acquire the selected carrier within a specified maximum time (set to 15 s), it then moves to the other carrier before performing another system selection by returning to the system determination substate. In case it is successful, it then moves to the sync channel acquisition substate.

Sync Channel Acquisition Substate

In this substate, the mobile station processes the sync channel message, which conveys information on system configuration and timing. Thus, it sets its code channel for the sync channel (which is always 32 for the sync channel—Walsh sequence W_{32}). If the mobile fails to receive a valid sync channel message within a specified maximum time (set to 1 s), it then returns to the system determination substate. If it is successful but the protocol revision level supported by the mobile station is less than the minimum protocol revision level supported by the base station, it then returns to the system determination substate. If the received sync channel message is valid and the protocol revision level restriction is satisfied, the mobile then moves to the timing change substate.

The following is the information conveyed by the sync channel message:

Protocol revision level
System identification
Network identification
Pilot PN sequence offset
Long code state
System time
Paging channel data rate
Number of leap seconds that have occurred since the start of the system time
Offset of local time from system time
Daylight saving time indicator

Timing Change Substate

In this substate, the mobile synchronizes its system timing and its long code timing to those of the CDMA system. This is achieved by means of the corresponding parameters obtained from the sync channel message, namely, pilot PN sequence offset, long code state, and system time. The exit of the timing change substate corresponds to the exit of the mobile station initialization state, from which the mobile station idle state is reached.

5.9.2 Mobile Station Idle State

The mobile station enters the mobile station idle state after the initialization procedures have been successfully completed within the mobile station initialization state. Upon entering the mobile station idle state, the mobile sets its code channel for the paging channel and sets the data rate. In this mobile station idle state, the mobile station performs a series of tasks as follows:

Paging channel monitoring procedures
Message acknowledgment procedures
Registration procedures
Idle handoff procedures
Response to overhead information operation
Page match operation
Mobile station order and message processing operation
Mobile station origination operation
Mobile station message transmission operation
Mobile station power-down operation

Paging Channel Monitoring Procedures

The paging channel is divided into 80-ms slots. Paging monitoring can be carried out in two modes: nonslotted mode and slotted mode. Paging or control messages addressed to the mobile operating in the nonslotted mode can be received in any slot. In such a mode of operation, the mobile station must monitor all slots. Paging and control messages may also be scheduled to be transmitted within certain assigned slots, and this characterizes the slotted mode of operation. In this latter case, because the mobile station is required to monitor only the assigned slots, it can conserve battery power. In any mode of operation, whenever a message is received the mobile station checks its integrity and whether or not the message is addressed to the mobile. If the timer set for the arrival of the message expires and no valid message arrives, the mobile declares loss of the paging channel. It is interesting to mention that the mobile station monitors the paging channel of one base station only.

Message Acknowledgment Procedures

Messages sent to the mobile station on the paging channel may or may not require an acknowledgment. The acknowledgment procedures provide a means of improving the reliability of message exchanges between base and mobile stations. If a message is received that requires an acknowledgment in addition to another message in response, the mobile station shall include the acknowledgment within the response. If only acknowledgment is required, the acknowledgment is then sent in a mobile station acknowledgment order.

Registration Procedures

The mobile station signalizes its presence within the system by means of a registration procedure. In the registration process, information such as location, status, identification, slot cycle, class mark, common air interface revision number, and others are passed over to the system so that the database relative to that mobile station is updated. The possible forms of registration supported by the system follow:

- *Power-up registration.* A registration occurs when the mobile station powers-on or changes from the analog to the CDMA system.
- *Power-down registration.* A registration occurs when the mobile station powers-off in a system within which it has been previously registered.
- *Timer-based registration.* A registration occurs when a timer expires. This causes the mobile station to register at regular time intervals.
- *Distance-based registration.* A registration occurs when the distance between the current base station and the base station in which the mobile station last registered exceeds a certain threshold.
- *Zone-based registration.* A registration occurs when the mobile station enters a new zone. A zone is defined as a group of base stations within a given system and network.
- *Parameter-change registration.* A registration occurs when the mobile station modifies any of the following stored parameters: the preferred slot cycle index, the station class mark, the call termination enabled indicator.
- *Ordered registration.* A registration occurs when the base station commands a mobile station to register.
- *Implicit registration.* A registration implicitly occurs whenever an origination message or a page response message is sent by the mobile station, from which the location of the mobile station can be inferred by the base.
- *Traffic channel registration.* A registration notification is sent to the mobile station when a traffic channel is assigned to it.

Idle Handoff Procedures

An idle handoff corresponds to a handoff occurring when the mobile station is still in the mobile station idle state, i.e., a call has not been established yet. While in the mobile station idle state, the mobile station continuously searches for the strongest pilot channel signal on the current CDMA frequency assignment whenever it monitors the paging channel. As it moves from the coverage area of the serving base station into a coverage area of another base station, it may find that the pilot channel signal from the new base station is stronger than that of the old base station and it determines that an idle handoff should occur. As already mentioned, the mobile station monitors the paging channel of one base station only. Therefore, soft handoff is not applicable in the mobile station idle state.

Response to Overhead Information Operation

The response to overhead information operation occurs whenever the mobile station receives an overhead message. The overhead messages on the paging channel follow:

System Parameters Message
Access Parameters Message
Neighbor List Message
CDMA Channel List Message
Extended System Parameters Message
Global Service Redirection Message

Whenever an overhead message is received on the paging channel, internally stored information is compared with the contents of these messages. If the comparison results in a match, the mobile station may ignore the message. If the comparison results in a mismatch, internal updates occur.

Page Match Operation

A page match operation is carried out by the mobile station whenever a page message is received on the paging channel. The page messages on the paging channel follow:

Page Message
Slotted Page Message
General Page Message

The records in these messages are processed according to a given procedure and a page match is declared whenever the parameters present in the records match those stored in the mobile station. Some of these parameters are IMSI, MIN, SID, and NID. If a page match is declared, the mobile station enters the update overhead information substate of the system access state with a page response indication within a specified time (set to 0.3 s) after the page message is received.

Mobile Station Order and Message Processing Operation

Except for the overhead messages and page messages all the other messages and orders addressed to the mobile station are processed during the mobile station order and message processing operation. The processing may result in an acknowledgment, in an order/message response, or in an acknowledgment and order/message response, as appropriate. The processing may also result in a reject order if any field of the message is outside its permissible range (message field not in valid range or message not acceptable in the present state). The messages and orders received during the mobile station order and message processing operation follow:

Abbreviated Alert Order
Audit Order
Authentication Challenge Message
Base Station Acknowledgment Order
Base Station Challenge Confirmation Order
Channel Assignment Message
Data Burst Message
Feature Notification Message
Local Control Order
Lock until Power-Cycled Order
Maintenance Required Order
Registration Accepted Order
Registration Rejected Order
Registration Request Order
Service Redirection Message
SSD Update Message
Unlock Order

The authentication challenge message causes the mobile station to respond with an authentication challenge response message and to enter the update overhead information substate of the system access state. The base station challenge confirmation order causes the mobile station to respond with an

SSD update confirmation order or SSD update rejection order and to enter the update overhead information substate of the system access state. The channel assignment message causes the mobile station to set its CDMA channel to the CDMA channel specified in the assignment, to tune to the new frequency assignment, to measure the strength of each pilot listed in the assignment, and to set the pilot PN offset to the corresponding offset of the strongest pilot in the list. The lock until power-cycled order causes the mobile station not to enter the system access state until it has received the unlock order or until after the next mobile station power-up. (The user must be notified of the locked condition.) On the other hand, the mobile station may depart from the mobile station idle state and move into the system determination substate of the mobile station initialization state with a lock indication, which permits its operation in the analog mode while locked. The maintenance required order causes the mobile station to remain in the locked condition, if it has previously received a lock until power-cycled order, or in the unlocked condition, otherwise. The registration rejected order causes the mobile station to enter the system determination substate with a registration-rejected indication. This same substate is reached if the service redirection message is received. The SSD update message causes the mobile station to enter the update overhead information substate of the system access state. The unlock order causes the mobile to change its condition to unlocked and to enter the system determination substate of the mobile station initialization state with an unlocked indication. All the other messages and orders are processed by the mobile station as previously described.

Mobile Station Origination Operation

The origination operation is performed when the mobile station is initiating a call. In such a case, the mobile station enters the update overhead information substate of the system access state with an origination indication within a specified time (set to 0.3 s).

Mobile Station Message Transmission Operation

The message transmission operation is performed when the mobile station is directed by the user to transmit a data burst message. In this case, the mobile station enters the update overhead information substate of the system access state with a message transmission indication within a specified time (set to 0.3 s). The support of this operation is optional.

Mobile Station Power-Down Operation

The power-down operation is performed when the mobile station is directed by the user to power off. In such a case, the mobile station updates the stored parameters and performs other registration procedures.

5.9.3 System Access State

The mobile station enters the system access state to perform those access procedures for which the access channel is required. An access channel is required whenever any of the following two types of messages are to be sent by the mobile: a response message and a request message. A response message is transmitted in response to a base station message. A request message is sent autonomously by the mobile station.

For any given message to be transmitted by the mobile station, the access channel is chosen pseudorandomly among all the access channels associated with the current paging channel and is seized by the mobile station by means of a random procedure. The process of sending one message and receiving or failing to receive an acknowledgment is known as *access attempt*. Each transmission of the same message on the access channel in an access attempt is known as *access probe*. Several access probes compose the *access probe sequence*. Therefore, the access probe sequence is a sequence of transmissions of the same message on the access channel in an access attempt. The access probes occur at increasing power levels, the first power level given by the nominal open-loop power level and the subsequent ones incremented by some given power step. Sequence or sequences of access probes occur as a consequence of an unsuccessful communication between mobile station and base station. The failure of communication is declared if after transmitting an access probe an acknowledgment is not received within a specified period of time.

The number of access probes within a probe sequence, limited to a maximum of 16, is set by the system operator. In the same way, the number of probe sequences within an access attempt, limited to a maximum of 16, is set by the system operator, but may be set differently for the two types of messages (response and request). An access attempt occurs on an access channel whose long code mask is given by a number that has been randomly generated within the range of zero to the number of access channels supported by the current paging channel. An access probe occurs at the start of an access channel slot and the timing between access probes of an access probe sequence is generated in a pseudorandom manner. More specifically, a back-off delay, given in numbers of time slots after the failure of communication is declared, is pseudorandomly generated within the range from zero to 1 + probe back-off slots. In the same way, the timing between access probe sequences is pseudorandomly generated. In particular, for the response-type message the number of slots to be waited until the next access probe sequence is initiated is obtained by a number randomly selected within the range from zero to 1 + access channel back-off. For the request-type message an additional delay is imposed by the use of a persistence test. The persistence test is carried out for each slot and consists of performing a pseudorandom test having as input the parameters characterizing the reason for the access attempt and the access overload class of the mobile station. If the test passes, the first access probe of the sequence begins in that slot; otherwise, the access probe sequence is deferred until at least the next slot. The persistence test is not required for the response-type message because the base station can control the rate at which it receives this type of message from the mobile station by controlling the rate at which it sends messages requiring response. The request-type message is generated autonomously by the mobile station and a persistence test imposes a further randomness in the rate at which it reaches the base station.

The transmit timing of each access attempt is determined by a PN randomizing procedure that consists of computing the number of PN chips by which the access attempt should be delayed. This number is determined from 0 to $2^{TR} - 1$ using a hash function that depends on the ESN, where TR denotes the time randomization. This delay adjustment includes the delays of the direct sequence spreading long code and of the quadrature spreading PN sequences. In effect, this procedure augments the apparent range from the mobile station to the base station. Therefore, it increases the probability of the base station to discern among mobile stations using the same access channel slot and encountered at similar distances from the base.

The system access state is composed of six substates as follows:

1. Update Overhead Information Substate
2. Page Response Substate
3. Mobile Station Order/Message Response Substate
4. Mobile Station Origination Attempt Substate
5. Registration Access Substate
6. Mobile Station Message Transmission Substate

These substates are illustrated in Figure 5.17.

Update Overhead Information Substate

In this substate, the mobile station monitors the paging channel to check for the current configuration messages. It compares sequence numbers to determine whether all the configuration messages are up-to-date. With this purpose, the mobile station receives at least one message containing the access message sequence field and, if necessary, waits for the access parameters message. The configuration messages are as follows:

System Parameters Message
Neighbor List Message
CDMA Channel List Message
Extended System Parameters Message
Global Service Redirection Message

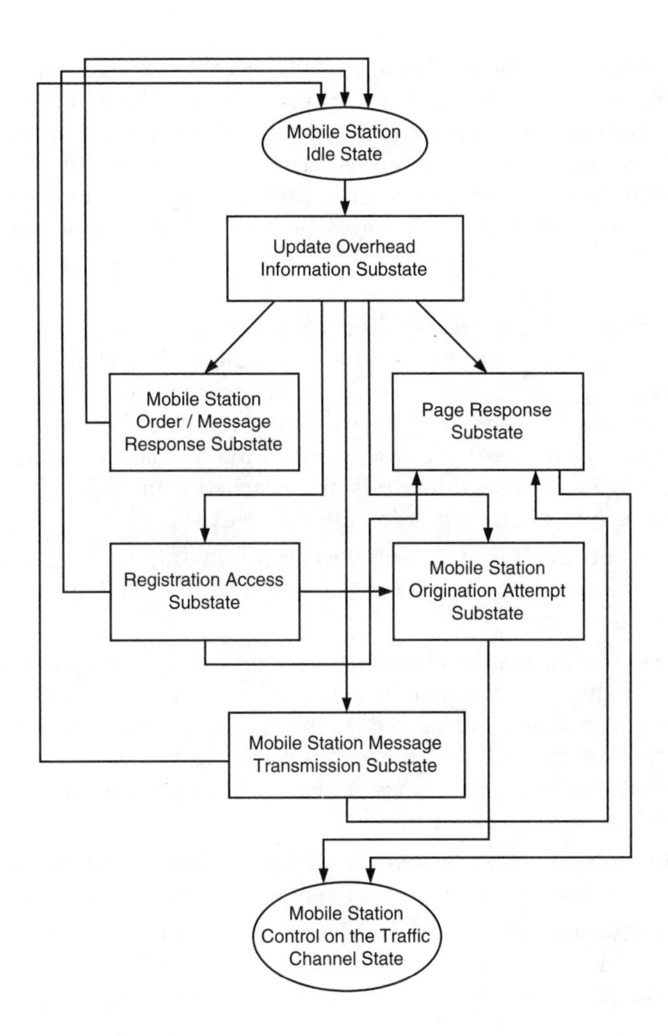

FIGURE 5.17
Mobile station access state.

The mobile station may also receive the general page message, the page message, and the slotted page message. These messages contain the access message sequence field.

Upon entering this substate, a timer is set to a specific value (4 s) while the mobile station monitors the paging channel and processes the received messages. If the timer expires while in this substate, the mobile station enters the system determination substate of the mobile initialization state indicating system loss. In the same way, if the paging channel is declared lost, it enters the mobile station idle state. The page messages cause the current access message sequence to be set to the received access message sequence. They also incite the mobile station to determine whether there is a page match. The timer is disabled when the mobile station finds that the stored configuration parameters are current and its current access message sequence and the received access message sequence match each other and are not null. The mobile station will leave this substate if any of the following conditions occur:

1. If a page match is declared or if the substate was reached with a page response indication. In this case, the mobile station checks if the message resulting in the page match was received on the current paging channel. In the affirmative case, it enters the page response substate. In the negative case, it enters the mobile station idle state. The page response substate is also reached from this substate if a page response retransmission indication exists.
2. If this substate was reached with an origination indication. In this case, the mobile station enters the mobile station origination attempt substate.
3. If this substate was reached with an order/message response indication. In this case, the mobile station checks if the message resulting in the response was received on the current paging channel. In the affirmative case, it enters the mobile station order/message response substate. In the negative case, it discards the response and enters the mobile station idle state.
4. If this substate was reached with a registration indication. In such a case, the mobile station enters the registration access substate.
5. If this substate was reached with a message transmission indication. In this case, the mobile station enters the mobile station message transmission substate.

The messages and orders received during the update overhead information substate follow:

System Parameters Message
Access Parameters Message
Neighbor List Message
CDMA Channel List Message
Extended System Parameters Message
Global Service Redirection Message
Lock until Power-Cycled Order

Except for the lock until power-cycled order, all the other messages have already been mentioned and they are processed by the mobile station as previously described. The lock until power-cycled order causes the mobile station to notify the user of the locked condition and to enter the system determination substate of the mobile initialization state with a lock indication. The mobile station does not move into the system access state until an unlock order is received.

Page Response Substate

In this substate, the mobile station sends a page response message in response to page messages (general page message, page message, or slotted page message) from the base station. The page response message is sent through the access attempt procedures, as already described. Therefore, the access attempt is ended upon the reception of an acknowledgment for it. While in this substate, the mobile station monitors the paging channel which may be declared lost if no valid message arrives. In such a case, it disables its transmitter and enters the mobile station idle state.

A series of messages and orders may be received while the mobile is this substate. These include:

Authentication Challenge Message
Base Station Challenge Confirmation Order
Channel Assignment Message
Data Burst Message
Feature Notification Message
Local Control Order
Lock until Power-Cycled Order
Maintenance Required Order
Registration Accepted Order
Registration Rejected Order
Release Order
Service Redirection Order
SSD Update Message

One of the important messages received in this substate is the channel assignment message. This message is exhibited in several modes, each of which causes the mobile station to perform a succession of different actions. These modes comprise the following: traffic channel assignment, paging channel assignment, acquire analog system, analog voice channel assignment. In the traffic assignment mode, upon receiving the traffic channel assignment message, the mobile station stores the appropriate parameters from the message and enters the traffic channel initialization substate of the mobile station control on the traffic channel state. In paging channel assignment mode, the mobile station sets its CDMA frequency to the new the CDMA frequency assignment, tunes to the new CDMA frequency assignment, measures the strength of the pilot listed in the assignment, and sets the pilot PN sequence offset of the strongest pilot in the neighbor set list. In the acquire analog mode and analog voice assignment mode, the mobile station enters the initialization task with a page response indication or waits for a page indication.

The lock until power-cycled order causes the mobile station not to enter the system access state until it has received the unlock order or until after the next mobile station power-up. (The locked condition must be notified to the user.) On the other hand, the mobile station may depart from the mobile station idle state and move into the system determination substate of the mobile station initialization state with a lock indication, which permits its operation in the analog mode while locked. The maintenance required order causes the mobile station to remain in the locked condition, if it has previously received a lock until power-cycled order, or in the unlocked condition, otherwise. The registration rejected order causes the mobile station to enter the system determination substate with a registration-rejected indication. This same substate is reached if the service redirection message is received. The release order causes the mobile station to enter the mobile station idle state or the system determination substate of the mobile station initialization state.

Note that the page response substate is reached from the update overhead information substate upon receiving the page messages. From this substate, besides the state transitions above described, the state to be reached is the mobile station control on the traffic channel state.

Mobile Station Order/Message Response Substate

In this substate, the mobile station sends a message to the base station in response to a message received from it. The response message is sent through the access attempt procedures, as already described. Therefore, the access attempt is terminated upon the reception of an acknowledgment for it. While in this substate, the mobile station monitors the paging channel, which may be declared lost if no valid message arrives. In this case, it disables its transmitter and enters the mobile station idle state. Except for the channel assignment message and the release order, all the other messages and orders accepted in the page response substate are also allowed in this mobile station order/message response substate and they are processed as already described.

Note that the mobile station order/message response substate is reached from the update overhead information substate upon receiving messages or orders requiring acknowledgment or response. From this substate, besides the state transitions above described, the state to be reached is the mobile station idle state.

Mobile Station Origination Attempt Substate

In this substate, the mobile station sends an origination message containing the dialed digits using the access procedures, as already described. Therefore, the access attempt is ended upon the reception of an acknowledgment for it. While in this substate, the mobile station monitors the paging channel, which may be declared lost if no valid message arrives. In this case, it disables its transmitter and enters the mobile station idle state. In addition to all the messages and orders accepted in the page response substate, the mobile origination attempt substate also allows the intercept order and the reorder order. Both the intercept order and the reorder order cause the mobile station to enter the mobile station idle state. All the other messages are processed as already described.

In case the mobile station is required to disconnect a call, the mobile station aborts any access attempt in progress and enters the system determination substate of the mobile station initialization state.

Note that the mobile station order/message response substate is reached from the update overhead information substate when the user initiates a call. From this substate, besides the state transitions described above, the state to be reached is the mobile station control on the traffic channel state.

Registration Access Substate

In this substate, the mobile station sends a registration message using the access procedures, as already described. Therefore, the access attempt is ended upon the reception of an acknowledgment for it. While in this substate, the mobile station monitors the paging channel, which may be declared lost if no valid message arrives. In this case, it disables its transmitter and enters the mobile station idle state. In addition to all the messages and orders accepted in the mobile station order/message response substate the registration attempt substate also allows the release order. In case the mobile station is required to originate a call, the mobile station aborts any access attempt in progress and enters the mobile station origination attempt substate. In this substate, the page messages may also be received. In case a page match is declared, any access attempt in progress is aborted and the mobile station enters the page response substate.

Note that the registration access substate is reached from the update overhead information substate when a registration is required. From this substate, besides the state transitions described above, the state to be reached is the mobile station idle state.

Mobile Station Message Transmission Substate

In this substate, whose support is optional, the mobile station sends a data burst message using the access attempt procedure, as already described. Therefore, the access attempt is terminated upon the reception of an acknowledgment for it. While in this substate, the mobile station monitors the paging channel, which may be declared lost if no valid message arrives. In this case, it disables its transmitter and enters the mobile station idle state. Except for the Feature Notification Message, all the messages and orders accepted in the mobile station order/message response substate are also allowed in the mobile station message transmission substate, and they are processed as already described. In this substate, the page messages may also be received. In case a page match is declared, any access attempt in progress is aborted and the mobile station enters the page response substate.

Note that the mobile station message transmission substate is reached from the update overload information substate when a data burst message is to be sent. From this substate, besides the state transitions described above, the state to be reached is the mobile station idle state.

5.9.4 Mobile Station Control on the Traffic Channel State

The mobile station enters the mobile station control on the traffic channel state departing from the system access state upon receiving the channel assignment message. In this state, the mobile station uses the forward traffic channel and reverse traffic channel to communicate with the base station.

The mobile station control on the traffic channel state is composed of five substates as follows:

1. Traffic Channel Initialization Substate
2. Waiting for Order Substate
3. Waiting for Mobile Station Answer Substate
4. Conversation Substate
5. Release Substate

These substates are described next and are illustrated in Figure 5.18.

Traffic Channel Initialization Substate

In this substate, the mobile station verifies if it can support the assigned CDMA channel and the assigned forward traffic code channel. In the negative case, the mobile station enters the system determination substate of the mobile initialization state with an error notification. In the affirmative case, the mobile station performs the following actions:

It tunes to the assigned CDMA channel.

It sets its code channel for the assigned forward traffic code channel.

It sets its forward and reverse traffic channel frame offsets to the assigned frame offset.

It sets its forward and reverse traffic channel long code masks to the public long code mask.

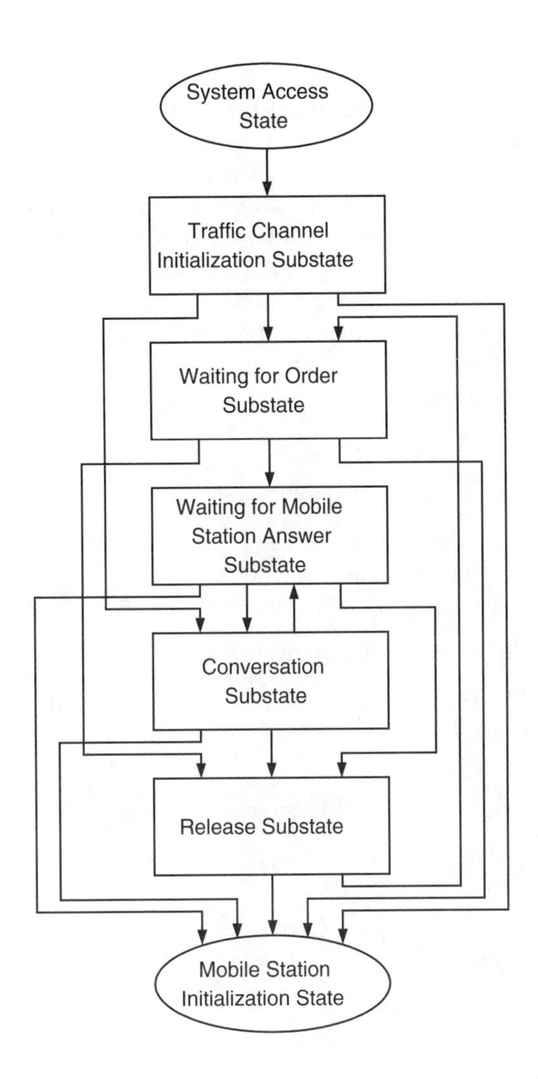

FIGURE 5.18
Mobile station control on the traffic channel states.

In this state, among others, the mobile monitors the forward traffic channel and performs pilot strength measurements. In case the mobile station does not receive a specified number of consecutive good frames (two good frames) or if an acknowledgment failure is declared, it enters the system determination substate of the mobile initialization state with a system lost indication. In the successful event, the mobile station moves to the waiting for order substate, if the call is terminated by the mobile station, or to the conversation substate, if the call is originated by the mobile station.

Waiting for Order Substate

In this substate, the mobile station waits for an alert with information massage, which commands the mobile station to alert the subscriber of an incoming call. Upon entering this substate, a timer is set to a specified value (5 s) and if during this period the mentioned message is not received the mobile station disables its transmitter and enters the system determination substate of the mobile station initialization state. If the message is received, then the mobile station moves to the waiting for mobile station answer substate. In case the mobile station is commanded by the user to power down, it enters the release substate with a power-down indication. The messages and orders received in this substate follow:

Alert with Information Message
Analog Handoff Direction Message
Audit Order
Authentication Challenge Message
Base Station Acknowledgment Order
Base Station Challenge Confirmation Order
Data Burst Message
Extended Handoff Direction Message
Handoff Direction Message
In-Traffic System Parameters Message
Local Control Order
Lock until Power-Cycled Order
Long Code Transition Request Order
Maintenance Order
Maintenance Required Order
Message Encryption Mode Order
Mobile Station Registered Message
Neighbor List Update Message
Parameter Update Order
Pilot Measurement Request Order
Power Control Parameters Message
Release Order
Retrieve Parameters Message
Service Option Control Order
Service Option Request Order

Service Option Response Order
Set Parameters Message
SSD Update Message
Status Request Order

The alert with information message causes the mobile station to alert the user and to enter the mobile station answer substate. The analog handoff direction message causes the mobile station to process the message and to enter the waiting for order substate. The base station challenge confirmation order causes the mobile station to respond with an SSD update confirmation order or an SSD update rejection order. The maintenance order causes the mobile station to move into the mobile station answer substate. The parameter update order causes the mobile station to send a parameter update confirmation order. The release order causes the mobile station to enter the release substate with a release indication. The set parameters message causes the mobile station to set its parameters according to the specification in the message, or to send a mobile reject order if these parameters are not acceptable. The status request order causes the mobile station to respond with a status message. All the other messages are processed as already described.

Waiting for Mobile Station Answer Substate

In this substate, the mobile station waits for the user to answer the mobile station terminated call. Upon entering this state, a timer is set to a specified value (65 s) and a series of tasks is performed. If the timer expires, the mobile station disables its transmitter and enters the system determination substate of the mobile station initialization state indicating loss of the system. The same happens if a loss of the forward traffic channel is declared or if the acknowledgment procedures fail. In case the mobile station is directed by the user to answer the call, a connect order is sent to the base station and the mobile station moves to the conversation substate. If a power-down is to be carried out upon the user's direction the mobile station enters the release substate with a power-down indication. All the messages accepted in the waiting for order substate are also allowed in the waiting for mobile station answer substate, and they are processed as already described.

Conversation Substate

In this substate, the primary traffic service option application of the mobile station exchanges primary traffic bits with the base station. This substate is reached from the traffic channel initialization substate, when the call is originated by the mobile, and from the waiting for mobile station answer substate, when the call is terminated by the mobile. A series of tasks is performed by the mobile station while in this substate. One such task includes the supervision of the forward traffic channel, in which if a loss is declared the mobile station moves to the system determination substate of the mobile station initialization state with a system loss indication. The same happens upon failure of the acknowledgment procedures. Still in the conversation substate, the mobile station may send the remaining dialed digits of a mobile-originated call for which not all digits had been sent at the call origination time. A data burst message may also be sent by the mobile station if directed by the user. From the conversation substate, the release substate is reached upon a call disconnection as directed by the user, in which case a release indication is set. In the same way, if directed by the user to power down, the mobile station enters the release state with a power-down indication. The messages allowed in this substate follow:

Alert with Information Message
Analog Handoff Direction Message
Audit Order
Authentication Challenge Message
Base Station Acknowledgment Order
Base Station Challenge Confirmation Order
Continuous DTMF Tone Order
Data Burst Message
Extended Handoff Direction Message
Flash with Information Message
Handoff Direction Message
In-Traffic System Parameters Message
Local Control Order
Lock until Power-Cycled Order
Long Code Transition Request Order
Maintenance Order
Maintenance Required Order
Message Encryption Mode Order
Mobile Station Registered Message
Neighbor List Update Message
Parameter Update Order
Pilot Measurement Request Order
Power Control Parameters Message
Release Order

Retrieve Parameters Message
Send Burst DTMF Message
Service Option Control Order
Service Option Request Order
Service Option Response Order
Set Parameters Message
SSD Update Message
Status Request Order

The support of the DTMF message or order is optional. All the messages are processed as already described.

Release Substate

In this substate, the mobile station confirms the call disconnect. Upon entering this state, a timer is set to a specified value (65 s) and a series of tasks is performed. If the timer expires, the mobile station disables its transmitter and enters the system determination substate of the mobile station initialization state indicating loss of the system. The same happens if a loss of the forward traffic channel is declared or if the acknowledgment procedures fail. Upon entering the release substate, a release order is sent by the mobile station in case there is a power-down indication, or a base station release indication, or a mobile station release indication. In the first case, the mobile station performs a power-down registration procedure. And in the second case, the mobile station disables its transmitter and enters the system determination substate of the mobile station initialization state with a release indication. The messages or orders allowed in the release substate follow:

Alert with Information Message
Base Station Acknowledgment Order
Data Burst Message
Extended Handoff Direction Message
Handoff Direction Message
In-Traffic System Parameters Message
Local Control Order
Lock until Power-Cycled Order
Maintenance Required Order
Mobile Station Registered Message
Neighbor List Update Message
Power Control Parameters Message

Release Order
Retrieve Parameters Message
Service Option Control Order
Status Request Order

All the messages are processed as already described.

5.10 Base Station Call Processing

This section describes the states and procedures a base station goes through to process a call. TIA/EIA/IS-95 specifies these states and procedures separately for mobile stations and base stations. Four types of procedures compose the base station call processing:

1. Pilot and Sync Channel Processing
2. Paging Channel Processing
3. Access Channel Processing
4. Traffic Channel Processing

5.10.1 Pilot and Sync Channel Processing

The pilot and sync channel processing refers to the actions performed by the base station for the transmission of the pilot and sync channels. As already mentioned, the pilot channel serves as a reference used by the mobile station for acquisition and timing purposes, and as phase reference for coherent demodulation. In the same way, the sync channel provides the mobile station with system configuration and timing information. For every CDMA channel supported by the base station there is a pilot channel and at most one sync channel, the latter conveying the sync channel message in a continual way.

5.10.2 Paging Channel Processing

The paging channel processing refers to the actions performed by the base station for the transmission of the paging channel. As already mentioned, the paging channel is monitored by the mobile station while in the mobile station idle state and system access state. As many as seven paging channels for each CDMA channel can be transmitted by the base station. In the same way, for each sync channel, the base station should transmit a paging channel.

And for each paging channel the base station should send valid paging channel messages.

The resources assigned to the mobile station, namely, CDMA channel, paging channel, and paging channel slots, are determined by the base station through the use of a hash function with the appropriate inputs. These inputs include the mobile station identification number or the international mobile subscriber number, the number of CDMA channels on which the base station transmits paging channels, the number of paging channels that the base station transmits on assigned CDMA channel, and the maximum number of paging channel slots (2048).

The transmission of messages on the paging channel is made more reliable by means of acknowledgment procedures involving the access channel. Such a mechanism is supported by adequately using the Layer 2 fields such as acknowledgment address type, acknowledgment sequence number, message sequence number, acknowledgment required, and valid acknowledgment. In the same way, a message may be sent several times to increase the probability that the message is received correctly. The retransmission of the same message should be concluded within a specific time (set to 2.2 s) after the first retransmission. If a message is to be retransmitted after this specified time, another message sequence number should be used.

Apart from the general page message, page message, and slotted page message, all the other messages on the page channel may be addressed to a specific mobile station, a specific mobile station identification number, or a specific IMSI. (These last two parameters can be active in more than one mobile station.) However, the page message and slotted page message can only be addressed to a specific mobile identification number, whereas the general page message can only be addressed to a specific IMSI.

System overhead messages are also sent on the paging channel and each occurs at least once per a specified time (set to 1.28 s). These messages include:

Access Parameters Message
CDMA Channel List Message
Neighbor List Message
System Parameters Message

A series of messages and orders directed to the mobile station may be sent by the base station. They include:

Abbreviated Alert Order
Audit Order
Authentication Challenge Message

Base Station Acknowledgment Order
Base Station Challenge Confirmation Order
Channel Assignment Message
Data Burst Message
Feature Notification Message
General Page Message
Intercept Order
Local Control Order
Lock until Power-Cycled Order
Maintenance Required Order
Page Message
Registration Accepted Order
Registration Rejected Order
Registration Request Order
Release Order
Reorder Order
Service Redirection Message
Slotted Page Message
SSD Update Message
Unlock Order

In addition, the data burst messages may be directed to broadcast addresses.

5.10.3 Access Channel Processing

Access channel processing refers to the actions performed by the base station as a consequence of the monitoring of the access channels. As already mentioned, the access channel is transmitted by the mobile station while in the system access state. Each access channel is associated with a paging channel. And as many as 32 access channels can be associated with a paging channel.

The communication between base station and mobile station is made more reliable by means of transmission of messages and acknowledgment procedures involving the paging channel and the access channel. Such a mechanism is supported by adequately using the Layer 2 fields such as acknowledgment address type, acknowledgment sequence number, message sequence number, acknowledgment requited, and valid acknowledgment.

A page response message sent over the access channel by the mobile station causes the base station to send a release order or a channel assignment message to the mobile station. In the same way, an origination message sent by the mobile station causes the base station to send an intercept order, or a reorder order, or a release order, or a channel assignment message. Finally, a registration message sent by the mobile station causes the mobile station to send a registration accepted order or a registration reject order. In all cases, an authentication procedure may be started so that the identity of the mobile station can be confirmed (the sets of shared secret data of both should match).

5.10.4 Traffic Channel Processing

The traffic channel processing refers to the actions taken by the base station while using the forward traffic channel and reverse traffic channel to communicate with the mobile station. As in the mobile station case, the traffic channel processing has five substates:

1. Traffic Channel Initialization Substate
2. Waiting for Order Substate
3. Waiting for Answer Substate
4. Conversation Substate
5. Release Substate

Some special functions and actions are performed by the base station while in traffic channel processing state. These include forward traffic channel power control, service option, acknowledgment procedure, message action time, and long code transition request processing. Forward traffic channel power control basically consists in reporting frame error rate statistics using power measurement report message. Service option refers to the several primary traffic services supported by the base station and mobile station. These service options may be negotiated between these two entities and the negotiation ends when they find a mutually acceptable service option or when one of them rejects a service option requested by the other. Acknowledgment procedure refers to the protocols used by the base station and the mobile station to facilitate the reliable exchange of messages between them. Message action time refers to the action times a message should require to take effect. There may be two types of action time: implicit or explicit. A message with an implicit action time should take effect no later than the first 80 ms boundary, relative to system time, occurring at least 80 ms after the end of the frame containing the last bit of the message. A message with an explicit action time should take effect when system time, in units of 80 ms modulo 64, becomes equal to the time specified within the message. Long code transition request processing refers to action taken by the base station when a code transition is requested. The transition may be to a private long code or to a public long code, as specified in the origination message.

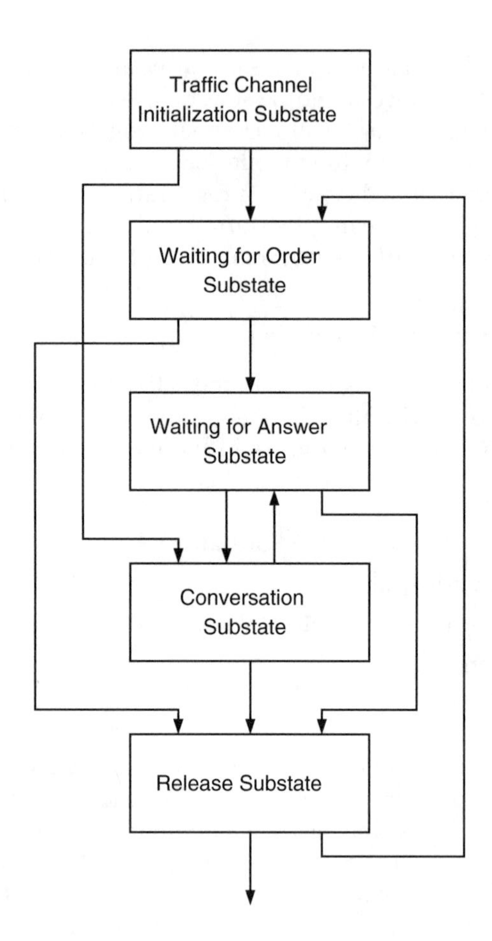

FIGURE 5.19
Base station traffic channel processing state.

The substates of the traffic channel processing state are illustrated in Figure 5.19 and are described next.

Traffic Channel Initialization Substate

In this substate, the transmission on the forward traffic channel is started by the base station while the acquisition of the reverse traffic channel is pursued. Upon entering this substate, the base station, among others, sets its forward traffic channel and reverse traffic channel long code masks to the public long code mask and its forward traffic channel and reverse traffic channel frame offsets to the frame offset assigned to the mobile station.

If the acquisition of the reverse traffic channel fails, the base station may either retransmit the channel assignment message on the paging channel and remain in this same substate or else it disables transmission on the forward traffic channel and discontinues the traffic channel for the mobile station. If the acquisition is successful, the base station sends the base station acknowledgment order requiring an acknowledgment. For call terminated by the mobile station it enters the waiting for order substate and for a mobile station–originated call, it moves to the conversation substate.

Waiting for Order Substate

In this substate, the base station sends an alert with information message, which commands the mobile station to alert the subscriber of an incoming call. Upon entering this substate, the base station processes the service option request specified in the page response message. While in this substate, the base station, among others, transmits the power control subchannel, performs forward traffic channel power control, acknowledgment procedures, and authentication procedures. As part of the service option tasks, it may process the received primary traffic bits or it may request a service option. It may also request a long code transition.

The base station may send the following messages or orders:

Alert with Information Message
Analog Handoff Direction Message
Audit Order
Authentication Challenge Message
Base Station Acknowledgment Order
Base Station Challenge Confirmation Order
Data Burst Message
Extended Handoff Direction Message
Handoff Direction Message
In-Traffic System Parameters Message
Local Control Order
Lock until Power-Cycled Order
Long Code Transition Request Order
Maintenance Order
Maintenance Required Order
Message Encryption Mode Order
Mobile Station Registered Message
Neighbor List Update Message
Parameter Update Order
Pilot Measurement Request Order

Power Control Parameters Message
Release Order
Retrieve Parameters Message
Service Option Control Order
Service Option Request Order
Service Option Response Order
Set Parameters Message
SSD Update Message
Status Request Order

After sending the alert with information message to the mobile station, the base station moves into the waiting for answer substate. And after sending the maintenance order it enters the waiting for answer substate.

The base station may receive the following messages or orders:

Base Station Challenge Order
Data Burst Message
Handoff Completion Message
Long Code Transition Request Order
Mobile Station Acknowledgment Order
Mobile Station Reject Order
Parameters Response Message
Parameter Update Confirmation Order
Pilot Strength Measurement Message
Power Measurement Report Message
Release Order
Request Analog Service Order
Request Narrow Analog Service Order
Request Wide Analog Service Order
SSD Update Confirmation Order
SSD Update Rejection Order
Service Option Control Order
Service Option Request Order
Service Option Response Order
Status Response Order

The release order causes the base station to send a release order to the mobile station and to enter the release substate. Or it may send an alert with information message and enter the waiting for answer substate. In case a loss of reverse traffic channel continuity is declared by the base station, the base station it sends a release order to the mobile station and moves into the release substate.

Waiting for Answer Substate

In this substate, the base station waits for a connect order from the mobile station. Most of the functions and actions performed by the base station while in the waiting for order substate are also performed in this substate. In the same way, the possible messages to be transmitted are the same. In addition to the possible messages to be received while in the waiting for order substate, the connect order may be received in the waiting for answer substate.

In case a loss of reverse traffic channel continuity is declared by the base station, it sends a release order to the mobile station and moves into the release substate. If it receives the connect order, it enters the conversation substate.

Conversation Substate

In this substate, the base station exchanges primary traffic bits with the mobile station primary traffic service option application. Upon entering this substate, the base station processes the service option request present in the origination message. While in this substate, the base station, among others, transmits the power control subchannel, performs forward traffic channel power control, acknowledgment procedures, and authentication procedures. As part of the service option tasks, it may process the received primary traffic bits or it may request a service option. It may also request a long code transition.

The base station may send the following messages or orders:

Alert with Information Message
Analog Handoff Direction Message
Audit Order
Authentication Challenge Message
Base Station Acknowledgment Order
Base Station Challenge Confirmation Order
Continuous DTMF Tone Order
Data Burst Message
Extended Handoff Direction Message
Flash with Information Message
Handoff Direction Message

In-Traffic System Parameters Message
Local Control Order
Lock until Power-Cycled Order
Long Code Transition Request Order
Maintenance Order
Maintenance Required Order
Message Encryption Mode Order
Mobile Station Registered Message
Neighbor List Update Message
Parameter Update Order
Pilot Measurement Request Order
Power Control Parameters Message
Release Order
Retrieve Parameters Message
Send Burst DTMF Message
Service Option Control Order
Service Option Request Order
Service Option Response Order
Set Parameters Message
SSD Update Message
Status Request Order

After sending the alert with information message to the mobile station, the base station moves into the waiting for answer substate. And after sending the maintenance order, it enters the waiting for answer substate.

The base station may receive the following messages or orders:

Base Station Challenge Order
Continuous DTMF Tone Order
Data Burst Message
Flash with Information Message
Handoff Completion Message
Long Code Transition Request Order
Mobile Station Acknowledgment Order
Mobile Station Reject Order
Origination Continuation Message
Parameters Response Message

Parameter Update Confirmation Order
Pilot Strength Measurement Message
Power Measurement Report Message
Release Order
Request Analog Service Order
Request Narrow Analog Service Order
Request Wide Analog Service Order
SSD Update Confirmation Order
SSD Update Rejection Order
Send Burst DTMF Control Order
Service Option Control Order
Service Option Request Order
Service Option Response Order
Status Response Order

The release order causes the base station to send a release order to the mobile station and to enter the release substate. Or it may send an alert with information message and enter the waiting for answer substate. In case a loss of reverse traffic channel continuity is declared by the base station, the base station sends a release order to the mobile station and moves into the release substate.

Release Substate

In this substate, the base station executes the tasks related to call disconnection. While in this substate, the base station, among others, performs forward traffic channel power control and acknowledgment procedures. It transmits null traffic channel data for a specified time, after which it stops transmitting on the forward traffic channel.

The base station may send the following messages or orders:

Alert with Information Message
Audit Order
Base Station Acknowledgment Order
Data Burst Message
Extended Handoff Direction Message
Handoff Direction Message
In-Traffic System Parameters Message
Local Control Order
Lock until Power-Cycled Order

Maintenance Order
Maintenance Required Order
Mobile Station Registered Message
Neighbor List Update Message
Parameter Update Order
Power Control Parameters Message
Release Order
Retrieve Parameters Message
Service Option Control Order
Status Request Order

After sending the alert with information message to the mobile station, the base station moves into the waiting for answer substate. And after sending the maintenance order, it enters the waiting for answer substate.

The base station may receive the following messages or orders:

Base Station Challenge Order
Connect Order
Continuous DTMF Tone Order
Data Burst Message
Flash with Information Message
Handoff Completion Message
Long Code Transition Request Order
Mobile Station Acknowledgment Order
Mobile Station Reject Order
Origination Continuation Message
Parameters Response Message
Parameter Update Confirmation Order
Pilot Strength Measurement Message
Power Measurement Report Message
Release Order
Request Analog Service Order
Request Narrow Analog Service Order
Request Wide Analog Service Order
SSD Update Confirmation Order
SSD Update Rejection Order
Send Burst DTMF Control Order

Service Option Control Order
Service Option Request Order
Service Option Response Order
Status Response Order

All these messages are processed as already described.

5.11 Authentication, Message Encryption, and Voice Privacy

TIA/EIA/IS-95 defines authentication as "the process by which information is exchanged between mobile station and base station for the purpose of confirming the identity of the mobile station." An authentication is said to be successful if the outcome of this process determines that both mobile station and base station possess identical sets of shared secret data (SSD). Message encryption is employed to protect sensitive information in certain fields of the signaling messages. Voice privacy is achieved with special codes allocated to a conversation for which privacy is required.

Key to the authentication process is the SSD. As already defined, SSD is a 128-bit term of information stored in the semipermanent memory of the mobile station. It is composed of two equal-sized parts: SSD_A, used to support the authentication procedures, and SSD_B, used to support voice privacy and message encryption. SSD is not accessible to the user. SSD is generated according to a given cryptographic algorithm and is initialized with mobile station–specific information, random data, and the A-key of the mobile station. The A-key is a 64-bit secret key known only to the mobile station and to its associated HLR/AC. It is stored in the permanent security and identification memory of the mobile station.

5.12 Authentication

The following procedures require authentication:

- Updating the SSD
- Mobile Station Registrations
- Mobile Station Originations
- Mobile Station Terminations
- Mobile Station Data Burst
- Unique Challenge-Response Procedure

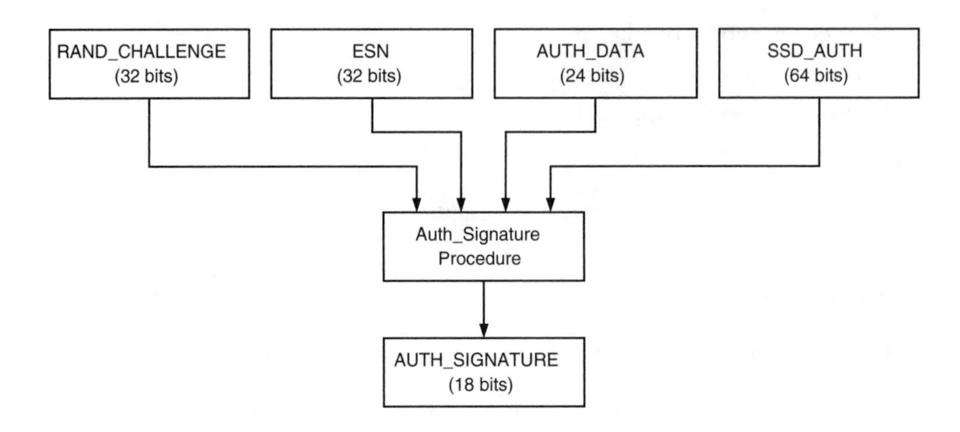

FIGURE 5.20
Computation of the authentication signature for authentication purposes.

All these procedures make use of a signature algorithm whose inputs may vary for each procedure. TIA/EIA/IS-95 names this signature algorithm Auth_Signature Procedure, its four inputs as RAND_CHALLENGE (32 bits), ESN (32 bits), AUTH_DATA (24 bits), and SSD_ AUTH (64 bits), and its output as AUTH_SIGNATURE (18 bits), as shown in Figure 5.20. In addition to the signature code, a further security mechanism is provided by controlling the content of an 8-bit memory register in the mobile station, the call-history register. The content of this register, named COUNT, is updated (incremented) every time the mobile station receives a parameter update message, transmitted on the paging channel or on the forward traffic channel by the base station. At the appropriate time, as required, the mobile station sends the content, COUNT, of this register to the base station, which is checked against the value stored at the base station. Note that, even when SSD is intercepted by an unauthorized terminal, the content of this register must be known so that this terminal may obtain access to the system. And such a content, COUNT, will be known in case this terminal intercepts all the parameter update messages directed to the authorized terminal.

5.12.1 Updating the Shared Secret Data

For security reasons, both base station and mobile station obtain a new SSD periodically. The SSD-update process starts when the base station sends an SSD-update message to the mobile station. This message may go through the paging channel, if there is no call in progress, or through the forward traffic channel, if a call is in progress. The RANDSSD field of the update message contains a 56-bit-long number, randomly generated at the HLR/AC.

Upon receiving the SSD-update message, the mobile station computes a new SSD by means of a SSD_Generation Procedure, for which the inputs are the information of the RANDSSD (56 bits), the ESN (32 bits), and the A-key (64 bits). The new 128-bit-long SSD is then equally split into the SSD_A and SSD_B.

The mobile station generates a 32-bit random number to compose the RANDBS field of the base station challenge order. This order, which also contains the ESN, the IMSI, and the new SSD_A, is sent to the base station either on the access channel or on the reverse traffic channel. With the new SSD_A and the RANDBS, both base station and mobile station use the Auth_Signature Procedure (signature algorithm) whose inputs (RAND_CHALLENGE, ESN, AUTH_DATA, and SSD_AUTH) are set as follows:

RAND_CHALLENGE = RANDBS

ESN = ESN

AUTH_DATA = IMSI (24 most significant bits)

SSD_AUTH = new SSD_A

The output AUTH_SIGNATURE of the signature algorithm is assigned to the AUTHBS field of the base station confirmation order, sent by the base station to the mobile station. The mobile station compares its internally generated 18-bit signature (internal AUTHBS) with the one received in this message order (received AUTHBS). If they match, an SSD-update confirmation order is sent; otherwise, a SSD-update rejection is transmitted. Note that none of the strategic information, namely, SSD and A-key, appears in the radio messages.

5.12.2 Mobile Station Registrations

A registration attempt occurs when the mobile station sends a registration message on the access channel. The fields in the registration message relevant to the authentication process are RANDC (eight most significant bits of RAND), COUNT, ESN, IMSI, and AUTHR. The values of RAND and COUNT are held in the mobile station memory; ESN and IMSI identify the terminal; and AUTHR is the result (AUTH_SIGNATURE) of the Auth_Signature Procedure (signature algorithm) whose inputs (RAND_CHALLENGE, ESN, AUTH_DATA, and SSD_AUTH) are set as follows:

RAND_CHALLENGE = RAND

ESN = ESN

AUTH_DATA = IMSI (24 most significant bits)

SSD_AUTH = SSD_A

Upon receiving the registration message, the base station computes the value of AUTHR using the same parameters and the same algorithm as used by the mobile station. It then compares the received values of RANDC, COUNT, and AUTHR with those internally stored RANDC, COUNT, and internally generated AUTHR. If any of the comparisons fail, the base station assumes that the registration attempt is unsuccessful. It then may initiate the unique challenge-response procedure or the process of updating the SSD.

5.12.3 Mobile Station Originations

A mobile station origination occurs when the mobile station attempts to originate a call by sending an origination message on the access channel. The fields in the origination message relevant to the authentication process are RANDC (eight most significant bits of RAND), COUNT, CHARi, ESN, IMSI, and AUTHR. The values of RAND and COUNT are held in the mobile station memory, CHARi contains the dialed digits, ESN and IMSI identify the terminal, and AUTHR is the result (AUTH_SIGNATURE) of the Auth_Signature Procedure (signature algorithm) whose inputs (RAND_CHALLENGE, ESN, AUTH_DATA, and SSD_AUTH) are set as follows:

RAND_CHALLENGE = RAND

ESN = ESN

AUTH_DATA = last six digits in the CHARi fields or IMSI (24 most significant bits) if fewer than six digits are included in the message

SSD_AUTH = SSD_A

Upon receiving the origination message, the base station computes the value of AUTHR using the same parameters and the same algorithm as used by the mobile station. It then compares the received values of RANDC, COUNT, and AUTHR with those internally stored RANDC, COUNT, and internally generated AUTHR. If any of the comparisons fail, the base station may deny service and may initiate the unique challenge-response procedure or the process of updating the SSD. If the comparisons are successful, the base station may start the channel assignment procedure. And after channel assignment, a parameter update order on the forward traffic channel may be issued to update the content of COUNT in the mobile station.

5.12.4 Mobile Station Terminations

A mobile station termination occurs when the mobile station responds to a page by sending a page response message on the access channel. The fields in the page response message relevant to the authentication process are RANDC (eight most significant bits of RAND), COUNT, ESN, IMSI, and AUTHR. The values of RAND and COUNT are held in the mobile station memory, ESN and IMSI identify the terminal, and AUTHR is the result (AUTH_SIGNATURE) of the Auth_Signature Procedure (signature algorithm) whose inputs (RAND_CHALLENGE, ESN, AUTH_DATA, and SSD_AUTH) are set as follows:

RAND_CHALLENGE = RAND

ESN = ESN

AUTH_DATA = IMSI (24 most significant bits)

SSD_AUTH = SSD_A

Upon receiving the page response message, the base station computes the value of AUTHR using the same parameters and the same algorithm as used by the mobile station. It then compares the received values of RANDC, COUNT, and AUTHR with those internally stored RANDC, COUNT, and internally generated AUTHR. If any of the comparisons fails, the base station may deny service and may initiate the unique challenge-response procedure or the process of updating the SSD. If the comparisons are successful, the base station may start the channel assignment procedure. And after channel assignment, a parameter update order on the forward traffic channel may be issued to update the content of COUNT in the mobile station.

5.12.5 Mobile Station Data Burst

A mobile station data burst may occur when the mobile station attempts to send a data burst message on the access channel. The fields in the data burst message relevant to the authentication process are RANDC (eight most significant bits of RAND), COUNT, CHARi, ESN, IMSI, BURST_TYPE, and AUTHR. The values of RAND and COUNT are held in the mobile station memory, CHARi contains the dialed digits, ESN and IMSI identify the terminal, BURST_TYPE directs the burst procedure, and AUTHR is the result (AUTH_SIGNATURE) of the Auth_Signature Procedure (signature algorithm) whose inputs (RAND_CHALLENGE, ESN, AUTH_DATA, and SSD_AUTH) are set as follows:

RAND_CHALLENGE = RAND

ESN = ESN

AUTH_DATA = IMSI (24 most significant bits) partially or totally replaced by up to six 4-bit digits in CHARi as indicated in BURST_TYPE

SSD_AUTH = SSD_A

Upon receiving the page response message, the base station computes the value of AUTHR using the same parameters and the same algorithm as used by the mobile station. It then compares the received values of RANDC, COUNT, and AUTHR with those internally stored RANDC, COUNT, and internally generated AUTHR. If any of the comparisons fails, the base station may deny service and may initiate the unique challenge-response procedure or the process of updating the SSD. If the comparisons are successful, the base station processes the message.

5.12.6 Unique Challenge-Response Procedure

The unique challenge-response procedure may be initiated by the base station upon failure of one of the following procedures: mobile station registration, mobile station originations, mobile station terminations, and mobile station data burst. The unique challenge-response procedure is carried out either on the forward traffic channel and on reverse traffic channel, if a call is in progress, or on the paging channel and on the access channel, otherwise. It is initiated by the base station with the transmission of the authentication challenge message to the mobile station. The field in the authentication challenge message relevant to the authentication procedure is the RANDU, which contains a 24-bit quantity randomly generated by the base station. Upon receiving the authentication challenge message, the mobile station carries out the Auth_Signature Procedure (signature algorithm) whose inputs (RAND_CHALLENGE, ESN, AUTH_DATA, and SSD_AUTH) are set as follows:

RAND_CHALLENGE = RANDU (24 bits) concatenated with IMSI (eight least significant bits)

ESN = ESN

AUTH_DATA = IMSI (24 most significant bits)

SSD_AUTH = SSD_A

The 18-bit output AUTH_SIGNATURE of the Auth_Signature Procedure is assigned to the AUTHU field of the authentication challenge response message, which is sent by the mobile station to the base station. Upon receiving the authentication challenge response message, the base station computes the value of AUTHU using the same parameters and the same algorithm as used by the mobile station. It then compares the received value of AUTHU with the internally generated AUTHU. If the comparison fails, the base station may deny service (further attempts, continuation of a call) to the mobile station or initiate the process of updating the SSD.

5.13 Message Encryption

Encryption may be applied to certain fields of selected messages running on the traffic channels. Message encryption can be carried out only after authentication is performed. It is controlled individually for each call. The ENCRYPTION field is present in all the reverse traffic channel messages. It indicates the encryption mode (encryption disabled or encrypt call control messages) prevalent when the message was created. The first message to indicate the initial encryption mode for the call is the channel assignment message and such information is conveyed in its ENCRYPT_MODE field. The following messages, with the ENCRYPT_MODE field conveniently set, can turn off message encryption:

- Extended Handoff Direction Message
- Handoff Direction Message
- Message Encryption Mode Order

After channel assignment, these same messages, with the ENCRYPT_MODE field conveniently set, can turn on message encryption. The control messages on the analog control channels can also be encrypted. In this case, the indication to enable or not encryption is conveyed in the MEM field of the analog handoff direction message.

5.14 Voice Privacy

Voice privacy is provided upon request. It is applied to information conveyed on the traffic channels only and after authentication is performed. It consists in assigning a private long code mask for PN spreading to replace the long code mask with which all calls are initiated. The request may occur during call setup or during traffic channel operation. In the first case, the voice privacy request is indicated in the PM field of the origination message or in the same field of the page response message sent by the mobile station. In the second case, the long code transition order is used either by the base station or by the mobile station. Two other messages, namely, extended handoff direction message and handoff direction message, sent by the base station, can implement voice privacy by indicating in their PRIVATE_LCM field that privacy is or is not being requested.

5.15 Roaming

For roaming purposes, a base station can be seen as a member of a network and a network as a member of a system. A system contains one or more (usually a few) networks and a network one or more (usually several) base stations. A network is identified by NID, the network identification. In the same way, a system is identified by SID, the system identification. The mobile station has a list of (SID, NID) pairs. The mobile station is considered as a home, i.e., nonroaming, mobile station if the base station with which it communicates has an (SID, NID) pair coinciding with one of the (SID, NID) pairs within the mobile station's list. Therefore, a subscriber may move from one network or system to another and still be considered a home subscriber. The mobile station is considered an NID roamer if the base station with which it communicates has an (NID, SID) pair of which only the SID is within the (NID, SID) list of the mobile station. In a similar way, it is considered a foreign SID roamer if its base station has an SID that does not belong to any of the (NID, SID) pairs of the mobile station.

5.16 Handoff

Handoff in a CDMA system differs from that of other systems in many respects. As in other systems, the terminal plays an important role in the handoff process by performing measurements and reporting the respective results to the base station. In this sense, the handoff is mobile-assisted (MAHO, or mobile assisted handoff). But as opposed to other systems, where the participation of the mobile station in the handoff process is restricted to measurements and reports, in the CDMA system a handoff can be initiated by the mobile station. The initiation of the handoff process can take place when the mobile station, by the analyses of the measurements, recognizes that a handoff might be necessary. The measurements are performed over the pilot channels and the parameter of interest in this case is the ratio between the pilot energy per chip, E_c, and the total received spectral density I_0 (noise and signals). Many other peculiarities of the handoff process in the CDMA system are detailed next.

5.16.1 Types of Handoff

In a hybrid CDMA–analog system with dual-mode operation, five types of handoff may be supported: soft handoff, softer handoff, soft softer handoff, CDMA to CDMA hard handoff, and CDMA to Analog handoff.

In soft handoff, the mobile station maintains communication simultaneously with the old base station and with one or more new base stations, provided that all base stations have CDMA channels with identical frequency assignment. Note that soft handoff constitutes a means of diversity for both forward traffic channel and reverse traffic channel paths. It is supposed to occur in the vicinities of the boundary of the cell, where the signal is presumably weaker. On the forward link, diversity is accomplished by means of the transmission of the same information through different forward traffic channels each of which is supported by the respective base station involved in the handoff process. In the mobile station, each forward traffic channel is correlated at one of the fingers of its RAKE receiver, and the received signals then are combined to yield a single-output bit stream. On the reverse link, diversity is achieved by means of the reception of the reverse traffic channel by the base stations involved in the handoff process. Each base station demodulates the received signal and sends the results to the mobile switching center (MSC) on a frame-by-frame basis. The MSC then selects the best received frame, i.e., the frame with the lower error rate, again on a frame-by-frame basis, and this is sent to the source decoder. Note that, in the forward link case, diversity is achieved at the expense of extra forward traffic channels especially assigned for soft handoff purposes. In the reverse link case, on the other hand, no extra reverse traffic channel is necessary. However, the base stations involved must provide for equipment to demodulate the received signal.

In softer handoff, the mobile station maintains communication simultaneously with two or more sectors of the same base station and certainly within the same CDMA channel of that base station. As with soft handoff, softer handoff also constitutes a means of diversity and is intended to occur in the vicinity of the boundaries of the sector coverage area. On the forward link, the mobile station processes the signal in the same way as for the soft handoff case. On the reverse link, however, because only one base station is involved, the received signals are demodulated and selected within the same base station and only one frame is transmitted to the MSC.

In soft softer handoff, the mobile station maintains communication simultaneously with two or more sectors of the same base station and with one or more base stations (or their sectors). The remarks about the previous situations also apply for this situation.

In CDMA to CDMA hard handoff, also known as D-to-D handoff, the communication with the old base station through a given CDMA channel is discontinued and a new communication with a new base station and necessarily through another CDMA channel is established. This may occur if the mobile station moves between base stations supporting different frequency assignments or different frame offsets.

In CDMA to analog handoff, also known as D-to-A handoff, the communication with the old base station through a given CDMA channel is discontinued and a new communication with a new base station and necessarily through an analog channel is established.

5.16.2 Handoff and Pilot Sets

The monitoring of pilot channels is decisive in the handoff process. Within a given CDMA frequency assignment, the mobile searches for pilots to detect CDMA channels. Each pilot is associated with the forward traffic channels in the respective forward CDMA channel. A pilot with sufficient signal strength and not associated with any of the forward traffic channels assigned to the mobile station, when detected by it, causes the mobile station to send a pilot strength measurement message to the base station. The base station, in turn, can direct the mobile station to perform a handoff. There are four sets of pilots over which the searches are carried out: active set, candidate set, neighbor set, and remaining set.

The active set contains the pilots (as many as six pilots) associated with the forward traffic channels currently assigned to the mobile station. The candidate set contains the pilots (as many as five pilots) that are not in the active set, but have sufficient strength to permit successful demodulation of the forward traffic channels associated with it. The neighbor set contains the pilots (at least 20 pilots) that are neither in the active set nor in the candidate set but are likely to be candidates to be moved to one of these sets. The remaining set contains all the pilots in the system that are not currently in any one of the other sets.

5.16.3 Handoff Parameters

As can be inferred, the pilots play a very important role in the entire handoff process. The search for pilots to detect the presence of CDMA channels to establish the various pilot sets is first step in the handoff process. For each pilot set a search window (range of PN offsets) is specified by the base station. Within the search window, which is centered at the earliest-arriving usable multipath component of the pilot, the mobile station searches for usable multipath components, those the mobile station can employ for the demodulation of the associated forward traffic channels. The window sizes are specified in the system parameters message in the fields to be described next.

For pilots within both the active set and candidate set the search window is in the SRCH_WIN_A field. For pilots within the neighbor set the window size information is conveyed in the SRCH_WIN_N field. And for the remaining set the window size in given in the SRCH_WIN_R field. There are 16 possible window sizes, numbered from 0 to 15. The window size fields convey one of these numbers. The actual window size, given in PN chips, as a function of these numbers, is shown in Table 5.1. Note that Table 5.1 gives the total window size, such that, because the search window is centered at the earliest-arriving usable pilot component, the search is carried out within the range ±PN chips/2. The window size should be set according to the propagation

TABLE 5.1

Search Window Sizes

SRCH_WIN_A SRCH_WIN_N SRCH_WIN_R	Window Size (PN Chips)	SRCH_WIN_A SRCH_WIN_N SRCH_WIN_R	Window Size (PN Chips)
0	4	8	60
1	6	9	80
2	8	10	100
3	10	11	130
4	14	12	160
5	20	13	226
6	28	14	320
7	40	15	452

conditions. It should be large enough to detect all usable multipaths and as small as possible to speed the search process.

According to the ratio between the pilot energy per chip E_c and the total received spectral density I_0 (noise and signals) accumulated within the search windows for each pilot in the pilot sets, these pilots will move from one set to another. Actions concerning the handoff process are taken, based on the parameters specified in the fields T_ADD (pilot detection threshold), T_DROP (pilot drop threshold), T_COMP (comparison threshold), T_TDROP (drop timer value) of some messages. T_ADD, T_DROP, and T_COMP are related to signal strength, and T_TDROP is associated with the time the signal remains under a certain level. T_ADD and T_DROP are given in units of -0.5 dB E_c/I_0; i.e., the threshold value is given by $-0.5 \times$ T_ADD. T_COMP is given in units of 0.5 dB; i.e., the threshold is given by $0.5 \times$ T_COMP. T_TDROP refers to an expiration time of a timer whose enabling is triggered whenever the strength of any pilot in the active set or in the candidate set drops below the value in T_DROP. The timer is considered expired within 10% of the expiration time values shown in Table 5.2. A handoff drop timer is maintained

TABLE 5.2

Handoff Expiration Time

Drop Timer Value	Expiration Time (s)	Drop Timer Value	Expiration Time(s)
0	0.1	8	27
1	1	9	39
2	2	10	55
3	4	11	79
4	6	12	112
5	9	13	159
6	13	14	225
7	19	15	319

for each pilot in the active set and in the candidate set. These parameters are further explained later in this section.

5.16.4 Handoff Messages

Five messages running on the forward traffic channel and two on the reverse traffic channel are associated with the handoff process. For the forward traffic channel:

- Pilot Measurement Request Order
- Handoff Direction Message
- Analog Handoff Direction Message
- Neighbor List Update Message
- Extended Handoff Direction Message

For the reverse traffic channel:

- Pilot Strength Measurement Message
- Handoff Completion Message

The pilot measurement request order, sent by the base station, causes the mobile station to send, within a certain time (0.2 s), a pilot measurement message.

The handoff direction message, sent by the base station, causes the mobile station to update the active set, the candidate set, and the neighbor set. It also causes the mobile station to discontinue the forward traffic channels not associated with pilots not listed in the message, to change the frame offset as specified in the message, and to use the long code mask as specified in the message. The mobile station may encrypt some fields of the message if specified, and perform soft handoff or CDMA to CDMA hard handoff as required. It also stores the values of the fields SRCH_WIN_A, T_ADD, T_DROP, T_COMP, T_TDROP.

The analog handoff direction message, sent by the base station, directs the mobile to perform a CDMA to analog handoff. If the mobile station has narrow analog capability (channel with one third of the conventional analog channel), a narrow analog channel may be specified.

The neighbor list updated message, sent by the base station, causes the mobile station to update the neighbor set with the pilots specified in the message. If the addition of a pilot to the set exceeds its maximum capacity, the pilots remaining longer in the set are replaced.

The extended handoff direction message, sent by the base station, causes the mobile station to perform actions as for the handoff direction message case.

In contrast to the handoff direction message, the extended handoff direction message includes the HARD_INCLUDED field, which indicates whether or not the mobile station should change the parameters relative to the hard handoff (FRAME_OFFSET, PRIVATE_LCM, ENCRYPT_MODE, NOM_PWR, BAND_CLASS, CDMA_FREQ).

The pilot strength measurement message is sent by the mobile station in two situations: (1) as a response to the pilot measurement request order or (2) autonomously, which requires acknowledgment. The autonomous transmission of such a message is triggered by the following events:

- The strength of a pilot in the neighbor set or in the remaining set is found to exceed the value in T_ADD.
- The strength of a pilot in the candidate set exceeds that of a pilot in the active set by $0.5 \times$ T_COMP dB and a pilot strength measurement message conveying such information has not been sent since the last arrival of the handoff direction message or the extended handoff direction message.
- The handoff drop timer of a pilot in the active set has expired and a pilot strength measurement message conveying such information has not been sent since the last arrival of the handoff direction message or the extended handoff direction message.

The handoff completion message is sent by the mobile station as a response to the handoff direction message or extended handoff direction message, i.e., after the actions required by these messages have been completed.

5.16.5 Pilot Sets Updating

The updating of the various pilot sets is triggered by a series of events as summarized next.

Active Set. The active set, with a maximum size of six pilots, is initialized when the mobile station is first assigned a forward traffic channel, in which case it shall contain only the pilot associated with the assigned channel. The reception of the handoff direction message or of the extended handoff direction message triggers the mobile station to replace the pilots in the active set with those listed in the message. All pilots in the active set have their strength continuously monitored by the mobile station.

Candidate Set. The candidate set, with a maximum size of five pilots, is initialized to contain no pilots. The update of the candidate set occurs as follows.

- A pilot is moved from the neighbor set or from the remaining set to the candidate set if the strength of this pilot exceeds T_ADD.

- A pilot from the active set is added to the candidate set if the received handoff direction message or extended handoff direction message does not contain that pilot and its handoff drop timer has not expired.
- A pilot is moved from the candidate set to the active set if the received handoff direction message or extended handoff direction message contains that pilot.
- A pilot is moved from the candidate set to the neighbor set if its handoff drop timer expires.
- A pilot is removed from the candidate set if, by adding another pilot in the candidate set, its maximum size is exceeded. In such a case, the pilot chosen to be deleted is the one with its handoff drop timer closest to the expiration time. If more than one pilot is found in this condition or if no pilot has its handoff drop timer enabled, then the pilot with the lowest strength is deleted.

Neighbor Set. The neighbor set, with a minimum size of 20 pilots, is initialized to contain all the pilots specified in the last-received neighbor list message when the mobile is first assigned a forward traffic channel. An aging mechanism is employed by the mobile station to keep in the set those pilots that were most recently detected. In addition to the aging mechanism, the update of the neighbor set occurs as follows.

- A pilot from the candidate set is added to the neighbor set if its handoff drop timer expires.
- A pilot from the active set is added to the neighbor set if the received handoff direction message or extended handoff direction message does not contain that pilot and its handoff drop timer has not expired.
- A pilot from the candidate set is added to the neighbor set if this pilot has been deleted from the candidate set because the number of pilots in this set has exceeded its maximum allowable number at the candidate set updating process.
- A pilot is deleted from the neighbor set if its strength exceeds T_ADD.
- A pilot is deleted from the neighbor set if the received handoff direction message or extended handoff direction message contains that pilot.
- A pilot is removed from the neighbor set if, by adding another pilot in the neighbor set, its maximum size is exceeded. In this case, the pilot chosen to be deleted is the one that has remained longer in the set. If more than one pilot is found in this condition, then the pilot with the lowest strength is deleted.

5.17 Power Control

To achieve the best performance, CDMA technology requires equalization of the signal strengths of the mobile station arriving at the base station. Ideally, because the forward link operates coherently, only the reverse link, which operates incoherently, requires power control. The near–far phenomenon effect is more relevant in multipoint-to-point transmission (mobile stations to base station) than in point-to-multipoint transmission. In the first case (multipoint-to-point), because the mobile stations may be at different distances from the base the various signals arriving at the base will have different strengths: at the base station, the signal strength of a mobile station near a base station is equivalent to a number of mobile stations away from the base station. Therefore, if no power control is exercised, the near–far phenomenon will drastically affect system capacity. In the second case (point-to-multipoint), and in theory, the various signals transmitted by the base will reach a given mobile station with the same power loss, thus maintaining power proportionality. In practice, however, both reverse link and forward link require power control, the reverse link for the reasons already outlined, and the forward link to compensate for poor reception conditions encountered by the mobile station. TIA/EIA/IS-95 specifies detailed power control algorithms for the reverse link. For the forward link, on the other hand, only an exchange of information between base station and mobile station is stipulated; specific procedures, however, are left to individual implementations.

5.17.1 Reverse-Link Power Control

Two independent means of reverse-link power control are specified by the TIA/EIA/IS-95 standard: open-loop power control and closed-loop power control.

Open-loop power control is so called because it is a purely mobile-controlled operation (the base station is not involved). The power adjustment is based on the pilot signal level measured at the mobile station: the larger the received power at the terminal, suggesting proximity between base station and mobile station, the smaller the transmitted power from the terminal, and vice versa. Additional open-loop operation adjustments are carried out as the mobile station attempts to transmit on the access channel in the process known as probing. As already explained in this chapter, each probe on the access channel is carried out at increasing power levels until successful access is accomplished. The initial transmission on the reverse traffic channel, therefore, accumulates the additional power due to the access probes. Note that open-loop power control assumes reciprocity between reverse link and forward link, i.e., it assumes that both links experience correlated fading.

Closed-loop power control is so called because it involves both base station and mobile station in the power adjustment process. By monitoring the reverse-link quality, the base station commands the mobile station to adjust its output power to achieve the desired grade of service. The command is given through the power control bits (PCBs) sent over the forward traffic channel, as explained previously. Closed-loop power control encompasses an inner loop control and an outer loop control. Only inner loop power control is specified by EIA/TIA/IS-95. In inner loop control, the commands to increase or to diminish the output power of the mobile station are given based on the fact that a given ratio of energy per bit and total noise density, E_b/N_o, has been established as the threshold for the required performance. In outer loop control, the E_b/N_o level is adjusted to give the minimum acceptable frame error rate (FER). Therefore, the output of the outer loop control process constitutes the input for the inner loop control process, with both processes interacting dynamically to give the minimum output power of the mobile station for a minimum E_b/N_o to yield the desired FER.

Because of the large frequency separation between forward channels and reverse channels, forward and reverse links fade independently. Therefore, for an efficient power control mechanism, both open-loop power control and closed-loop power control must interact. Open-loop power control may be considered coarse-tuning and closed-loop power control fine-tuning of the overall reverse-link power adjustment process.

The temporal response of the mobile station to open-loop control is intentionally made nonlinear. If the mobile station perceives a sudden increase of the signal strength of a pilot, it immediately (within microseconds) responds with a proportionally reduced output power. However, if the opposite occurs (a sudden decrease in received signal strength), it responds with a slow (within milliseconds) but proportional increase of its output power. The rationale for this lies in the fact that, apart from fading, a higher received power is a better estimate of the average link loss. In addition, if an improvement in the radio path is found, a decrease of the output power of the mobile station is imperious, so that undue interference and consequent decrease in capacity are avoided. In the same way, a sudden worsening of the radio path followed by an immediate increase of the output power of the mobile station may cause undue interference, and consequent decrease in capacity, because such a worsening may be due to fading occurring only on the forward link. Note that this rule benefits the whole system performance to the detriment of the single user.

Regarding closed-loop power control, the temporal response of the mobile station is immediate. The 800-bit/s PCB rate implies that once at each 1.25 ms an action is taken with respect to the output power of the mobile station. A 0 in PCB causes a 1-dB increase and a 1 causes a 1-dB decrease in the overall power output. During soft handoff, however, there may be conflicting commands with different base stations involved in the handoff instructing the mobile station to act differently. The mobile station will power up if all the base stations involved in the handoff command it to do so; if at least one base station commands a power-down, a power-down shall be done.

The overall power control formula is given by

$$\begin{aligned}\text{mean output power} = {} & \text{constant} \\ & + \text{open loop adjustment} \\ & + \text{closed loop adjustment (dBm)}\end{aligned} \tag{5.4}$$

The parameters considered in the Equation 5.4 are specified in the access parameters message and are obtained by the mobile station prior to transmitting. The *constant* in Equation 5.4 is given by the sum of the values of NOM_PWR (-8 dB $\leq$ NOM_PWR $\leq$ 7 dB, nominally 0 dB), INIT_PWR (-16 dB $\leq$ INIT_PWR $\leq$ 15 dB, nominally 0 dB), and -73. The figure -73 is obtained as follows.[1] The link budget equation, in decibels, for the reverse link can be simplistically written as

$$\begin{aligned}\text{base received SNR} = {} & \text{mobile transmitted power} \\ & - \text{propagation losses} \\ & - \text{reverse noise and interference}\end{aligned} \tag{5.5}$$

For the forward link:

$$\begin{aligned}\text{mobile received SNR} = {} & \text{base transmitted power} \\ & - \text{propagation losses} \\ & - \text{forward noise and interference}\end{aligned} \tag{5.6}$$

Assuming that the propagation losses are the same for forward and reverse links, then manipulating Equations 5.5 and 5.6 leads to

$$\begin{aligned}\text{mobile transmitted power} = {} & (\text{base received SNR} \\ & + \text{forward and reverse noise and interference} \\ & + \text{base transmitted power}) \\ & - \text{mobile received power}\end{aligned} \tag{5.7}$$

For base-received SNR = -13 dB, forward and reverse noise and interference = -100dBm, and base transmitted power = 40 dBm (10 W), the terms between parentheses in Equation 5.7 yield -73 dBm.

NOM_PWR is the adjustment to give the correct received power at the base station if INIT_PWR is 0 dB. And INIT_PWR provides the adjustment to the first access channel probe with the aim of providing at the base station a received power somewhat less than the required signal power, which is a conservative measure to compensate for the partially decorrelated path losses between forward and reverse links. The open-loop adjustment in Equation 5.4 is given by the sum of the mean input power and the accumulated increase in power due to the channel access probes. The increase power step is given in PWR_STEP (0 dB $\leq$ PWR_STEP $\leq$ 7 dB). If n channel access probes are carried out, then $(n-1) \times$ PWR_STEP is the total power increase at the end of the sequence. The closed-loop adjustment in Equation 5.4 is given by the net value of the increase and decrease in power given by the PCBs. If n_0 is the number of bits 0 received and n_1 is the number of bits 1 received in the PCBs, then the mentioned net value is $(n_0 - n_1)$. The final open-loop and closed-loop equation is

$$\begin{aligned}\text{mean output power} = {} & -73 + \text{NOM\_PWR} + \text{INIT\_PWR} \\ & - \text{mean input power} + (n-1)\text{PWR\_STEP} \\ & + (n_0 - n_1) \qquad \text{(dBm)}\end{aligned} \tag{5.8}$$

5.17.2 Forward-Link Power Control

As mentioned previously, specific procedures for forward-link power control algorithms are not defined by EIA/TIA/IS-95. And these are left to individual implementations. Some directions, on the other hand, are to be followed. They vary for the Rate Set 1 (9.6, 4.8, 2.4, and 1.2 kbit/s) and for the Rate Set 2 (14.4, 7.2, 3.6, and 1.8 kbit/s) configurations. For both configurations, the quality of the reverse link is monitored by the mobile station and its condition is reported back to the base station. In particular, for the Rate Set 1 the quality of the forward link is assessed by its FER. The FER is then reported to the base station in the power measurement report message either periodically (at each 250 ms or less) or if it exceeds a certain threshold. Based on this report, the base station increases its power by a certain amount, nominally 0.5 dB, limited to ±6 dB about the nominal power. For the Rate Set 2, the forward-link quality is assessed by means of the occurrence of a frame erasure. A frame erasure causes the mobile station to appropriately set an erasure indicator bit in the corresponding frame of the reverse link for the pertinent action by the base station. Note that the power tracking performance of the Rate Set 2 configuration (once at each 20 ms, the duration of a frame) is much faster than that of the Rate Set 1 configuration.

5.18 Call Procedures

This section illustrates some simplified call procedures. Here, the abbreviations MS, for Mobile Station, and BS, for Base Station, are used.

5.18.1 Mobile Station Origination

The following sequence of events occurs in an MS call origination.

1. MS detects user-originated call.
2. MS sends the origination message, using the access channel and the access procedure.
3. BS detects the origination message and performs the authentication verification process. (The steps concerning the authentication procedure are not detailed here.) BS sets up the forward traffic channel.
4. BS sends an acknowledgment order on the paging channel.
5. BS sends a sequence of 1s and 0s (the null traffic channel data) on the designated forward traffic channel.
6. BS sends the channel assignment message containing the channel information (ESN, code channel, CDMA frequency assignment, frame offset). This is carried out on the paging channel.
7. MS detects the channel assignment message. It tunes to the assigned forward traffic channel and detects the null traffic channel data.
8. MS sends the traffic channel preamble using the reverse traffic channel.
9. BS receives the traffic channel preamble. It then sends an acknowledgment order using the forward traffic channel.
10. MS detects the acknowledgment order. It then sends the null traffic channel data using the forward traffic channel.
11. BS detects the null data and sends the service connect message using the forward traffic channel.
12. MS receives the service connect message. If the MS can fulfill the service requirements specified in this message, it then sends a service connect completion message using the forward traffic channel.
13. BS detects the service connect completion message and sends the alert with information message using the forward traffic channel. This message contains information on the ring-back tone.
14. MS detects the alert with information message and applies the ring-back tone, as required.
15. BS is informed about the off-hook condition of the called party. It sends the alert with information message on forward traffic channel. This message commands the ring-back tone off.
16. MS detects the alert with information message and removes the ring-back tone.
17. Bidirectional conversation begins.

5.18.2 Mobile Station Termination

The following sequence of events occurs in an MS call termination.

1. BS detects user-terminated call. It uses the paging channel to send a page message or a slotted page message, with information that identifies the user.
2. MS detects the (slotted) page message. It sends a page response message using the access channel and the access procedure.
3. BS performs the authentication procedure. It sets up the forward traffic channel.
4. BS sends a sequence of 1s and 0s (the null traffic channel data) on the designated forward traffic channel.
5. BS sends the channel assignment message containing the channel information (ESN, code channel, CDMA frequency assignment, frame offset). This is carried out on the paging channel.
6. MS detects the channel assignment message. It tunes to the assigned forward traffic channel and detects the null traffic channel data.
7. MS sends the traffic channel preamble using the reverse traffic channel.
8. BS receives the traffic channel preamble. It then sends an acknowledgment order using the forward traffic channel.
9. MS detects the acknowledgment order. It then sends the null traffic channel data using the forward traffic channel.
10. BS detects the null data and sends the service connect message using the forward traffic channel.
11. MS receives the service connect message. If the MS can fulfill the service requirements specified in this message, it then sends a service connect completion message using the forward traffic channel.
12. BS detects the service connect completion message and sends the alert with information message using the forward traffic channel. This message contains information on the ring tone.
13. MS detects the alert with information message and applies the ring tone, as required.
14. MS detects hook-off condition of the user (user has answered). It stops ringing and sends a connect order using the forward traffic channel.
15. BS detects the connect order.
16. Bidirectional conversation begins.

5.18.3 Call Disconnect

A call disconnect procedure can be initiated by the MS or BS. The following sequence of events occurs in a call procedure. The communication between MS and BS is carried out on the traffic channels.

1. MS (or BS) sends a release order message.
2. BS (or MS) receives the release order message and sends a release order message.
3. MS (or BS) receives the release order.
4. MS (or BS) enters the system determination substate of the initialization state.

5.19 EIA/TIA/IS-95B

The capacity of CDMA systems is rather different for forward and reverse links. It was initially believed that the reverse link would be the capacity-limiting link. This belief was supported by the fact that the reverse link operates asynchronously and the multiple users accessing the base station constitute multiple sources of interference. As for the forward link, because of its synchronous mode of operation, a small number of interferers exist that affect the link performance. These assumptions turned out to be incorrect and, in practice, the forward link constitutes the capacity-limiting link. Three main reasons can be cited for this.

1. The interference on the reverse link is provoked by a large number of low-power transmitters (mobile stations). According to the law of large numbers, this tends to be statistically stable. The interference on the forward link is provoked by a small number of high-power transmitters (base stations). This is critical at the borders of the cells where mobile stations may have equivalent radio paths to their serving base as well as to interfering bases.
2. The use of soft handoff may alleviate the interference at the cell borders, but at the expense of the use of additional forward traffic channels in the involved base stations.
3. The use of fast and accurate power control mechanisms for the reverse link enhanced reverse-link capacity. Forward-link power control, on the other hand, did not enjoy the same advantage, the slow power control mechanisms greatly compromising forward-link capacity.

Several innovations have been included in the evolutionary revisions of cdmaOne. In particular, in EIA/TIA/IS-95B these improvements have brought the forward-link capacity to parity with the reverse-link capacity. This section describes the main innovations implemented in EIA/TIA/IS-95B.

5.19.1 Increase in the Transmission Rate

In EIA/TIA/IS-95B, each traffic channel is composed of one fundamental traffic channel and as many as seven supplemental traffic channels. The maximum number of fundamental traffic channels and supplemental traffic channels is still limited by the total number of Walsh sequences (64) minus the number of the remaining channels, namely, pilot channel, sync channel, and paging channels. Rate Set 1 and Rate Set 2 are kept, but they now include eight channel multiplex options each. For Rate Set 1 these multiplex options are numbered 1, 3, 5, . . . , 15, and for Rate Set 2 the multiplex options are numbered 2, 4, 6, . . . , 16. The increase in the transmission rate is achieved by aggregating up to seven supplemental traffic channels to the fundamental traffic channel.

In Rate Set 1, the transmission rate for each channel is set to 9.6 kbit/s, whereas this is 14.4 kbit/s in Rate Set 2. Table 5.3 illustrates the possible total transmission rates for the various multiplex options.

The multiplex options concern both forward link and reverse link. Primary data (voice) and secondary data (fax, image, etc.) may be transmitted through one fundamental traffic channel aggregated with supplemental traffic channels. Signaling data, on the other hand, use only the fundamental traffic channel.

TABLE 5.3

Possible Transmission Rates for the Various Multiplex Options

Multiplex Option (Rate Set 1)	Multiplex Option (Rate Set 2)	Supplemental Channels	Rate Set 1 (kbit/s)	Rate Set 2 (kbit/s)
1	2	0	1.2, 2.4 4.8, 9.6	1.8, 3.6 7.2, 14.4
3	4	1	19.2	28.8
5	6	2	28.8	43.2
7	8	3	38.4	57.6
9	10	4	48.0	72.0
11	12	5	57.6	86.4
13	14	6	67.2	100.8
15	16	7	76.8	115.2

In the reverse link, the fundamental traffic channel and the supplemental traffic channels are transmitted with the same frame offset. Each carrier for each reverse supplemental traffic channel is shifted by a fixed phase with respect to the carrier of the fundamental traffic channel.

5.19.2 Power Control

A new power control algorithm has been implemented in EIA/TIA/IS-95B. In this new algorithm an interference correction factor as well as the inclusion of supplemental traffic channels are taken into account. The new formula for the power control with all factors included is

$$\begin{aligned}\text{mean output power} = {} & -73 + \text{NOM\_PWR} + \text{INIT\_PWR} \\ & - 16 \times \text{NOM\_PWR\_EXT} \\ & + \text{interference correction} \\ & + 10 \times \log(1 + \text{reverse\_supplemental\_channels}) \\ & - \text{mean input power} + (n-1)\text{PWR\_STEP} \\ & + (n_0 - n_1) \quad (\text{dBm})\end{aligned} \tag{5.9}$$

The parameter NOM_PWR_EXT is equal to zero for Band Class 0 (800 MHz) mobile stations. The *interference correction* takes into account the E_c/I_0 level of the active pilot, measured over a span of the last 500 ms. It is given as

$$\text{interference correction} = \min\{\max\{-7 - \text{ECIO}, 0\}, 7\} \tag{5.10}$$

where ECIO is E_c/I_0 level (dB) of the active pilot. Note from Equation 5.10 that no correction is implemented for E_c/I_0 within the range from 0 dB down to −7 dB. From −7 dB down to −14 dB an increase of 1 dB per −1 dB decrease in the E_c/I_0 is implemented. Below −14 dB a constant correction of 7 dB is implemented.

Note from Equation 5.10 that an increase in the power is performed as the number of supplemental traffic channel increases.

5.19.3 Soft Handoff Criteria

The main innovation in the soft handoff criteria is the definition of new threshold levels to include or remove the pilots from the various lists. In addition to the well-known parameters, namely, T_ADD, T_COMP, T_DROP, and T_TDROP, three new parameters are defined: ADD_INTERCEPT, DROP_INTERCEPT, and SOFT_SLOPE.

The inclusion of a pilot into the candidate set follows the same procedure as described earlier, i.e., if the E_c/I_0 level of a pilot exceeds T_ADD, this pilot is included within the candidate set. As opposed to EIA/TIA/IS-95A, in which the inclusion or removal of pilots into the active set is based on fixed thresholds, in EIA/TIA/IS-95B such an inclusion is based on a dynamic threshold, which varies depending on the E_c/I_0 levels of the active pilots. An inclusion is allowed if

$$10 \log \mathrm{PS} \geq \frac{\mathrm{ADD\_INTERCEPT}}{2} + \frac{\mathrm{SOFT\_SLOPE}}{8} 10 \log \sum \mathrm{PS}_i$$

where PS is the E_c/I_0 level of the candidate pilot and PS_i is the E_c/I_0 level of the active pilot i. Similarly, a pilot is removed from the active set if

$$10 \log \mathrm{PS} \leq \frac{\mathrm{DROP\_INTERCEPT}}{2} + \frac{\mathrm{SOFT\_SLOPE}}{8} 10 \log \sum \mathrm{PS}_i$$

Note that the admission criterion is now more stringent because it is no longer based on the minimum signal level, but on the mean signal level of the active pilots. Therefore, in EIA/TIA/IS-95B case, the pilots in the active set present signal levels higher than those of the EIA/TIA/IS-95A case.

It can be said that the new soft handoff criteria render the system more flexible with the thresholds for inclusion and removal of pilots set dynamically. Therefore, unnecessary handoffs are avoided leading to an increase in the system capacity.

5.19.4 Hard Handoff

The hard handoff process here concerns the D-to-D handoff, in which a handoff between CDMA carriers occurs. In EIA/TIA/IS-95A, a D-to-D handoff is carried out via the handoff direction message and the process is known as break-before-make. In break-before-make, the call is interrupted before the acquisition of the new carrier. If the acquisition is successful, the call continues; otherwise, the call is interrupted. In EIA/TIA/IS-95B, this uncomfortable situation is circumvented by means of the inclusion of several innovations.

Two new pilot lists have been created:

1. Candidate Frequency Neighbor, which is a list containing the pilots of the CDMA candidate carrier.
2. Candidate Frequency Search Set, which is a list containing the pilots of the CDMA candidate carrier to be searched under the command of the base station.

Four new messages have also been included:

1. Candidate Frequency Search Request Message (forward direction). This message is used to command the mobile station to perform either a periodic or single search for pilots of the candidate carrier.
2. Candidate Frequency Response Message (reverse direction). This message is used by the base station to determine an appropriate period for the search for pilots of the candidate carrier.
3. Candidate Frequency Control Message (forward direction). This message is used to direct the mobile station to perform a single search, to initiate a periodic search, or to end a periodic search for the pilot of the candidate carrier.
4. Candidate Frequency Search Report Message (reverse direction). This message contains the measures of the pilot signal strengths of the candidate carrier.

Several hard handoff scenarios can be described, depending on the mode with which the search is performed and on the exchange of messages. In all these scenarios, the main purpose is to avoid the break-before-make situation. For this purpose, the configuration of the current pilot set lists is saved, the D-to-D handoff is tried, and if it is not successful the saved configuration is restored and no handoff is performed.

5.19.5 Idle Handoff

EIA/TIA/IS-95B allows the execution of idle handoffs in situations not permitted by EIA/TIA/IS-95A. These situations concern the access procedures, as follows:

Access Handoff. Access handoff may occur in two situations: while the mobile station is waiting for an answer from the base station or before sending a response to the base station. The aim is to have the mobile station utilize a paging channel with a better signal level.

Access Probe Handoff. Access probe handoff may occur when the mobile station declares loss of the paging channel during an access attempt. Therefore, the mobile station is permitted to carry out a handoff between access probes.

5.19.6 Conclusions

A number of innovations have been implemented in the evolved version of EIA/TIA/IS-95A, the EIA/TIA/IS-95B, to increase capacity and provide for higher data rate transmissions.

5.20 Summary

TIA/EIA/IS-95 supports a direct sequence spread spectrum technology with 1.25-MHz band duplex channels. Coexistence of analog and digital systems implies that dual-mode mobile stations are able to place and receive calls in any system and, conversely, all systems are able to place and receive calls from any mobile station. Handoff operations in such a scenario require some attention. A mobile station may initiate a call in the CDMA system and, while the call is still in progress, it may migrate to the analog system if required. A number of innovations have been introduced in the CDMA system as compared with earlier cellular systems. Soft handoff is certainly a great novelty. In soft handoff, handoff from one base station to another occurs in a smooth manner. Another innovation introduced in the CDMA system is the use of GPS receivers at the base stations. GPSs are utilized so that base stations are synchronized, a feature vital to the soft handoff operation. Vocoders at variable rates are specified to accommodate different voice activities aiming at controlling interference levels, thus increasing system capacity. Sophisticated power control mechanisms are used so that the full benefit of the spread spectrum technique is achieved. The first CDMA systems were employed under the TIA/EIA/IS-95A specifications. The *A* version of the specifications evolved to TIA/EIA/IS-95B, in which new features related to higher data rate transmissions, soft handoff algorithms, and power control techniques have been introduced.

References

1. Lee, J. S. and Miller, L. E., *CDMA Systems Engineering Handbook*, Artech House, Boston, 1998.
2. Mobile Station–Base Station Compatibility Standard for Dual-Mode Wideband Spread Spectrum Cellular System (TIA/EIA-95-A), 1995.
3. Mobile Station–Base Station Compatibility Standard for Dual-Mode Spread Spectrum Cellular System (TIA/EIA-95-B), 1988.

Part III

Wireless Data

6

Wireless Data Technology

6.1 Introduction

The Internet is without doubt one of the major driving forces behind new developments in the area of telecommunications networks. Wireless communication has gained universal acceptance, with the number of wireless subscriptions already exceeding the number of fixed lines in many countries. The merger of the two technologies—Internet and wireless—is an obvious step and so is the increase of the demand for mobile access to Internet applications. To support efficiently the several types of traffic generated by the great variety of applications, the resultant network must provide for packet data services.

Within this framework, three data technologies applied to wireless networks appear as alternatives to be used for packet applications in wireless systems: General Packet Radio Service (GPRS), TIA/EIA/IS–95B, and High Data Rate (HDR). This chapter describes the basic architectures of GPRS and HDR and outlines throughputs at the application level that may be achieved by these technologies. TIA/EIA/IS-95B has already been explored in Chapter 5, and only its achievable data rates are repeated in this chapter.

6.2 General Packet Radio Service

The discussions about GPRS started in the early 1990s with the focus on applications related to road transport telematics and financial services. The initial work aimed at the GSM community. More recently, however, with the widespread use of end-user applications, such as Web browsing and e-mail, the Internet has become the dominant driving force in the standardization of GPRS, and technologies other than GSM also incorporate GPRS. Based on packet-switched principles, GPRS provides optimized access to the Internet while reusing to a large extent existing wireless infrastructure. Connection to the Public Land Mobile Network (PLMN) is based on the Internet protocol (IP), and, on the air interface, the resources are assigned to the terminals on a per-IP packet basis.

The introduction of GPRS within the mobile radio network allows the following:

- Both circuit-switched services and packet-switched services
- Better use of radio resources
- Efficient setup time and access time
- Connection to other packet data networks
- Services based on quality of service (QoS) requirements
- Volume-based charging
- Point-to-point and point-to-multipoint services

Five different parameters characterize the QoS profiles, namely, precedence, reliability, maximum bit rate, mean bit rate, and delay for packets of 128 octets. Precedence concerns the priority for transmission, with the priority ranked as high, normal, or low. Reliability concerns the packet loss probability, which can be fixed according to the needs (e.g., 10^{-9}, 10^{-4}, 10^{-2}, etc.). Maximum bit rate concerns the maximum transmission rate, with the transmission rate ranging from 8 kbit/s to 2 Mbit/s. Mean bit rate concerns the mean transmission rate, which is specified to be within the range 0.22 bit/s to 111 kbit/s. Delay for packets of 128 octets concerns the maximum permitted delay, with the delay specified according to the traffic classes. Four traffic classes are defined: conversational (Class 1), streaming (Class 2), interactive (Class 3), and background (Class 4). They are specified in terms of the maximum mean delay and maximum delay for 95% of the time. Specifically, the maximum mean delay and maximum delay, in seconds, are stipulated, respectively, as 0.5 and 1.5 for Class 1 traffic, 5 and 25 for Class 2 traffic, 50 and 250 for Class 3 traffic, and best effort for Class 4.

For a detailed specification of GPRS, the reader is referred to References 1 through 9.

6.2.1 Network Architecture

The GPRS network architecture is heavily based on the GSM architecture which, in turn, has inspired the ANSI-41 architecture. As such, GPRS can be seen as an extension of GSM or, equivalently, of ANSI-41. In addition to the

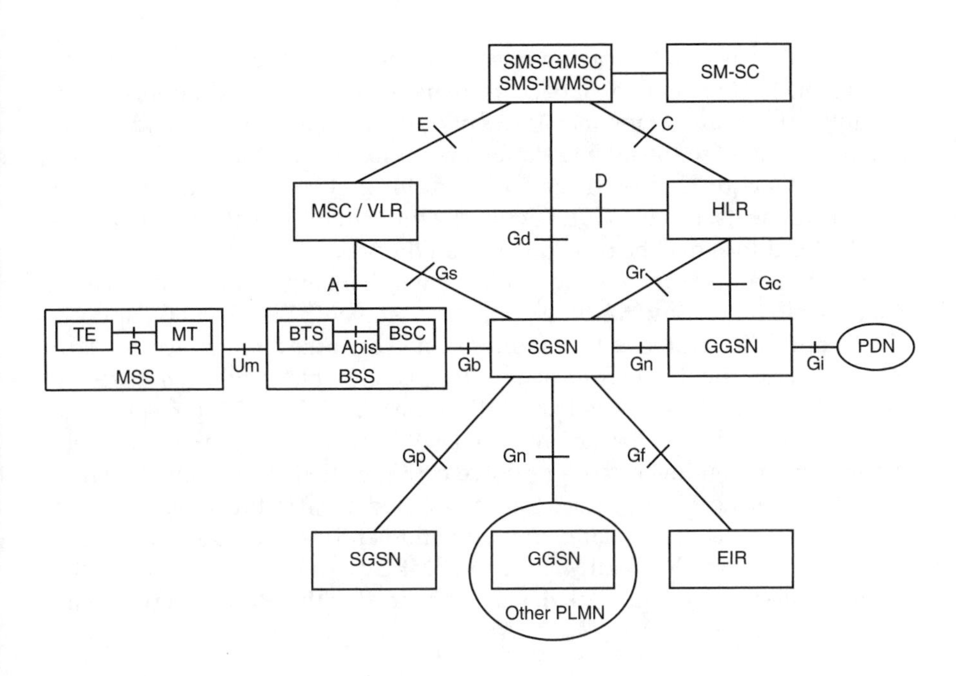

FIGURE 6.1
GPRS architecture.

well-known elements of these standards, a new logical network node, named GPRS support node (GSN), is introduced to provide independent packet routing and transfer within the PLMN. One such node is the gateway GPRS support node (GGSN) and the other is the serving GPRS support node (SGSN). The remaining elements are those as already described in Chapter 4: mobile station subsystem (MSS), which comprises the terminal equipment (TE), mobile terminal (MT), terminal adapter (TA), and subscriber identity module (SIM); base station subsystem (BSS), which comprises the base transceiver station (BTS) and base station controller; mobile switching center (MSC); visitor location register (VLR); home location register (HLR); equipment identity register (EIR); gateway MSC (GMSC); and short message service center (SMSC). Figure 6.1 depicts the GPRS network architecture with its respective blocks and interfaces (TA and SIM are not shown here). Next described are the GPRS-relevant functionalities. (For simplicity MSS shall be referred to as MS.)

The GGSN acts as a logical interface with the external packet data network (PDN), which includes the IP PDN or X.25/X.75 packet-switched PDN. For an external IP network GGSN is seen as an ordinary IP router serving all IP addresses of the MSs. In addition, it may include firewall and packet-filtering mechanisms and it provides means of assigning the correct SGSN to the MS depending on its location.

The SGSN acts as a logical interface with the radio access network, being responsible for the delivery of packets to the correct BSS. In addition, ciphering, authentication, session management, mobility management, and logical link management to the mobile station are performed by the SGSN.

The MS is equipped with the GPRS protocol stack providing means of connecting the user to the GPRS network. Both circuit-switched and packet-switched facilities may be provided within the MS.

The BSS, as already mentioned, contains two elements, namely, the base station controller (BSC) and the base station transceiver (BTS). The BSC supports all relevant GPRS protocols for communication over the air interface. In addition, it includes the packet control unit (PCU) logical block, which may reside physically within the BSC itself, the BTS, or the SGSN. The PCU is responsible for tasks concerning the packet-switched calls, including setup, supervision, disconnection, handover, radio resource configuration, and channel assignment. As far as the GPRS protocols are concerned, the BTS functions as a relay station performing tasks related to modulation and demodulation.

The MSC, GMSC, VLR, HLR, EIR, and SMSC are functional entities of the initial circuit-switched network that are enhanced with GPRS subscriber data and routing information.

6.2.2 Protocol Architecture

Figure 6.2 shows the GPRS protocol stack to the application layer according to the International Organization for Standardization/Open Systems Interconnection (ISO/OSI) Reference Model. It can be seen that GPRS supports

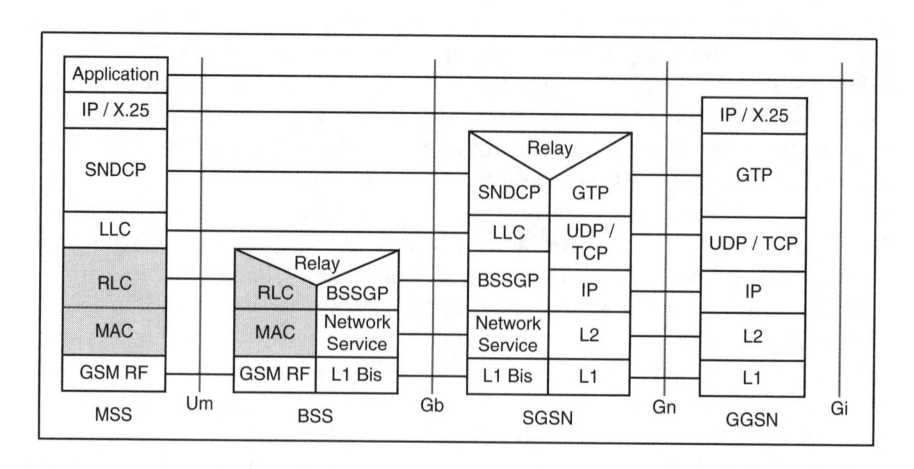

FIGURE 6.2
GPRS protocol stack.

applications based on IP and X.25 as well as other data protocols. Within the GPRS network, say, between two GSNs nodes (e.g., GGSN, SGSN) the GPRS tunnel protocol (GTP) is used. The GTP encapsulates the protocol data units (PDUs) at the originating GSN and decapsulates them at the destination GSN. The GTP PDU is then routed over the IP-based GPRS backbone network using either the transmission control protocol (TCP), for X.25-based applications, or the user data protocol (UDP), for IP applications. The entire process, as described, is known as *GPRS tunneling*. Between the serving SGSN and the MS, the Subnetwork Dependent Convergence Protocol (SNDCP) is used to map the network layer characteristics onto the underlying network. It provides functionalities such as multiplexing of network layer messages onto a single virtual logical connection, encryption, segmentation, and compression. The logical link control (LLC) provides the logical link between the MS and the SGSN and performs tasks such as ciphering, flow control, sequence control, and error control. The LLC is used by the SNDCP to transfer network layer PDUs; it is also used by the SMS protocol to transfer SMS messages; and it gives support to GPRS mobility management to transfer control data. The radio link control/multiple access control (RLC/MAC) layer is located within the PCU. It provides means for the transfer of LLC PDUs using a shared medium between multiple MSs and the network, with this medium the GPRS radio interface. In particular, the RLC layer is responsible for the segmentation and reassembling of LLC PDUs. It may operate in two modes, namely, *acknowledged* or *unacknowledged*, in accordance with the requested QoS. The acknowledged mode of operation provides for detection and recovery of transmission errors, whereas the unacknowledged mode of operation provides for retransmission procedures for uncorrectable data blocks. The MAC layer operates between the MSs and the BSS (BTS, more specifically). It is responsible for the signaling procedures concerning radio channel access. It performs contention resolution between access attempts, arbitration between multiple service requests from different MSs, and medium allocation to individual users in response to service requests. In particular, procedures are defined that allow one MSs to utilize several physical channels simultaneously or several MSs to share one physical channel. The physical layer provides means for information transfer over the physical channel between the MSs and the network. Tasks performed by the physical layer include forward error correction coding, interleaving, detection of physical link congestion, modulation and demodulation of physical waveforms, and others. The BSS GPRS protocol (BSSGP) operates between the BSS and the SGSN conveying routing and QoS-related information.

6.2.3 Data Flow and Data Structure

The network layer PDUs (or packets) received from the network layer are transmitted across the GPRS air interface using the LLC protocol. This is

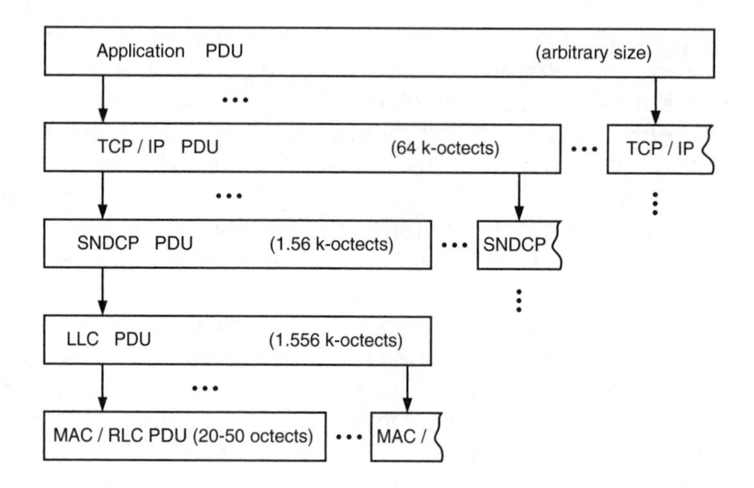

FIGURE 6.3
Segmentation and encapsulation in GPRS (the figures in parentheses show the maximum PDU size without the header).

accomplished by segmenting and encapsulating each PDU into several other PDUs. The segmentation and encapsulation process results in considerable header and signaling overhead such that, in general, 20 to 30% of the GPRS air interface capacity is spent on protocol overhead.[10] Starting with an arbitrary PDU size for any given application, successive segmentations and encapsulations lead to IP PDUs of 64 kbytes, then to SNDCP PDUs of 1.56 kbytes, then to LLC PDUs of 1.556 kbytes, and finally to MAC/RLC PDUs of 20 to 50 bytes,[10] as illustrated in Figure 6.3. The RLC/MAC layer provides means for information transfer over the physical layer of the GPRS radio interface. In particular, the RLC is responsible for the transmission of data blocks across the air interface and for the backward error correction procedures, which consist of selective retransmission of uncorrectable blocks (selective ARQ, S-ARQ). In S-ARQ, blocks in error are retransmitted until a complete frame is successfully transferred across the RLC layer, in which case the error-free frame is forwarded to the LLC layer.

Given an information block—a segment—resulting from the segmentation of an LLC PDU as illustrated in Figure 6.3, a radio block containing such information is configured for transmission. The processes involved in this configuration include aggregation of overhead information to the segment, convolutional encoding of the resulting block, and puncturing of bits of the encoded message.

The aggregation of information includes a block header, a block check sequence, and tail bits. The convolutional encoding is specified according to four different coding schemes (CS), namely, CS-1 (code rate of 1/2), CS-2 (code rate of 2/3), CS-3 (code rate of 3/4), and CS-4 (code rate of 1). The resulting radio block, in all cases, will always accommodate 456 bits and the radio block is sent in four time slots. The four time slots use the normal burst and are arranged in four consecutive frames, the slots and frames following the GSM specifications. The block header is different for control blocks and user data blocks. In the first case, only an MAC header is included. In the second case, both an MAC header and an RLC header are included. The MAC header consists of uplink state flag (USF), block type indicator (TI), and power control (PC) fields. The USF field, appearing in any given downlink channel, is used to grant the MS, to which it has been assigned permission to use the corresponding uplink channel after a connection is established. The TI field identifies the message type. The PC field controls the transmission power. Several types of messages are identified. Among others, these messages include:

- Packet Data Block (PDB)
- Packet Control Acknowledgment (PCA)
- Packet Channel Request (PCR)
- Packet Resource Request (PRR)
- Packet Uplink Assignment (PUA)
- Packet Uplink ACK/NACK (PUAck/NAck)
- Packet Downlink ACK/NACK (PDAck/NAck)
- Packet Paging Request
- Packet Paging Response

The use of some of these messages is illustrated later in this chapter.

A given network PDU is identified by a temporary flow identifier (TFI), which is a data field included in all messages belonging to that particular PDU. Figure 6.4 shows the radio block structures for user data and control messages. It also illustrates the coding, encapsulation, and transmission processes. In Figure 6.4, the numbers indicate the bits in each field. Table 6.1 shows the GPRS coding schemes and the respective throughput rate.

As already mentioned, GPRS employs the same frame structure as GSM. Each frame has a duration of 4.615 ms and consists of eight equal-duration time slots, with each time slot 0.577 ms long. The basic data transmission unit is called *radio block* and it is used for the transmission of the RLC/MAC PDUs. A radio block uses four time slots arranged in four consecutive frames. Therefore, a radio block has a duration of 2.3075 ms. The structure of a multiframe containing 52 frames is defined so that every 13th frame, known as an idle burst, is used for purposes other than data transmission, such as measurements. Effectively, within a multiframe (52 frames) only 48 frames are used for

TABLE 6.1

GPRS Coding Schemes

Code Scheme	Code Rate	RLC Data (octects)	Throughput Rate *octets/20 ms* (kbit/s)
CS-1	$\frac{1}{2}$	20	8
CS-2	$\frac{2}{3}$	30	12
CS-3	$\frac{3}{4}$	36	14.4
CS-4	1	50	20

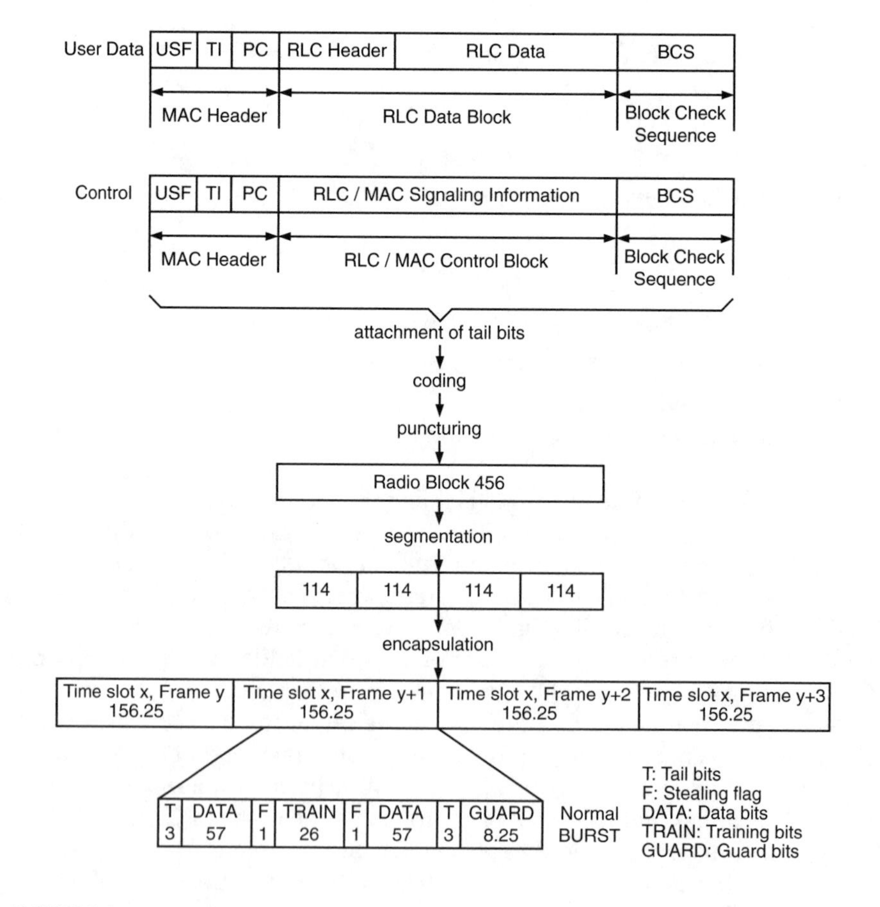

FIGURE 6.4
GPRS radio block structures: user data and control messages; coding, encapsulation, and transmission processes.

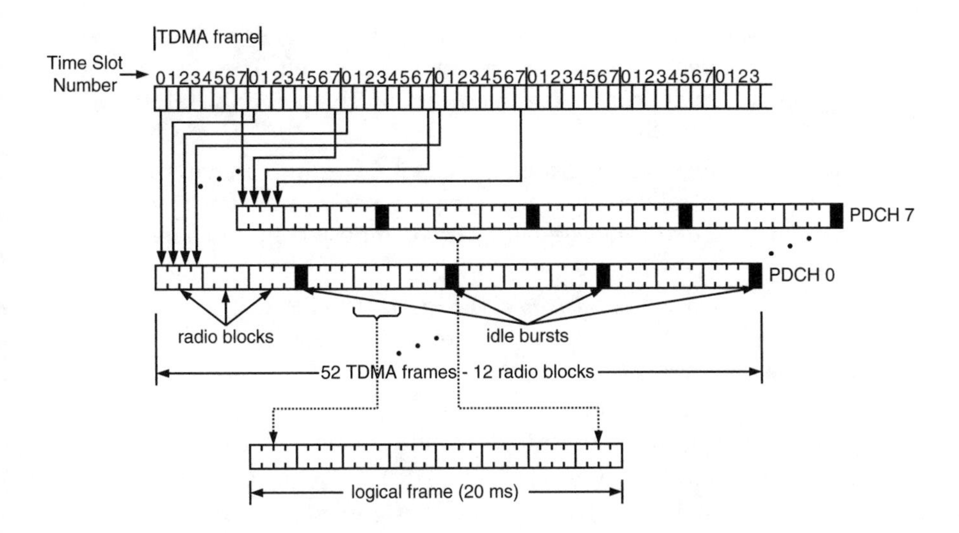

FIGURE 6.5
GSM PDCH, radio blocks, and logical frames.

data transmission purposes, these leading to $48 \div 4 = 12$ radio blocks within a multiframe. Therefore, the mean transmission time per radio block is $(52 \times 4.615 \text{ ms}) \div 12 = 20$ ms. The set of four frames, which on average is 20 ms long, defines a data structure constituting a logical frame. Thus, within a logical frame an access is identified by a channel—named the packed data channel (PDCH)—and by the TDMA frame. This is illustrated in Figure 6.5.

6.2.4 Physical Channels and Logical Channels

As an extension of conventional cellular systems, GPRS uses the same frequency bands of the cellular systems, both sharing the same physical channels. Physical channels in GPRS are time slots within a given carrier frequency. The time slots then can be assigned either to circuit-switched calls or to packet-switched data.

The physical channel assigned to a circuit-switched call is called a traffic channel (TCH), whereas the physical channel assigned to a packet-switched data is called a packet data channel (PDCH). PDCHs and TCHs, sharing the same pool of physical channels, can be allocated dynamically according to the capacity-on-demand principles. In such a case, the resources are assigned depending on the required service and on the respective QoS.

Three groups of logical channels compose the set of PDCHs, as follows: packet broadcast control channel (PBCCH), packet common control channel (PCCCH), and packet traffic channel (PTCH).

- *Packet Broadcast Control Channel.* The PBCCH group contains only one channel, the packet broadcast control channel itself. The PBCCH is a downlink channel conveying overhead information to all GPRS terminals within a cell.
- *Packet Common Control Channel.* The PCCCH group conveys all the necessary control signaling for initiating packet transfer as well as user data and dedicated signaling. It is composed of four channels as follows:
  1. *Packet Random Access Channel (PRACH).* The PRACH is an uplink channel used by the MS to initiate packet transfers or to respond to paging messages.
  2. *Packet Paging Channel (PPCH).* The PPCH is a downlink channel used to page an MS prior to downlink packet transfer.
  3. *Packet Access Grant Channel (PAGCH).* The PAGCH is a downlink channel used to convey resource assignments to an MS in the packet transfer establishment phase.
  4. *Packet Notification Channel (PNCH).* The PNCH is a downlink channel used to send a point-to-multipoint-multicast (PTM-M) notification to a group of MSs prior to a PTM-M packet transfer. Such a notification has the form of a resource assignment.
- *Packet Traffic Channel.* The PTCH group is responsible for packet transfers and for the control of such transfers. It is composed of two channels as follows:
  1. *Packet Data Transfer Channel (PDTCH).* The PDTCH works in the uplink and downlink directions. It is used for data transfer purposes. Any given MS may use one or more PDTCHs for individual packet transfers.
  2. *Packet Associated Control Channel (PACCH).* The PACCH works in the uplink and the downlink directions. It conveys signaling information (e.g., acknowledgment, power control information) and resource (re)assignment messages related to a given MS. It is associated with one or several PDTCHs assigned to that MS.

The GPRS logical channels are depicted in Figure 6.6.

6.2.5 Medium Access

The GPRS radio interface comprises asymmetric and independent uplink and downlink channels. Therefore, an uplink PDCH occupying a certain time slot may convey information from a given MS, whereas a downlink PDCH

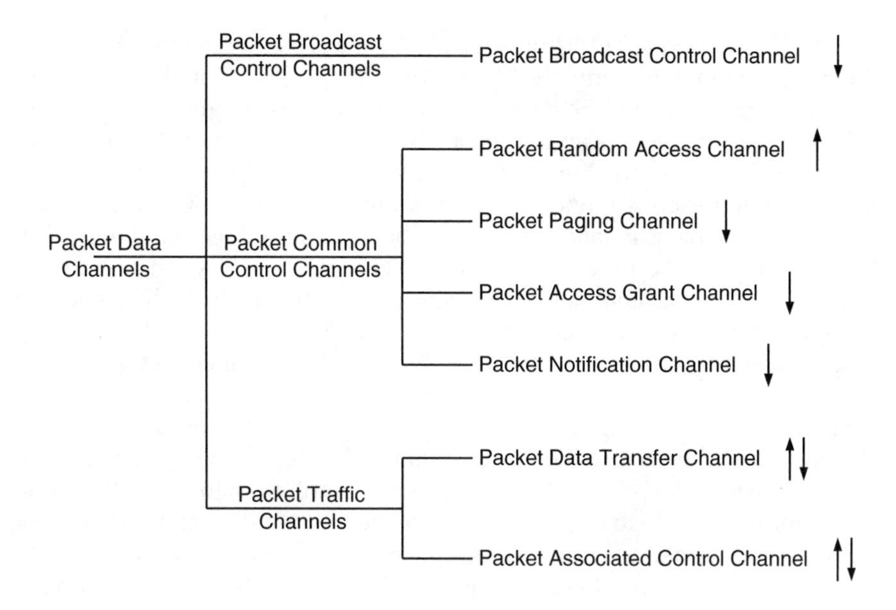

FIGURE 6.6
GPRS logical channels. The downward arrow indicates the downlink direction; the upward arrow indicates the uplink direction.

occupying the same time slot may convey information to a different MS. The assignment of PDCH resources in both downlink and uplink directions is controlled by the BSC, with the assignment carried out on a temporary basis to support the packet-switched principle.

Data transfer is accomplished by means of temporary block flow (TBF). A TBF corresponds to a virtual connection between an MS and the BSS supporting unidirectional transfer of LLC PDUs on packet data physical channels. Such a virtual connection is maintained throughout the duration of the data transfer, and the data transfer comprises all the blocks identified by the same TFI (temporary flow identifier). A TFI is unique in each direction and is assigned by the BSS. It is 7 bits long for the uplink and 5 bits long for the downlink. A TBF can be of the closed-ended type (CET) or of the open-ended type (OET). In a CET TBF, the amount of data to be transferred is negotiated at the initial access. In the OET TBF, an arbitrary amount of data can be transferred. A TBF is terminated after the completion of the data transfer, when the TFI is released. Such a resource allocation scheme is certainly appropriate for packet transfer. It renders GPRS very flexible, allowing the TBF duration to vary in accordance with the amount of data to be transferred.

Downlink multiplexing is achieved by assigning each data transfer (TBF) a unique TFI. Therefore, data streams destined to different MSs can share the same resources because each data stream belonging to a given TBF is identified by a unique TFI. Thus, several MSs may be assigned the same set of downlink channels but only the blocks with the specific TFI in these channels will be accepted by the MS to which this TFI has been assigned.

Uplink multiplexing is accomplished by indicating in the downlink channels, via USFs univocally identifying the wanted MSs, which MSs are allowed to transmit in the corresponding uplink channels. The USF is 3 bits long, implying that up to eight data transfers can be multiplexed on one channel. (In fact, USF = 111 is reserved; thus, only seven data transfers can actually be multiplexed on each channel.[11]) Therefore, data streams from different MSs can share the same uplink channels on an noncontention basis. In that case, several MSs may be assigned the same set of uplink channels but only those allowed to use these channels can transmit. This, grossly speaking, defines what is known as a centralized in-band polling, a scheme used by the base station to poll the desired MS. In a centralized in-band polling scheme, the base station sets the USF field in one or more downlink channels to the value that identifies the MS to be polled. Because the MS monitors all the downlink channels that are paired with the uplink channels assigned to it, it will identify its USF in the downlink channels whenever it appears. It shall transmit on the corresponding uplink channels in the next logical frame. The maximum number of time slots (channels) an MS may use is determined by its multislot class.

With such a flexible packet transmission scheme, downlink channels may transmit data to a given MS while a different MS may be using the corresponding uplink channels to transmit its data. This mechanism is illustrated in Figure 6.7. In Figure 6.7, in the downlink, the figures within each PDCH indicate, respectively, the USF for the given MS and the MS for which the data packet is transmitted. In the uplink, the number within the PDCH identifies the MS from which the data packet is transmitted. Note that channels 1 and 2 (identified as PDCH0 and PDCH1) of frame 1 in the downlink contains the USF assigned to MS1 and, therefore, MS1 can use the corresponding channels

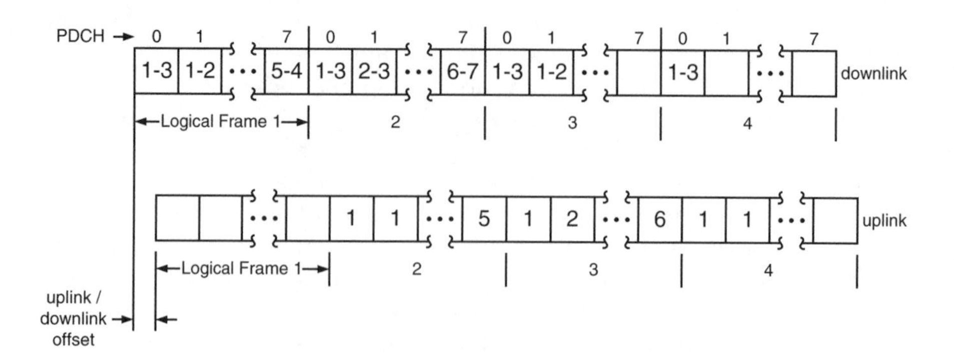

FIGURE 6.7
Downlink and uplink multiplexing.

of frame 2 in the uplink. In the same channels (channels 1 and 2 of frame 1), in the downlink, the packet data are transmitted to MS3 and MS2, respectively. Note also that while in channel 1 of frame 2 the permission to transmit is given to MS1, in channel 2 of the same frame this is given to MS2 (recall that in the previous frame this was given to MS1). Other multiplexing situations are illustrated in Figure 6.7.

6.2.6 Data Transfer Procedure

Two types of access procedures are permitted for data transfer: one-phase access and two-phase access. In the one-phase procedure, data transfer and overhead procedures, such as service negotiation and mobile verification, occur concurrently within the same access. In the two-phase procedure, data transfer occurs only after the overhead procedures are accomplished. The one-phase procedure, therefore, can be faster than two-phase whenever an ongoing service negotiation is acceptable from the system and application point of view. The choice for one or another procedure is left to the system operator.

As detailed previously, once a TBF is established a very efficient data transfer procedure is accomplished and no concurrent accesses occur. On the other hand, the initial uplink access, at the establishment of a TBF, may involve contention. GPRS uses a slotted-ALOHA-based random-access procedure for contention resolution. Initial access is certainly a critical phase in establishment of TBF and for this reason a special access burst format is specified. In such a format, five fields are defined as follows. The first field contains 8 bits (bits numbered from 0 to 7) corresponding to the extended tail bits. The second field contains 41 bits (bits numbered from 8 to 48) corresponding to the extended synchronization bits. The name *extended* refers to the bits in excess of 26, as 26 is the number of bits of the synchronization field in the normal burst. As already explained in Chapter 4, such an extension is provided because at the initial access the MS may not have any timing information. The third field contains 36 bits (bits numbered from 49 to 84) corresponding to the encrypted information content. Before coding and encryption, these bits total 8 or 11 in number. The information provided by these bits includes:

- Indication of the reason for the access, including one-phase access request, two-phase access request, page response, measurement report, and others
- Indication of the MS class and the requested radio priority
- Indication of the number of blocks to be transmitted

The fourth field contains 3 bits (bits numbered from 85 to 87) corresponding to the tail bits. The fifth field contains 68.25 bits (bits numbered from 88 to 156) corresponding to the extended guard period. Here, again, *extended* refers to the bits in excess of 8.25, as 8.25 is the number of bits of the guard period field in the normal burst. These bits, together with the extended synchronization bits, help in the initial access when no timing information may be available.

In addition to the special burst format used for the initial access, an initial access also differs from an ordinary data transfer, in which the data block occupies four time slots, by using only one time slot. Therefore, if a PDCH is assigned one time slot per frame, there are four access opportunities in every logical frame (20 ms).

6.2.7 Mobile-Originated Data Transfer

This subsection illustrates mobile-originated data transfer. The MS sends a packet channel request message to the BSS over the packet random access channel. In this message, among others, the access type (one-phase or two-phase) is specified. The BSS responds with an packet uplink assignment message over the packet access grant channel.

- *One-Phase Procedure*. In the one-phase procedure, the packet uplink assignment message conveys the information regarding the resource to be assigned to the MSs. This includes carrier, time slot, USF, TFI, open-ended or closed-ended TBF, packet-associated control channel, etc. Note that, at this point, the network has not yet been informed about the MS identity or about the requested service. This is accomplished when the MS communicates with the BSS using an extended RLC header in the first few blocks to include, for example, the MS temporary logical link identifier (TLLI). The network then decodes the TLLI and sends an acknowledgment to the MS in the packet uplink Ack/NAck message. After the successful reception of this message by the MS the contention resolution phase is completed on both sides (network and terminal). Figure 6.8 illustrates a simplified one-phase access for uplink data transfer procedure. A TBF is terminated with the packet control acknowledgment message.
- *Two-Phase Procedure*. In the two-phase procedure, the packet uplink assignment message conveys the information regarding a single block allocation. This includes the time slot number, TFI, power control parameters, and the packet-associated control channel. The MS then sends a packet resource request over the packet-associated control channel detailing the information regarding the resource and the service to be requested. Such information includes the MS TLLI, peak throughput class, radio priority, measurement report, and open-ended or closed-ended TBF. The BSS responds with a packet uplink assignment message indicating the time slot, TFI, USF, TLLI, and others. Data transfer then begins. Figure 6.9 illustrates a simplified two-phase access for the uplink data transfer procedure. A TBF is terminated with the packet control acknowledgment message.

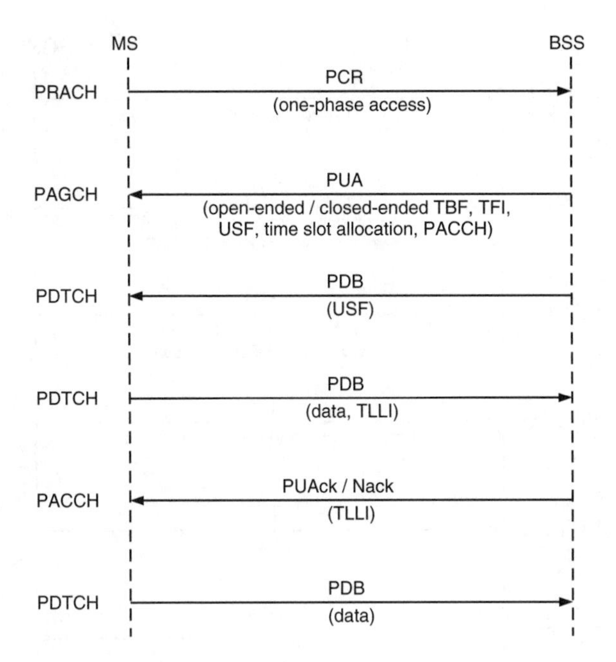

FIGURE 6.8
A simplified one-phase access for uplink data transfer procedure.

6.2.8 Mobile-Terminated Data Transfer

This subsection comments on mobile-terminated data transfer. The BSS sends a packet paging request message over the packet paging channel. The MS responds to the page with a packet channel request message using the packet random-access channel. From this point on, the message flow proceeds in a way very similar to the packet access procedure, as described in the previous subsections. A TBF is terminated with the packet control acknowledgment message.

6.2.9 Throughput Performance

Throughput performance can be analyzed from two standpoints: system and user. The system throughput is used to assess the system capacity. This is the throughput as perceived from the operator's viewpoint. The user throughput concerns the data rate or the delay as perceived by the user when a service is requested. In theory, for one carrier, which supports eight time slots per TDMA frame, the throughput can be estimated directly from Table 6.1 for each code scheme by multiplying the Throughput Rate column by eight (or by seven if one USF is reserved, as explained previously). In that case, CS-1 yields 64 kbit/s, CS-2 yields 96 kbit/s, CS-3 yields 115 kbit/s, and CS-4 yields

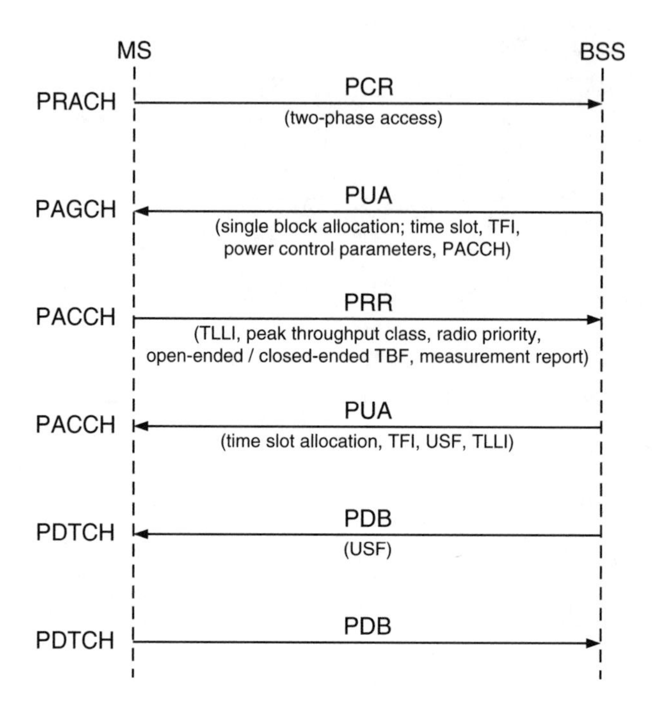

FIGURE 6.9
A simplified two-phase access for uplink data transfer procedure.

160 kbit/s. Nevertheless, if the overhead of all protocols and the phenomena involved in the transmission are accounted for, then the actual throughputs will be substantially smaller than these. Note that a great many parameters are involved in such a complex system. Therefore, a precise determination of the achievable throughputs is a rather intricate task. Many works (e.g., References 10, 12, and 13) attempt to estimate the GPRS throughput by means of simulation. Although the figures may vary from one work to another due to the different simulation platforms utilized, the main conclusions are similar.

The figures to be shown here are extracted from Reference 10. In Reference 10, the simulator models only one cell and one carrier frequency, and mobility and handover scenarios are excluded. In addition, the traffic streams comprise those used for typical Internet applications such as Web browsers, e-mail transfer, and file transfer. The number of PDCHs considered is eight and the carrier-to-interference ratio (C/I) varies depending on the code scheme utilized. These values are chosen in such a way that each code scheme operates to provide for low block error rates and good balance between forward error correction and automatic repeat request. The terminals operate in multislot mode (MSM), in which case they are allowed to make use of more than one time slot. Specifically, in the simulations of Reference 10 the MSM allows the use of one, two, or four time slots (MSM1, MSM2, MSM4, respectively, as named here). The TCP segment size is specified to have 536 bytes and the mean Internet packet loss rate allowed is 2%. Given the asymmetry of the Web traffic, and because the uplink network load is not the critical aspect to be considered, the investigations tend to concentrate on the downlink performance. Details of the model and parameters used in the simulation may be found in Reference 10.

System Throughput

The system throughput saturates as the number of Web sessions increases, with the saturation occurring at 44 kbit/s for CS-1 ($C/I = 12$ dB), 66 kbit/s for CS-2 ($C/I = 16$ dB), 75 kbit/s for CS-3 ($C/I = 20$ dB), and 100 kbit/s for CS-4 ($C/I = 24$ dB). If these figures are compared to the theoretical throughputs we notice that losses of 31.25, 31.25, 34.9, and 37.5% for the respective code schemes are found. The multislot mode feature affects the system throughput only for low traffic load condition (say, fewer than 20 Web users). Better performances are achieved by the modes with a greater number of slots. For heavy traffic load (say, more than 30 Web users) the multislot feature does not affect the system throughput substantially, for all PDCHs are found fully loaded most of the time. The channel conditions, as expected, play a decisive role in the system performance. CS-1, the most robust code scheme, yields the best performance only for $C/I < 6$ dB, an unrealistic operation condition. CS-2 yields the best performance for 6 dB $< C/I <$ 10 dB. CS-3 yields the best performance for 10 dB $< C/I <$ 17 dB. Finally, CS-4 performs better for $C/I > 17$ dB, i.e., for very good channel conditions. It has also been observed that the system throughput increases almost linearly with the increase of the number of PDCHs, which suggests that better data capacity can be achieved with the addition of transceivers within a cell.

User Throughput

The user throughput is measured in terms of packet bit rate, the object size (number of bits) divided by the time taken for its transmission (delay). Note that this is measured during the activity period of the user. Because of the packet transmission principle, in which the resources are allotted dynamically and only during the period of activity of the users, idle phases of some users can be used for the transmission of data of those users that are active during these phases. Here, again, it is observed that the higher the traffic load the less the influence of the multislot mode feature on the user throughput; the smaller the traffic load, the greater the influence of the multislot mode feature on the user throughput.

For example, for ten Web users MSM1 yields a user throughput that increases almost linearly from approximately 5 kbit/s for CS-1 to 12.5 kbit/s for CS-4; for MSM2 the increase is from 7.5 to 21 kbit/s; for MSM4 the increase is from 10 to 31 kbit/s. On the other hand, for high load (say, more than 40 Web users) the user throughput is almost constant for all the multislot modes, still increasing slightly from 1 kbit/s for CS-1 to 4 kbit/s for CS-4 (for all multislot modes). Specifically, for CS-2 and for ten Web users 8, 15, and 23 kbit/s are achieved, respectively, for MSM4, MSM2, and MSM1. As the number of users increases, for all the multislot modes the throughput decreases almost linearly, yielding the same throughput of 4.5 kbit/s for 40 Web users, decreasing further to 2.5 kbit/s for 60 users. In the case of 20 ongoing Web sessions, for example, each MSM4 user achieves a mean packet bit rate of 15 kbit/s. Another interesting measure is the cumulative distribution of the packet bit rate for the several conditions. It is found that for low packet bit rates the distributions for different numbers of users are almost coincident. This is explained by the fact that small objects present a relatively large header and protocol overhead (e.g., TCP connection setup) resulting in low packet rate. In addition, for lost IP packets, time-consuming retransmissions are necessary. For fewer than 20 users the great majority of the Web objects are received with data rates greater than 10 kbit/s. Large objects are usually received with higher data rates because the header and the relative protocol overhead in this case are smaller compared with those for small objects.

6.2.10 GPRS—Summary

GPRS technology constitutes an efficient solution for data services. It can be aggregated into existing 2G cellular systems. It has been optimized to support several classes of traffic allowing wireless users to access the Internet and carry out ordinary conversational connections. GPRS permits dynamic and flexible radio resource allocation, this resulting in efficient multiplexing and high throughput. Higher throughputs can still be accomplished by keeping GPRS core network and introducing new air interfaces. For example, in EDGE (enhanced data rate for global evolution) a higher-level modulation (8-PSK), in addition to the traditional GMSK, is introduced. This by itself may increase the data rate to 384 kbit/s and the spectrum efficiency to 0.5 bit/Hz/base.

6.3 EIA/TIA/IS-95B

EIA/TIA/IS-95B was explored in Chapter 5. Here only the main data transmission features of this technology are summarized. In EIA/TIA/IS-95B, each traffic channel is composed of one fundamental traffic channel and as many as seven supplemental traffic channels. The maximum number of fundamental traffic channels and supplemental traffic channels is still limited by the total number of Walsh sequences (64) minus the number of the

TABLE 6.2

Possible Transmission Rates for the Various Multiplex Options

Multiplex Option Rate Set 1	Multiplex Option Rate Set 2	Supplemental Channels	Rate Set 1 (kbit/s)	Rate Set 2 (kbit/s)
1	2	0	1.2, 2.4 4.8, 9.6	1.8, 3.6 7.2, 14.4
3	4	1	19.2	28.8
5	6	2	28.8	43.2
7	8	3	38.4	57.6
9	10	4	48.0	72.0
11	12	5	57.6	86.4
13	14	6	67.2	100.8
15	16	7	76.8	115.2

remaining channels, i.e., pilot channel, sync channel, and paging channels. Rate Set 1 and Rate Set 2 are kept, but they include now eight channel multiplex options each. For Rate Set 1 these multiplex options are numbered 1, 3, 5, 15, and for Rate Set 2 the multiplex options are numbered 2, 4, 6, 16. The increase in the transmission rate is achieved by aggregating as many as seven supplemental traffic channels to the fundamental traffic channel.

In Rate Set 1, the transmission rate for each channel is set to 9.6 kbit/s, whereas this is 14.4 kbit/s in Rate Set 2. Table 6.2 illustrates the possible total transmission rates for the various multiplex options.

The multiplex options concern both forward link and reverse link. Primary data (voice) and secondary data (fax, image, etc.) may be transmitted through one fundamental traffic channel aggregated with supplemental traffic channels. Signaling data, on the other hand, use only the fundamental traffic channel.

In the reverse link, the fundamental traffic channel and the supplemental traffic channels are transmitted with the same frame offset. Each carrier for each reverse supplemental traffic channel is shifted by a fixed phase with respect to the carrier of the fundamental traffic channel.

6.4 High Data Rate

The introduction of CDMA systems into the modern wireless market posed new challenges in the design and planning of cellular networks. The 2G CDMA technology was conceived primarily for two-way conversational speech, which requires strict adherence to symmetry and small latencies (less than 100 ms). Circuit-switched principles easily solve the latency question but, depending on the access technology, full symmetry requires more elaborate solutions. Downlink and uplink in 2G CDMA systems are inherently asymmetric with the reverse direction (uplink) initially believed to be the capacity-limiting link (bottleneck). This surmise turned out to be incorrect: the forward direction (downlink) constitutes the actual bottleneck. (The reader is referred to Chapter 5 for more information on this.) The evolutionary revisions of cdmaOne (IS-95-B and cdma2000), which implement fast power control in the forward link and introduce new criteria for enhanced soft handoff, have considerably improved the forward-link performance, bringing its capacity to parity with that of the uplink. Although this may suffice for voice applications, which are the main target of 2G systems, link symmetry is by no means a solution for high-speed data applications. On the contrary, the downlink demand is likely to be several times greater than the uplink demand, for example, whenever Internet applications are the focus.

cdmaOne provides an excellent solution for systems where low-rate channels share a common bandwidth. For a small number of high-rate data users sharing the same bandwidth, however, better solutions may be found. A CDMA system providing for mixed low-rate data service (with voice the primary service in this category) and high-rate data service requires an innovative solution. HDR is certainly one such innovative solution.

6.4.1 HDR Solution

HDR are stand-alone systems, designed to be highly interoperable with IS-95 and multicarrier (MC) CDMA systems, with the design leveraging in many ways the lessons learned from the development and operation of IS-95 networks. Along with the interoperability feature, HDR systems can reuse large portions of components and designs already implemented in IS-95 products. Dual-mode terminals, therefore, can be implemented in a compact and cost-effective manner, allowing users to have access to both voice services, through an IS-95 frequency carrier, and data services, through an HDR frequency carrier.

The first fundamental design choice in HDR is to separate low-rate data services (voice) from high-rate data services by using nonoverlapping spectrum allocations for these different services. In essence, the HDR solution for a CDMA system in which both types of services are provided uses separate carriers for the respective services. An IS-95 RF carrier is used for voice applications and a separate HDR RF carrier is utilized for data applications. Within an HDR RF carrier, forward and reverse links are designed to operate in distinct modes. Packet transmissions on the HDR forward link are time-division-multiplexed and occur at full power and at variable rates, with packets having variable durations (packets have variable sizes). Packet transmissions on the HDR reverse link are code-division-multiplexed and occur at variable rates, but with packets having fixed durations (packets have fixed sizes).

In HDR, the access terminal (user terminal) and the access point (base station) jointly determine the highest rate a subscriber can support at any instant. This is accomplished by means of the deployment of a combination of techniques based on channel measurement, channel control, and interference suppression and mitigation. In particular, each access terminal assesses the quality of the signals (signal-to-noise ratio, or SNR) received from neighboring access points. The best access point is chosen and its SNR, or equivalently the supportable data rate value, is indicated to the corresponding access point. Note that this may change continuously because of the nonstationary condition of the wireless channel (motion of the user, change of the propagation conditions, etc.). Therefore, the monitoring of the SNRs of the various access points and the selection of the best access point must be performed continuously, which has a direct implication for the supportable data rate. Packet transmissions on the forward link, then, occur with data rates and packet durations that vary in accordance with the user channel conditions. Hence, it may be said that the HDR forward-link supports dynamic data rates. Packet transmissions on the reverse link, in turn, occur with variable rates but fixed packet duration. The HDR reverse link does not support dynamic data rates.

By always making use of the best SNR available, interference is reduced. Moreover, because data services tolerate latencies better than voice services, robust error-correction coding techniques (for example, turbo coding) can be employed. Therefore, the system may operate at lower ratio of energy-per-bit to noise (E_b/N_0) and hence lower SNR and higher interference levels. HDR designers[14] claim that the HDR peak data rate may reach 2457.6 kbit/s on the forward link and 307.2 kbit/s on the reverse link.

6.4.2 Network Architecture

The basic elements of an HDR system are the access terminal and the access point. Several network architectures may be selected to accommodate these elements and provide for wireless packet data communication. In Reference 15, the choice fell on the most ubiquitous packet data network, namely, the Internet. Basically, as far as air interface is concerned, the entire network may be split into the *user side* and the *network side*.

The user side—the user terminal—contains two functional elements: data terminal (e.g., a computer) and access terminal. These functional elements may physically reside in two devices or in one device. The functional partitioning method in the first case is referred to as *network model*, whereas in the second case this is known as *relay model*. In the network model, the two elements may be connected by means of any standard interface (e.g., Ethernet,

PCMCIA, universal serial bus), with the responsibility of implementing the protocol stack for a reliable communication with the network side left to the access terminal. In the relay model, the two elements may be connected by means of an appropriate interface (e.g., RS-232, universal serial bus), but with some of the implementation of the stack protocol also left to the data terminal.

The network side—the access network—implements many of the functions of a cellular network, but on a decentralized basis. It contains a number of elements that may reside in the access point, as follows: transceiver, controller, network access server, network interface, routers, and the several IP servers, stations, and centers such as simple network management protocol (SNMP) based network management center (NMC) stations, domain name system (DNS) servers, secure hypertext transfer protocol (HTTPS)-based customer care center (CCC) servers, and others. The transceiver and the controller are responsible for providing frequency, power output, modulation, channel structure, message fragmentation, message reassembling, multiplexing, and other functions typical of lower layers. The network access server implements those functions related to higher layers. The network interface is responsible for the implementation of the protocols and interfaces necessary for a connection to an IP network and a backhaul network. Figure 6.10 depicts a possible network architecture for an HDR system implementation.

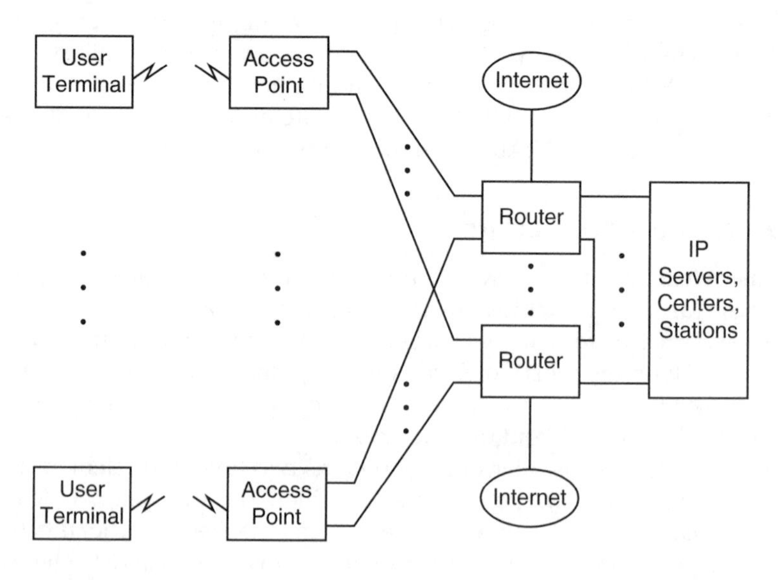

FIGURE 6.10
HDR architecture.

6.4.3 Protocol Architecture

The HDR protocol stack layers and the respective default protocols are shown in Figure 6.11.[14] Next, these layers and protocols are briefly described.

- *Physical Layer (PHYL).* The PHYL is responsible for the provision of the channel structure, frequency, power output, modulation, and encoding specifications for both the forward channels and reverse channels.
- *Multiple Access Control Layer (MACL).* The MACL is responsible for the definition of the procedures used to transmit and receive over the PHYL. Four default protocols governing the operation of the various channels are defined within the MACL as follows (the channels are described later in this chapter):

  Control Channel MAC Protocol (CMP). The CMP contains rules concerning the control channel. It governs the following: access network transmission over the control channel; packet scheduling over the control channel; access terminal acquisition of the control channel; access terminal packet reception over the control channel; and inclusion of the access terminal address into the packets to be transmitted.

  Access Channel MAC Protocol (AMP). The AMP contains rules concerning the access channel. It governs the following: access terminal transmission timing; power characteristics; and rate selection.

  Forward Traffic Channel MAC Protocol (FMP). The FMP contains rules concerning the forward traffic channel. In particular, the FMP supports both variable operation and fixed operation of the

Layers	Protocols			
APPL	HMP	PPP	-	
	SLP	RLP		
STRL	STP			
SSSL	SBP		SCP	
CNNL	LMP	ISP	DSP	CSP
	PCP	RDP		OMP
SCTL	-			
MACL	CMP	FMP	AMP	RMP
PHYL	Forward and Reverse Channels			

FIGURE 6.11
HDR protocol stack layers (default protocols).

forward traffic channel. It governs tasks related to the data rate control channel including access terminal data rate control channel transmission and access network data rate control channel reception.

Reverse Traffic Channel MAC Protocol (RMP). The RMP contains rules concerning the reverse traffic channel. It governs the following: communication between the access terminal and the access network for the acquisition of the reverse traffic channel; and reverse traffic channel data rate selection.

- *Security Layer (SCTL).* The SCTL is simply a relay layer, responsible for transferring packets between its adjacent layers. It does not provide any services.
- *Connection Layer (CNNL).* The CNNL is responsible for the control of the state of the HDR airlink and for the prioritization of the traffic sent over this layer. In particular, the access terminal may be found in one of three possible states, namely, initialization state, idle state, and connected state. In the initialization state, the access terminal has yet to acquire the network. In the idle state, the access terminal has already acquired the network but the connection is still closed. In the connected state, the access terminal has an open connection with the network. Within CNNL, seven default protocols are defined as follows.

Air Link Management Protocol (LMP). The LMP is responsible for the activation of the protocols that perform the different actions according to the states at which the access terminal is found. Each state is governed by one protocol. Therefore, three different state protocols, the next three described below, are defined.

Initialization State Protocol (ISP). The ISP is activated by the LMP performing the actions associated with the HDR network acquisition.

Idle State Protocol (DSP). The DSP is activated by the LMP performing actions associated with access terminal location, connection opening, and access terminal power conservation.

Connection State Protocol (CSP). The CSP is activated by the LMP performing actions associated with connection closing and link management between access terminal and access network.

Packet Consolidation Protocol (PCP). The PCP is responsible for the consolidation and prioritization of packets for transmission.

Route Update Protocol (RDP). The RDP is responsible for actions associated with access terminal location and link maintenance between access terminal and access network.

Overhead Messages Protocol (OMP). The OMP is responsible for broadcasting essential parameters over the control channel.

- *Session Layer (SSSL).* The SSSL is responsible for providing protocols used to negotiate a session between the access terminal and the access network. A session is defined as a state shared by the access terminal and the access network in which the following information is included: access terminal unicast address; airlink protocols for communication between the access terminal and the access network; configuration settings for airlink session protocols; and estimate of the access terminal location. Within SSSL, two default protocols are defined as follows.

  Session Control Protocol (SCP). The SCP is responsible for the following: provision of means for session negotiation; provision of airlink protocols; provision of configuration parameters; and closing of the session.

  Session Boot Protocol (SBP). The SBP is responsible for providing the following: unicast address assignment to the access terminal; and means to negotiate the type of SCP to be used during the session.

- *Stream Layer (STRL).* The STRL accommodates the stream protocol (STP). It is responsible for the provision of multiplexing of application streams for one access terminal. It is also responsible for configuration messages that map applications to streams.
- *Application Layer (APPL).* The APPL contains four default protocols as follows.

  HDR Messaging Protocol (HMP). The HMP is responsible for routing messages to protocols.

  Signaling Link Protocol (SLP). The SLP is responsible for the provision of best effort and reliable message delivery for the HMP, message fragmentation, and duplicate detection and retransmission for messages using reliable delivery.

  Radio Link Protocol (RLP). The RLP is responsible for providing an octet stream service with a satisfactorily low erasure rate so that higher layer protocols (e.g., TCP) may operate adequately. The RLP makes use of a NAck-based scheme and of an efficient retransmission mechanism due to sequencing of octets rather than sequencing of frames. In that case, complex segmentation and reassembling tasks are avoided in case retransmitted frames cannot fit into the available payload information field. Error rates in the order of 10^{-6} may be delivered by RLP.

Point-to-Point Protocol (PPP). The PPP is responsible for providing an octet stream to convey transmission of packets between the access terminal and the access network.

6.4.4 Channels and Channel Structure

The HDR solution for a CDMA system in which voice and packet services are provided uses separate carriers for the respective services. An IS-95 RF carrier is used for voice applications and a separate HDR RF carrier is used for data applications. In contrast to the IS-95 system, in which the forward-link power is shared by all users, in HDR, there is no power sharing and packet transmissions are time-division-multiplexed and occur at full power. Packet transmissions on the forward link occur with data rates and packet lengths that vary in accordance with the user channel conditions. The HDR reverse link, on the other hand, enjoys a solution similar to that of the IS-95, enhanced to accommodate higher data rates. Packet transmissions on the reverse link, in turn, occur with variable data rates but with fixed packet lengths. In HDR, both forward link and reverse link use a pilot-aided, coherent demodulation, as opposed to IS-95 in which only the forward link presents such a feature. Several channels and subchannels, six on the forward link and seven on the reverse link, are defined, as illustrated in Figure 6.12.

Forward-Link Channel Structure

The HDR forward link contains four groups of time-division-multiplexed channels: pilot channel, medium access channels, traffic channel, and control channel. The medium access channels are composed of the forward activity channel, reverse activity channel, and reverse power control channel.

- *Pilot Channel.* The pilot channel is a common pilot transmitted at full power by each access point at predetermined intervals. It can be used to estimate the channel conditions (SNR) and is received only in the presence of pilots from other access points, and is not affected by interference of data transmission.
- *Forward Activity Channel.* The forward activity channel indicates the activity of the forward link. It is used to help the access point and the access terminal determine the optimum data rate for packet transmission.
- *Reverse Activity Channel.* The reverse activity channel indicates the activity of the reverse link. It is used to help the access point and the access terminal determine the optimum data rate for packet transmission.

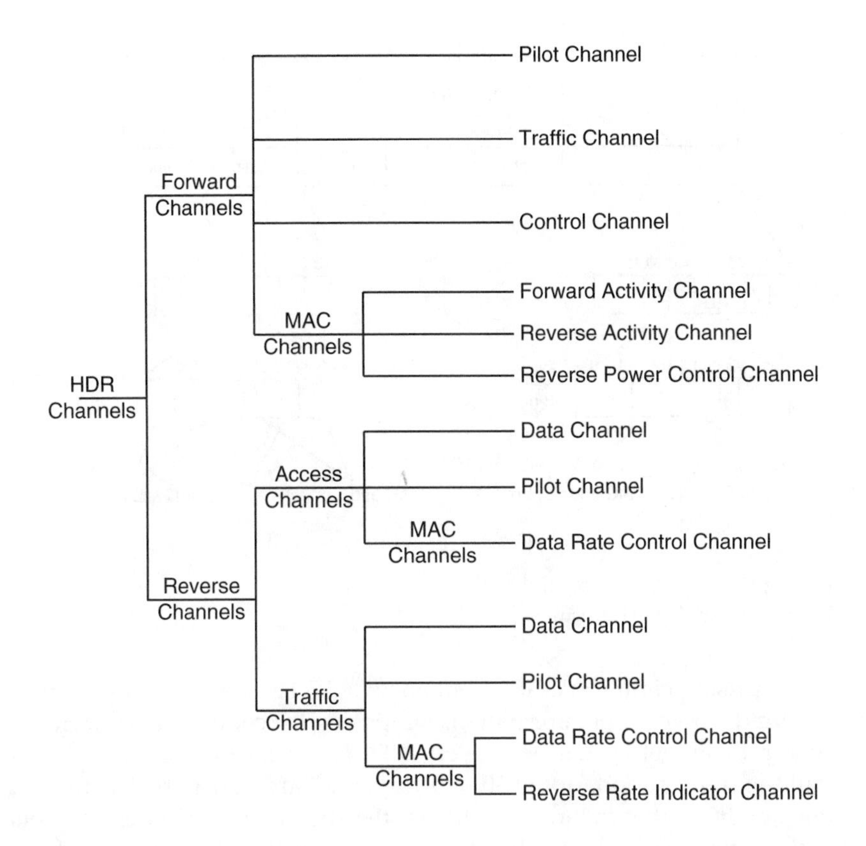

FIGURE 6.12
HDR channels and subchannels.

- *Reverse Power Control Channel.* The reverse power control channel is used for reverse-link power control purposes.
- *Traffic Channel.* The traffic channel is used for packet data transmission purposes.
- *Control Channel.* The control channel combines the functions of the IS-95 sync and paging overhead channels. It is time set aside on the forward link at a given fixed rate, as explained next.

The HDR forward-link channel structure is based on 26.67 ms frames aligned to the PN rolls of the zero-offset PN sequences, a frame containing 2^{15} chips in accordance with the 1.2288 Mchip/s transmission rate. A frame is composed of 16 1.67 ms slots; each slot, therefore, encompasses 2048 chips. For control purposes, a frame is split into two half frames, each half frame

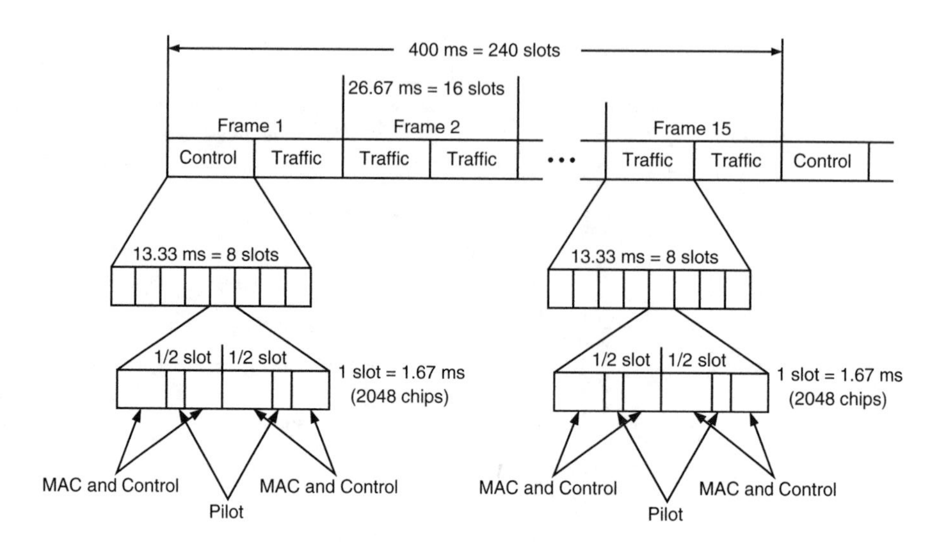

FIGURE 6.13
HDR forward-link channel structure.

then comprising eight slots. At a frequency of 2.5 Hz, one such half frame is transmitted with control information, composing the control channel. Hence, the control channel is sent once every 400 ms for a duration of 13.33 ms. Within the same period, of the 30 half frames 29 are transmitted with data, composing the traffic channels. Note that the 400 ms encompasses 240 slots, eight dedicated to the control channel and 232 to the traffic channels. Each slot is further split into two half slots, each of which contains a pilot burst—the pilot channel. The pilot channel has a duration of 78.125 μs, or 96 chips, and is centered at the midpoint of the half slot. The remainder of the half slot is shared by the medium access control channels and either the control channel or the traffic channel, as appropriate. Whereas the traffic channels are transmitted at a variable bit rate ranging from 38.4 to 2457.6 kbit/s, the control channel is transmitted at a fixed rate of 76.8 kbit/s. Figure 6.13 illustrates the HDR forward-link channel structure.

Reverse-Link Channel Structure

The HDR reverse link contains two groups of code-multiplexed channels: traffic channel group and access channel group. The traffic channel group contains the pilot channel, the medium access control channel, and the data channel. The access channel group contains the pilot channel, the medium access control channel, and the data channel. Within the traffic channel group, the medium access control channel includes the reverse rate indicator channel and the data rate control channel. Within the access channel group, the medium access channel includes only the reverse rate indicator channel. In both the traffic channel group and the access channel group, the pilot channel and the medium access control channels are time-multiplexed and separated from the data channel by means of codes.

- *Pilot Channel (Traffic Channel group).* The pilot channel is used for coherent demodulation of the (traffic) data channel.
- *Reverse Rate Indicator Channel (Traffic Channel group).* The reverse rate indicator channel indicates the rate at which the (traffic) data channel is sent.
- *Data Rate Control Channel (Traffic Channel group).* The data rate control channel indicates the required rate at which data information is sent on the (traffic) data channel.
- *Data Channel (Traffic Channel group).* The data channel conveys user data for data transfer purposes.
- *Pilot Channel (Access Channel group).* The pilot channel is used for coherent demodulation of the (access) data channel.
- *Reverse Rate Indicator Channel (Access Channel group).* The reverse rate indicator channel indicates the rate at which the (access) data channel is sent.
- *Data Channel (Access Channel group).* The data channel conveys data concerning user access.

The HDR reverse-link channel structure is based on fixed-duration packets, a deviation from the forward-link channel structure in which packets have variable sizes. The reverse-link structure consists of 32 1.67-ms slots, with a total duration of 53.33 ms. Each slot comprises 2048 chips and is further split into 32 64-chip subslots. The frames are multiples of 26.67 ms and traditional IS-95 power control mechanisms and soft handoffs are supported. Whereas the traffic data channel supports data rates ranging from 4.8 to 307.2 kbit/s, the data rate on the access data channel is fixed at 9.6 kbit/s. In both the traffic channel group and access channel group, the pilot channel and the data rate control channel are time-division-multiplexed at a 50% duty cycle. The reverse rate indicator channel, which is not included within the access channel group, appears as the preamble of the reverse-link frame. More specifically, the reverse rate indicator channel is time-multiplexed with the pilot channel and date rate control channel, occupying only one subslot (64 chips). Figure 6.14 illustrates the HDR reverse-link channel structure.

6.4.5 Medium Access

As already mentioned, HDR employs a time-shared forward link, and no forward power sharing is used, such that, when being served, an access terminal

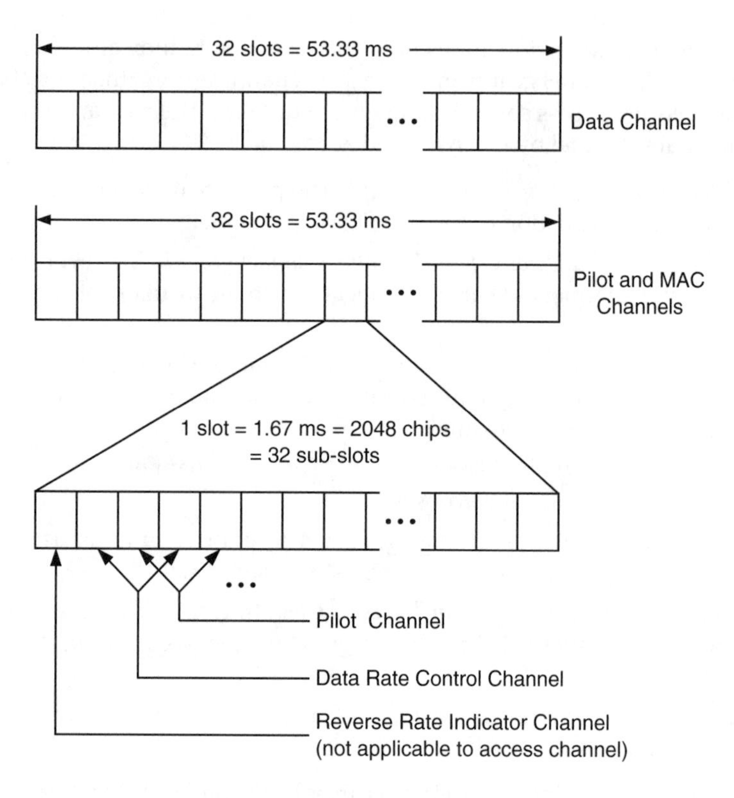

FIGURE 6.14
HDR reverse-link channel structure.

receives full power from the access point. This provides high peak rates for users within good coverage areas. The transmission of the pilot bursts by the access points at their maximum power and at predetermined time intervals allows the access terminals quickly to accomplish an accurate estimate of the channel conditions expressed in terms of the effective SNRs received from different access points. The highest SNR can be selected and mapped onto a value representing the maximum data rate that can be supported for a given error rate performance. Such information is sent by the access terminal to the best-serving access point on the data rate request channel and updated on a slot-by-slot (1.67 ms) basis. Only the best-serving access point transmits to the respective access terminal at the requested data rate. Note that this differs from the IS-95 solution in which more than one base station may serve one terminal.

Users in the *connected* state are assigned Walsh covers on which to look for the preamble that indicates whether or not the data in the corresponding slot is addressed to the respective user. Although 32 Walsh covers are available, only 29 can be assigned to the users, with the remaining three used for other purposes. Therefore, a maximum of 29 simultaneous users can actually be in the connected state, actively transmitting or receiving packets. A larger number of users may be supported depending on the channel activity. For bursty traffic, which is usually the case in Internet applications, some users may not be actively using the link. In this case, the connection becomes *dormant* and more users may be included within the system. Note that the total throughput of an HDR forward link is shared by all active users, with users allowed to transmit packets of variable lengths. Therefore, the smaller the number of active users, the higher the achievable throughput per user.

The HDR system makes use of a scheduling algorithm to maximize its throughput. The scheduling algorithm takes advantage of two facts as follows. In general, different users require different services, therefore different data rate requirements. In addition, the channel conditions vary from user to user. Note that the different services in connection with the channel variability lead to time-varying requested data rates. Basically, the scheduling algorithm makes use of the proportional fairness scheduler (PFS) in which only those users near their peaks of requested rates are allowed to be served, and the disadvantaged users are not then served. PFS maximizes the product of the throughputs delivered to all of the users such that the increase of the throughput of one user by a given proportion is accomplished at the expense of the decrease of the cumulative throughput of all users by the same proportion. More specifically, the PFS is implemented as follows. The instantaneous data rate requests on the data rate request channel are filtered over a certain number of slots (100 slots as specified in Reference 14) by means of the moving average algorithm. The instantaneous data rate requests are compared with the filtered average and only those data rates requests above the filtered average will be served. It is claimed[14] that a substantial increase in the throughput is obtained with the application of such an algorithm. "PFS takes advantage of the channel variation over the short term to increase throughput but maintain the grade of service fairness over longer periods."[14]

6.4.6 Throughput Performance

HDR forward and reverse links operate in different modes, as already mentioned. The forward link supports dynamic data rates, whereas the reverse link offers fixed data rates for a given connection. Table 6.3 shows the theoretical HDR forward-link throughput. Note that different modulations schemes are used. In particular, QPSK is used to achieve 38.4 through 1228.8 kbit/s, 8-PSK for 1843.2 kbit/s, and 16-QAM for 2457.6 kbit/s. Based on simulation studies and according to Reference 14 the required SNRs, expressed in terms of received signal energy density and the total nonorthogonal single-sided noise density (intercell interference, thermal noise, and nonorthogonal

TABLE 6.3

HDR Forward-Link Throughput

Data Rate (kbit/s)	Packet Length (bits)	Packet Duration (ms)	Number of Slots	Modulation Scheme
38.4	1024	26.67	16	QPSK
76.8	1024	13.33	8	QPSK
102.4	1024	10.00	6	QPSK
153.6	1024	6.67	4	QPSK
204.8	1024	5.00	3	QPSK
307.2	1024	3.33	2	QPSK
614.4	1024	1.67	1	QPSK
921.6	3072	3.33	2	QPSK
1228.8	2048	1.67	1	QPSK
1843.2	3072	1.67	1	8-PSK
2457.6	4096	1.67	1	16-QAM

intracell interference), for the various rates are shown in Table 6.4. Table 6.5 shows the theoretical HDR forward-link throughput. Note that BPSK is used to achieve 4.8 through 153.6 kbit/s and QPSK for 307.2 kbit/s. The access terminal is granted free permission to transmit at 4.8, 9.6, and 19.2 kbit/s. In order for the access terminal to transmit at rates above 19.2 kbit/s, permission must be granted by the access point, and this will depend on the link activity.

In Reference 14, the average throughput (bits/sector) of an HDR forward-link sector for a system consisting of two three-sector cells and one omni-cell site is reported. The throughput has been observed to increase with the increase of the number of users, a consequence of the use of the scheduling algorithm in which the more-advantaged users are preferentially served. The increase, however, seems to reach a saturation point around ten users per

TABLE 6.4

HDR Forward-Link Throughput and SNR

Data Rate (kbit/s)	SNR (dB)
76.8	–9.5
102.4	–8.5
153.6	–6.5
204.8	–5.7
307.2	–4.0
614.4	1.0
921.6	1.3
1228.8	3.0
1843.2	7.2
2457.6	9.5

TABLE 6.5

HDR Reverse-Link Throughput

Data Rate (kbit/s)	Packet Length (bits)	Packet Duration (ms)	Number of Slots	Modulation Scheme
4.8	256	53.33	32	BPSK
9.6	512	53.33	32	BPSK
19.2	1024	53.33	32	BPSK
38.4	2048	53.33	32	BPSK
76.8	4096	53.33	32	BPSK
153.6	8192	53.33	32	BPSK
307.2	8192	53.33	32	QPSK

sector. Single-antenna and dual-antenna access terminals with maximal ratio combining have been tested. In the first case (single-antenna access terminal), the average throughput is reported to be 750 kbit/s for stationary users and 500 kbit/s for nonstationary users. In the second case (dual-antenna access terminal), the figures are reported to be 1.05 and 900 Mbit/s, respectively. Results for indoor stationary users have been obtained by means of simulation; the average throughput is 650 kbit/s for a single-antenna access terminal and 1.15 Mbit/s for a dual-antenna access terminal.

6.4.7 Handoff Features

In HDR systems, the downlink communication follows the *best server* rule. In other words, the access terminal communicates with only one access point, the one presenting the highest SNR. Therefore, as opposed to IS-95 systems, in which the terminal may combine signals from different base stations for soft handoff purposes, in HDR, only one access point—the best server—is selected at any instant, with the best server selection occurring rapidly from one access point to another. Note that, whereas a variability in user throughput or in user latency is unacceptable for voice services, for packet data services this is quite legitimate. And this is the reason soft handoff is a mandatory feature in IS-95 systems and totally dispensable in HDR networks. The uplink communication in HDR systems enjoys the same feature as that of the IS-95 networks. In other words, the access terminal may communicate with more than one access point, and a frame selection procedure is carried out to have the best frame selected.

Assuming dual-mode terminals, several handoff scenarios between HDR and IS-95—or its evolved 3G version (the so-called multicarrier or MC)—systems are envisaged. For ease of notation, both IS-95 and MC are here referred to as simply MC.

- *HDR system and MC system with nonoverlapped service areas.* Assume that the access terminal is found within an area served by either an HDR system or an MC System. As it moves toward the other system, a handoff may be performed in the active mode or in the dormant mode. This occurs on the airlink side and on the network side. On the airlink side, the access terminal is informed that it is going out of range and that it should look for pilots from both systems. The system with the strongest pilot is then chosen to host the subscriber. On the network side, the access terminal performs the registration, as required.
- *HDR system and MC system with overlapped service areas.* Assume that the access terminal is found within an area served by both an HDR system and an MC system. Consider that a page for an incoming voice service instance from the MC system is received while the access terminal is exchanging data within the HDR system. The following may be carried out: the access terminal may continue the data call on the MC system; or the access terminal may abandon the HDR data services instance, handoff to the MC system, and continue with the voice service only. The access terminal is able to detect the paging because it monitors the MC system forward common channel even though a data connection is in process.

6.4.8 HDR Summary

HDR technology constitutes an efficient solution for data services that can be aggregated into existing 2G CDMA cellular systems. It is designed to be highly interoperable with CDMA networks, the design leveraging in many ways the lessons learned from the development and operation of IS-95 networks. HDR systems separate low-rate data services (voice) from high-rate data services by using nonoverlapping spectrum allocations for these different services. In essence, the HDR solution for a CDMA system in which both types of services are provided uses separate carriers for the respective services. An IS-95 RF carrier is used for voice applications and a separate HDR RF carrier is used for data applications. Within an HDR RF carrier, forward and reverse links are designed to operate in distinct modes. Packet transmissions on the forward link are time-division-multiplexed and occur at full power and at variable rates, with packets having variable durations (packets have variable sizes). Hence, it may be said that HDR forward link supports dynamic data rates. Packet transmissions on the reverse link are code-division-multiplexed and occur at variable rates, but with packets having fixed durations (packets have fixed sizes). The HDR reverse link does not support dynamic data rates. HDR designers[14] claim that the HDR peak rate may reach 2457.6 kbit/s on the forward link and 307.2 kbit/s on the reverse link.

6.5 Summary

The demand for mobile access to Internet applications has impelled designers to look for efficient solutions that provide for packet data services. Within this framework, three data technologies applied to wireless networks appear as good alternatives to be used for packet applications in wireless systems: GPRS, TIA/EIA/IS-95B, and HDR. GPRS technology aimed initially at the GSM community. More recently, however, with the widespread use of end-user applications, such as Web browsing and e-mail, the Internet has become the dominant driving force in the standardization of GPRS, and technologies other than GSM also incorporate GPRS. Based on packet-switched principles, GPRS provides optimized access to the Internet while reusing to a large extent existing wireless infrastructure. TIA/EIA/IS-95B makes full use of the TIA/EIA/IS-95A platform and provides data rates compatible to GPRS. HDR technology, in turn, is designed to be highly interoperable with CDMA systems, the design leveraging in many ways the lessons learned from the development and operation of IS-95 networks. The first fundamental design choice in HDR is to separate low-rate data services (voice) from high-rate data services by using non-overlapping spectrum allocations for these different services. In essence, the HDR solution for a CDMA system in which both types of services are provided uses separate carriers for the respective services. An IS-95 RF carrier is used for voice applications and a separate HDR RF carrier is used for data applications. The HDR forward link supports dynamic data rates, whereas the HDR reverse link operates with packets with fixed lengths.

References

1. Digital Cellular Telecommunications System (Phase 2+), General Packet Radio Service, Channel Coding, Stage 2, GSM 05.03 v. 6.0.0, January 1998.
2. Digital Cellular Telecommunications System (Phase 2+), General Packet Radio Service, GPRS Tunneling Protocol across Gn and Gp Interface, Stage 2, GSM 09.60 v. 6.2.2, November 1998.
3. Digital Cellular Telecommunications System (Phase 2+), General Packet Radio Service, Logical Link Control (LLC) layer specification, GSM 04.60 v. 6.3.0, March 1999.
4. Digital Cellular Telecommunications System (Phase 2+), General Packet Radio Service, Radio Link Control/Medium Access Control (RLC/MAC), GSM 04.64 v. 6.3.0, March 1999.

5. Digital Cellular Telecommunications System (Phase 2+), General Packet Radio Service, Subnetwork Dependent Convergence Protocol (SNDCP), GSM 04.65 v. 6.3.0, March 1999.
6. Digital Cellular Telecommunications System (Phase 2+), General Packet Radio Service, Multiplexing and Multiple Access on the Radio Path, GSM 05.02 v. 6.6.0, March 1999.
7. Digital Cellular Telecommunications System (Phase 2+), General Packet Radio Service (GPRS), Service Description, Stage 2, GSM 03.60 v. 6.3.0, April 1999.
8. Digital Cellular Telecommunications System (Phase 2+), General Packet Radio Service, Overall Description of the GPRS Radio Interface, Stage 2, GSM 03.64 v. 6.2.0, May 1999.
9. Digital Cellular Telecommunications System (Phase 2+), General Packet Radio Service, Service Description, Stage 1, GSM 02.60 v. 6.2.1, August 1999.
10. Kalden, R., Meirick, I., and Meyer, M., Wireless internet access based on GPRS, *IEEE Personal Commun.*, 8–18, April 2000.
11. Qiu, X., et al., RLC/MAC design alternatives for supporting integrated services over EGPRS, *IEEE Personal Commun.*, 20–33, April 2000.
12. Meyer, M., TCP Performance over GPRS, *IEEE Wireless Commun. Networking Conf.*, New Orleans, LA, September 1999.
13. Hoff S., Meyer, M., and Sachs, J., Analysis of General Packet Radio Service (GPRS) of GSM as access to the Internet, ICUPC'98, Florence, Italy, October 1998.
14. 1× High Data Rate (1×HDR) Airlink Overview, Qualcomm, Inc., April 28, 2000, Revision 3.1.
15. Bender, P. et al., CDMA/HDR: a bandwidth efficient high speed wireless data service for nomadic users, *IEEE Personal Commun.*, 20–33, April 2000.

Part IV

3G Systems

7

IMT-2000

7.1 Introduction

The conception of third-generation (3G) wireless systems is embodied by the International Mobile Telecommunications-2000 (IMT-2000). IMT-2000 standards and specifications have been developed by various standards organizations worldwide under the auspices of the International Telecommunications Union (ITU). 3G telecommunication services target both mobile and fixed users, with the access provided via a wireless link. A wide and ambitious range of user sectors, radio technology, radio coverage, and user equipment is covered by IMT-2000. In essence, a 3G system must provide for:

- Multimedia services, in circuit-mode and packet-mode operations
- User sectors such as private, public, business, residential, local loop, and others
- Terrestrial-based and satellite-based networks
- Personal pocket, vehicle-mounted, or any other special terminal

Moreover, global roaming and virtual home environment (VHE) are supported by 3G systems. Global roaming capability allows users to roam across different wireless networks, and the VHE characteristic provides roamers with the set of services and features of their home network. The key features of IMT-2000, as specified in Reference 1, include:

- High degree of commonality of design worldwide
- Compatibility of services within IMT-2000 and fixed networks
- High quality
- Small terminal for worldwide use
- Capability for multimedia applications and wide range of services and terminals

IMT-2000 may be implemented as a *stand-alone network* or as an *integral part of the fixed networks*. In the first case, gateways and interfacing units must be provided so that internetworking operations can be supported. Note that this has been the standard solution for the internetworking operation between conventional public land mobile and fixed networks and also between conventional mobile networks of different technologies. In the second case, the fixed network must accommodate the functionality specific to the mobile networks, such as location registration, paging, handover, and others. This is feasible with the development of IN (intelligent network) capability and the use of exchanges with ISDN (Integrated Service Digital Network) and B-ISDN (Broadband-ISDN) capabilities.

3G systems provide for modularity in terms of both capacity and functionality. A start-up network may gradually evolve from a simple and small 3G system to a large network with complex 3G applications. In this sense, the IMT-2000 functional model is developed to be flexible to accommodate a wide range of 3G applications while meeting the necessary requirements of quality of service (QoS).

As opposed to the fixed QoS (defined as *best effort*) in the 1G and 2G wireless systems, 3G systems provide for flexible QoS. Users and their applications can be assigned a default QoS with negotiations for a suitable QoS occurring as desired. Resources are then allotted depending on a series of parameters such as QoS profiles authorized by subscription, system load, propagation conditions, type of traffic (which is dependent on the application), and others.

In summary, the key goals for 3G networks are the following:

- Universal adoption of a core set of standards for the air interface
- Promotion of global roaming
- Efficient support of a wide range of data services including multimedia

In practical terms, 3G systems are deployed with two main objectives:

1. To support packet data services with speed and quality available on fixed networks
2. To provide Internet access

7.2 Some Definitions

This section lists the definitions of some terms[3] that are used in a more specific sense in the context of this chapter.

- *Access Link*. An access link constitutes an aggregation of logical channels supporting a connection link between the mobile terminal and the core network. It comprises two segments: the access radio link and the BS approach link.
- *Access Radio Link*. An access radio link is the radio portion of the access link.
- *Association between an IMT-2000 Terminal/User and the Network (Terminal/User Association)*. A terminal/user association corresponds to a logical association between the IMT-2000 terminal/user and the network. Such an association is established at the first outgoing or incoming call of the terminal/user at the idle state. It is used by the network to identify the IMT-2000 terminal/user among all the IMT-2000 terminals/users having control relationships with it. It is kept active until all the calls and connections on the terminal are released.
- *Authentication Data*. Authentication data encompass authentication parameters and authentication information as used for authentication purposes.
- *Authentication Information*. The authentication information constitutes the information used for user authentication (e.g., the triplet challenge, response, and ciphering key).
- *Authentication Parameter*. An authentication parameter constitutes secret data utilized for individual user authentication (e.g., authentication key, shared secret data).
- *Bearer*. A bearer is a communication path between adjacent nodes associated with a given connection.
- *Bearer Control*. The bearer control supports the node-to-node control of network resources providing for an end-to-end carriage of information.
- *BS Approach Link*. A BS approach link is the segment of the access link which, together with the access radio link, composes the access link.
- *Call*. A call is defined as an end-to-end logical association between two or more parties. It is associated with a service request.
- *Call Control*. A call control constitutes a set of functions (service negotiation, setup, modification, and release) performed to process a call as a result of a service request. One call may trigger connections ranging from none to several connections.
- *Camp on a Cell*. A mobile terminal is said to "camp on a cell" when it lodges temporarily within a cell after the cell selection/reselection process is completed.
- *Connection*. A connection consists of an end-to-end association of network resources or entities to provide means for a transfer of

information between points within the network. A connection is established by the union of connection links.

- *Connection Control.* A connection control constitutes a set of functions performed to set up, maintain, and release a communication path between users or between a user and a network entity.
- *Connection Link.* A connection link constitutes a part of a connection between two connection control functions.
- *Diversity Convergence Point.* A diversity converging point is the point into which physical bit streams merge to form one logical information stream.
- *Diversity Branch.* A diversity branch corresponds to a path diverging from the diversity point.
- *Diversity Handover.* Diversity handover is a handover that uses macrodiversity techniques.
- *Diversity Link.* A diversity link is a set of diversity paths.
- *Diversity Path.* A diversity path is a branch of a diversity link.
- *Handover Branch.* A handover branch constitutes an access radio link branch that takes part in a handover.
- *Handover Path.* A handover path constitutes a sequence of link elements that take part in a handover.
- *Link Element.* A link element is a logical channel connecting two adjacent functional nodes.
- *Macrodiversity.* Macrodiversity is a kind of diversity in which the radio paths of redundant information are separated by large distances as compared to the propagated radio wavelength. In the uplink, the communication is characterized by a point-to-multipoint connection with the mobile terminal signal captured by several base stations. In the downlink, the communication is characterized by a multipoint-to-point connection with the signals from several base stations captured by a single mobile terminal. The macrodiversity techniques include base station diversity, diversity handover, simulcast, and others.
- *Radio Resource.* The radio resource is the portion of the radio spectrum available within the cell.

7.3 Frequency Allocation

The World Administrative Radio Conference 1992 (WARC 1992), Torremolinos, Spain, identified two frequency bands, a total of 230 MHz, to be used for IMT-2000 applications. The identified frequency bands spanned from 1885 to 2025 MHz and from 2110 to 2200 MHz. Later on, the World Radio Conference 2000 (WRC 2000), Istanbul, Turkey, identified three other frequency bands, an additional 519 MHz for IMT-2000 applications. These new bands spanned from 806 to 960 MHz, from 1710 to 1885 MHz, and from 2500 to 2690 MHz. These bands are intended to be available on a global basis for countries wishing to implement IMT-2000 systems. Note that, by allowing these bands to overlap with those bands already used for other wireless services, the 2G cellular systems included, a high degree of flexibility is provided with the operators able to migrate toward IMT-2000 according to the market and other national considerations.

7.4 Features and Services

The features and services provided by IMT-2000 systems are defined as *service capabilities* in the ITU-T Recommendations. An extensive list of features and services of IMT-2000 networks, divided in several categories, are listed in Reference 1 under the item *Capability Set 1*. This section presents the primary capabilities. In the descriptions that follow, the terms *system* and *family member* are used interchangeably; these concepts are better defined in the next section.

The capability categories and the respective capabilities are described next.

- *Existing Capability*. This category recommends that the 2G core fixed and mobile services and capabilities be kept or possibly enhanced.
- *Long-Term Objectives*. This category traces the long-term objectives of IMT-2000 and establishes that distinct improvement over 2G systems in the areas of voice, data messaging, image, and multimedia be supported. In particular, enhanced roaming, increased data rates, and multimedia and Internet wireless services are targeted.
- *Bearer Capability*. The following constitute some of the bearer capabilities:

  Terrestrial access with BER $\leq 10^{-6}$ for transmission rates of at least:

  1. 144 kbit/s in vehicular radio environment.
  2. 384 kbit/s in outdoor to indoor and pedestrian radio environment.
  3. 2048 kbit/s in indoor office environment, for both circuit and packet services.

  Data rates for satellite access varying with the operating environment and with the type of terminal and ranging from 9.6 to 144 kbit/s.

  QoS negotiated real time/non-real time basis and according to the delay characteristics, maximum acceptable bit error rate, and bit rate/throughput.

Packet services on radio as well as on fixed interfaces.

Communication configurations for bidirectional point-to-point and point-to-multipoint services, the latter comprising broadcast and multicast capabilities.

Types of communication including connectionless-oriented network service and connection-oriented network service.

Access links of the symmetric and asymmetric type.

Fixed and variable bit rate traffic.

- *Access Network Capability*. The following constitute some of the access network capabilities:

  Packet services with negotiable parameters such as bit rate, delay tolerance, and reliability classes (probability of data loss, out of sequence delivery, etc.).

  Constant bit rate with timing and variable bit rate with timing for connection-oriented network, constant bit rate without timing and variable bit rate without timing for connectionless-oriented network, and efficient link layer recovery.

  Radio resource control capabilities including radio channel quality monitoring, macrodiversity monitoring, channel allocation, and power control.

  Fixed wireless access.

- *Core Network Capability*. The following constitute some of the core network capabilities:

  Constant bit rate with timing and variable bit rate with timing for connection-oriented network, constant bit rate without timing and variable bit rate without timing for connectionless-oriented network, and efficient link layer recovery.

  Handling of voice, data, and video by means of both circuit and packet communication.

  Internetworking with ISDN, B-ISDN, X.25 PDN (Packet Data Network), IP (Internet protocol) networks, and PSTN (Public Switched Telephone Network).

  Terminal mobility, personal mobility, and service mobility.

  Internetworking with IP networks and provision of Internet-type services.

  Global roaming and service interoperability between IMT-2000 family members.

  Support of packet-switched and circuit-switched operations.

  Support of evolved family member network architecture.

Support of open interfaces with intelligent network servers, and others.

- *Network Capabilities—Call Control*. The following constitute some of the network capabilities concerning call control:

  Multiple simultaneous calls per terminal or directory number, multimedia calls, call internetworking procedures, emergency calls, and priority calls.

  Geographic positioning of a terminal or user.

  Multiconnection calls independent of connection characteristics.

- *Network Capabilities—Security Procedures*. The following constitute some of the network capabilities concerning security procedures:

  User authentication and ciphering, service-dependent authentication and ciphering meconiums, and user–network mutual authentication.

  Terminal identification.

  Prevention of fraudulent (unauthorized) use.

  Privacy of data and user message.

- *Network Capabilities—Resource Allocation*. The following constitute some of the network capabilities concerning resource allocation:

  Allocation based on QoS, overload control, mixed services configurations, and route optimization.

- *Network Capabilities—Numbering and Addressing*. The following constitute some of the network capabilities concerning numbering and addressing:

  Support of numbering and addressing portability.

  Support of conventional and advanced addressing and numbering plans.

  Identity management for terminal, international mobile user, subscriber ISDN, multicast group.

- *Network Capabilities—Charging and Accounting*. The following constitute some of the network capabilities concerning charging and accounting:

  Standardized billing and charging user profiles and new charging mechanisms based, for example, on the traffic volume, QoS, time, etc.

  Real-time charging, third-party charging, prepaid billing, location-dependent billing and charging, and real-time access to billing information.

Charging information generation for circuit-switched calls, packet data transmission, and signaling traffic.

- *Network Capabilities—Roaming*. The following constitute some of the network capabilities concerning roaming:

  Mobility and global roaming including location management, user registration and cancellation, security and authentication database management and control, and others.

  Ability to supplement mobility management and authentication control with intelligent network–type service logic.

- *Network Capabilities—Service Portability*. The following constitute some of the network capabilities concerning service portability:

  The capabilities under service portability basically concern the VHE feature. VHE, as previously defined, provides roamers with the set of services and features of their home network.

- *Network Services/Features—Handover*. The following constitute some of the network services/features concerning handover:

  Support of hierarchical cell structure with call transfer and handover across cell layers and location management within multiple cell layers.

- *Network Services/Features—Service Provisioning*. The following constitute some of the network services/features concerning service provisioning:

  Provision for over-the-air service such as support for voice and data, and security and authentication.

- *Network Services/Features—Quality of Service*. The following constitute some of the network services/features concerning QoS:

  Support of QoS based on subscription, or on negotiation during a service invocation, or on renegotiation during a service session.

  Provision for QoS equivalent to that of the wired access.

  Fulfillment of minimum delay requirements.

- *Network Services/Features—Supplemental Support*. The following constitute some of the network services/features concerning supplemental support:

  Support of cordless telephone access, virtual private networks, IP-based services, satellite access, media transparency, and operator services.

- *Network Services/Features—Terminals and User Interface Modules*. The following constitute some of the network services/features concerning terminals and user interface modules:

Provision for a network model to support uploading and downloading of user profiles, data information capabilities, software configurable terminals, future enhancements in software-defined radios.

Support of mobiles and user interface modules with downloading capabilities over the air for data and applications.

Support of multiple calls on a single terminal and multiple registration of one user on several terminals for different services.

- *Network Services/Features—Packet Transfer Control.* The following constitute some of the network services/features concerning packet transfer control.

  Support of registration/authentication, static and dynamic address assignment, sleep mode for battery power conservation, optimal packet routing, multiprotocol, data compression, internetworking, location identification, load balancing across RF channels, multiple simultaneous address registrations on a single terminal, priority access, and multimedia sessions.

As mentioned before, the set of capabilities listed above corresponds to what is called Capability Set 1 in the ITU-T Recommendations. Future capability sets with enhanced and new capabilities shall be built upon the previous capability sets and shall support backward and forward compatibility between capability sets.

7.5 Traffic Classes

Four classes of traffic are defined to be supported by 3G systems: conversational, streaming, interactive, and background.

- *Conversational Class.* The conversational class is the most familiar class of traffic. It preserves the time relation between information entities of the stream and demands a constant and short end-to-end delay. An example application is voice.
- *Streaming Class.* The streaming class preserves the time relation between information entities of the stream. This class of traffic works within a small range of delays and throughput rates. It is characterized by applications for which the processing of the traffic can be started for presentation to the user before the whole file is transmitted to the subscriber. An example application is streaming video.
- *Interactive Class.* The interactive class requests response pattern and preserves payload content. This class of traffic works with an

intermediate range of delays and throughput. It is used by online applications in which a subscriber is allowed to interact with a server. Example applications include Web browsing, E-commerce, games, and location-based services.

- *Background Class.* The background class is characterized by applications for which the destination is not expecting the data within a certain time. This class of traffic preserves payload content. It works within a wide range of throughput rates and is very tolerant of delays but relatively intolerant of errors. Example applications include background download of e-mails, short messaging services, and file downloads.

7.6 IMT-2000 System and IMT-2000 Family

Two important concepts are formally defined within the IMT-2000 philosophy: *IMT-2000 System* and *IMT-2000 Family.*

An IMT-2000 System comprises a set of subsystems, entities, and interfaces designed to perform actions and interactions to provide its users with IMT-2000 capabilities as defined in the IMT-2000 Capability Set. Four functional subsystems, i.e., user identity module (UIM), mobile terminal (MT), radio access network (RAN), and core network (CN), compose an IMT-2000 System, as described next.

- *User Identity Module.* The UIM functional subsystem performs functions to support user security and user services. It may be implemented either as a removable physical card for an MT or it may be integrated into the physical MT.
- *Mobile Terminal.* The MT subsystem performs functions to support communication with the UIM and with the RAN. In addition, it supports user mobility and user services.
- *Radio Access Network.* The RAN subsystem performs functions to support communication with the MT and with the CN. It provides means for exchanging information between MT and CN, acting as a bridge, router, and gateway as required.
- *Core Network.* The CN subsystem performs functions to support communication with the RAN and with other CNs. It provides means to support user mobility and user services.

The implementation of each functional subsystem may require one or more physical platforms and such an implementation may appear in a number of arrangements. Figure 7.1 depicts the architecture of IMT-2000. In Figure 7.1,

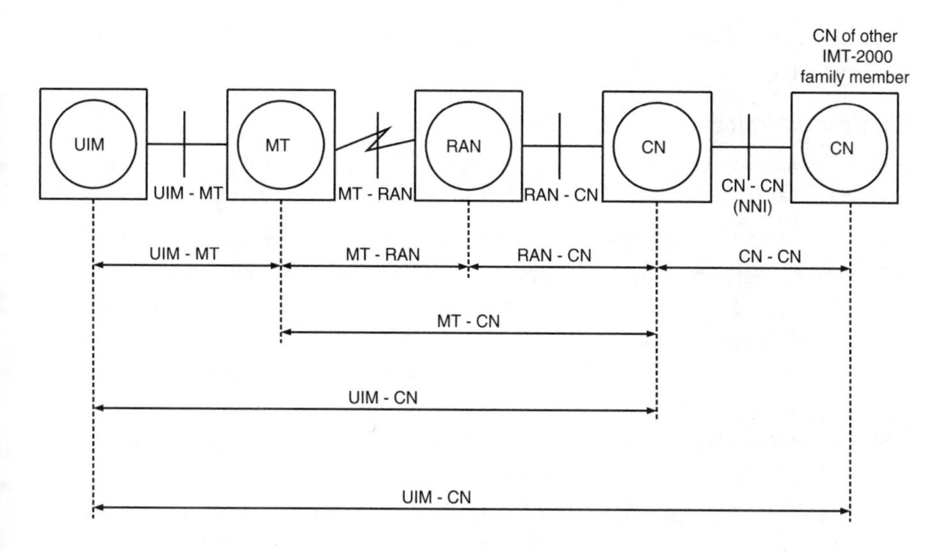

FIGURE 7.1
IMT-2000 architecture. Physical and functional interfaces of an individual IMT-2000 family member.

the circles indicate the functional subsystems and the squares represent the (set of) physical platforms where the respective functional subsystem resides. Note that any system functionally structured as described above and providing for the capabilities as defined in IMT-2000 Capability Set constitutes an IMT-2000 System. This gives rise to very different implementations leading to different IMT-2000 Systems. To accommodate all these systems under the same 3G principles the IMT-2000 Family concept emerged as a necessity.

"The IMT-2000 Family is a federation of IMT-2000 Systems providing IMT-2000 capabilities to its users as identified in IMT-2000 Capability Set."[1] An IMT-2000 family member is therefore an IMT-2000 System. Thus, the UIM, MT, RAN, and CN functional subsystems along with their internal processes, interactions, and communication may be specific to each IMT-2000 System (family member). On the other hand, any family member should be able to provide for support of users of other member systems in a roaming service offering and to yield a consistent set of service offerings based on IMT-2000 Capability Sets and interfaces.

7.6.1 Interfaces

The several interfaces identified for standardization within IMT-2000 are depicted in Figure 7.1. The vertical bars between the squares in Figure 7.1 indicate the physical interfaces between the respective physical platforms where the functional subsystems reside. The two-sided arrows between the functional subsystems in Figure 7.1 indicate the functional interfaces between the respective subsystems.

Physical Interfaces

Four physical interfaces, as shown in Figure 7.1, are identified in an individual IMT-2000 family member: UIM–MT, MT–RAN, RAN–CN, and CN–CN.

- *UIM–MT Interface.* The UIM–MT physical interface refers to the interface between the user-removable UIM and the MT. It is defined in terms of physical specifications including size, contacts, electrical specifications, protocols, and others. It supports the UIM–MT and UIM–CN (home and visited) functional communications.
- *MT–RAN Interface.* The MT–RAN physical interface is the radio interface between MT and RAN. It supports the UIM–CN, MT–RAN, and MT–CN functional communications.
- *RAN–CN Interface.* The RAN–CN physical interface provides means of connecting a RAN to different CNs or a CN to different RANs. Access technologies such as fixed radio, cordless terminal, satellite, and wireline may also be supported by this interface. It supports the UIM–CN, MT–CN, and RAN–CN functional communication.
- *CN–CN Interface.* The CN–CN interface is also referred to as network-to-network interface (NNI) in the ITU-T Recommendations. It provides means to interconnect different CNs supporting the UIM–CN, MT–CN, RAN–CN, and the several CN–CN functional communications.

Functional Interfaces

Seven functional interfaces, as shown in Figure 7.1, are identified in an individual IMT-2000 family member: UIM–MT, UIM–CN (home), UIM–CN (visited), MT–RAN, MT–CN, RAN–CN, CN–CN.

- *UIM–MT Interface, UIM–CN (home) Interface, and UIM–CN (visited) Interface.* The communication between the UIM and the MT conveys information to be processed within the MT itself or to be transferred to the CN (home or visited). Examples of classes of information flowing through these functional interfaces include:

  UIM access control (e.g., transfer of personal identification number for authentication purposes)

  Identity management (e.g., transfer of internationally unique subscriber identity)

  Authentication control (e.g., transfer of challenges and responses in the authentication process)

  Service control (e.g., transfer of user service profiles)

Man–machine interface control (e.g., transfer of user-specific man–machine interface configuration)

A UIM–MT functional communication provides means for the exchange of family-specific information between these two entities. Exchange of family-specific information is also supported by a UIM–CN (home) functional interface. A UIM–CN functional communication can support services (e.g., profile services and data services) requiring software download over a pseudotransparent data path established across the network.

- *MT–RAN Interface and MT–CN Interface.* The MT–RAN functional communication supports data protection and resource management. The MT–CN functional communication supports call control and mobility management, whose information flows transparently through the RAN.
- *RAN–CN Interface.* The RAN–CN functional communication supports bearer traffic (e.g., voice, data), control information (e.g., call, mobility), data security information, management information, and others.
- *CN–CN Interface.* The CN–CN functional communication comprises a set of three functional communications: Serving_CN-Home_CN functional communication, Serving_CN-Transit_CN functional communication, and CN–CN functional communication for Packet Data. These functional communications support information exchange between home, serving, and transit CNs, as required, for purposes such as:

Authentication control, subscriber-specific service control (for the VHE capability), and location information

Establishment of mobile-terminated calls for a roaming user as well as call and service control (e.g., call setup, negotiation of service capabilities)

Mobility management and delivery of bearer packet data

7.6.2 Global Roaming

The IMT-2000 architecture is conceived so that global roaming and VHE, two IMT-2000 key features, be fully supported. Global roaming is envisaged basically as two alternatives: *terminal mobility* and *UIM portability*. The global roaming terminal mobility alternative concerns the roaming of UIM and MT jointly. In this case, a UIM provided by the home network is used with an MT that is available both in the home network and in the visited network. The global roaming UIM portability alternative concerns the roaming of an UIM. In this case, a UIM provided by the home network is used with an MT provided by the visited network. These situations are illustrated in

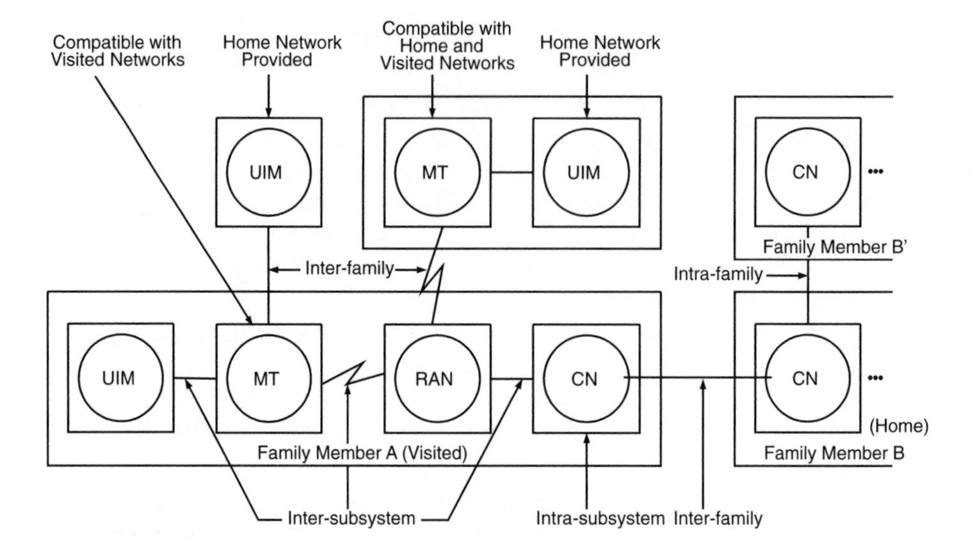

FIGURE 7.2
Global roaming.

Figure 7.2. Other features related to global roaming include *routing/addressing*, *service portability*, and *user profile accessibility/transportability*.

The global roaming routing/addressing feature concerns the ability of the network to address and route communications and services not to a geographic location or to a physical device but to roaming IMT-2000 users. The global roaming service portability feature concerns the ability of the network to grant subscribers access to customized services within and between any IMT-2000 systems. The global roaming user profile accessibility/ transportability feature concerns the ability of the network to handle (access, transfer, download, modify) the user's service profile independently of its location.

Figure 7.2 also illustrates the various interconnections between subsystems and the possible signaling streams within and between IMT-2000 subsystems and families. These interconnections and signaling are classified as being of the kind:[2] intra-subsystem, intersubsystem, intrafamily, and interfamily.

- *Intrasubsystem*. Intrasubsystem signaling is characterized by signaling flowing within a specific subsystem. For example, the signaling within a CN of a family member remains within the extent of the family member. This is outside the scope of the ITU-T standardization.
- *Intersubsystem*. Intersubsystem signaling is characterized by signaling flowing between two subsystems, either contained within the same or within different IMT-2000 family members. In the first case, the entity responsible for its specification is the respective family member

supplier, whereas in the second case the body responsible for its specification is the ITU-T. Interconnections between subsystems such as UIM–MT, MT–RAN, RAN–CN are of the intersubsystem type.

- *Intrafamily.* Intrafamily signaling is characterized by signaling flowing between the same IMT-2000 family members and its specification remains within the family member scope. In such a case ITU-T proves the framework for commonality. An interconnection between the CN of family member B and the CN of family member B′ is of the intrafamily type.
- *Interfamily.* Interfamily signaling is characterized by signaling flowing between two subsystems contained in different IMT-2000 family member systems. The body responsible for its specification is ITU-T. An interconnection between the CN of family member A and the CN of family member B is of the interfamily type.

In the global roaming context, it is informative to define more formally the types of logical networks required to support the interconnections:

- *Destination Network.* A destination network is the target network for an outgoing call from an IMT-2000 user.
- *Home Network.* A home network is the network of subscription of the IMT-2000 user. It maintains location and service profile information of the user on a permanent basis.
- *Interrogating Network.* An interrogating network is the network that requests routing data retrieval from the home network.
- *Previously Visited Network.* A previously visited network is the network visited by an IMT-2000 user immediately before entering the visited (serving) network.
- *Supporting Network.* A supporting network is the network that provides services to the IMT-2000 user. These services include, for example, logic programs and data for IN supplementary services.
- *Visited (Serving) Network.* A visited network is the network serving an active IMT-2000 user.

7.7 Specific Functions

To provide its users with IMT-2000 capabilities, as defined in IMT-2000 Capability Set, several network functions are identified. However, only those functions necessary to support the services chosen to be provided by a given family member need to be implemented within the respective family member.

Some of these functions are listed in Reference 3, where they are grouped according to their relations with the overall service and network capabilities. This section summarizes these functions.

7.7.1 Overall System Access Control Functions

The access to services and facilities of IMT-2000 are supported by functions controlling the means by which a user is connected to an IMT-2000 network. A system access may be initiated by the user (mobile-originated call) or by the network (mobile-terminated call). The following functions are identified:

- *System Access Information Broadcasting.* This function is responsible for providing the mobile terminal with configuration information to enable it to camp on a cell, register, or initiate and receive calls.
- *System Access Information Monitoring and Analysis.* This function enables the mobile terminal to monitor and analyze the system access–related information transmitted by the network. Upon processing such information, the mobile terminal should be able to camp on a suitable cell, belonging to a suitable network or network operator.
- *Cell Selection in Idle Mode.* This function enables the mobile terminal to control the tracking of the active cell while in the idle mode. This process involves tasks such as knowledge of the system configuration, signal quality monitoring, acquisition and selection of surrounding cells, and others.
- *Cell Selection in Packet Data Transfer Mode.* This function enables the mobile terminal to select the best cell when engaged in a packet data transaction. This process involves tasks such as signal quality monitoring, detection of traffic condition within the candidate cells, and others.

7.7.2 Radio Resource Management and Control Functions

Functions related to allocation and maintenance of physical channels are grouped under this item. The following functions are identified:

- *Synchronization Control.* This function performs tasks to ensure correct synchronization between the mobile terminal and the network.
- *Access Radio Link Setup and Release.* This function is responsible for tasks that take part in the processing of the setup and the release of a connection upon request of a given entity. These tasks also include the management and the maintenance of the radio link of such a connection.

- *Reservation and Release of Physical Channels*. This function performs reservation or release of physical channels in situations such as radio link setup or release requests, handover requests, service requests, and macrodiversity requests.
- *Allocation and Deallocation of Physical Channels*. This function is responsible for allocating or deallocating the physical channels that have been reserved.
- *Packet Data Transfer over Radio*. This function is responsible for tasks dealing with data transfer. In general, it provides for packet access control over radio channels, packet multiplexing over common physical radio channels, packet discrimination within the mobile terminal, error detection and correction, flow control procedures, and load balancing across RF channels.
- *RF Power Control*. This function controls the levels of transmitted power of both the mobile terminal and the base station upon assessing the radio channel quality.
- *RF Power Setting*. This function constitutes an intrinsic part of any power control scheme. It is used to adjust the output power of the transmitter.

7.7.3 Random-Access Functions

Functions related to random access are grouped under this item. The following functions are identified:

- *Random-Access Initiation*. This function initiates the specified random-access procedure when the mobile terminal accesses the network.
- *Random-Access Detection and Handling*. This function detects the random-access initiation attempt carried out by the mobile terminal. It handles the random access by responding to the access attempt, by activating procedures for collision resolution, by generating requests for resources allocation, etc.

7.7.4 Radio Resource Request Acceptability Functions

Functions related to radio resource request acceptability are grouped under this item. The following functions are identified:

- *Radio Resource Request Acceptability Information Setting*. This function checks the availability of radio resources within a cell.
- *Radio Resource Request Acceptability Information Broadcasting*. This function broadcasts, within a cell, the radio resource request acceptability information.

- *Radio Resource Request Acceptability Judgment*. This function uses the information on the availability of radio resources to initiate certain procedures. In the network side, the concerned procedures are related to the allocation of resources on request. On the mobile terminal side, the concerned procedures are related to the permission to initiate a service request.

7.7.5 Channel Coding Function

This item contains only one function as follows:

- *Radio Channel Source and Error Protection Coding and Decoding*. This function is related to the procedures used to protect the data transmission against errors. This is achieved by the use of convolutional codes for error detection and correction, of cyclic redundancy check codes for error detection, and of interleaving for error protection.

7.7.6 Handover Function

Functions related to handover are grouped under this item. The following functions are identified:

- *Radio Channel Quality Estimation*. This function monitors the radio channel conditions performing measurements to estimate parameters such as received signal strength, bit error rate, transmission range, Doppler spread, synchronization status, type of propagation environment (high speed, low speed, satellite, etc.), and others.
- *Cell Selection in Dedicated Mode*. This function makes it possible for a mobile terminal engaged in a circuit-switched connection to select the best cell for access or handover operations. The selection is based on the signal quality monitoring and on the traffic condition of the candidate cells.
- *Quality of Service Assessment*. This function concerns the overall QoS assessment to recommend or not resource reallocation, use of macrodiversity, or handover execution. QoS metrics may include radio signal quality, throughput, delay, and others.
- *Resource Reallocation*. This function carries out the reallocation of those resources that are being used. This is carried out to accomplish better performances.
- *Handover Decision*. This function generates a request for new or for different resources. A handover decision may be initiated by the mobile terminal or by the network.
- *Macrodiversity Control*. This function controls both the distribution of replicas of information through diversity branches and the

combination of them, as appropriate. This is activated upon request from the handover decision function or from the QoS assessment function.

- *Handover Execution*. This function performs tasks such as reservation/activation of the new radio and wireline resources required for handover and final switching from the old to the new resources.
- *Handover Completion*. This function releases the resources that are no longer needed.
- *Handover Trigger*. This function recognizes cells to be added to or deleted from the access link.

7.7.7 Location Management and Geographic Position–Finding Functions

Functions related to terminal paging, location data management, registration and deregistration, and location registration are grouped under this item. The following functions are identified:

- *Paging Decision and Control*. This function is responsible for identifying the location area of the mobile terminal, for determining the status of the mobile terminal (e.g., busy, idle, active) to execute or not the paging, and for processing the paging response by the mobile terminal.
- *Paging Execution*. This function carries out the paging of the mobile terminal.
- *Paging Detection*. This function detects a paging and responds to it.
- *Location Data Management Initiation*. This function requests service features related to location management, including terminal location updating, detach, and attach. This may be carried out on a periodic or demand basis.
- *Location Data Management*. This function is responsible for procedures controlling the feature services related to location management. Examples of tasks include location information updating, obtaining information on the reachability of the mobile terminal, and others. The location of the mobile terminal may be recognized with different accuracy (e.g., location unknown, multiple cell accuracy, cell area accuracy).
- *Terminal-Initiated Location Update*. This function enables the mobile terminal to inform the network of its presence.
- *Network-Initiated Location Update*. This function enables the network to request the mobile terminal to identify itself to the network.

- *Location Registration for Call and Service Delivery*. This function provides for an interaction between serving (visited) and home networks to enable service transparency. In this case, service profile downloading from the home network to the serving network may occur. Several scenarios may be considered. For standardized IMT-2000 services, if download occurs, the visited network will use the downloaded profile locally without requesting further information from the home network; if download does not occur, the visited network will have to request instructions from the home network.
- *Geographic Position Determination*. This function provides the network and the mobile terminal with means to determine the position of the mobile terminal, by taking into account the dynamic information available (e.g., signal strengths, time of arrival of the signals, angle of arrival of the signals).
- *Geographic Position Notification*. This function enables the notification of the position of the mobile terminal to the authorized entities.

7.7.8 Mobile Call Handling Functions

Functions related to call handling and routing are grouped under this item. The following functions are identified:

- *Service Feature Analysis*. This function checks the compatibility between requested services, current subscription, and terminal capabilities.
- *Provision of Terminal Capability Information*. This function provides the network with the information on the capabilities of the terminal.
- *Negotiation of Data Rates and Quality of Service*. This function handles the negotiation of data rates and QoS between the mobile terminal and the network.
- *Access Restrictions*. This function checks for restrictions to services according to subscription options, authorization, and network conditions.
- *Request Routing Information*. This function activates the routing information handling function for mobile-terminated calls.
- *Routing Information Handling*. This function provides the network with the necessary routing information for the call to be established.
- *Routing of Packet-Data Mobile Communication Service*. This function provides means to support the routing of packet-data communication service. The routing may be carried out on a dial-up basis or on a service subscription-based manner. In the first case, the user is allowed to establish an on-demand path of choice by providing the network with

the packet handler "address" as destination address information. In the second case, the user is allowed to specify the packet-handling function of choice at the time of subscription with the home IMT-2000 family member.

- *Connections and Address Management of Mobile Data Packets*. A number of functions under this item provide means to support point-to-point and point-to-multipoint services, dynamic and static address management, Internet control message protocol, packet filtering, and data privacy.
- *Handling of Multimedia Calls*. This function provides the system with the ability to support multimedia services. This includes the ability to provide bearers with flexible QoS, parallel calls, and point-to-multipoint calls.
- *Management of Circuit-Switched and Packet-Switched Communications*. This function supports circuit-switched and packet-switched communication paths simultaneously.

7.7.9 Data Coding and Compression Functions

Functions related to data coding, signaling compression, and data compression are grouped under this item. The following functions are identified:

- *Data Coding*. This function implements voice coding or compression of image data.
- *Signaling Compression*. This function implements the compression of signaling information.
- *Data Compression*. This function implements the compression of network protocol packet contents.

7.7.10 Network Intelligence and Service Control Functions

Functions related to the network intelligence and service control in general are grouped under this item. The following functions are identified:

- *Support of UPT Users*. This function grants users with a UPT (universal personal telecommunications) number the ability to access telecommunication services from any terminal.
- *Support of Service Portability, Supplementary Services, Virtual Home Environment, and Global Roaming*. A number of functions provide the network and the users with the ability to have transparent access to their subscribed services while roaming.
- *Support of IN*. Several functions provide the network and the users with the ability to support IN procedures.

7.7.11 User Privacy and Network Security Functions

Functions related to ciphering, authentication, fraud/abuse control, and identity management are grouped under this item. The following functions are identified:

- *Confidentiality Control.* This function is closely related to the authentication mechanisms. This is regarded as a centralized function. It provides the required information for the radio channel ciphering/deciphering functions.
- *Physical Radio Channel Ciphering/Deciphering.* These are pure computational functions used to protect the radio-transmitted data against nonauthorized third parties.
- *Ciphering Execution Control.* This function obtains ciphering information from the confidentiality control function and triggers ciphering and deciphering on the physical radio channel.
- *Authentication Data Management.* This function controls and manages the authentication information used in the network.
- *User Authentication Processing.* This function triggers the user authentication procedure and processes the results.
- *Network Fraud/Abuse Control.* This function provides for mechanisms to protect the user and the network from fraudulent use of the system. This involves a combination of real-time and non-real-time analyses of events monitored by the networks. Monitored events typically include call addressing, geographic position, subscriber identity, network element address, and supplementary service invocation.
- *Identity Management.* A number of functions are used to prevent a permanent user identity from eavesdropping over the radio interface. Confidentiality of the identity is kept by performing a series of functions such as:

  Assigning a temporary identity to the mobile terminal

  Periodically updating the assigned temporary identity

  Retrieving a permanent user identity from another network

7.7.12 Emergency Services Functions

Functions related to emergency services are grouped under this item. The following functions are identified:

- *Identification of Emergency Calls.* This function is responsible for recognizing calls to emergency services.
- *Emergency Services Calls Handling.* This function is responsible for tasks that handle calls to emergency services. These tasks include, for

example, the provision for priority access to identified emergency services calls, the use of different charging procedures, the provision for enhanced capabilities (e.g., such as call back), the provision for enhanced information about the users of such services, including location or geographic position, and others.

7.7.13 Charging Functions

Functions related to charging are grouped under this item. The following functions are identified:

- *Circuit-Switching Information Generation*. This function collects charging information parameters related to circuit-switching connections. This is carried out at instants such as call setup, during a call, and at call release. Charging parameters include call duration, used bandwidth, invoked service instances, time-dependent charging rate, access charge rate, and registration charge rate.
- *Packet Data Information Generation*. This function collects charging information parameters related to packet data transfer connections. This is carried out at such instants as registration or attach, packet transfer, and at deregistration or detach. Charging parameters include number of exchanged packets, packet data rate or used bandwidth, average packet size, access charge rate, and registration charge rate.
- *Charging Processing*. This function processes the parameters received from the circuit-switching information generation function and packet data information generation function to yield parameters that are relevant to the billing and accounting functions.

7.7.14 Support of Users Function

This item contains only one function, as follows:

- *IMT-2000 Personal Mobility*. This function supports the mobility of UIM devices among mobile terminals. Note that UIM portability is possible only for situations in which UIM and mobile terminals are physically separated. Because IMT-2000 supports both integrated UIM–MT and separated UIM–MT this function constitutes an optional capability. The personal mobility capability enables the user to access the telecommunication services specified within the user's profile from any IMT-2000 mobile terminal that accepts the portable UIM.

7.7.15 Subscriber Data Management Functions

Functions related to subscriber data management perform tasks concerning the following: update and deletion of subscriber data within the serving network; fault recovery of subscriber data; and control of supplementary services (SS) data. The functions related to updating and deletion of subscriber data may be activated by events, such as change of subscription of basic or supplementary services (carried out by the operator); change of subscriber data (carried out by the operator); application, change, or removal of call barring (carried out by the operator); change of data concerning the subscriber's SS using a subscriber procedure (carried out by the subscriber). The functions related to fault recovery of subscriber data permits the recovery from faulty situations (e.g., invalid subscriber data, missing location data). The functions related to control of SS data enable the subscriber, or the network on behalf of the subscriber, to control the SS. The following functions are identified:

- *Insert Subscriber Data.* This function updates certain subscriber data (e.g., change of the subscriber data associated with the subscriber's bearer services, SS, VHE services, regional subscription, etc.).
- *Delete Subscriber Data.* This function deletes certain subscriber data (e.g., deletion of basic services, deletion of SS, or deletion of VHE services).
- *Reset.* This function is used to announce to serving networks that a failure within the home network has occurred.
- *Restore Data.* This function is used by the serving network to inform the home network that the information associated with the provided IMUI (international mobile user identity) is not valid or not available.
- *SS Data Handling.* This function updates, retrieves, or deletes the SS data.
- *SS Activation.* This function is used to activate or to deactivate the SS upon request of the subscriber.
- *SS Password Protection.* This function is activated to protect operations on SS requiring restricted access (e.g., registration of a password).
- *SS Subscriber Data.* This function enables the exchange of subscriber data between the subscriber and the serving network.
- *SS Invocation.* This function invokes the subscriber SS.

7.7.16 Messaging Service Management Functions

Functions used to support short message service (SMS) and application data delivery service (ADDS) are grouped under this item. SMS and ADDS make use of a message center acting as a store-and-forward center. SMS is a feature supporting limited-size messages sent to or from the mobile terminal. ADDS is a feature supporting general-purpose wireless data delivery (application/teleservice messages between service users). The following functions are identified:

- *Message Delivery*. This function is responsible for delivering the message to the mobile terminal.
- *Message Handling*. This function is responsible for receiving the message from the terminal and forwarding it to the message center.
- *Message Alerting*. This function is used to alert the home network about a message being stored in the message center for a mobile terminal.

7.7.17 Software-Configurable Terminals Functions

The functions supporting the software-configurable terminals capability are responsible for tasks controlling the mechanisms that allow applications to interact and operate with any mobile terminal (MT). These applications together with the related data may reside within the UIM, within the MT, within an external device, or they may be downloaded by the CN.

- *Capability Profile Exchange*. This function supports the exchange of service capability information between UIM, MT, and CN. For example, services capability of one functional subsystem may be provided to another functional subsystem (e.g., from MT to UIM or to CN, from UIM to MT or to CN, and from CN to MT or to UIM).
- *Application Data Transfer*. This function supports the exchange of applications and associated data between UIM, MT, and CN. For example, data and applications of one subsystem may be provided to another subsystem (e.g., from MT to UIM or to CN, from UIM to MT or to CN, and from CN to MT or to UIM).
- *Proactive Applications*. This function supports mechanisms through which applications can initiate actions to be taken by the MT. These may include: "display text from UIM or CN to MT; send short message; set up a voice call to a number held by the UIM, MT, or an external device; set up a data call to a number with bearer capabilities held by the UIM, MT, or external device; send a supplementary service control or service data; play tones in earpiece; initiate a dialogue with the user; provide local information from the MT to the UIM or to the CN; provide help information on each command involved in the dialogue with the user."[3]
- *Screening Service by UIM*. This function, if activated, confers the UIM the ability to allow, bar, or modify the call, the SS operation, or the service data operation at the time anyone of these services is requested.

- *Security*. This function provides those applications designed with the features in this capability with data confidentiality, data integrity, data sender validation, or any subset of these.

7.8 Network Architecture

The functions implementing the IMT-2000 capabilities are distributed among or grouped into functional entities (FEs). These FEs are then grouped into physical entities (PEs) which, in turn, are distributed among the functional subsystems composing an IMT-2000 system, namely, UIM, MT, RAN, and CN. The FEs are detailed in the next section. The PEs implementing the functional subsystems are described as follows.

The UIM functional subsystem is implemented by means of the UIM PE. The MT functional subsystem is implemented by means of the MT PE. The RAN functional subsystem is implemented by means of the base station (BS) and the radio network controller (RNC) PEs. The CN functional subsystem is implemented by means of the following PEs: authentication center (AC), drift MSC (DMSC), gateway location register (GLR), gateway MSC (GMSC), home location register (HLR), intelligent peripheral (**IP**),* mobile switching center (MSC), packet data gateway node (PDGN), packet data serving node (PDSN), serving control point (SCP), serving data point (SDP), and visitor location register (VLR).

7.8.1 Physical Entities—Reference Model

The PEs composing the functional subsystems are briefly described next. A detailed specification of these PEs can be obtained by defining the set of FEs composing each PE, which are described in the next section.

- *UIM*. The UIM is the only PE composing the UIM functional subsystem. It contains an FE that supports user security and user services.
- *MT*. The MT is the only PE composing the MT functional subsystem. It contains FEs that support call control agent, connection control agent, mobile control, mobile geographic position, mobile radio transmission and reception, packet service control agent, and radio access control agent.

* This chapter uses the bold notation **IP** for intelligent peripheral, and the normal notation for IP for Internet protocol.

- *BS*. The BS is one of the PEs composing the RAN functional subsystem. It contains FEs that support radio frequency transmission and reception and system access information broadcast.
- *RNC*. The RNC is one of the PEs composing the RAN functional subsystem. It contains FEs that support radio access control, connection control, geographic positioning, and access link relay.
- *AC*. The AC is one of the PEs composing the CN functional subsystem. It contains an FE that supports authentication management.
- *DMSC*. The DMSC is one of the PEs composing the CN functional subsystem. It contains an FE that supports access link relay.
- *GLR*. The GLR is one of the PEs composing the CN functional subsystem. It contains FEs that support authentication management and location management. The GLR is an optional node between the VLR and the HLR used to enhance the subscriber location data across network boundaries. For subscribers roaming away from the home IMT-2000 network, the GLR functions as an HLR toward the VLR and as a VLR toward the HLR. The roaming process for subscribers within the home IMT-2000 network does not involve the GLR.
- *GMSC*. The GMSC is one of the PEs composing the CN functional subsystem. It contains FEs that support call control, connection control, and service switching. The GMSC is used to contact the HLR for routing information so that the call can be routed to the corresponding visited MSC.
- *HLR*. The HLR is one of the PEs composing the CN functional subsystem. It contains FEs that support location management.
- *IP*. The IP is one of the PEs composing the CN functional subsystem. It contains an FE that supports specialized resources used for IN services, mobile multimedia services, application delivery services, and packet data transfer services.
- *MSC*. The MSC is one of the PEs composing the CN functional subsystem. It contains FEs that support access link relay, call control, connection control, geographic positioning control, service access control, and service switching.
- *PDGN*. The PDGN is one of the PEs composing the CN functional subsystem. It contains an FE that supports packet service gateway control.
- *PDSN*. The PDSN is one of the PEs composing the CN functional subsystem. It contains FEs that support geographic position control and packet service control.

TABLE 7.1

Reference Points and Interfaces

	G/VLR	GMSC	HLR	MSC	MT	PDGN	RNC	SCP	SDP
AC			F					I5	
BS					U		Q		
DMSC				E1			L		
G/VLR	D2	B2	D1	B1		V4		I4	
GMSC	B2	E3	C1	E2				I2	
HLR	D1	C1		C2		V2		I3	
IP		G2		G1				J	
MSC	B1	E2	C2				P	I1	
MT									
PDGN	V4		V2						
PDSN	V3		V1			N	M	I6	
RNC				P			R		
SCP	I4	I2	I3	I1				S	K
SDP								K	W
UIM					H				

- *SCP*. The SCP is one of the PEs composing the CN functional subsystem. It contains FEs that support service control and service data.
- *SDP*. The SDP is one of the PEs composing the CN functional subsystem. It contains an FE that supports service data.
- *VLR*. The VLR is one of the PEs composing the CN functional subsystem. It contains FEs that support authentication management and location management.

7.8.2 Interface—Reference Points

A number of reference points (34 all together) defining the interfaces between the several PEs within an IMT-2000 network are specified. Table 7.1 shows, in matrix form, the PEs and the existing reference points (interfaces) between these PEs.

7.9 Physical Entities and Functional Entities

The network functional model described in Reference 3 specifies a number of FEs identified to support the IMT-2000 capabilities. Two approaches for the allocation of functionalities related to call control and connection control are possible: (1) integrated call control and connection control FEs and (2) separated call control and connection control FEs. The distribution of FEs into the PEs are depicted in Figure 7.3 where Tx signifies Transit Exchange and the other abbreviations are as already previously defined. The Reference Model for IMT-2000 with integrated call control and connection control FEs is obtained in Figure 7.3 by eliminating the FEs shown in gray. The Reference Model for IMT-2000 with separated call control and connection control FEs is obtained in Figure 7.3 by including the FEs shown in gray.

The next subsections briefly describe the PEs in terms of their respective FEs. A detailed description of these FEs can be found in Reference 3.

7.9.1 User Identity Module

The following constitutes the FE within the UIM PE:

- *UIMF, User Identification Management Module.* The UIMF incorporates functions that are responsible for tasks related to the identification, authentication, and service handling in the UIM. These functions store user identification information that is used by the network and by the service provider to identify both the user and the MT. They also store location management-related information and security- and privacy-related information. The UIMF interacts with other FEs to exchange application information, to provide serving system selection information on location area identity, service availability, service preferences, to generate ciphering key, and to perform authentication.

7.9.2 Mobile Terminal

The following constitute the FEs within the MT PE:

- *CCAF′, Call Control Agent Function (enhanced).* The CCAF′ interfaces user with network call control functions providing service access for users. Tasks performed by CCAF′ include interaction with the user to establish, maintain, modify, and release a call or other service instance.
- *CnCAF, Connection Control Agent Function.* The CnCAF is an FE that is implemented for the situation in which call and connection control functions are separated. In this situation it interfaces the call control agent function with the radio access control agent function. It interacts with other FEs to establish, maintain, modify, and release connections. It also maintains information on the connection state.
- *MCF, Mobile Control Function.* The MCF is responsible for the service access control logic and processing at the mobile side of the radio interface. It interacts with other FEs to establish and release the association between an MT and the network; initiate location and MT status management; set up, maintain, modify, and release a signaling connection

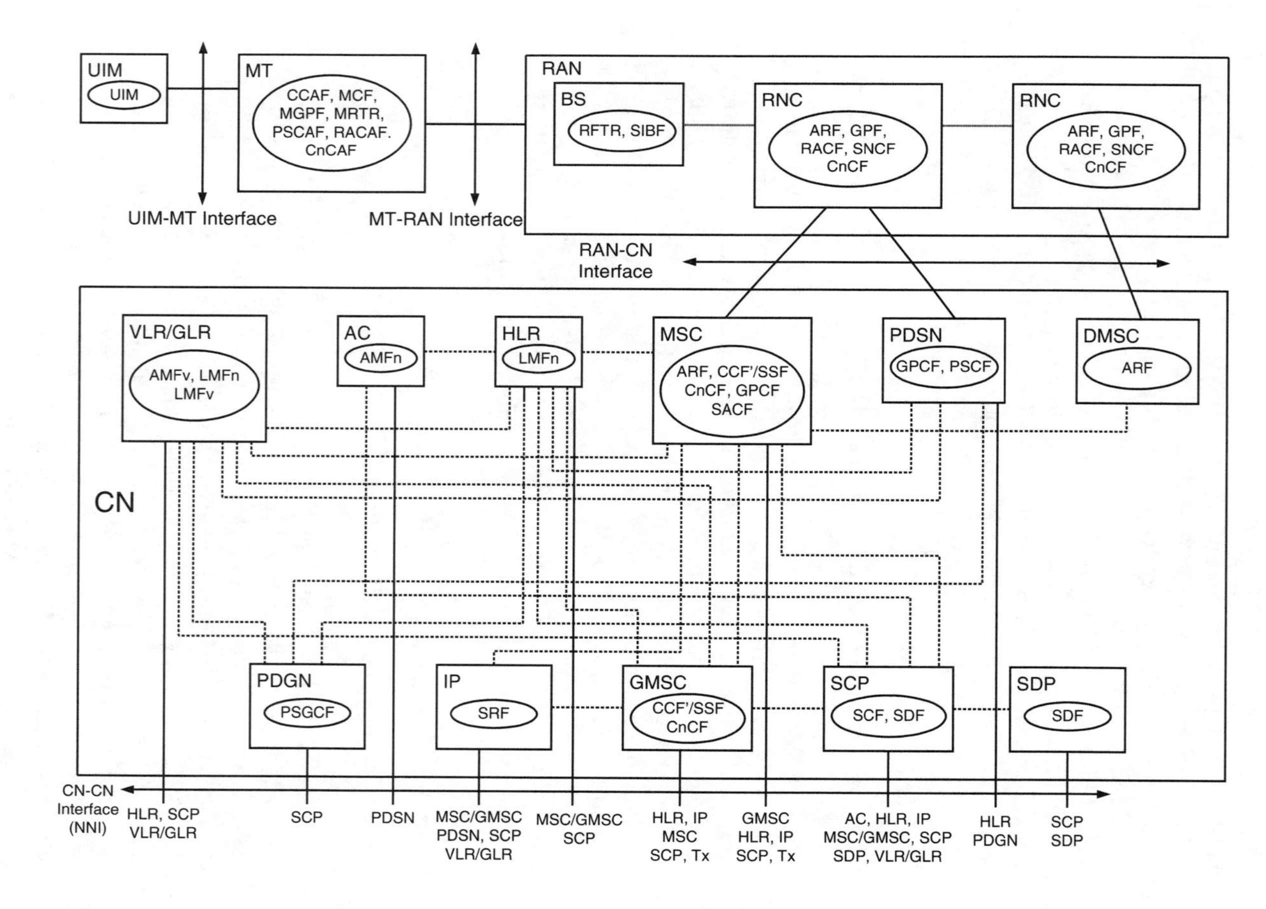

FIGURE 7.3
IMT-2000 reference model.

or an access channel; retrieve user identification information, location management–related information, and security- and privacy-related information; handle paging; manage unique authentication challenge activities; perform serving system selection.

- *MGPF, Mobile Geographic Position Function*. The MGPF, on the MT side, is responsible for tasks leading to the determination of the geographic position of the MT. Based on information received from the radio receiver, or from the GPS, or even from the network, it performs the computations to determine the geographic coordinates of the MT. It interacts with other FEs to perform RF measurements or to modify the transmitted RF signals so that the network can perform these measurements.
- *MRTR, Mobile Radio Transmission and Reception*. The MRTR performs functions related to the control of the interconnection and adaptation of the access radio link to the rest of the MT. These functions include tasks such as maintenance of the state of an access radio link, radio channel ciphering and deciphering, estimation of the quality of the radio channels in the active cell and in the neighboring cells, setting the RF power, cell selection in the idle mode, and others.
- *PSCAF, Packet Service Control Agent Function*. The PSCAF is responsible for the provision of the packet service control agent functionality in the MT. Tasks performed by the PSCAF include control of the transport of data packets across the radio interface according to the chosen packet service; control of the dynamic allocation of resources for the transport of data across the radio interface; support of authentication and of service control handling; provision of location management information to another FE.
- *RACAF, Radio Access Control Agent Function*. The RACAF is the mobile-side agent responsible for the association and access link control between the MT and the network. Tasks performed by the RACAF include interaction with other FEs to allocate and reallocate physical channels for a branch of an access radio link and to set up, maintain, modify, and release the access radio link; execution of handover decision; detection of paging; initiation of handover; RF power control; judgment of service acceptability; control of measurements of the quality of radio channels; the selection of cells in the idle mode; and others.

7.9.3 Base Station

The following constitute the FEs within the BS PE.

- *RFTR, Radio Frequency Transmission and Reception*. The RFTR is responsible for the control of the interconnection and adaptation of the access radio link corresponding to the BS approach link. Tasks performed by the RFTR include estimation of the quality of the radio channel; setting of the RF power; detection and handling of random access; control of the interconnection of branches of an access radio link and macrodiversity in a handover situation; realization of measurements to determine the geographic position of the MT; modification of the transmitted RF; and others.

- *SIBF, System Access Information Broadcast Function*. The SIBF is responsible for the tasks involving the broadcast of system access information. Such information may be provided to the SIBF by means of an operation-and-maintenance function.

7.9.4 Radio Network Controller

The following constitute the FEs within the RNC PE.

- *ARF, Access Link Relay Function*. The ARF is responsible for the control of a branch of a BS approach link. Tasks performed by an ARF include interaction with FEs set up, release, maintain the state of, and obtain routing instructions for a branch of a BS approach link.
- *GPF, Geographic Position Function*. The GPF is responsible for tasks concerning the geographic positioning of the MT. These tasks include RF measurements, modification of the RF signals, synchronization, reporting of the results, and others.
- *CnCF, Connection Control Function*. The CnCF is an FE that is implemented for the situation in which call and connection control functions are separated. It is responsible for the control of the connection processing, interacting with other FEs to establish, maintain, modify, and release connection instances.
- *RACF, Radio Access Control Function*. The RACF is responsible for the control of the association and access link between an MT and the network. Tasks performed by the RACF include interaction with other FEs to allocate and reallocate physical channels for a branch of an access radio link and to set up, maintain, modify, and release a branch of a BS approach link and associated access radio link; execution of handover decision; execution of handover; completion of handover; RF power control; judgment of service acceptability; message delivery; scheduling of the delivery of messages; and others.
- *SNCF, Satellite Network Control Function*. The SCNF is responsible for the dynamic control of the configuration of the radio network resources, particularly in satellite networks. Other networks may not require this FE.

7.9.5 Authentication Center

The following constitutes the FE within the AC PE:

- *AMFh, Authentication Management Function (home)*. The AMFh is responsible for functions related to authentication and confidentiality within the AC. Tasks performed by the AMFh include storage of authentication data; validation of received authentication data; user authentication; confidentiality control; authentication parameters control; and others.

7.9.6 Drift MSC

The following constitutes the FE within the DMSC PE:

- *ARF, Access Link Relay Function*. The ARF, as previously described, is responsible for the control of a branch of a BS approach link. Tasks performed by an ARF include interaction with FEs to set up, release, maintain the state of, and obtain routing instructions for a branch of a BS approach link.

7.9.7 Gateway Location Register

The following constitute the FEs within the GLR PE:

- *AMFv, Authentication Management Function (visitor)*. The AMFv, like the AMFh within the AC, is responsible for functions related to authentication and confidentiality within the GLR and the VLR. Tasks performed by the AMFv include storage of authentication data; validation of received authentication data; user authentication; confidentiality control; authentication parameters control; and others.
- *LMFh, Location Management Function (home)*. The LMFh is responsible for the support of location management, mobility management, activation status management, and identity management.
- *LMFv, Location Management Function (visitor)*. The LMFv, like the LMFh, is responsible for the support of location management, mobility management, activation status management, and identity management.

7.9.8 Gateway MSC

The following constitute the FEs within the GMSC PE:

- *CCF′, Call Control Function (enhanced)*. The CCF′ is responsible for the provision of call processing control and connection processing control. Tasks performed by CCF′ include the interaction with other FEs to establish, maintain, modify, and release a call or connection instances; set up and release an access link; support IN services; receive routing and profile information for mobile calls; and others.

- *CnCF, Connection Control Function*. The CnCF is an FE that is implemented for the situation in which call and connection control functions are separated. It is responsible for the control of the connection processing, interacting with other FEs to establish, maintain, modify, and release connection instances.
- *SSF, Service Switching Function*. The SSF is associated with the CCF′ providing functions to enable the interaction between CCF′ and SCF.

7.9.9 Home Location Register

The following constitutes the FE within the HLR PE:

- *LMFh, Location Management Function (home)*. The LMFh, as already described, is responsible for the support of location management, mobility management, activation status management, and identity management.

7.9.10 Intelligent Peripheral

The following constitutes the FE within the **IP** PE:

- *SRF, Specialized Resource Function*. The SRF is responsible for the provision of specialized resources necessary for the support of IN services, mobile multimedia services, application delivery services, and packet data transfer services.

7.9.11 Mobile Switching Center

The following constitute the FEs within the MSC PE:

- *ARF, Access Link Relay Function*. The ARF, as previously described, is responsible for the control of a branch of a BS approach link. Tasks performed by an ARF include interaction with FEs to set up, release, maintain the state of, and obtain routing instructions for a branch of a BS approach link.
- *CCF′, Call Control Function (enhanced)*. The CCF′, as previously described, is responsible for the provision of call processing control and connection processing control. Tasks performed by CCF′ include the

interaction with other FEs to establish, maintain, modify, and release a call or connection instances; set up and release an access link; support IN services; receive routing and profile information for mobile calls; and others.

- *CnCF, Connection Control Function*. The CnCF, as previously described, is an FE that is implemented for the situation in which call and connection control functions are separated. It is responsible for the control of the connection processing, interacting with other FEs to establish, maintain, modify, and release connection instances.
- *GPCF, Geographic Position Control Function*. The GPCF is the network-side agent responsible for the provision of the overall control for the GPF function. Tasks performed by the GPCF include reception of and response to requests to determine the geographic position of an MT; establishment, maintenance, and release of a service instance for a geographic positioning request; interaction with other FEs to request that RF measurements be performed or transmitted RF signals be modified; and others.
- *SACF, Service Access Control Function*. The SACF is responsible for the provision of processing and control of call-related and call-unrelated services. The SACF interact with other FEs to perform tasks, including establishment and release of association between an IMT-2000 terminal/user and the network; provision of routing information for establishment of calls and mobility management functionality; provision of IN services; set up and release of an access link; set up and release of a branch of a BS approach link; detection and handling of paging response; request of paging execution; control of ciphering execution; provision of paging strategy; support of location-based services; management of temporary routing numbers for roamers; control of supplementary services; delivery of messages; and others.
- *SSF, Service Switching Function*. The SSF, as previously described, is associated with the CCF′ providing functions to enable the interaction between CCF′ and SCF.

7.9.12 Packet Data Gateway Node

The following constitutes the FE within the PDGN PE:

- *PSGCF, Packet Service Gateway Control Function*. The PSGCF is responsible for the provision of the gateway control functionality within the CN. Tasks performed by the PSGCF include interaction with other packet data networks; routing and relaying of data packets between other packet data networks and the PSCF; and others.

7.9.13 Packet Data Serving Node

The following constitute the FEs within the PDSN PE:

- *GPCF, Geographic Position Control Function*. The GPCF, as already described, is the network-side agent responsible for the provision of the overall control for the geographic position finding function. Tasks performed by the GPCF include reception of and response to requests to determine the geographic position of an MT; establishment, maintenance, and release of a service instance for a geographic positioning request; interaction with other FEs to request that RF measurements be performed or transmitted RF signals be modified; and others.
- *PSCF, Packet Service Control Function*. The PSCF is responsible for the provision of the control of packet service functionality within the CN. The PSCF interacts with other FEs to perform tasks, including provision of updates on the packet data service; transfer of user data between external packet data networks and the MT; authentication; confidentiality control, user authentication, and parameter update for packet data services; initiation of dynamic allocation of radio resources for the transport of data packets across the radio interface; ciphering execution control; location management; and others.

7.9.14 Service Control Point

The following constitute the FEs within the SCP PE:

- *SCF, Service Control Function*. The SCF encompasses the IN service control functionality in the network. It interacts with other FEs to perform tasks, including provision of mobility management–related IN services; provision of call-related and call-unrelated IN services; provision of specialized resources for IN services.
- *SDF, Service Data Function*. The SDF provides a logical data view to the SCF, hiding from it the real data implementation. Tasks performed by the SDF include storage of service data; consistency checks on data; application data management; exchange of service data; and generation and management of user service-related data.

7.9.15 Service Data Point

The following constitutes the FE within the SDP PE:

- *SDF, Service Data Function*. The SDF, as already described, provides a logical data view to the SCF, hiding from it the real data implementation. Tasks performed by the SDF include storage of service data; consistency checks on data; application data management; exchange of service data; and generation and management of user service-related data.

7.9.16 Visitor Location Register

The following constitute the FEs within the VLR PE:

- *AMFv, Authentication Management Function (visitor).* The AMFv, like the AMFh within the AC, is responsible for functions related to authentication and confidentiality within the GLR and VLR. Tasks performed by the AMFv include storage of authentication data; validation of received authentication data; user authentication; confidentiality control; authentication parameters control; and others.
- *LMFh, Location Management Function (home).* The LMFh is responsible for the support of location management, mobility management, activation status management, and identity management.
- *LMFv, Location Management Function (visitor).* The LMFv, like the LMFh, is responsible for the support of location management, mobility management, activation status management, and identity management.

7.10 Functional Entities and Their Interrelations

The FEs, as described in the previous section, implement all of the functions required to support the capabilities specified in Reference 1 and cited in this chapter. However, only those functions necessary to support the services chosen to be offered by a given family member need be implemented.

Broadly speaking, the FEs can be bundled into two groups: one containing FEs that deal with *radio access control* and another containing FEs that deal with *communication and service control*. The radio access control–related FEs are dependent on the transmission technology, whereas the communication and service control–related FEs are independent of the transmission technology. Within the communication and service control group a subgroup containing FEs that deal with packet data service can be identified.

The interrelation among these FEs and the respective groups are illustrated in Figure 7.4.

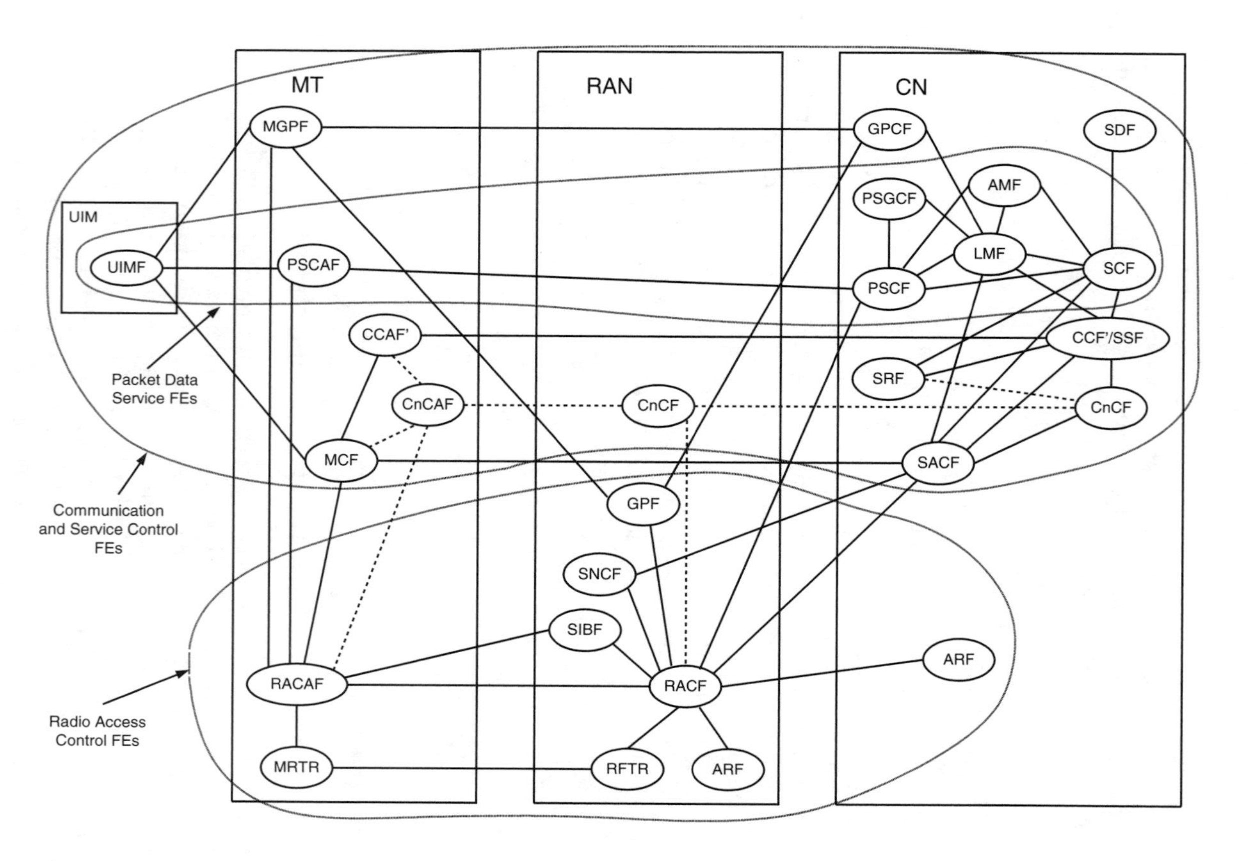

FIGURE 7.4
FEs and their interrelations.

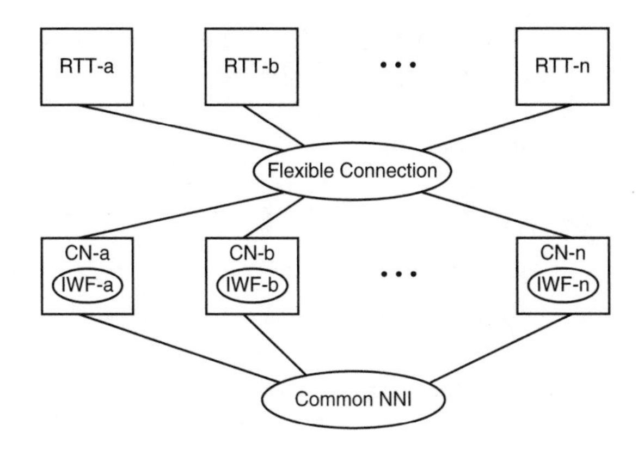

FIGURE 7.5
IMT-2000 interconnection model.

7.11 Application of the IMT-2000 Family Member Concept

In a multinetwork environment, effective internetworking can be achieved by using a common NNI (CN–CN interface). With this solution, global roaming and VHE can be fully supported. In addition, benefits from existing investment can be derived. The common NNI provides the means for CNs of different family members to be interconnected with the common NNI communicating with each family member through a particular interworking function (IWF). In this case, only one IWF per family member is required. This solution provides for transparency and modularity with changes in one family member not affecting other family members and with new family members easily accommodated in the network. This interconnection model and the family member concept are illustrated in Figure 7.5. In Figure 7.5, the radio transmission technology (RTT) comprises all but the CN subsystems of a family member. Note in Figure 7.5 that, *in theory*, any RTT can be accommodated into any CN of different family members. This is possible by means of a flexible connection between radio modules and CNs based on operator needs.[2]

7.11.1 Radio Transmission Technology

IMT-2000 provides access to a wide range of telecommunications services via several radio links. It consists of both terrestrial and satellite component radio interfaces. This subsection is basically concerned with the terrestrial component of IMT-2000. Five radio interfaces of the terrestrial component of IMT-2000 are recommended by the ITU Radio Communication Assembly. These radio interfaces are identified as follows:

IMT-2000 CDMA Direct Spread
IMT-2000 CDMA Multicarrier
IMT-2000 CDMA TDD
IMT-2000 TDMA Single-Carrier
IMT-2000 FDMA/TDMA

The IMT-2000 CDMA direct spread radio interface is called universal terrestrial radio access (UTRA), frequency division duplex (FDD), or wideband CDMA (WCDMA). As is implicit in its name, the radio access scheme of the IMT-2000 CDMA direct spread is direct-sequence CDMA, with the information transmitted with a chip rate of 3.84 Mchip/s, spread over a bandwidth of approximately 5 MHz. The radio protocol is flexible so that different services such as speech, data, and multimedia can be simultaneously required by a user and multiplexed into a single carrier. The radio bearer services provide for real-time and non-real-time service support. QoS can be adjusted in terms of delay, bit error rate, and frame error rate. The radio interface for this technology has been specified with the aim of achieving harmonization with the TDD component for maximum commonality. Maximum commonality is accomplished by carefully harmonizing important parameters in the physical layer and developing a common set of protocols in the higher layers for both FDD and TDD.

The IMT-2000 CDMA multicarrier radio interface is called cdma2000, consisting of the 1X and 3X components, as explained next. It is a wideband spread spectrum radio interface designed to meet the requirements of the 3G wireless systems as well as the requirements of the 3G evolution of the 2G TIA/EIA-95-B family standards. The physical layer of cdma2000 supports RF channel bandwidths of $N \times 1.25$ MHz, where N is the spreading rate number. N has initially been specified to be equal to 1 and 3, but is easily extendable to 6, 9, and 12. The radio interface encompasses radio configurations to support the specific spreading rates. In particular, six radio configurations for the reverse link and nine radio configurations for the forward link are specified for spreading rates 1 and 3. Spreading rate 1 corresponds to the 1X component and spreading rate 3 corresponds to the 3X component.

The IMT-2000 CDMA TDD radio interface is called UTRA time division duplex (TDD), and TD-SCDMA (time division-synchronous code division multiple access). In fact, TD-SCDMA is referred to as UTRA TDD low chip rate option, and UTRA TDD represents the UTRA TDD high chip rate option. The radio access scheme for both technologies is direct-sequence CDMA, with the chip rate and the bandwidths varying for these technologies: UTRA

TDD operates with chip rate of 3.84 Mchip/s with the information spread over approximately 5 MHz bandwidth; TD-SCDMA operates with chip rate of 1.28 Mchip/s with the information spread over approximately 1.6 MHz bandwidth. The radio bearer services provide for real-time and non-real-time service support. QoS can be adjusted in terms of delay, bit error rate, and frame error rate. The radio interface for this technology has been specified with the aim of achieving harmonization with the TDD component for maximum commonality. TD-SCDMA has a significant commonality with UTRA TDD.

The IMT-2000 TDMA single-carrier radio interface is called universal wireless communications 136 (UWC-136). It is a TDMA radio interface designed to meet the requirements of 3G wireless systems as well as the requirements of the 3G evolution of the 2G TIA/EIA-136 standard. UWC-136 was developed aiming at a maximum commonality between TIA/EIA-136 (designated 136) and GSM GPRS. The evolution from the TIA/EIA-136 technology toward a 3G system is accomplished by means of four strategies: 136^+, 136HS outdoor, 136HS indoor, and 136EHS. The 136^+ strategy is based on the enhancement of voice and data capabilities of the 30-kHz channels by the inclusion of the GPRS technology. The 136HS outdoor strategy is based on the addition of a 200-kHz carrier component, as in EDGE (enhanced data rate for global evolution), accommodating data rates of 384 kbit/s in high-mobility applications. The 136HS indoor strategy is based on the addition of 1.6-MHz carrier component, accommodating data rates of 2 Mbit/s in low-mobility applications. The 136EHS strategy is based on the inclusion of an additional alternative 200-kHz carrier (EDGE) component for high-speed data. The specification of the combined strategies composes the UWC-136 radio interface. The evolution of such a technology toward the 3G applications has been governed by the compatibility principle, which facilitates the use of the already deployed 136 platforms. Commercially effective evolution and global roaming are further strengthened by providing for flexible spectrum allocation and compatibility between 136EHS and GSM EGPRS (enhanced GPRS), where, in the latter, the aggregation of carriers leads to higher data rates.

The IMT-2000 FDMA/TDMA radio interface is called digital enhanced cordless telecommunications (DECT). It is a TDMA radio interface with TDD. The carrier spacing is 1.728 MHz, with the equipment allowed to use four-level and/or eight-level modulation in addition to two-level modulation. The RF bit rates, therefore, depend on the modulation scheme and are specified as 1.152, 2.304 , and 3.456 Mbit/s, for the two-level, four-level, and eight-level modulation schemes, respectively. Symmetric and asymmetric connections, connection-oriented and connectionless data transport, as well as variable bit rates up to 2.88 Mbit/s per carrier are supported by the standard. The radio interface is a high-capacity digital technology, with the cell radii ranging from a few meters to several kilometers, as required. Therefore, the standard supports applications ranging from simple residential cordless telephones to large systems for a wide scope of telecommunications services, including fixed wireless access.

7.11.2 Core Network

The CN for IMT-2000 must comply with the CN capabilities as described in the section on features and services in this chapter. Five CNs accommodating the radio interfaces of the terrestrial component of IMT-2000 are identified as follows:

Evolved GSM (MAP) CN
Evolved ANSI-41 CN
ANSI-41 UWC-136 CN
IP-Based CN
Future CN

The first three CNs—evolved GSM (MAP), evolved ANSI-41, and ANSI-41 UWC-136—as is implicit in their names, are evolutions of those CNs largely used for the 2G systems. Therefore, as these networks have been oriented primarily toward voice applications, the packet data functionalities must be added to implement the 3G capabilities. This can be achieved by the combination of circuit-switched and GPRS packet-switched networks. An all-IP-based CN is certainly an emerging alternative. Other CNs may also be accommodated within IMT-2000.

7.11.3 Radio Transmission Technologies and Core Networks

The IMT-2000 CDMA direct-spread and IMT-2000 TDD radio interfaces have been developed with the aim of interconnecting with the evolved GSM (MAP) CN. The resulting 3G network is known as the universal mobile telecommunications system (UMTS). In UMTS, the UIM subsystem is called USIM (UMTS subscriber identify module); the MT is called ME (mobile equipment); the RAN is called UTRAN (UTRA network or UMTS terrestrial RAN); the equivalent of the BS within UTRAN is called Node B and that of RNC is also the RNC. On the other hand, in the development of these interfaces, specifications also include the necessary capabilities for operation with an evolved ANSI-41 CN. Conversely, the IMT-2000 CDMA multicarrier radio interface has been developed to interconnect with the evolved ANSI-41 CN. On the other hand, in the development of this interface, specifications also include the necessary capabilities for operation with an evolved GSM (MAP) CN.

7.12 Toward 3G

7.12.1 An Overview

The evolution of the 2G systems toward a truly 3G network includes an intermediate step—the so-called 2.5G systems. GSM evolves toward the GSM Phase 2^+ in which provision for high-data-rate transmission constitutes the main target. In this case, two technologies are included: HSCSD and GPRS. HSCSD stands for high-speed circuit-switched data and provides for high-data-rate services by means of circuit-switched technology. It increases the data rate in GSM channels from 9.6 to 14.4 kbit/s and allows users to be assigned an asymmetric number of channels within uplink and downlink. GPRS stands for general packet radio service and provides for high-data-rate services by means of packet-switched technology. It overlays a packet-switched radio network on the wireless circuit-switched model. An extremely primitive version of GPRS is the CDPD technology. CDPD stands for cellular digital packet data and appeared as a solution for packet data transmission overlaying the analog AMPS. It was then implemented within ANSI-136 and IS-95, but not very successfully. GPRS was originally conceived for GSM systems, but ANSI-136 has also adopted this solution. Such a migration, together with the adoption of similar vocoders, shows a clear convergence between the two standards, GSM and ANSI-136. The new generation of this TDMA system is referred to as TDMA-136^+. GSM phase 2^+ evolves toward EDGE in which higher data rates are achieved. EDGE stands for enhanced data rates for global evolution and replaces GMSK modulation scheme with 8-PSK, which doubles normal bit rates and triples peak bit rates. It initially aimed at the GSM evolution and was previously spelled out as enhanced data rate for GSM evolution. GSM also evolves toward WCDMA. TDMA-136^+ evolves toward UWC-136, which includes EDGE. IS-95 evolves toward IS-95B, in which handoff procedures and forward power control procedures have been drastically improved and packet data transmission have been introduced. The next step of the evolution, the IS-95C or, equivalently, IS-2000, constitutes a primitive version of the multicarrier cdma2000.

To develop standards with a more global and collaborative scope, the wireless industry has created three partnership projects: 3GPP, 3GPP2, and G3G. The concept of a partnership project was pioneered by the European Telecommunications Standards Institute (ETSI) early in 1998 with the proposal to create the Third Generation Partnership Project (3GPP) focusing on GSM-based technology. Although discussions did take place between ETSI and the ANSI-41 community with a view at consolidating collaboration efforts for all

ITU family members, in the end it was deemed appropriate that a parallel group, the Third Generation Partnership Project 2 (3GPP2), should be established. 3GPP2 is a collaborative 3G telecommunications standards-setting project comprising North American and Asian interests with two main objectives: the development of global specifications for ANSI/TIA/EIA-41 "Cellular Radio Telecommunication Intersystem Operations" network evolution to 3G, and the development of global specifications for the radio transmission technologies (RTTs) supported by ANSI/TIA/EIA-41. G3G has been created with the aim of harmonizing the proposals of the two other groups. These partnership projects are supported by officially recognized standards development organizations (SDOs).

3GPP gathers the following SDOs:

- ARIB—Association of Radio Industries and Businesses (Japan)
- CWTS—China Wireless Telecommunication Standards Group (China)
- ETSI—European Telecommunications Standard Institute (Europe)
- T1—Standards Committee T1 Telecommunications, sponsored by the Alliance for Telecommunications Industry Solutions and accredited by ANSI for interoperability standards (North America)
- TTA—Telecommunications Technology Association (Korea)
- TTC—Telecommunications Technology Committee (Japan)

3GPP2 is a collaborative effort between the following SDOs:

- ARIB—Association of Radio Industries and Businesses (Japan)
- CWTS—China Wireless Telecommunication Standards Group (China)
- TIA—Telecommunications Industry Association (United States, Canada, Mexico)
- TTA—Telecommunications Technology Association (Korea)
- TTC—Telecommunications Technology Committee (Japan)

For any given operator, the already installed 2G plant—the legacy system—constitutes a key factor to be accounted for in the introduction of new 3G networks. Backward compatibility is facilitated and investment is minimized if similar platforms are used for 2G and 3G systems. This is feasible for those systems making use of GSM and ANSI-41 platforms, which constitute the great majority of the installed systems. The Japanese Pacific (or Personal) Digital Cellular (PDC) is certainly an exception, for it is based on neither of these two platforms. Interoperability between 2G and 3G systems in this case will have to rely on an infrastructure specially built for such an aim.

7.12.2 Network Architecture

Among the CNs for IMT-2000, as identified previously, the IP-based CN is certainly the one more open to innovative proposals. Any IP-based CN would have to support both stream and best effort services and the MT the IP-based clients. The advent of the IP technology and the growth in data traffic impelled the wireless industries toward the development of an all-IP mobile network. The two working groups, namely, 3GPP and 3GPP2, have different approaches concerning the IP network architecture[4] but a convergence between these approaches is possible.

3GPP IP-Based Network Architecture

The main feature of the 3GPP IP-based network is that it is supported by the GPRS technology and a clear separation between service control and connection control exists.[4] Essentially, the 3GPP IP-based network architecture has a core packet network that is overlaid with call control functions and gateway functions to support voice over IP (VoIP) and other multimedia services. The packet network is based on GPRS technology and the functions are provided via the Internet Engineering Task Force (IETF). The usual RAN is used to provide the radio access, as necessary.

The function providing for VoIP capability is named the call state control function (CSCF), which is a function analogous to that used for call control in a circuit-switched environment. Besides the usual call control functions, the CSCF also performs service switching functions, address translation functions, and vocoder negotiation functions. The communication between the MT and CSCF is carried out by means of a protocol that supports VoIP. The communication between this CN and the PSTN and other legacy networks is provided by a gateway MSC. In the same way, roaming to 2G wireless networks is supported by roaming gateway functions. A GPRS serving node, the SGSN, makes use of ordinary GSM registration and authentication procedures, thus rendering itself dependent on access technology. An HLR, enhanced to support IP services, is also used. A GPRS gateway node (GGSN) anchors the IP address of a data terminal making use of the packet network. A foreign agent (FA) is incorporated into the GGSN to provide mobility of the data terminal throughout other networks.

3GPP2 IP-Based Network Architecture

The main characteristic of the 3GPP2 IP-based network is that it is an all-IP network built on the CDMA 2G and 3G air interface data services.[4] Both the 3G high data rates and the works in IETF on mobile IP have been used to enhance the network architecture to provide IP capabilities. The use of globally accepted IETF protocols facilitates internetworking and roaming and provides private network access via a mobile IP tunnel with IP security. By investing in the synergies of Internet technologies, a single network for all services can be utilized. An all-IP network provides for an end-to-end IP connectivity, distributed control and services, and gateways to legacy networks. The IP connectivity reaches all the way to the IP-based RAN (IP-B-RAN), which is a router-based IP node containing radio control functions such as power control, soft handoff frame selection, etc. Other control functions, namely, call/session control, FA, mobility management, and gateway functions, reside within the managed IP network. Gateway functions such as roaming gateway and PSTN gateway for roaming to 2G networks and internetworking with the PSTN, respectively, are also provided.

For a given data session, the MT attaches to the FA, which establishes a mobile IP tunnel to the home agent (HA) for registration purposes. The HA then accesses the home authorization, authentication, and accounting (HAAA) server for authentication purposes. The HA now anchors the IP address of the MT for the duration of the session. The data device attached to the MT may be handed over to any other access device supporting mobile IP.

An Intermediate Step

The architecture of a truly all-IP network is significantly different from that of legacy wireless systems. The descriptions given previously of an all-IP network are rather general and the actual implementation of it is subject to interpretations, because the requirements are not fully developed.[5] Packet and IP access can be introduced in legacy systems in a graceful fashion taking advantage of new technology built to be adapted within these systems. Replacement of existing blocks within existing networks, then, may occur gradually, as required. The GSM community makes the transition to a 3G network in three steps, as follows. In the first step, a packet-switched network—the GPRS—is superimposed to the network, and the access blocks—base station controller and mobile station subsystem—are kept virtually unchanged, except for the addition of a packet control unit (PCU) in both. In the second step, the base station subsystem is replaced by UTRAN—the UMTS terrestrial RAN. In the third and final step, the complete UMTS handsets replace the GSM handsets.

7.13 Summary

IMT-2000 embodies the concepts of 3G networks. IMT-2000 standards and specifications have been developed by various standards organizations worldwide under the auspices of the International Telecommunications Union. Mobile and fixed users are targeted and a wide and ambitious range of user sectors, radio technology, radio coverage, and user equipment is covered by IMT-2000. In essence, a 3G system must provide for multimedia services, in circuit-mode and packet-mode operations, for user sectors such as private, public, business, residential, local loop, and others, for terrestrial-based and satellite-based networks, for personal pocket, vehicle-mounted, or any other special terminal. Global roaming and virtual home environment constitute two of the key features of IMT-2000. A start-up network may gradually evolve from a simple and small 3G system to a large network with complex 3G applications. In this sense, the IMT-2000 functional model is developed to be flexible to accommodate a wide range of 3G applications while meeting the necessary requirements of QoS.

Two important concepts are formally defined within the IMT-2000 philosophy: *IMT-2000 System* and *IMT-2000 Family*. An IMT-2000 System comprises a set of subsystems, entities, and interfaces designed to perform actions and interactions to provide its users with IMT-2000 capabilities as defined in the IMT-2000 Capability Set. Four functional subsystems, i.e., user identity module, mobile terminal, radio access network, and core network, compose an IMT-2000 System. Any system functionally structured as described and providing for the capabilities as defined in the IMT-2000 Capability Set constitutes an IMT-2000 System. This gives rise to the possibility of different implementations leading to different IMT-2000 Systems. To accommodate all these systems under the same 3G principles the IMT-2000 Family concept emerged as a necessity. "The IMT-2000 Family is a federation of IMT-2000 Systems providing IMT-2000 capabilities to its users as identified in IMT-2000 Capability Set."[1] An IMT-2000 family member is therefore an IMT-2000 System.

IMT-2000 consists of both terrestrial and satellite component radio interfaces. Five radio interfaces of the terrestrial component of IMT-2000 are recommended by the ITU Radio Communication Assembly. These radio interfaces are identified as follows:

IMT-2000 CDMA Direct Spread
IMT-2000 CDMA Multicarrier
IMT-2000 CDMA TDD
IMT-2000 TDMA Single-Carrier
IMT-2000 FDMA/TDMA

The CNs for IMT-2000 must comply with the CN capabilities as described in the section on features and services. Five CNs accommodating the radio interfaces of the terrestrial component of IMT-2000 are identified as follows:

Evolved GSM (MAP) CN
Evolved ANSI-41 CN
ANSI-41 UWC-136 CN
IP-Based CN
Future CN

To develop standards with a more global and collaborative scope, the wireless industry has created three partnership projects: 3GPP, 3GPP2, and G3G. The 3GPP (3rd Generation Partnership Project) is a consortium for the development of 3G standards for GSM-based systems. The 3GPP2 is a consortium for the development of 3G standards for IS-95-based CDMA systems. The G3G has been created with the aim of harmonizing the proposals of the two other groups. The advent of IP technology and the growth in data traffic impelled the wireless industries toward the development of an all-IP mobile network. The two working groups 3GPP and 3GPP2 have different approaches concerning the IP network architectures,[4] but convergence of these approaches is possible.

References

1. ITU-T Recommendation Q.1701: Framework for IMT-2000 Networks, March 1999.
2. Supplement to ITU-T Recommendation Q.1701 Framework for IMT-2000 Networks—Roadmap to IMT-2000 Recommendations, Standards and Technical Specifications, June 2000.
3. ITU-T Recommendation Q.1711: Network Functional Model for IMT-2000, March 1999.
4. Patel, P. and Dennett S., The 3GPP and 3GPP2 movements toward an all-IP mobile network, *IEEE Personal Commun.*, 62–64, August 2000.
5. 3GPP TR 23.922, Architecture for an All IP Network, December 1999.

8

UTRA

8.1 Introduction

The IMT-2000 radio interface for direct-sequence code division multiple access is referred to as universal terrestrial radio access (UTRA) or wideband CDMA (WCDMA). It has two modes of operation: frequency division duplex (FDD) and time division duplex (TDD). The radio interface specifications for both UTRA FDD and UTRA TDD have been developed with the strong objective of harmonization of these two components to achieve maximum commonality.[1,2] In this case, important physical parameters and higher-layer protocols are common to both technologies. The core network (CN) specifications are based on an evolved GSM-MAP and capabilities are included so that operation with an evolved ANSI-41 based CN is possible. The radio interfaces are defined in such a way a wide range of services including speech, data, and multimedia can be simultaneously used by a subscriber and multiplexed on a single carrier. Therefore, circuit-switched services (e.g., PSTN-based and ISDN-based networks) and packet-switched services (e.g., IP-based networks) are efficiently supported, and real-time and non-real-time operations employing transparent or nontransparent data transport are specified. The quality-of-service (QoS) is an important feature and is specified to be adjusted in terms of parameters such as delay, bit error rate, frame error rate, and others. Although the FDD and TDD components have been designed to achieve maximum commonality, some basic differences between them exist, and they are detailed in this chapter. One of them concerns macrodiversity applied to soft handover situations. Whereas this constitutes an intrinsic feature of UTRA FDD, macrodiversity is not necessary in UTRA TDD because of the TDMA component.

In UTRA FDD, the information is spread over approximately 5 MHz bandwidth with a chip rate of 3.84 Mchip/s. In UTRA TDD, two options are specified. In one of them, the information is spread over approximately 5 MHz bandwidth with a chip rate of 3.84 MHz. In the other, the information is spread over approximately 1.6 MHz with a chip rate of 1.28 Mchip/s, and this alternative is mainly based on smart antenna technology. The first option is referred to as simply UTRA TDD, whereas the second option is known as time division-synchronous code division multiple access (TD-SCDMA) or 1.28 Mchip/s TDD. This chapter, for didactic purposes only, refers to the first option as UTRA TDD-3.84 and to the second as UTRA TDD-1.28. When no specific reference is made to any particular technology, the description in this chapter is applicable to the three technologies, namely, UTRA FDD, UTRA TDD-3.84, and UTRA TDD-1.28. Special attention, however, is given to UTRA FDD.

The integration of user equipment, UTRA, and a CN results in a 3G system known as the universal mobile telecommunications system (UMTS). Although this chapter gives an overview of UMTS, it is mainly concerned with the UTRA component. For information on the other elements and 3G services the reader is referred to Chapter 7. The descriptions that follow are based on the technical specifications of References 8 through 18.

8.2 Network Architecture

The overall architecture of UMTS is shown in Figure 8.1. In accordance with the IMT-2000 specifications, three basic blocks compose UMTS: user equipment (UE); UMTS terrestrial radio access network (UTRAN); and core network (CN). These elements are connected among themselves and to external networks, as appropriate.

- *User Equipment*. The UE provides means for the user to access the system. It consists of the mobile equipment (ME) and of the UMTS subscriber identity module (USIM):

  Mobile Equipment. The ME is the radio terminal performing functions to support mobile radio access to the system.

  UMTS Subscriber Identity Module. The USIM is a smartcard performing functions to support user security and user services. It holds the subscriber identity and some subscription information, performs authentication procedures, and stores authentication and encryption keys.

- *UMTS Terrestrial Radio Access Network*. The UTRAN performs functions to support communication with the MT and with the CN. It

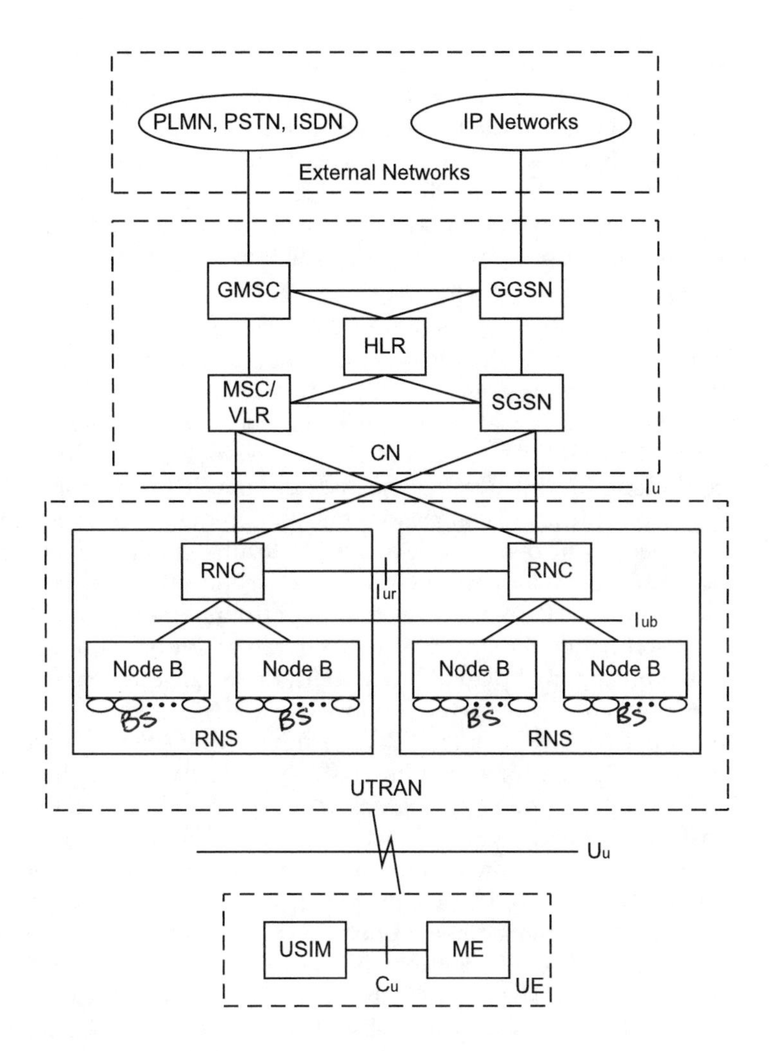

FIGURE 8.1
UMTS architecture.

provides means for exchanging information between MT and CN, acting as a bridge, router, and gateway, as required. It consists of a set of radio network subsystems (RNS), each of which contains a radio network controller (RNC) and one or more entities known as Node B:

Node B. Node B is an entity supporting radio frequency transmission and reception and system access information broadcast. It performs channel coding and interleaving, rate adaptation, spreading, radio resource management operation, as in inner loop power control, and others. Node B logically corresponds to the well-known entity base station. (The term *Node B* is one of those provisional names, initially adopted in the UMTS standardization process, that now is permanent.) One Node B can handle one or more cells, indicated as ellipses in Figure 8.1. Whenever applicable, Node B may comprise an optional combining and splitting functions to support macrodiversity within Node B itself.

Radio Network Controller. The RNC supports radio access control, connection control, geographic positioning, and access link relay. It owns and controls the radio resources in its domain constituting the service access point for all services UTRAN provides to the CN. It logically corresponds to the well-known entity base station controller (BSC). An RNC controlling one Node B is identified as controlling RNC (CRNC). In such a case, the CRNC responds for the load control and congestion control of its own cells, executing admission control and code allocation for the new radio links to be established within those cells. In case a connection between one UE and the UTRAN involves resources from more than one RNS (in a soft handover situation, for example), the RNCs implicated in the process will play different logical roles for this connection. One of them will act as the serving RNC (SRNC) and the other as the drift RNC (DRNC). The SRNC terminates the link for transport of user data; the corresponding radio access network application part signaling; and signaling protocol between UE and UTRAN. Tasks performed by SRNC include the mapping of radio access bearer parameters into air interface transport channel parameters, handover decision, outer loop power control, and all the basic radio resource management operations. The SRNC may serve as CRNC of some Node Bs. Note that only one SRNC is assigned for one connection. The DRNC is the RNC, other than the SRNC, controlling the cells used by the subscriber. Whenever applicable, the DRNC may comprise optional combining and splitting functions to support macrodiversity between different Node Bs. It acts as a relay between the Node B and the SRNC whenever appropriate. One UE may have zero or more DRNCs. One physical RNC may contain the functionalities of CRNC, SRNC, and DRNC.

- *Core Network*. The CN performs functions to support communication with the UTRAN and with other CNs. It provides means to support user mobility and user services. Several elements, as described in Chapter 7, may compose the CN. In particular, the elements for circuit-switched services are those of the 2G systems, whereas the elements for packet-switched services are those of the general packet data service (GPRS) technology. The basic elements of a CN, however, are the mobile switching center/visitor location register (MSC/VLR), home location register (HLR), gateway MSC (GMSC), serving GPRS support node (SGSN), gateway GPRS support node (GGSN). The elements for intelligent network (IN) services are also part of CN, but are not considered here. The reader is referred to Chapter 1 for the concepts and details of IN.

Mobile Switching Center/Visitor Location Register. The MSC performs the switching functions and coordinates the calls and routing procedures within the network. The VLR is a database containing a copy of the service profile of the visiting subscriber as well as information on the location of the subscriber within the system. The combination MSC/VLR serves the UE in its current location for circuit-switched services.

Home Location Register. The HLR is a database residing within the home system and contains a list of the home subscribers with their respective service profiles. In the HLR, these subscribers are associated with information records relevant to the call management. Both permanent and temporary data are held within the HLR, the former constituting data that are only modified for administrative reasons, and are kept for every call, and the latter comprising data that are modified to accommodate the transient status of the subscribers' parameters, and can be changed from call to call. The HLR also stores the UE location for the purposes of routing incoming transactions (calls, short messages) to the UE.

Gateway MSC. The GMSC supports call control, connection control, and service switching. The GMSC is used to contact the HLR for routing information so that the call can be routed to the corresponding visited MSC or any external circuit-switched network. Incoming and outgoing circuit-switched connections are served by the GMSC.

Serving GPRS Support Node. The SGSN acts as a logical interface to the UTRAN, responsible for the delivery of packets to the correct Node B. In addition, ciphering, authentication, session management, mobility management, and logical link management to the mobile station are performed by the SGSN. Its functions are similar to those of MSC/VLR but for packet services only.

Gateway GPRS Support Node. The GGSN acts as a logical interface to the external packet data network (PDN), which includes the IP PDN or X.25/X.75 packet-switched PDN. For an external IP network, GGSN is seen as an ordinary IP router. In addition, it may include firewall and packet-filtering mechanisms. It also provides means of assigning the correct SGSN to the UE depending on its location. Its functions are similar to those of GMSC but for packet services only.

As shown in Figure 8.1, connections to external networks include those with switched-circuit services, such as PLMN, PSTN, ISDN, and those with packet-switched services, such as the Internet. The internal functionalities of the UMTS logical network elements are not specified in detail. On the other hand, the various interfaces between these elements are defined; the main open interfaces are the Cu interface, Uu interface, Iu interface, Iur interface, and Iub interface,[3] as shown in Figure 8.1. The open interfaces allow the operators to set up their equipment with elements acquired from different manufacturers.

- *Cu Interface.* This is the interface between USIM and ME and is defined in terms of physical specifications including size, contacts, electrical specifications, protocols, and others. This interface follows the standard format for smartcards.
- *Uu Interface.* This is the radio interface between ME and UTRAN, which is the main subject of this chapter.
- *Iu Interface.* This is the interface between UTRAN and CN. It is presented in two instances, namely, Iu circuit switched (Iu CS) and Iu packet switched (Iu PS). Iu CS connects UTRAN to the circuit-switched domain of the CN, whereas the Iu PS connects UTRAN to the packet-switched domain of the CN. Some of the functions supported by Iu include:

  Relocation of SRNS functionality from one RNS to another without changing the radio resources and without interrupting the user data flow

  Relocation of SRNS from one RNS to another with a change of radio resources for hard handover purposes

  Setup, modification, and clearing of radio access bearer

  Release of all resources from a given Iu instance related to the specified UE, this including the RAN-initiated case

  Report of unsuccessfully transmitted data

  Paging

  Management of the activities related to a specific UE–UTRAN connection

  Transparent transfer of UE–CN signaling messages

  Implementation of the ciphering or integrity feature for any given data transfer

Management of overload

Reset of the UTRAN side and/or CN side of Iu

Reporting of the location of a given UE

Framing of data into segments of predefined sizes according to the adaptive multirate codec speech frames or to the frame sizes derived from the data rate of a circuit-switched data call.

- *Iur Interface.* This is the interface between RNCs of different RNSs. It can be conveyed over physical direct connection between RNCs or via any suitable transport network. Iur was initially designed to support inter-RNC soft handover. More features, however, have been added and four distinct functions are provided. These functions are defined in terms of four modules as follows: support of the basic inter-RNC mobility (Iur1); support of dedicated channel traffic (Iur2); support of common channel traffic (Iur3); and support of global resource management (Iur4).[3]

  Iur1. The functions offered in Iur1 include support of SRNC relocation; support of inter-RNC cell and UTRAN registration area update; support of inter-RNC packet paging; and reporting of protocol errors.

  Iur2. The functions offered in Iur2 include establishment, modification, and release of the dedicated channel in DRNC due to hard handover and soft handover in the dedicated channel state; setup and release of dedicated transport connections across Iur; transfer of dedicated channel traffic transport blocks between SRNC and DRNC; management of the radio links in DRNS via dedicated measurement report and power setting procedures.

  Iur3. The functions offered in Iur3 include setup and release of the transport connection across Iur for common channel data streams separation of the MAC layer between SRNC and DRNC; flow control between the separated MAC layers.

  Iur4. The functions offered in Iur4 include transfer of cell measurements between two RNCs; transfer of Node B timing information between two RNCs.

- *Iub Interface.* This is the interface between Node B and RNC. This interface supports all the procedures for the logical operation and maintenance (O&M) of Node B, such as configuration and fault management. It also supports all the signaling through dedicated control ports for the handling of a given UE context, after a radio link has been set up for this UE. More specifically, the following functions are performed: setup of the first radio link for one UE; cell configuration; initialization and reporting of cell or Node B specific measurements; fault management; handling of access channels and page channels; addition, release, and configuration of radio links for one UE context; handling of dedicated and shared channels; handling of softer combining; initialization and reporting of radio link–specific measurement; radio link fault management.

8.3 Protocol Architecture

A general protocol model, as depicted in Figure 8.2, is defined for all UTRAN terrestrial interfaces. The protocol architecture is modularly composed of layers and planes that are logically independent of each other. Two main layers are defined, namely, radio network layer (RNL) and transport network layer (TNL). RNL contains all visible UTRAN-related issues, whereas TNL is composed of standard transport technology selected to be used for UTRAN. Four planes are defined: control plane (CP), user plane (UP), transport network control plane (TNCP), and transport network user plane (TNUP).

CP is responsible for all UMTS-specific control signaling, comprising the application protocol and the signaling bearer. UP is responsible for transmission and reception of all user-related information, such as coded voice, in a voice call, or packets, in an Internet connection, comprising the data stream and the data bearer. TNCP performs functions related to control signaling within TNL, with the corresponding transactions carried out between CP and UP. It isolates CP from UP so that the communication between the application protocol, in CP, and the data bearer, in UP, is intermediated by the access link control application part (ALCAP) in TNCP. ALCAP is specific for the particular UP technology. In such a case, the application protocol can be completely

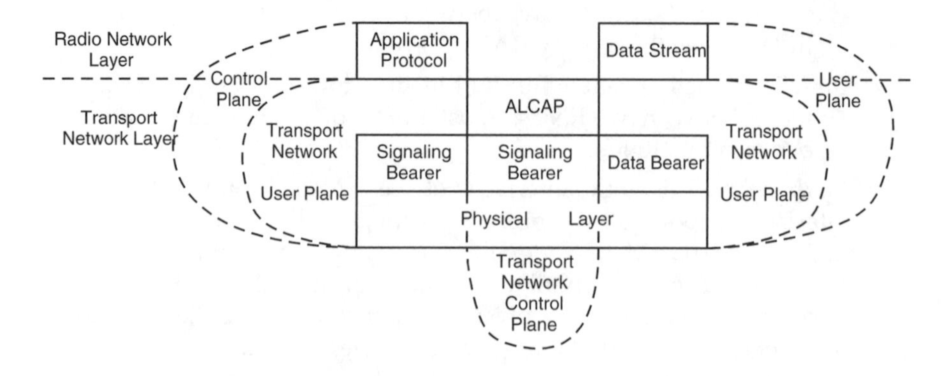

FIGURE 8.2
Protocol architecture for UTRAN terrestrial interface.

independent of the technology selected for the data bearer. For preconfigured data bearers, however, no ALCAP signaling transactions are necessary, in which case TNCP becomes dispensable. It must be emphasized that ALCAP is not used for setting up the signaling bearer for the application protocol during real-time operations. In addition, the signaling bearer for ALCAP and for the application protocol may be of different types. The signaling bearer for ALCAP is always set up by O&M actions. TNUP is responsible for the transport of user-related signaling and information comprising the data bearer, in UP, and the signaling bearer, in CP. The data bearer is controlled by TNCP during real-time operations, whereas the signaling bearer, in UP, is set up for O&M actions.

The protocols and functions within each layer and plane are shown in Table 8.1, where RNL User Plane in the second row refers to the Transport

TABLE 8.1

Protocol for the Various Interfaces

	IuCS	IuPS	Iur	Iub
RNL Control Plane	RANAP	RANAP	RNSAP	NBAP
RNL User Plane	IuUPP	IuUPP	DCHFP CCHFP	DCHFP RACHFP FACHFP PCHFP DSCHFP USCHFP
TNL User Plane	SCCP MTP3b SSCF-NNI SSCOP AAL5 ATM	SCCP MTP3b M3UA SSCF-NNI SCTP SSCOP IP AAL5 ATM	SCCP MTP3b M3UA SSCF-NNI SCTP SSCOP IP AAL5 ATM	SSCF-NNI SSCOP AAL5 ATM
TNL Control Plane	Q.2630.1 Q.2150.1 MTP3b SSCF-NNI SSCOP AAL5 ATM	—	Q.2630.1 Q.2150.1 MTP3b M3UA SSCF-NNI SCTP SSCOP IP AAL5 ATM	Q.2630.1 Q.2150.1 SSCF-NNI SSCOP AAL5 ATM
TNL User Plane	AAL2 ATM	GTP-U UDP IP AAL5 ATM	AAL2 ATM	AAL2 ATM

Network User Plane indicated in the left side of Figure 8.2, and RNL User Plane in the fourth row refers to the Transport Network User Plane indicated in the right side of Figure 8.2. The protocols in Table 8.1 are explained next.

8.3.1 Radio Network Layer

As mentioned before, the radio network layer contains application protocol, in CP, and data stream, in UP. Note that the application protocol is RANAP (RAN application part) for IuCS and IuPS, RNSAP (RNS application part) for Iur, and NBAP (Node B application part) for Iub. RANAP, RNSAP, and NBAP are signaling protocols whose functions are basically those already mentioned in the description of IuCS, IuPS, Iur, and Iub. The data stream comprises the IuUPP (Iu User plane protocol) for IuCS and IuPS, DCHFP (dedicated channel frame protocol) and CCHFP (control channel frame protocol) for Iur, and DCHFP (dedicated channel frame protocol), RACHFP (random-access channel frame protocol), FACHFP (forward access channel frame protocol), PCHFP (paging channel frame protocol), DSCHFP (downlink shared channel frame protocol), and USCHFP (uplink channel frame protocol) for Iub. IuUPP conveys user data related to radio-access bearer (RAB).

It may operate either in the support mode or in the transparent mode. In the first case, the protocol frames the user data into segment data units of predefined size and performs control procedures for initialization and rate control. In the second case, the protocol performs neither framing nor control and is applied to RABs not requiring such features. The various frame protocols, namely, DCHFP, CCHFP, RACHFP, FACHFP, PCHFP, DSCHFP, and USCHFP, handle the respective channels DCH (dedicated channel), CCH (control channel), RACH (random-access channel), FACH (forward-access channel), PCH (paging channel), DSCH (downlink shared common channel), and USCH (uplink shared common channel), which are described later in this chapter.

8.3.2 Transport Network Layer

As mentioned before, the RNL comprises the signaling bearer and the data bearer, in TNUP, ALCAP, and signaling bearer, in TNCP. In TNCP, ALCAP is implemented by means of Q.2630.1, and the adaptation is carried out by Q.2150.1. A number of broadband signaling system 7 (BB SS7) protocols are selected to implement the lower layers in CP and UP: SCCP (signaling connection control part), MTP3b (message transfer part), and SAAL-NNI (signaling ATM adaptation layer for network-to-network interfaces). SAAL-NNI is, in fact, split into SSCF (service-specific coordination function), SSCOP (service-specific oriented protocol), and AAL 5 (ATM adaptation layer type 5) layers. SSCF and SSCOP layers respond for the signaling transport in ATM networks, whereas AAL5 is responsible for the segmentation of data to compose the ATM cells. AAL2 (ATM adaptation layer type 2) deals with transfer of a service data unit with variable bit rate, transfer of timing information, and indication of lost or errored information not recovered by type 2. As an alternative to some BB SS7-based signaling bearers, an IP-based signaling bearer is specified. They consist of M3UA (SS7 MTP3—user adaptation layer), SCTP (simple control transmission protocol), and IP (Internet protocol), and they are shown side-by-side with MTP3b, SSCF-NNI, and SSCOP in Table 8.1. The multiplexing of packets on one or several AAL5 predefined virtual connections (PVC) is provided by GTP-U (user plane part of the GPRS tunneling protocol), which is responsible for identifying individual data flows. The data flow uses UDP (user datagram protocol) connectionless transport and IP addressing. Note that all planes share a common ATM (asynchronous transfer mode) transport. The physical layer constitutes the interface to the physical medium (optical fiber, radio link, copper cable) and can be implemented with standard off-the-shelf transmission technologies (SONET, STM1, E1).

8.4 Radio Interface Protocol Architecture

The handling of the radio bearer services is performed by the radio interface protocols. Generally speaking, the UTRA radio interface protocol architecture follows very closely the ITU-R protocol architecture as described in Reference 4. The basic radio interface architecture encompassing the blocks and protocols that are visible in UTRAN is illustrated in Figure 8.3. Only three layers, specifically, Layer 3, network layer, represented by its lowest sublayer; Layer 2, data link layer; and Layer 1, physical layer, are of interest. The higher-layer signaling, namely, mobility management and call control, belong to the CN and are not described here. Note that Layer 3 and part of Layer 2 are partitioned into CP and UP. The blocks in Figure 8.3 represent the instances of the respective protocol and peer-to-peer communication are provided by service access points (SAPs); some are shown in this figure in the form of ellipses.

Layer 3 contains no elements in this radio interface for UP. In its CP, on the other hand, it encompasses the radio resource control (RRC) that offers services to the nonaccess stratum (higher layers) through SAPs, with these SAPs used by the higher-layer protocols in the UE side and by Iu RANAP in the UTRAN side. RRC encapsulates higher-layer signaling (mobility management, call control, session management, etc.) into RRC messages to be transmitted over the radio interface. The control interfaces between RRC and the lower layers are used to convey information and commands to perform

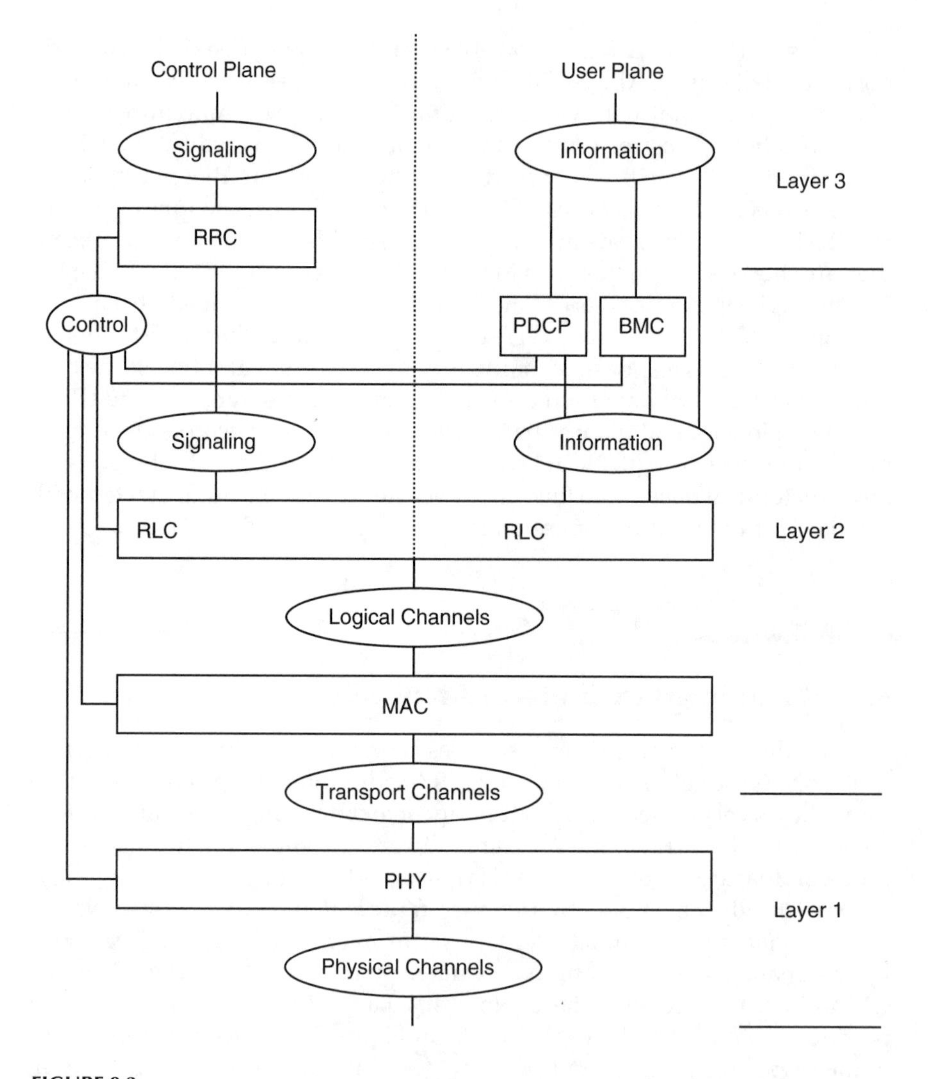

FIGURE 8.3
The basic radio interface architecture encompassing the blocks and protocols that are visible in UTRAN.

configuration of the lower-layer protocol entities (logical channels, transport channels, physical channels), to perform measurements, to report results of measurements, and others.

Layer 2 is split into several sublayers, such as packet data convergence protocol (PDCP), broadcast/multicast control (BMC), radio link control (RLC), and medium access control (MAC). BMC is used to convey messages originated from the cell broadcast center, including messages related to the short message services. PDCP is responsible for header compression and is specific to the packet-switched domain only. RLC provides services to RRC, on the CP side, and (radio bearers) to PDCP, BMC, and other higher layers, on the UP. These services are provided by means of SAPs and are called signaling radio bearers in the CP and radio bearers for services that do not use either PDCP or BMC in the UP. MAC offers services to RLC by means of SAPs.

Layer 1 contains the physical layer (PHY), which provides services to MAC by means of SAPs.

The main SAPs shown in Figure 8.3 are logical channels, transport channels, and physical channels. (The physical channels are not usually shown as SAPs in the ITU documents; however, here they are described as such only for didactic purposes.) The logical channels constitute the SAPs between RLC and MAC and are characterized by the type of information transferred. They are grouped into control channels and traffic channels. The transport channels constitute the SAPs between MAC and PHY and are characterized by how the information is transferred over the radio interface. The SAPs generated by PHY are the physical channels to be transmitted over the air. They may be defined in terms of carrier frequency, scrambling code, channelization code, relative phase, time slot, frame, and multiframe.

Figure 8.4 illustrates how higher-layer service data units (SDUs) and protocol data units (PDUs) are segmented and multiplexed to transport blocks

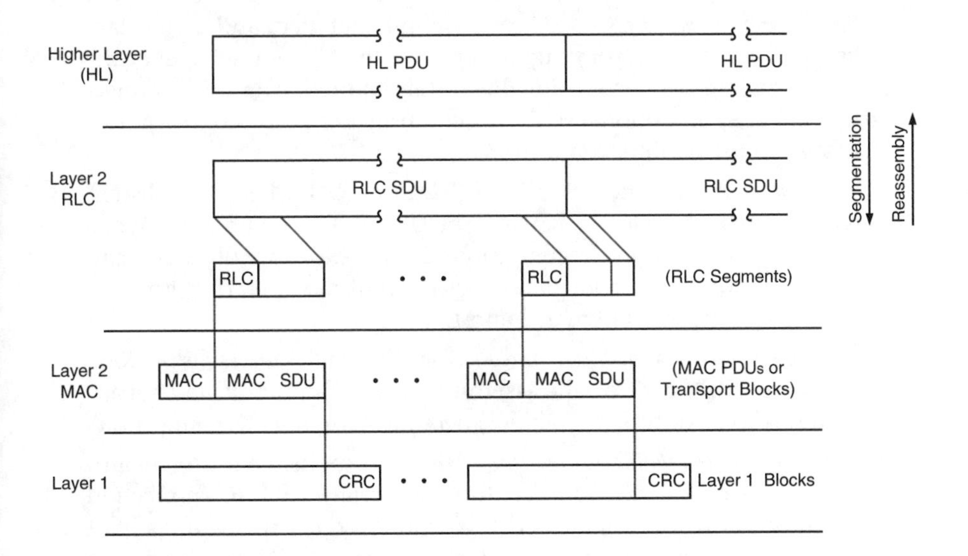

FIGURE 8.4
Data flow for a given service (nontransparent RLC and nontransparent MAC).

from Layer 3 through Layer 2 to be further treated by Layer 1. In Figure 8.4, Layer 2 is assumed to operate in the nontransparent mode, in which case protocol overhead is added to higher-layer PDUs. Next, the several blocks are detailed.

8.4.1 Layer 3

As already mentioned, only the lowest sublayer of Layer 3, represented by the RRC protocol, is visible in UTRAN.

Radio Resource Control

RRC is responsible for the CP signaling of Layer 3 between the UEs and the radio interface. It interacts with the upper layers, such as those for the CN, and performs functions as described next. (The last four functions are performed in UTRA TDD only.)

- *Broadcast of System Information*. RRC performs system information broadcasting from the network to all UEs. Such information is provided by both the nonaccess stratum (CN) and the access stratum. In the first case, the information may be cell specific and is transmitted on a regular basis, whereas in the second case the information is typically cell specific.
- *Establishment, Maintenance, and Release of an RRC Connection between the UE and the RAN*. An RRC connection is initiated, and then established, by a request from higher layers at the UE side whenever the first signaling connection for the UE is required. The RRC connection includes an optional cell reselection, an admission control, and a Layer 2 signaling link connection.
- *Establishment, Reconfiguration, and Release of RABs*. RRC establishes, reconfigures, and releases RABs of the UP on request of higher layers. The establishment and reconfiguration operations involve the realization of admission control and selection of parameters describing the RAB processing in Layer 2 and Layer 1.
- *Assignment, Reconfiguration, and Release of Radio Resources for the RRC Connection*. RRC controls the assignment, reconfiguration, and release of the radio resources (e.g., codes) necessary for an RRC connection.
- *RRC Connection Mobility Functions*. RRC is responsible for the evaluation, decision, and execution of functions related to RRC connection mobility during an established RRC connection. This includes handover, preparation of handover to other systems, cell reselection, and cell/paging area update procedures. These processes are based, for example, on measurements carried out by the UE.

- *Paging/Notification*. RRC is able to broadcast paging information from the network to selected UEs as well as to initiate paging during an established RRC connection.
- *Control of Requested QoS*. RRC is responsible for the accomplishment of the requested QoS for the access bearers. This includes the allocation of the necessary radio resources for the specific purpose.
- *UE Measurement Reporting and Control of Reporting*. RRC is responsible for the control of the measurements (what and how to measure and how to report) performed by the UEs. It is also responsible for reporting the measurements from the UEs to the network.
- *Outer Loop Power Control*. RRC controls the setting of the target signal-to-interference ratio of the closed-loop power control.
- *Control of Ciphering*. RRC establishes procedures for setting (on/off) ciphering between the UE and the RAN.
- *Initial Cell Selection and Reselection in Idle Mode*. RRC is responsible for the selection of the most suitable cell based on the idle mode measurements and cell selection criteria.
- *Arbitration of the Radio Resource Allocation between Cells*. RRC provides means of ensuring an optimal performance of the overall RAN capacity.
- *Contention Resolution (UTRA TDD)*. RRC reallocates and releases radio resources in the occurrence of collisions, as indicated by lower layers.
- *Slow DCA (UTRA TDD)*. RRC performs allocation of radio resources based on long-term decision criteria (slow dynamic channel allocation, or slow DCA).
- *Timing Advance Control (UTRA TDD-3.84)*. RRC controls the operation of timing advance.
- *Active UE Positioning (UTRA TDD-1.28)*. RRC determines the position of the active UE according to the information received from the physical layer.

8.4.2 Layer 2

Layer 2 is split into several sublayers including PDCP, BMC, RLC, and MAC.

Packet Data Convergence Protocol

PDCP is responsible for the transmission and reception of network PDUs. It provides for the mapping from one network protocol to one RLC entity. In the same way, it provides for compression (in the transmitting entity) and decompression (in the receiving entity) of redundant network PDU control information (header compression/decompression). It is present only in the UP and its tasks are related to services within the PS domain only.

Broadcast/Multicast Control

BMC is responsible for broadcast/multicast transmission services on the radio interface for common user data, appearing only in the UP. The following functionalities are handled by BMC: storage of cell broadcast messages, scheduling of BMC messages, transmission of BMC messages to UEs, delivery of broadcast messages to the upper layer, monitoring of traffic volume, and request of radio resource for cell broadcast services.

Radio Link Control

RLC provides for the establishment and release of RLC connections as well as for QoS setting and notification to higher layers in case of unrecoverable errors. It is responsible for data transfer, which may occur in three possible modes, depending on the type of service: transparent mode, unacknowledged mode, and acknowledged mode. The service provided by RLC in the CP is called signaling radio bearers, whereas in the UP this is called radio bearer. The communication of the RLC entities with the upper layer is provided by means of SAPs: TM-SAP, for the transparent mode; UM-SAP, for the unacknowledged mode; and AM-SAP, for the acknowledged mode. The communication with the lower layer (MAC) is provided by means of SAPs known as transport channels.

Transparent Mode. In the transparent mode, higher-layer PDUs are transmitted without the inclusion of extra protocol information, with the exception of the segmentation and reassembling functionalities.

Unacknowledged Mode. In the unacknowledged mode, higher-layer PDUs are transmitted to the peer entity without guaranteeing delivery. The following characteristics are applicable to the unacknowledged mode.

- *Immediate delivery*. RLC delivers an SDU to the higher-layer receiving entity upon arrival of the SDU at the receiver.
- *Unique delivery*. RLC delivers each SDU to the higher-layer receiving entity only once, using duplication detection function.
- *Detection of erroneous data*. RLC delivers to the higher-layer receiving entity only those SDUs that are free from transmission errors. This is achieved by means of the use of sequence-number check functions.

Acknowledged Mode. In the acknowledged mode, higher-layer PDUs are transmitted to the peer entity with guaranteed delivery and both in-sequence and out-of-sequence deliveries are supported. In case of unsuccessful transmission, the involved entity is notified. The following characteristics are applicable to the acknowledged mode.

- *Error-free delivery.* RLC delivers error-free SDUs to the higher layers, which is accomplished by means of retransmission of data blocks whenever erroneous data are detected.
- *Unique delivery.* RLC delivers each SDU to the higher-layer receiving entity only once, using duplication detection function.
- *In-sequence delivery.* RLC delivers SDUs to the higher-layer receiving entity keeping the same order as that of SDUs submitted by the higher-layer entity to RLC.
- *Out-of-sequence delivery.* RLC delivers SDUs to the higher-layer receiving entity without the need to keep the same order as that of SDUs submitted by the higher-layer entity to RLC.

Medium Access Control

MAC handles data streams directed to it from RLC and RRC and provides an unacknowledged transfer mode to the upper layers. The communication between RLC and MAC is carried out through SAPs known as logical channels. In the same way, the communication between MAC and PHY is handled through SAPs known as transport channels. More specifically, the following tasks are performed by MAC.

- Mapping of the different logical channels onto the appropriate Transport Channels
- Selection of appropriate transport formats for the transport channels on the instantaneous source bit rate basis
- Multiplexing of PDUs into transport blocks to be treated by PHY
- Demultiplexing of PDUs from transport blocks delivered by PHY
- Handling of priority issues for services to one UE according to information from higher layers and PHY (e.g., available transmit power level)
- Handling of priority between UEs by means of dynamic scheduling to improve spectrum efficiency
- Monitoring of traffic volume to be used by RRC so that, based on the detected volume, a reconfiguration of radio bearers or transport channels may be triggered.

8.4.3 Layer 1

Layer 1 contains the physical layer that provides services to MAC by means of SAPs known as transport channels. The SAPs generated by PHY are the physical channels to be transmitted over the air. They are defined in terms of carrier frequency, code, and relative phase for the UTRA FDD, and in terms of carrier frequency, code, time slot, and multiframe for UTRA TDD. The following functionalities are included within PHY. We note that some functionalities are particular to one or another UTRA technology, whereas others are common to the three of them. Next the functionalities in Layer 1 are cited. In the list that follows, those functionalities particular to any given technology will be appropriately indicated.

- Error detection on transport channels and indication to higher layers
- Forward error control encoding/decoding of transport channels
- Multiplexing of transport channels and demultiplexing of coded composite transport channels
- Rate matching
- Mapping of coded composite transport channels onto physical channels
- Power weighting and combining of physical channels
- Modulation and spreading/demodulation and despreading of physical channels
- Frequency and time (chip, bit, slot, frame) synchronization
- Closed-loop power control
- Radio frequency processing
- Macrodiversity distribution/combining and soft handover execution (UTRA FDD)
- Macrodiversity distribution/combining and handover execution (UTRA TDD-1.28)
- Support of timing advance on uplink channels (UTRA TDD)
- Radio characteristics measurements including frame error rate, signal-to-interference ratio, interference power levels, direction of arrival (UTRA TDD-1.28) and indication to higher layers
- Subframe segmentation (UTRA TDD-1.28)
- Random-access process (UTRA TDD-1.28)
- Dynamic channel allocation (UTRA TDD-1.28)
- Handover measurements (UTRA TDD-1.28)
- Uplink synchronization (UTRA TDD-1.28)

- Beam-forming for both uplink and downlink—smart antennas (UTRA TDD-1.28)
- UE location/positioning (UTRA TDD-1.28)

The physical layer is described in more depth in several sections of this chapter.

8.5 Logical Channels

Logical channels are SAPs located between RLC and MAC. They provide data transfer services between these two entities. A logical channel is characterized by the type of information it conveys. Broadly speaking, these channels can be grouped into control channels and traffic channels. The logical channels of the first group are used to convey CP information, whereas those in the second group are used to convey UP information.

The following are the logical channels of the control channel group:

- *Broadcast Control Channel (BCCH).* The BCCH is a downlink channel used to broadcast system control information.
- *Common Control Channel (CCCH).* The CCCH is a bidirectional channel used to convey control information between the network and the UEs.
- *Dedicated Control Channel (DCCH).* The DCCH is a point-to-point bidirectional channel used to convey dedicated control information between the network and a UE. Such a channel is established during the RRC connection establishment procedure.
- *Paging Control Channel (PCCH).* The PCCH is a downlink channel used to transmit paging information.

The following are the logical channels of the traffic channel group:

- *Control Traffic Channel (CTCH).* The CTCH is a point-to-multipoint unidirectional channel used to convey dedicated user information to all UEs or to a specified group of UEs.
- *Dedicated Traffic Channel (DTCH).* The DTCH is a point-to-point channel that can appear both in the downlink and uplink and is used to convey information to one UE.
- *Shared Channel Control Channel (SHCCH)—UTRA TDD only.* The SHCCH is a bidirectional channel used to convey information to and from shared transport channels.

8.6 Transport Channels and Indicators

Transport channels are SAPs located between MAC and PHY. They provide data transfer services between these two entities. A transport channel is characterized by how the information is transferred over the radio interface and by the type of information it conveys. Broadly speaking, these channels can be grouped into common channels and dedicated channels. A transport channel of the first group is shared by all UEs or by a group of UEs and contains an address field for address resolution purposes. The transport channel of the second group is used by a single UE and is defined by the physical channel; therefore, no specific address is needed for the UE.

The following are the transport channels of the common channel group:

- *Broadcast Channel (BCH)*. The BCH is a downlink transport channel used to broadcast system- and cell-specific information. Typically, the BCH conveys information such as the available random-access codes, available access slots, types of transmit diversity methods, etc. The BCH is transmitted over the entire cell.
- *Forward Access Channel (FACH)*. The FACH is a downlink transport channel used to convey control information to a UE whose location is known to the system. The FACH may also carry short packet data. There may exist one or more FACHs within a cell, one of them with a bit rate low enough to be detected by all UEs. The FACH does not make use of fast power control mechanisms. An in-band identification information is included to ensure correct reception of the message.
- *Paging Channel (PCH)*. The PCH is a downlink transport channel used to convey control information relevant to the paging procedure. A paging message may be transmitted to one or more (up to a few hundred) cells, depending on the system configuration.
- *Random-Access Channel (RACH)*. The RACH is an uplink transport channel used to convey control information from the UE to the network, typically for random-access purposes (requests to set up a connection). It may also be used to carry short packet data from a UE with messages lasting no longer than one or two frames.
- *Common Packet Channel (CPCH)—UTRA FDD*. The CPCH is an uplink transport channel used to convey short to medium-sized packet data from a UE. This is, in fact, an extension of the RACH; the main differences are the use of fast power control mechanisms, the longer

duration of transmission (several frames), and the channel status monitoring for collision detection purposes.

- *Downlink Shared Channel (DSCH).* The DSCH is a shared downlink transport channel used to convey dedicated user data and control information to several users. The DSCH resembles in many aspects the FACH; the main differences are the use of fast power control as well as variable bit rate on a frame-by-frame basis. It does not need to be received within the whole cell area. It can employ different modes of transmit antenna diversity methods and is associated with a downlink dedicated channel.
- *Uplink Shared Channel (USCH)—UTRA TDD only.* The USCH is a shared uplink transport channel used to convey dedicated user data and control information from several users. This is, in fact, an extension of the RACH; the main differences are the use of fast power control mechanisms, the longer duration of transmission (several frames), and the channel status monitoring for collision detection purposes.

The following is the transport channel of the dedicated channel group:

- *Dedicated Channel (DCH).* The DCH is the only transport channel within the dedicated channel group. It is an uplink or downlink transport channel conveying user information and control information. Note that, as opposed to 2G systems in which these two types of information are conveyed by different channels (traffic channel and associated control channel), the DCH carries both service data (e.g., speech frames) and higher-layer control information (e.g., handover commands, measurement reports). This is accomplished because variable bit rate and service multiplexing are supported. The DCH supports the following: fast power control mechanisms; fast data rate changed on a frame-by-frame basis; transmission to specific location within the cell by use of adaptive antennas; and soft handover (UTRA FDD).

Indicators, on the other hand, are low-level signaling entities that do not use information blocks of the transport channels. They are of the Boolean or three-valued type and are transmitted directly on the physical channels known as indicator channels. The following are the specified indicators: acquisition indicator (AI), access preamble indicator (API), channel assignment indicator (CAI), collision detection indicator (CDI), page indicator (PI), and status indicator (SI).

8.7 Physical Channels and Physical Signals

Physical channels may be defined in terms of carrier frequency, scrambling code, channelization code, relative phase, time slot, subframe, frame, and multiframe. Physical signals, on the other hand, are entities with the same basic attributes as physical channels but with no transport channels or indicators mapped to them. They may be associated with physical channels to support functionalities of the physical channels. This section briefly describes the functionalities of the physical channels, the details of their structures and specific functionalities being given in a later subsection.

8.7.1 UTRA FDD Physical Channels

The following are the physical channels of UTRA FDD.

- *Common Pilot Channel (CPICH).* The CPICH is a downlink physical channel—unmodulated code channel—used as a phase reference for the other downlink physical channels.
- *Synchronization Channel (SCH).* The SCH is a downlink physical channel used for cell search purposes.
- *Primary Common Control Physical Channel (P-CCPCH).* The P-CCPCH is a downlink physical channel conveying control information at 30 kbit/s (constant bit rate).
- *Secondary Common Control Physical Channel (S-CCPCH).* The S-CCPCH is a downlink physical channel conveying control information at a variable bit rate.
- *Acquisition Indicator Channel (AICH).* The AICH is a downlink physical channel used to convey the acquisition indicator for the random-access procedure. It is used to indicate the reception, at the base station, of the random access channel signature sequence.
- *Paging Indicator Channel (PICH).* The PICH is a downlink physical channel used to convey page indicators to indicate the presence of a page message on the PCH.
- *Physical Downlink Shared Channel (PDSCH).* The PDSCH is a downlink physical channel used to convey data and control information on a common basis. It is used in association with the downlink dedicated channel (downlink DCH) on which the information needed to decode the PDSCH is carried to the UE.
- *Access Preamble Acquisition Indicator Channel (AP-AICH), Collision-Detection/Channel-Assignment Indicator Channel (CD/CA-ICH), CPCH Status Indicator Channel (CSICH).* The AP-AICH, CD/CA-ICH, and

CSICH are downlink physical channels used for the uplink common packet channel (CPCH) procedure. The CPCH procedure is described later in this chapter.

- *Physical Random-Access Channel (PRACH).* The PRACH is an uplink physical channel used to convey control information for access purposes. It also carries short user packets from the UE.
- *Physical Common Packet Channel (PCPCH).* The PCPCH is an uplink physical channel used to convey short to medium-sized user packets. This is a fast-setup and fast-release channel. It is handled similarly to the RACH reception by the physical layer at the base station.
- *Dedicated Physical Data Channel (DPDCH).* This is an uplink and downlink physical channel used to convey user information.
- *Dedicated Physical Control Channel (DPCCH).* This is an uplink and downlink physical channel used to convey control information.

8.7.2 UTRA TDD Physical Channels

The following are the physical channels of UTRA TDD. When no specific reference is made to any particular technology, the physical channels as described below are applicable to the two TDD technologies, UTRA TDD-3.84 and UTRA TDD-1.28.

- *Downlink Pilot Time Slot (DwPTS)—UTRA TDD-1.28.* The DwPTS is a downlink physical channel used as a phase reference for the other downlink physical channels.
- *Uplink Pilot Time Slot (UpPTS)—UTRA TDD-1.28.* The UpPTS is an uplink physical channel used as a phase reference for the other uplink physical channels.
- *Synchronization Channel (SCH).* The SCH is a downlink physical channel used for cell search purposes.
- *Primary Common Control Physical Channel (P-CCPCH).* The P-CCPCH is a downlink physical channel conveying control information at 30 kbit/s (constant bit rate).
- *Secondary Common Control Physical Channel (S-CCPCH).* The S-CCPCH is a downlink physical channel conveying control information at a variable bit rate.
- *Paging Indicator Channel (PICH)—UTRA TDD-3.84.* The PICH is a downlink physical channel used to convey page indicators to indicate the presence of a page message on the PCH.
- *Physical Random-Access Channel (PRACH).* The PRACH is an uplink physical channel used to convey control information for access purposes. It also carries short user packets from the UE.

- *Physical Downlink Shared Channel (PDSCH).* The PDSCH is a downlink physical channel used to convey data and control information on a common basis. It is used in association with the downlink dedicated channel (downlink DCH) on which the information needed to decode the PDSCH is carried to the UE.
- *Physical Uplink Shared Channel (PUSCH).* The PUSCH is an uplink physical channel used to convey short to medium-sized user packets. This is a fast-setup and fast-release channel handled similarly to RACH reception by the physical layer at the base station.
- *Dedicated Physical Channel (DPCH).* This is an uplink and downlink physical channel used to convey user information and control information.

8.8 Mapping of Channels

Figure 8.5 illustrates the possible mapping of logical channels, transport channels, and physical channels. In Figure 8.5, AP-CPCH indicates the four downlink physical channels used for CPCH access procedure.

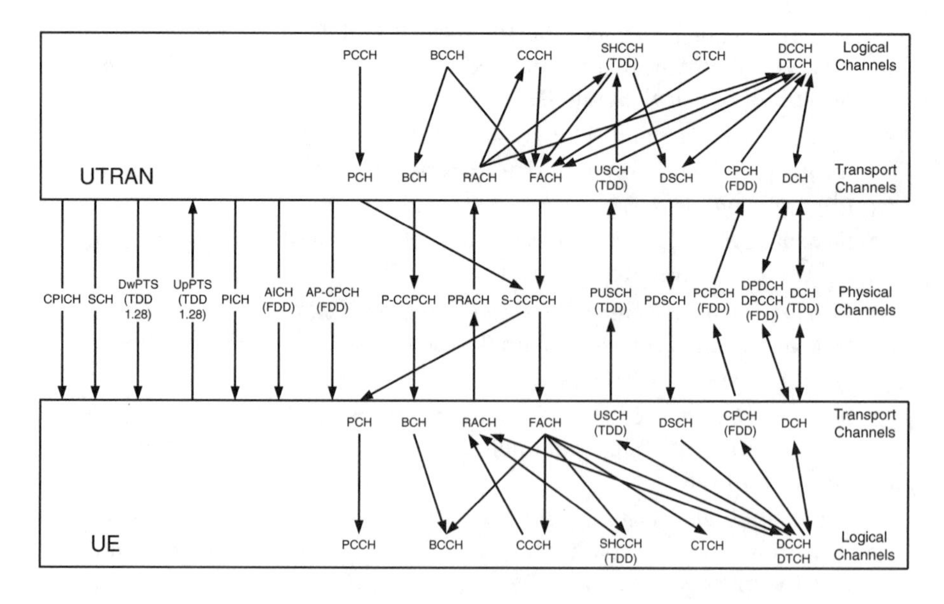

FIGURE 8.5
Mapping of logical channels, transport channels, and physical channels.

8.9 Physical Layer Transmission Chain

Figures 8.6 and 8.7 illustrate the physical layer transmission chain for UP data, respectively, for uplink and downlink. Each chain starts at the transport channel level and ends at the physical channel level. Note how several transport channels may be multiplexed onto one or more dedicated physical data channels. The transmission chain is identical in both directions for UTRA TDD and slightly different for UTRA FDD.

In the UTRA FDD uplink direction, the data stream is continuous with services multiplexed dynamically. The symbols are sent with equal power for all services, which implies that the relative rates for the different services must be adjusted to balance the power level requirements for the channel symbols. In the UTRA FDD downlink direction, the data stream is given discontinuous transmission (DTX) capability. In this case, DTX indication bits are inserted, but not transmitted over the air, and they inform the transmitter the bit positions at which the transmission should be turned off. The insertion point of the DTX indication bits depends on whether fixed or flexible positions of the transport channels in the radio frame is used. The decision whether fixed or flexible positions are used during a connection is the responsibility of UTRAN. There are two possible DTX insertion stages. The first insertion occurs only if the positions of the transport channels in the radio frame are fixed. In the second insertion, the indication bits are placed at the end of the radio frame. In such a case, because an interleaving step occurs, DTX will be distributed over all slots after the second interleaving. The transmission chain illustrated in Figures 8.6 and in 8.7 is detailed next.

The transport block, received from higher layers, is given error detection capability by means of cyclic redundancy check (CRC) attachment. Depending on the service requirements, as signaled from higher layers, the CRC attachment can take the length of 0, 8, 12, 16, or 24 bits, with the cyclic generator polynomials for the nonzero lengths, respectively, given by

$$
\begin{aligned}
G_{CRC8}(D) &= D^8 + D^7 + D^4 + D^3 + D + 1 \\
G_{CRC12}(D) &= D^{12} + D^{11} + D^3 + D^2 + D + 1 \\
G_{CRC16}(D) &= D^{16} + D^{12} + D^5 + 1 \\
G_{CRC24}(D) &= D^{24} + D^{23} + D^6 + D^5 + D + 1
\end{aligned}
$$

All transport blocks in a transmission time interval (TTI) are serially concatenated. If the size of a TTI, measured in number of bits, is larger than the maximum size of a code block under consideration, then code block

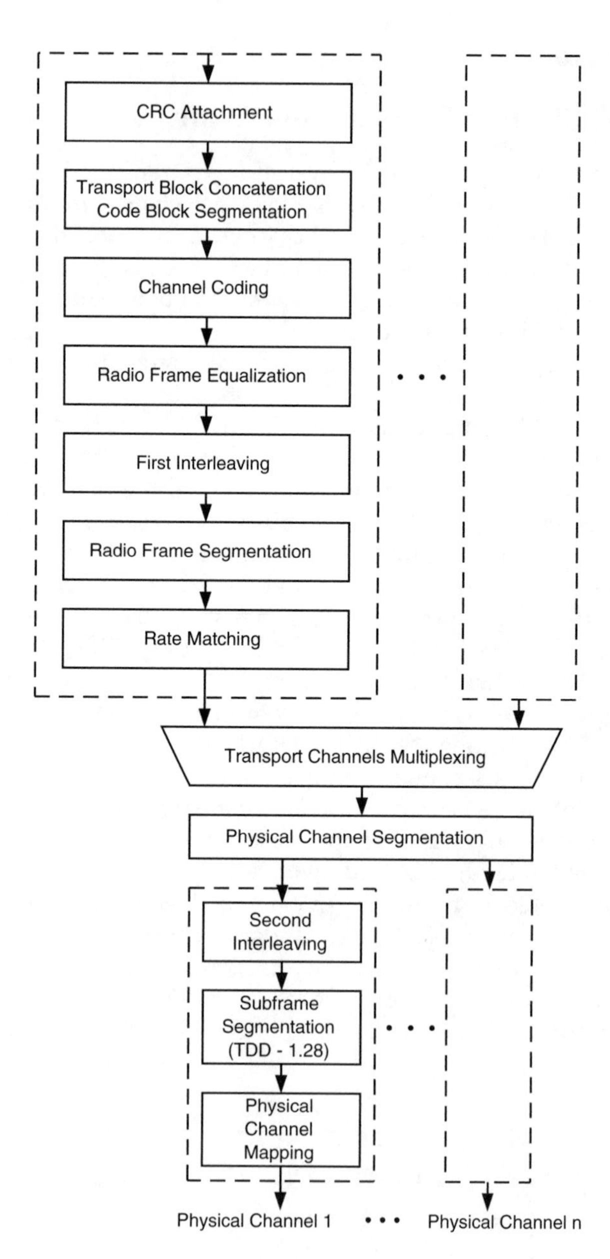

FIGURE 8.6
Physical layer transmission chain for UP data for uplink.

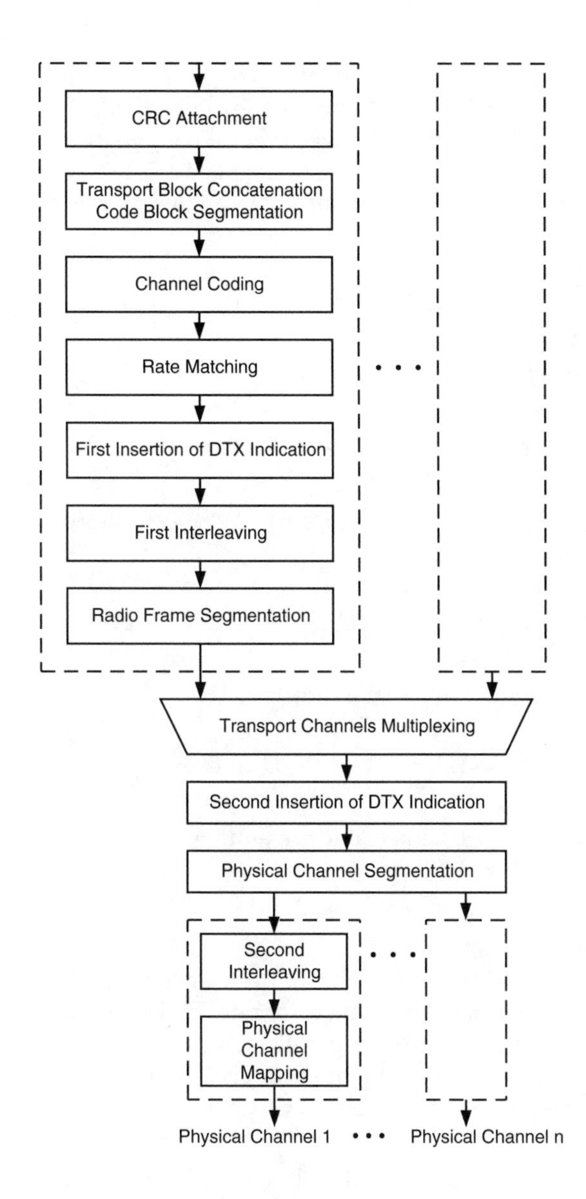

FIGURE 8.7
Physical layer transmission chain for UP data for downlink.

segmentation is carried out after the concatenation has been performed. The maximum size of the code blocks varies depending on the coding scheme used in the transport channel.

After concatenation/segmentation, the code blocks are delivered to the channel coding block, where the following coding schemes can be applied to the transport channels: convolutional coding, turbo coding, or no coding. Real-time services make use of forward error correction (FEC) encoding, whereas non-real-time services make use of a combination of FEC and automatic repeat request (ARQ). The possible convolutional coding rates are 1/2 or 1/3 whereas the turbo code rate is 1/3. In the 1/2 rate case, the generator polynomials are

$$G_0(D) = D^8 + D^6 + D^5 + D^4 + 1$$

for output 0 ($G_0(D) = 561$, in octal form), and

$$G_1(D) = D^8 + D^7 + D^6 + D^5 + D^3 + D + 1$$

for output 1 ($G_1(D) = 753$, in octal form). In the 1/3 rate case, the generator polynomials are

$$G_0(D) = D^8 + D^6 + D^5 + D^3 + D^2 + D + 1$$

for output 0 ($G_0(D) = 557$, in octal form),

$$G_1(D) = D^8 + D^7 + D^5 + D^4 + D + 1$$

for output 1 ($G_1(D) = 663$, in octal form) and

$$G_2(D) = D^8 + D^7 + D^6 + D^3 + 1$$

for output 2 ($G_2(D) = 711$, in octal form). Turbo encoding is used for data services requiring quality of service—bit error rate (BER)—within the range from 10^{-3} to 10^{-6}. For such a purpose, parallel concatenated convolutional code (PCCC) with eight-state constituent encoders is used. The transfer function of the eight-state constituent code for PCCC is

$$G(D) = \left[1, \frac{1 + D + D^3}{1 + D^2 + D^3}\right]$$

The initial value of the shift registers of the PCCC encoders is set to all zero. Its output is punctured to produce coded bits corresponding to the desired 1/3 code rate.

The radio frame equalization function concerns the frame size equalization, which ensures that the output can be segmented into equal-sized segments to be transmitted in the TTI of the respective transport channel. Frame size equalization is carried out by means of padding the input bit sequence and is performed only in the uplink.

The first interleaving, or interframe interleaving, performs a block interleaving with intercolumn permutation and possible interleaving depths of 10, 20, 40, and 80 ms.

After the first interleaving is performed, and if the transmission time interval is longer than 10 ms, the bit sequence is segmented and mapped onto consecutive equal-sized radio frames. Note that, following radio frame size equalization in the uplink and rate matching in the downlink, the input bit sequence length is guaranteed to have an integer multiple of the number of radio frames in the respective TTI. When the TTI is longer than 10 ms, the input bit sequence is segmented and mapped onto an integer number of consecutive frames. The number of bits on a transport channel may vary between different TTIs. In the downlink, for example, the transmission is interrupted if the number of bits is smaller than the maximum.

To guarantee that, after transport channels multiplexing, the total bit rate is identical to the total channel bit rate of the allocated dedicated physical channels, rate matching is performed. Rate matching is used to accommodate the number of bits to be transmitted within the frame. This is accomplished by repeating or puncturing bits of the transport channels. The rate-matching attribute is semistatic and can only be changed through higher-layer signaling. In the uplink direction, rate matching is a dynamic operation varying on a frame-by-frame basis. The multiplexing of several transport channels onto the same frame involves a rate-matching operation to guarantee the use of all symbols. In this case, a decrease of a symbol rate of a transport channel implies an increase of the symbol rate of another transport channel. Note that, by adjusting the rate-matching attribute, the quality of different services can be adjusted so that a near-equal symbol power level requirement is reached.

After rate matching, and at every 10 ms, each radio frame from each transport channel is delivered to the transport channel multiplexing block. These radio frames are serially multiplexed to yield a coded composite transport channel (CCTrCH).

In case more than one physical channel is required for the transmission of the CCTrCH, a physical channel segmentation is performed so that data are evenly distributed among the respective physical channels.

A second interleaving is then applied jointly to all data bits transmitted during one frame, or separately within each time slot (for UTRA TDD) onto which the CCTrCH is mapped. The selection of the second interleaving scheme is controlled by a higher layer.

The final step in this transmission chain is the mapping of bits from the second interleaver onto physical channels.

The multiplexing of different transport channels onto one CCTrCh and the mapping of one CCTrCH onto physical channels follow some basic rules:

- Transport channels multiplexed onto one CCTrCH should have coordinated timings. This means that transport blocks arriving on different transport channels of potentially different transmission time intervals shall have aligned transmission time instants. If the possible transmission time instants are multiple of Δt ms, then a transport channel of $k \times \Delta t$ ms will occupy k transmission time intervals starting at the allowed transmission time instants.
- Different CCTrCHs cannot be mapped onto the same physical channel.
- One CCTrCH is mapped onto one or several physical channels.
- Dedicated transport channels and common transport channels cannot be multiplexed into the same CCTrCH.
- Among the common transport channels only FACH and PCH may belong to the same CCTrCH.
- A CCTrCH carrying a BCH shall not carry any other transport channel.
- A CCTrCH carrying a RACH shall not carry any other transport channel.

Note that there are two types of CCTrCH: CCTrCH of the dedicated type, corresponding to the result of coding and multiplexing of one or more DCHs; and CCTrCH of the common type, corresponding to the result of the coding and multiplexing of a common channel, namely, RACH and USCH in the uplink, and DSCH, BCH, FACH, or PCH in the downlink.

8.10 Channel and Frame Structures

In this section, the frame structure and the channel structure of UTRA are described. Special attention is given to the UTRA FDD technology, where a detailed description of these structures is provided. These items are only superficially explored for the UTRA TDD technologies.

8.10.1 UTRA FDD Uplink Physical Channels

This subsection describes the various UTRA FDD uplink physical channels.

Dedicated Physical Data Channel and Dedicated Physical Control Channel

Two dedicated uplink physical channels are specified: dedicated physical data channel (DPDCH) and dedicated physical control channel (DPCCH). These two channels are code-multiplexed within each frame on an in-phase and quadrature basis (I/Q code multiplexing). DPDCH is used to convey the DCH transport channel, whereas DPCCH is used to carry Layer 1 control information. The number of DPDCHs may range from zero to six on each radio link with a possible spreading factor ranging from 256 to 4. On the other hand, there is only one DPCCH on each link with a fixed spreading factor of 256. Note, therefore, that the DPCCH data rate is fixed, whereas the DPDCH data rate may vary on a frame-by-frame basis.

The information conveyed by the DPDCH corresponds to the higher-layer information, including user data. The Layer 1 control information conveyed by the DPCCH includes pilot bits, used to support channel estimation for coherent detection; transmit power control (TPC) bits, used to carry power control commands for the downlink power control; feedback information (FBI) bits, used to support techniques requiring feedback from the UE to the UTRAN access point, including closed-loop mode transmit diversity and site selection diversity transmission; and transport format control information (TFCI) bits, used to inform the receiver about the instantaneous transport format combination (rate information of the transport channels mapped onto the DPDCH that is being multiplexed with the DPCCH carrying such control information). The presence or not of TFCI within the uplink dedicated physical channels characterizes, respectively, the several simultaneous services or the fixed-rate services on the channel.

The radio frame is 10 ms long and contains 15 slots. Given that the transmission rate is 3.84 Mchip/s, the total number of chips is 38,400 per radio frame and 2560 per slot. A slot corresponds to one power control period and is 0.66666... ms long. The frame structure of the uplink dedicated physical channels is shown in Figure 8.8. The number of bits of the DPDCH $N_{\text{bit/slot}}$ varies in accordance with the seven possible slot formats, ranging from 10 bits/slot, for slot format 0, to 640 bits/slot, for slot format 6. More specifically,

$$N_{\text{bit/slot}} = 10 \times 2^{8-\log_2(\text{spreading\_factor})}$$

where the spreading factor is

$$\text{spreading factor} = 2^{8-(\text{slot\_format})}$$

or, equivalently,

$$N_{\text{bit/slot}} = 10 \times 2^{\text{slot\_format}}$$

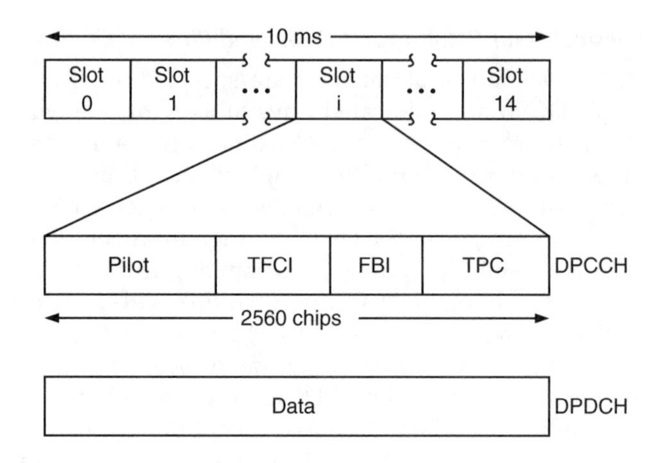

FIGURE 8.8
Frame structure of the uplink dedicated physical channels (DPCCH and DPDCH).

where the slot_format assumes the values 0, 1, 2, 3, 4, 5, or 6. Table 8.2 shows the DPDCH fields (the data rate shown corresponds to that immediately before spreading).

As for the DPCCH, the number of bits per slot, the channel bit rate, the symbol rate, and the spreading factor are constant and equal to 10 bits/slot, 15 kbit/s, 15 kbit/s, and 256, respectively. The number of pilot bits, TPC bits, TFCI bits, and FBI bits vary in accordance with the slot format, but their sum is constant and equal to 10 always.

The uplink dedicated physical channels may operate on a multicode basis, in which case several parallel DPDCHs are transmitted using different channelization codes. Even operating on a multicode basis, only one DPCCH per radio link is used.

TABLE 8.2
Uplink DPDCH Fields

Slot Format	Spreading Factor	Bits per Slot	Channel Bit Rate (kbit/s)
0	256	10	15
1	128	20	30
2	64	40	60
3	32	80	120
4	16	160	240
5	8	320	480
6	4	640	960

Physical Random-Access Channel

The PRACH is used to convey the RACH. The access policy is based on a slotted-ALOHA algorithm with fast acquisition indication. Defined are 15 access slots per two radio frames (15 access slots per 20 ms), spaced 5120 chips apart, i.e., each access slot has 5120 chips. The UE may start the random-access transmission at the beginning of any access slot, with the information on what access slots are available for random access transmission given by higher layers. The structure of the random-access transmission consists of one or more 4096-chip preambles and a 10- or 20-ms message part. The preamble is composed of 256 repetitions of a signature of length 16 chips, with a maximum of 16 available signatures. The message part of 10 ms duration consists of 15 slots, each of which contains 2560 chips. Each slot is split into two parts, one that conveys the RACH information and another that conveys the Layer 1 control information. The message part of 20 ms is composed of two 10-ms message parts. The data part consists of

$$N_{\text{bits}} = 10 \times 2^{\text{slot\_format}}$$

where the slot_format assumes the values 0, 1, 2, or 3, corresponding to spreading factors, respectively, of 256, 128, 64, and 32. The control part consists of eight known pilot bits, for channel estimation and coherent detection purposes, and two bits for TFCI, which corresponds to a spreading factor of 256. Note that the number of TFCI bits in the random-access message is 30 (2 bits/slot $\times$ 15 slots/frame). The data part and the control part are code-multiplexed and transmitted in parallel. The frame structure of the PRACH is shown in Figure 8.9. Table 8.3 shows the PRACH fields (the data rate shown corresponds to that immediately before spreading).

Physical Common Packet Channel

The PCPCH is used to convey the CPCH. The access policy is based on the digital (or data) sense multiple access with collision detection (DSMA-CD) algorithm with fast acquisition indication. Transmission from the UE may be started at the beginning of time intervals, defined with respect to the frame boundary of the received BCH of the current cell. The access transmission of the PCPCH uses one or more 4096-chip access preambles, one 4096-chip collision detection preamble, a 0 or 8-slot DPCCH power control preamble, and a variable-length message part. The access preamble part and the collision detection preamble part are similar to the RACH preamble part and use preamble signature sequences. The power control preamble presents a slot structure containing 10 bits per slot defined by the pilot bits, the TPC bits, the TFCI bits, and the FBI bit, with a channel bit rate of 15 kbit/s and a spreading factor of 256. The message part consists of one or more 10-ms frames, with the number of frames defined by higher layers. A frame is split into 15 2560-chip

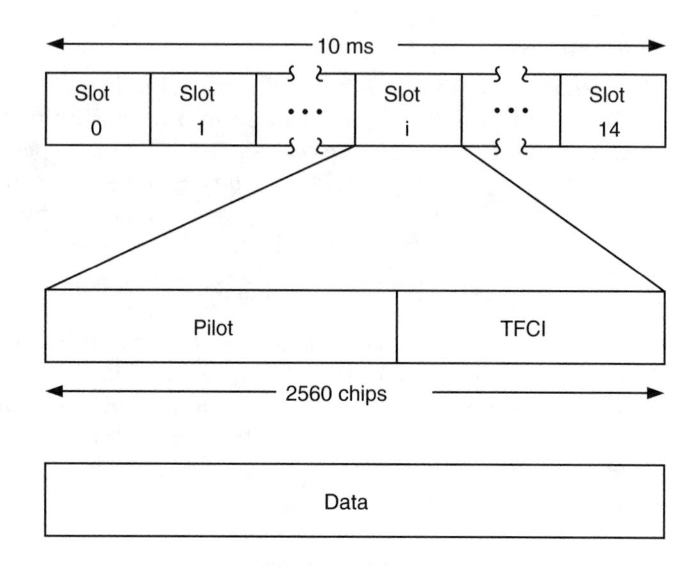

FIGURE 8.9
Frame structure of PRACH.

slots, each slot containing a data part and a control part. The data part and the control part are code-multiplexed and transmitted in parallel. The slot structure of Table 8.2 applies for the data part and that of the power control preamble applies for the control part.

8.10.2 UTRA FDD Downlink Physical Channels

This subsection describes the various UTRA FDD downlink physical channels.

TABLE 8.3
PRACH Fields

Slot Format	Spreading Factor	Bits per Slot	Channel Bit Rate (kbit/s)
0	256	10	15
1	128	20	30
2	64	40	60
3	32	80	120

Dedicated Physical Data Channel and Dedicated Physical Control Channel

Two dedicated downlink physical channels are specified: dedicated physical data channel and dedicated physical control channel. The DPDCH is used to convey the DCH transport channel, whereas the DPCCH is used to carry Layer 1 control information represented by the pilot bits, TFCI bits, and TPC bits. These two channels are time-multiplexed.

A radio frame is 10 ms long and contains 15 slots. Given that the transmission rate is 3.84 Mchip/s, the total number of chips is 38,400 per radio frame and 2560 per slot. A slot corresponds to one power control period and is 0.6666... ms long. The frame structure of the downlink dedicated physical channels is shown in Figure 8.10. The total number of bits per slot $N_{\text{bit/slot}}$ ranges from 10 to 1280 with the spreading factor varying from 512 to 4. More specifically,

$$N_{\text{bit/slot}} = 10 \times 2^{9-\log_2(\text{spreading\_factor})}$$

where the spreading factor is

$$\text{spreading\_factor} = 2^{9-(\text{slot\_format})}$$

or, equivalently,

$$N_{\text{bit/slot}} = 10 \times 2^{\text{slot\_format}}$$

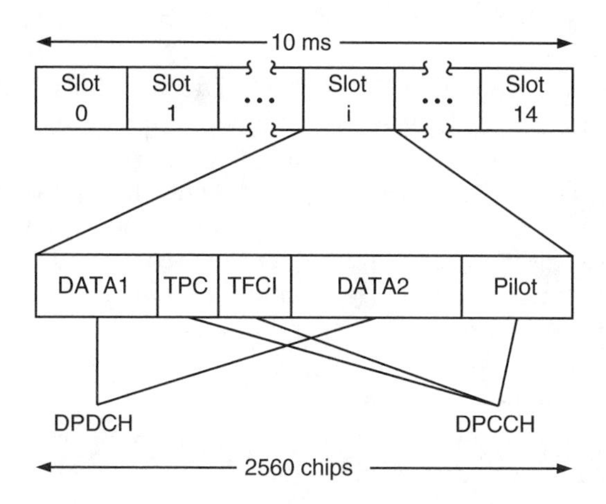

FIGURE 8.10
Frame structure of the downlink dedicated physical channels (DPCCH and DPDCH).

TABLE 8.4

Downlink DPCH and DPCCH Fields

Spreading Factor	Bits per Slot	DPDCH Bits per Slot	DPCCH Bits per Slot	Channel Bit Rate (kbit/s)
512	10	2;4	8;6	15
256	20	4;6;8;10;12;14;16	16;14;12;10;8;6;4	30
128	40	16;20;24;26;28;30;32;34	24;20;16;14;12;10;8;6	60
64	80	52;56;60;64	28;24;20;16	120
32	160	120;132;140	40;28;20	240
16	320	280;288	40;32	480
8	640	576;600;608	64;40;32	960
4	1280	1216;1240;1248	64;40;32	1920

where the slot_format assumes the values 0, 1, 2, 3, 4, 5, 6, or 7. Table 8.4 shows the DPDCH and DPCCH fields. The channel symbol rate is half of the channel bit rate, when both rates are obtained immediately before spreading.

Multicode transmission may also be employed in the downlink, in which one CCTrCH is mapped onto more than one parallel downlink channels using the same spreading factor. The control part of the channel, i.e., the DPCCH, is transmitted only on the first downlink channel. On the additional downlink channels, the time period of the DPCCH is filled with DTX bits. It is also possible to have several CCTrCHs mapped onto different downlink channels and transmitted to the same UE. In this case, different spreading factors can be used. As before, the control information is sent only on the first channel while DTX bits are transmitted during the corresponding time period for the other channels.

There is also the downlink DPCCH (DL-DPCCH) for CPCH, which is a special case of dedicated physical channels using a spreading factor of 512. The frame structure of DL-DPCCH is similar to that shown in Figure 8.10, but only the fields TPC, TFCI, pilot, and CCC (CPCH control commands) are present. The CCC bits are used to support CPCH signaling such as Layer 1 command (start of a message indicator) and higher layer command (emergency stop command). The channel bit rate, the channel symbol rate, and the number of bits per slot are, respectively, 15, 7.5, and 10 bits.

Common Pilot Channel

The CPICH is an unmodulated code channel transmitted at a fixed rate of 30 kbit/s (scrambling factor of 256), which conveys a predefined bit/symbol sequence. The frame structure of CPICH consists of 15 2560-chip slots and is 10 ms long. A slot is composed of 20 bits, or equivalently 10 symbols. Two types of

CPICHs are provided: the primary CPICH (P-CPICH) and secondary CPICH (S-CPICH). P-CPICH presents the same channelization code always, whereas S-CPICH may have an arbitrary channelization code of spreading factor equal to 256. P-CPICH is scrambled by the primary scrambling code (PSC), whereas S-CPICH is scrambled by either the PSC or secondary scrambling code (SSC). Each cell has one P-CPICH, but zero or more S-CPICHs. P-CPICH is broadcast throughout the cell, whereas S-CPICH may be transmitted over the entire cell or only over a specific part of it.

P-CPICH is used as a phase reference for SCH, P-CCPCH, AICH, PICH, AP-AICH, CD/CA-ICH, CSICH, and S-CCPCH carrying PCH. Moreover, by default, P-PCICH is the phase reference for S-CCPCH carrying FACH only and downlink DPCH, but this default condition can be changed by higher-layer signaling.

S-CPICH may be used as a phase reference for S-CCPCH carrying FACH only and/or downlink DPCH, but this is not the default condition.

Primary Common Control Physical Channel

The P-CCPCH is a fixed-rate channel transmitted at 30 kbit/s (scrambling factor of 256) used to convey the BCH transport channel. The frame structure of P-CCPCH consists of 15 2560-chip slots and is 10 ms long. A slot is composed of 20 bits but the initial two bits (256 chips) are not transmitted. The corresponding time interval is used to multiplex the primary SCH and secondary SCH. The P-CCPCH is transmitted over the entire cell.

Secondary Common Control Physical Channel

The S-CCPCH is a variable-rate channel used to convey FACH and PCH, these channels appearing in the same S-CCPCH or separate S-CCPCHs. The frame structure of S-CCPCH consists of 15 2560-chip slots and is 10 ms long. Each slot contains control and data information; the control part comprises the TFCI bits and the pilot bits. The total number of bits per slot $N_{\text{bit/slot}}$ ranges from 20 to 1280 with the spreading factor varying from 256 to 4. Accordingly, the channel bit rate varies from 30 to 1920 kbit/s, with the channel symbol rate half the channel bit rate. More specifically,

$$N_{\text{bit/slot}} = 20 \times 2^{8-\log_2(\text{spreading\_factor})}$$

where the spreading factor is

$$\text{spreading\_factor} = 2^{8-(\text{slot\_format})}$$

or, equivalently,

$$N_{\text{bit/slot}} = 20 \times 2^{\text{slot\_format}}$$

where the slot_format assumes the values 0, 1, 2, 3, 4, 5, or 6.

S-CCPCH carrying FACH may be transmitted over part of a cell (narrow lobe), if required. S-CCPCH may include TFCI or not, which is determined by UTRAN. The TFCI value in each radio frame determines the transport format combination of the FACHs and/or PCHs currently in use, and this is negotiated at each FACH/PCH addition/removal.

Synchronization Channel

The SCH is used for cell search purposes. It consists of the primary SCH (P-SCH) and secondary SCH (S-SCH), which are code-multiplexed and transmitted in parallel. The frame structure of SCH consists of 15 2560-chip slots and is 10 ms long. Of these 2560 chips in each slot, only the first 256 chips are used for SCH and the corresponding time interval is shared by P-CCPCH, as already mentioned. P-SCH consists of a modulated code transmitted once every time slot. Such a code, named primary synchronization code (PSC), is the same for every cell within the system. S-SCH consists of a sequence of 15 256-chip modulated codes, the secondary synchronization codes (SSC). Each 256-chip block is transmitted within each respective slot of the frame. An SSC is selected from a set of 16 codes of length 256. The sequence on S-SCH indicates the scrambling code group, out of 64 possible groups, to which the downlink scrambling code of the cell belongs.

Physical Downlink Shared Channel

The PDSCH is used to convey DSCH and is allocated on a radio frame basis. The following illustrates the possible mode of operation of PDSCH. Different PDSCHs may be allocated to different UEs on a code multiplex basis; multiple parallel PDSCHs, with the same spreading factor, may be allocated to a single UE; PDSCHs allocated to the same UE, but on different radio frames, may have different spreading factors. Each PDSCH is associated with the downlink DPCH, which does not necessarily have the same spreading factor, nor is it frame-aligned with PDSCH. All relevant Layer 1 control information concerning PDSCH is conveyed by the DPCCH part of the associated DPCH. The frame structure of PDSCH consists of 15 2560-chip slots and is 10 ms long. Each slot contains data information only, with the control information provided by the associated downlink PDCH. The total number of bits per slot $N_{\text{bit/slot}}$ ranges from 20 to 1280 with the spreading factor varying from 256 to 4. Accordingly, the channel bit rate varies from 30 to 1920 kbit/s, with the channel symbol rate half the channel bit rate. More specifically,

$$N_{\text{bit/slot}} = 20 \times 2^{8-\log_2(\text{spreading\_factor})}$$

where the spreading_factor is

$$\text{spreading\_factor} = 2^{8-(\text{slot\_format})}$$

or, equivalently,

$$N_{\text{bit/slot}} = 20 \times 2^{\text{slot\_format}}$$

where the slot_format assumes the values 0, 1, 2, 3, 4, 5, or 6. The UE receiving the associated DPCH is informed whether or not there are data to be decoded on the DSCH by means of either TFCI or higher-layer signaling.

Acquisition Indicator Channel

The AICH is used to convey the acquisition indicators that correspond to a signature on PRACH. The frame structure of AICH consists of 15 consecutive 5120-chip access slots and is 20 ms long. An access slot contains an acquisition indicator, which consists of 32 real-valued symbols (4096 chips), and a part, consisting of 1024 chips, reserved for possible use by CSICH or by another physical channel to be specified. AICH runs at a fixed rate with a spreading factor of 256 and has as a phase reference the P-CPICH.

CPCH Access Preamble Acquisition Indicator Channel

The AP-AICH is used to convey AP acquisition indicator of CPCH; an AP acquisition indicator corresponds to an AP signature transmitted by UE. The frame structure of AICH consists of 15 consecutive 5120-chip access slots and is 20 ms long. An access slot contains an AP acquisition indicator, which consists of 32 real-valued symbols, and a part, consisting of 1024 chips, reserved for possible use by CSICH or by another physical channel to be specified. The AP-AICH runs at a fixed rate with a spreading factor of 256 and has as a phase reference the P-CPICH.

CPCH Collision Detection/Channel Assignment Indicator Channel

The CD/CA-ICH is used to convey CD indicator, if CA is not active, or CD indicator/CA indicator simultaneously if CA is active. (CA is described later in this chapter.) The frame structure of CD/CA-ICH consists of 15 consecutive 5120-chip access slots and is 20 ms long. An access slot contains the CD indicator/CA indicator, which consists of 32 real-valued symbols, and a part, consisting of 1024 chips, reserved for possible use by CSICH or by another physical channel to be specified. The AP-AICH runs at a fixed rate with a spreading factor of 256 and has as a phase reference the P-CPICH.

Paging Indicator Channel

The PICH is used to convey the paging indicators and is always associated with an S-CCPCH onto which the PCH transport channel is mapped. The frame structure of PICH is 10 ms long and consists of 300 bits, with 288 used for paging indication and 12 bits that are reserved for possible future application. It runs at a fixed rate with a spreading factor of 256.

CPCH Status Indicator Channel

The CSICH runs in association with the physical channel used for transmission of CPCH AP-AICH, making use of the same channelization and scrambling codes. The frame structure of CSICH consists of 15 40-bit access slots and is 20 ms long. An access slot contains a part with no information and another consisting of the status indicator. The part with no information corresponds to 4096 chips and is reserved for use by AICH, AP-AICH, or CD/CA-ICH. The part containing the status indicator is 8 bits long. The CSICH runs at a fixed rate with spreading code of 256, having as a phase reference the P-CPICH.

8.10.3 UTRA TDD-3.84

UTRA TDD-3.84 timing is based on a multiframe structure. A multiframe contains *n* 10-ms-long frames, a frame encompassing 15 2560-chip slots. The time slots may be allocated either to the uplink or to the downlink, as required by the different environments and deployment scenarios. In particular, at least one time slot must be allocated to the uplink and at least one time slot must be allocated to the downlink. Note that a highly asymmetric frame configuration may be encountered for a given application. The time slot structure—the burst structure—contains two data fields separated by a midamble field, which is used for both channel equalization (training sequence) and coherent detection at the receiver. TFCI field is also incorporated to indicate the multiplexing mode within the frame. Whenever applicable, TPC may also be included. Moreover, a guard period is provided at the end of the burst for timing alignment purposes. The midambles reduce the user payload information, comprising training sequences of different users. The timing structure of UTRA TDD-3.84 is illustrated in Figure 8.11. UTRA TDD-3.84 allows for the use of multicode and multiple time slots. Next described are some of the main UTRA TDD-3.84 physical channels.

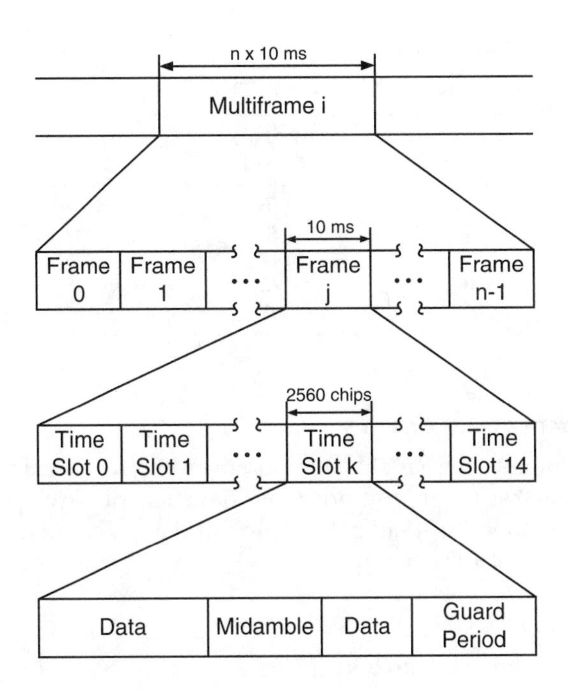

FIGURE 8.11
UTRA TDD-3.84 timing structure.

Dedicated Physical Channel

The DPCH is used to convey user information and control information. Two burst formats may be used for a slot conveying DCH. In burst format 1, each of the two data fields (payload information) contains 976 chips and these fields are separated by a 512-chip midamble. In burst format 2, each of the two data fields (payload information) contains 1104 chips and these fields are separated by a 256-chip midamble. In both cases, the guard period length is 96 chips, which, at 3.84 Mchip/s, corresponds to 25 μs (cell radius of 3.75 km). The symbols resulting from the modulation process are spread with a channelization code of length varying from 1 to 16 in the uplink and 1 and 16 in the downlink. The number of symbols per data field as a function of the spreading factor in each of the two burst formats is shown in Table 8.5.

The choice for burst format 1 or burst format 2 depends on the application and on the number of allocated users per time slot. The actual number of bits per burst depends on the modulation scheme utilized. For a QPSK modulation, the actual number of bits per data field is four times the figures shown in Table 8.5.

TABLE 8.5

Symbols per Data Field in TDD-3.84 Burst

Spreading Factor	Symbols Burst 1	Symbols Burst 2
1	976	1104
2	488	552
4	244	276
8	122	138
16	61	69

Physical Random Access Channel

The PRACH conveys the RACH and uses spreading factor values of 8 and 16. The PRACH burst structure comprises one data field of 976 chips, a midamble of 512 chips, another data field of 880 chips, and a guard period of 192 chips. The guard period of 192 corresponds to 50 μs (a cell radius of 7.5 km).

Synchronization Channel

The SCH is used for cell search purposes and occupies one or two time slots within a 10-ms (15-slot) frame. In the case of a structure with two slots, the slots containing SCH are separated by seven other slots. In the corresponding slot, a time offset is introduced so that distinct offsets may be used to identify the cells within the SCH slot. SCH contains four synchronization sequences constituted by one primary code and three secondary codes, each of which is 256 chips long with the four of them transmitted simultaneously. The following information is conveyed by SCH: base station code group (5 bits); frame position within the interleaving period (1 bit); slot position within the frame (1 bit); primary CCPCH locations (3 bits). An SCH sequence is used to decode the following: frame synchronization; time offset; midamble; spreading code set of the base station; spreading codes; and location of the BCCH.

Downlink Shared Channel and Uplink Shared Channel

The DSCH and USCH are allocated on a temporary basis and present the same structure as that of the dedicated channels. In DSCH, the TFCI may indicate which terminals need to decode the channel. In USCH, higher-layer signaling is used and, in practice, the channel is not shared on a frame-by-frame basis.

Common Control Physical Channel

The CCPCH is similar to the DPCH and can be mapped onto any downlink slot including P-SCH slots. The information for the mapping is conveyed by the primary BCH. In fact, once synchronization is achieved, the timing and the coding of the primary BCH are already known.

8.10.4 UTRA TDD-1.28

UTRA TDD-1.28 timing is based on a multiframe structure. A multiframe contains n 10-ms-long frames, a frame encompassing two 5-ms subframes. A subframe contains seven 675-μs time slots (normal time slots) and three special time slots: DwPTS (downlink pilot), G (guard period), and UpPTS (uplink pilot). A physical channel may be allocated one or more time slots on a continuous or discontinuous mode. In the first case, the allocation occurs in every frame. In the second case, the allocation occurs in a subset of all frames. One time slot (time slot 0) among the seven available is always allocated as downlink whereas another (time slot 1) is allocated as uplink. The remaining time slots may be allocated as downlink or uplink, as required. Note that a highly asymmetric frame configuration may be encountered for a given application. Downlink and uplink time slots are separated by a switching point but only two switching points are possible within each subframe. The time slot structure—the burst structure—contains two data fields separated by a midamble field, which is used for both channel equalization (training sequence) and coherent detection at the receiver. A guard period is provided at the end of the burst for timing alignment purposes. The midamble reduces the user payload information, comprising training sequences of different users. The timing structure of UTRA TDD-1.28 is illustrated in Figure 8.12, where time slots 2 and 3 are allocated as uplink and time slots 4, 5, and 6 are allocated as downlink. UTRA TDD-1.28 allows for the use of multicode and multiple time slots. Next, some of the UTRA TDD-1.28 physical channels are described.

Dedicated Physical Channel

The DPCH is used to convey user information and control information. The burst format is composed of two data fields (payload information), both amounting to 704 chips and separated by a 144-chip midamble, and a 16-chip guard period (12.5 μs, cell radius of 1.875 km). The symbols resulting from the modulation process are spread with a channelization code of length 16 in the downlink and varying from 1 to 16 in the uplink. The number of symbols per data field as a function of the spreading factor is shown in Table 8.6.

The burst type used allows the transmission of TFCI both in downlink and uplink. For each user the TFCI information is transmitted once every 10-ms frame. The transmission of TPC is negotiated at the call setup and can be renegotiated during the call. For each user the TPC is transmitted once every 5-ms subframe.

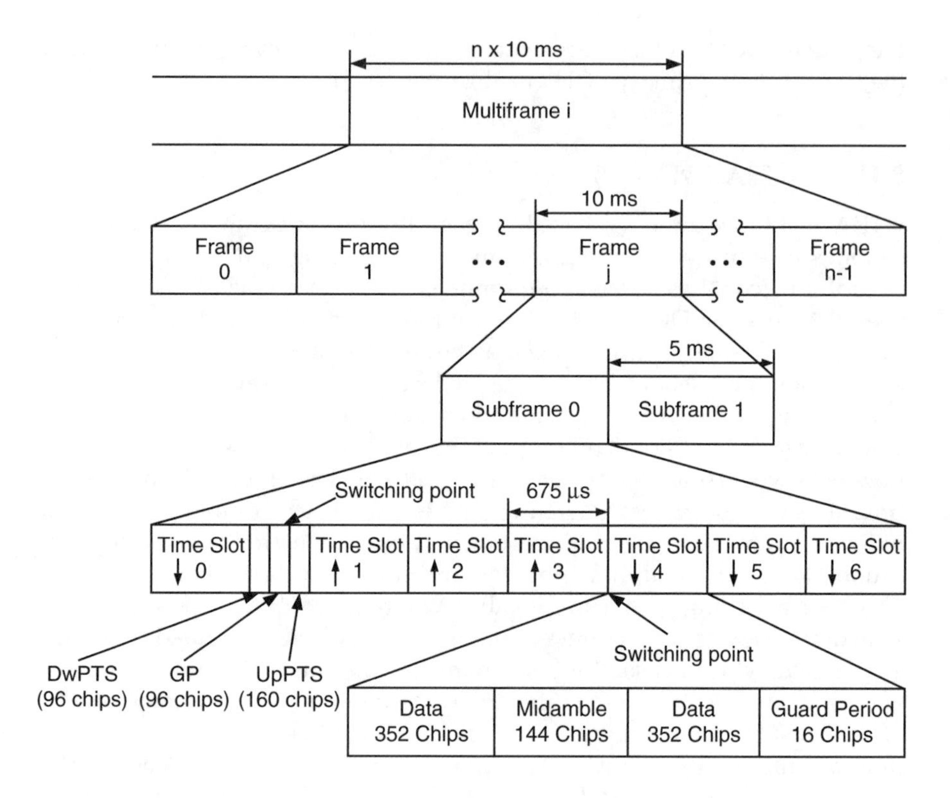

FIGURE 8.12
UTRA TDD-1.28 timing structure.

Primary Common Control Physical Channel

The P-CCPCH conveys the BCH. There are two P-CCPCHs: P-CCPCH1 and P-CCPCH2. They are mapped onto the first two code channels of time slot 0 and run at a spreading factor of 16.

TABLE 8.6

Symbols per Data Field in TDD-1.28 Burst

Spreading Factor	Symbols per Data Field
1	352
2	176
4	88
8	44
16	22

Secondary Common Control Physical Channel

The S-CCPCH conveys the PCH and FACH. There are two S-CCPCHs, namely, S-CCPCH1 and S-CCPCH2, which are mapped onto two code channels with spreading factor of 16. P-CCPCH and S-CCPCH can be time-multiplexed on time slot 0 or they can be allocated on any other downlink time slots. The time slot and the codes to be used are broadcast via cell information.

Physical Random Access Channel

The PRACH conveys the RACH. It uses a spreading factor of either 16 or 8. Its configuration (time slot number and spreading codes) is broadcast through BCH.

Synchronization Channel

There are two types of synchronization channels: DwPTS and UpPTS. DwPTS in each subframe is designed for both downlink pilot and SCH. It is transmitted omnidirectionally or sectorially at full power levels. It is composed of 96 chips (75 μs), 64 of which are used for synchronization purposes and 32 of which comprise the guard period. The PN code set is used to distinguish cells. UpPTS in each subframe is designed for both uplink pilot and SCH. It is composed of 150 chips (125 μs), 125 of which are used for synchronization purposes and 32 of which comprise the guard period. The PN code set is designed to distinguish UEs in the access procedure. In the condition of air registration or random access, the UE transmits the UpPTS first and, after an acknowledgment from the network, it then transmits RACH.

Physical Uplink Shared Channel

The PUSCH uses the same burst structure as that of DPCH. Information concerning power control, timing advance, or directive antenna settings are derived from FACH or DCH. TFCI may be transmitted within PUSCH.

Physical Downlink Shared Channel

The PDSCH uses the same burst structure as that of DPCH. Information concerning power control or directive antenna settings are derived from FACH or DCH. TFCI may be transmitted within PDSCH.

Paging Indicator Channel

The PICH conveys the paging indicators. It is always transmitted at the same reference power level and the same antenna pattern as for P-CCPCH. The same burst format is used for PICH throughout the cell.

8.11 Spreading and Modulation

Once the physical channels are formatted, they are ready for transmission. The transmission of the physical channels consists of two processes: *spreading* and *modulation*. Spreading consists of the following operations: channelization and scrambling. The channelization operation transforms each symbol into a number of chips, given by the spreading factor, thus increasing the signal bandwidth. In the scrambling operation, the resulting spread signal is further multiplied by a scrambling code. These procedures are the same for UTRA FDD, UTRA TDD-3.84, and UTRA TDD-1.28, with the peculiarity that in UTRA TDD the midamble part is not spread. The chip sequence generated by the spreading process is QPSK-modulated. In the modulation process the real part and the imaginary part of the spread signal are independently pulse shaped and multiplied by the in-phase carrier and the quadrature carrier, respectively. The pulse-shaping method uses a root-raised cosine filtering with a roll-off factor of 0.22. Figure 8.13 illustrates the transmission process in UTRA. The spreading process to be detailed here concerns UTRA FDD. In the spreading process, the binary physical channels to be spread are represented by real-valued sequences, in which the binary value 0 is mapped onto the real value +1 and the binary value 1 is mapped onto the real value −1.

8.11.1 Uplink Spreading

DPCCH/DPDCH Spreading

One DPCCH and as many as six parallel DPDCHs can be transmitted simultaneously as follows. The DPCCH is spread to the chip rate by a control channelization code (C_c), while each DPDCH is spread to the chip rate

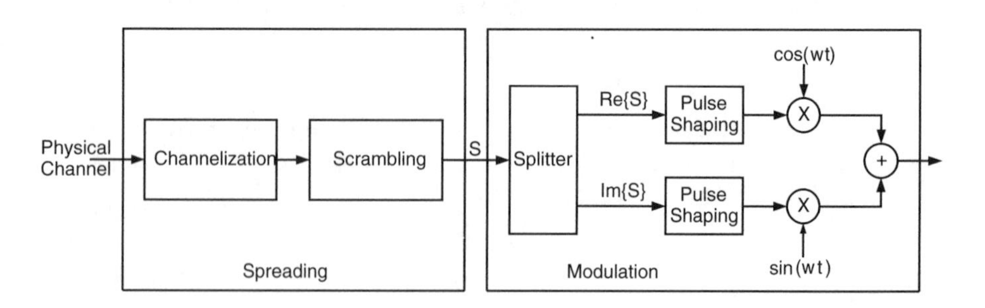

FIGURE 8.13
UTRA transmission: spreading and modulation.

by a specific data channelization code ($C_{d,i}$, $1 \leq i \leq 6$). The real-valued spread signals are multiplied by a gain factor (G_c for DPCCH and G_d for all the DPDCHs). The ratio G_c/G_d is quantized into 4-bit words. Therefore, 16 gain ratios can be represented and they are $15/15, 14/15, \ldots, 0/15$. The weighted streams of real-valued chips on the in-phase (I) and quadrature (Q) branches are added and treated as a complex-valued stream of chips. The complex-valued stream of chips is scrambled by the complex-valued scrambling code $S_{DPCH,n}$, which is aligned with the radio frames (the first scrambling chip starts at the beginning of a radio frame). Figure 8.14 illustrates the DPCCH/DPDCH spreading process.

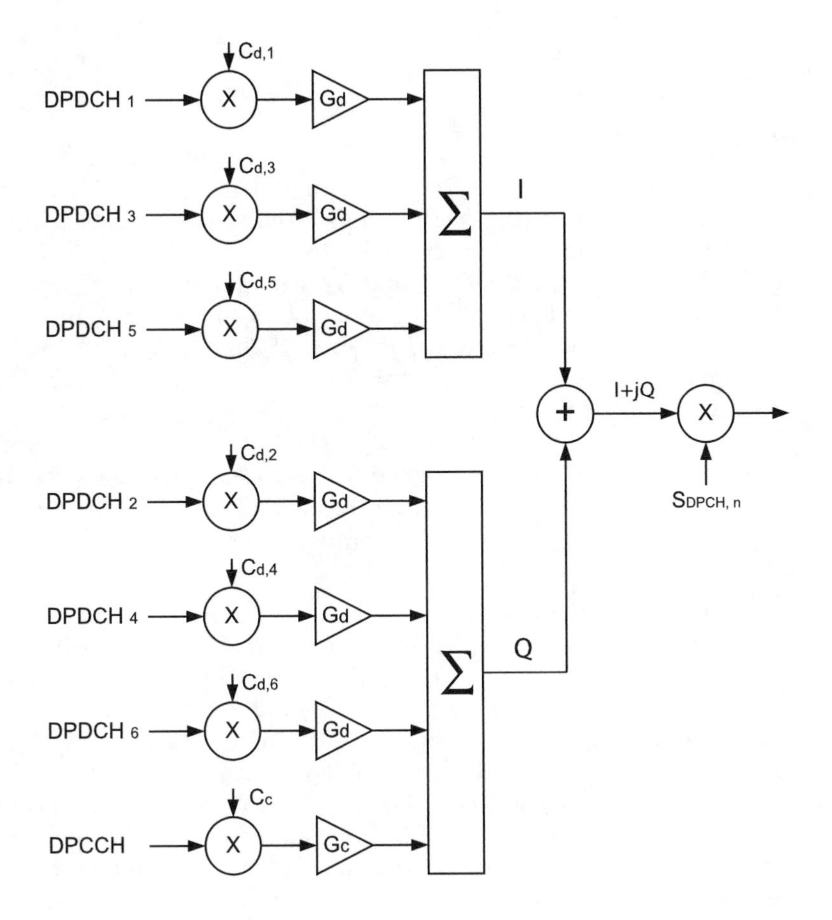

FIGURE 8.14
Uplink DPCCH/DPDCH spreading.

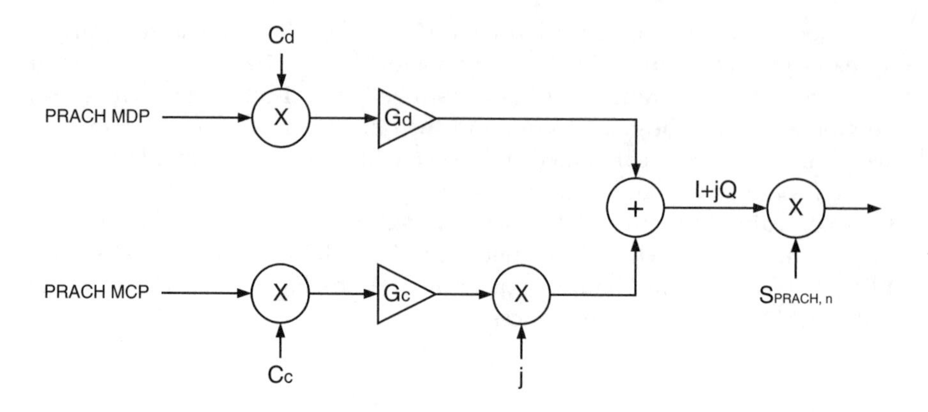

FIGURE 8.15
PRACH spreading.

PRACH Spreading

The PRACH message part is divided into a data part (PRACH MDP) and a control part (PRACH MCP). These two parts have different channelization codes and gain factors. The channelization code and the gain factor for PRACH MDP are, respectively, C_d and G_d, whereas these are C_c and G_c for PRACH MCP. The ratio G_c/G_d is quantized into 4-bit words. Therefore, 16 gain ratios can be represented and they are $15/15, 14/15, \ldots, 0/15$. The weighted streams of real-valued chips on the in-phase (I) and quadrature (Q) branches are added and treated as a complex-valued stream of chips. And the complex-valued stream of chips are scrambled by the 10-ms complex-valued scrambling code $S_{\mathrm{PRACH},n}$, which is aligned with the 10-ms message part radio frames (the first scrambling chip starts at the beginning of a message radio frame). Figure 8.15 illustrates the PRACH spreading process.

PCPCH Spreading

The PCPCH message part is divided into a data part (PCPCH MDP) and a control part (PCPCH MCP). These two parts have different channelization codes and gain factors. The channelization code and the gain factor PCPCH MDP are, respectively, C_d and G_d, whereas these are C_c and G_c for PCPCH MCP. The ratio G_c/G_d is quantized into 4-bit words. Therefore, 16 gain ratios can be represented and they are $15/15, 14/15, \ldots, 0/15$. The weighted streams of real-valued chips on the in-phase (I) and quadrature (Q) branches are added and treated as a complex-valued stream of chips. The complex-valued stream of chips is scrambled by the 10-ms complex-valued scrambling code $S_{\mathrm{PCPCH},n}$, which is aligned with the 10-ms message part radio frames (the first

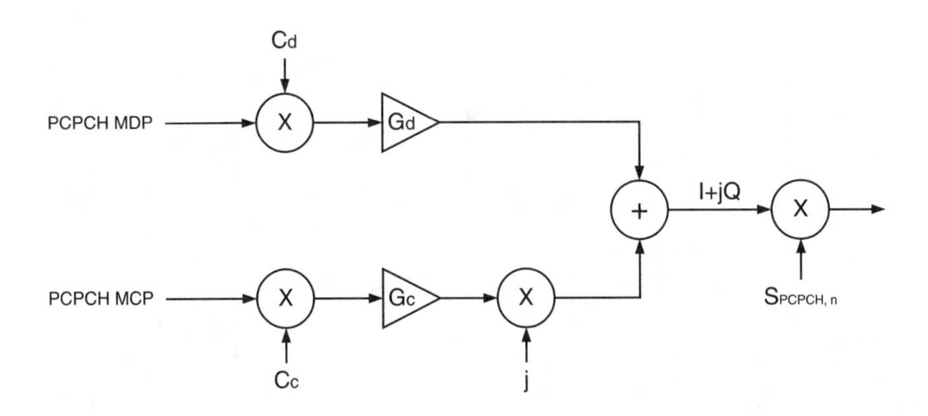

FIGURE 8.16
PCPCH spreading.

scrambling chip starts at the beginning of a message radio frame). Figure 8.16 illustrates the PCPCH spreading process.

8.11.2 Downlink Spreading

Except for SCH, the spreading operation is identical for all downlink physical channels, namely, P-CCPCH, S-CCPCH, CPICH, AICH, PICH, PDSCH, and downlink DPCH. Except for the AICH symbols, the nonspread physical channel symbols can take on the values +1, −1, and 0, the null value indicating discontinuous transmission. As for AICH, its symbol values depend on the exact combination of acquisition indicators.

The spreading process for these channels (except SCH) is as follows. Each pair of two consecutive symbols is serial-to-parallel converted, and the even-numbered and odd-numbered symbols are mapped onto I and Q branches, respectively. Each branch is then spread to the chip rate by the same channelization code C_m. The stream of real-valued chips on the in-phase and quadrature branches is added and treated as a complex-valued stream of chips. The complex-valued stream of chips is scrambled by the complex-valued scrambling code $S_{dw,n}$. The scrambling code for any downlink channel is always applied with its alignment referred to that of the scrambling code used for P-CCPCH. The scrambling code for P-CCPCH is aligned with the P-CCPCH frame boundary (the first complex chip of the spread P-CCPCH frame is multiplied with chip number zero of the scrambling code).

The downlink channels are then combined, with each downlink channel weighted by a different gain factor G_i. The synchronism channels P-SCH and

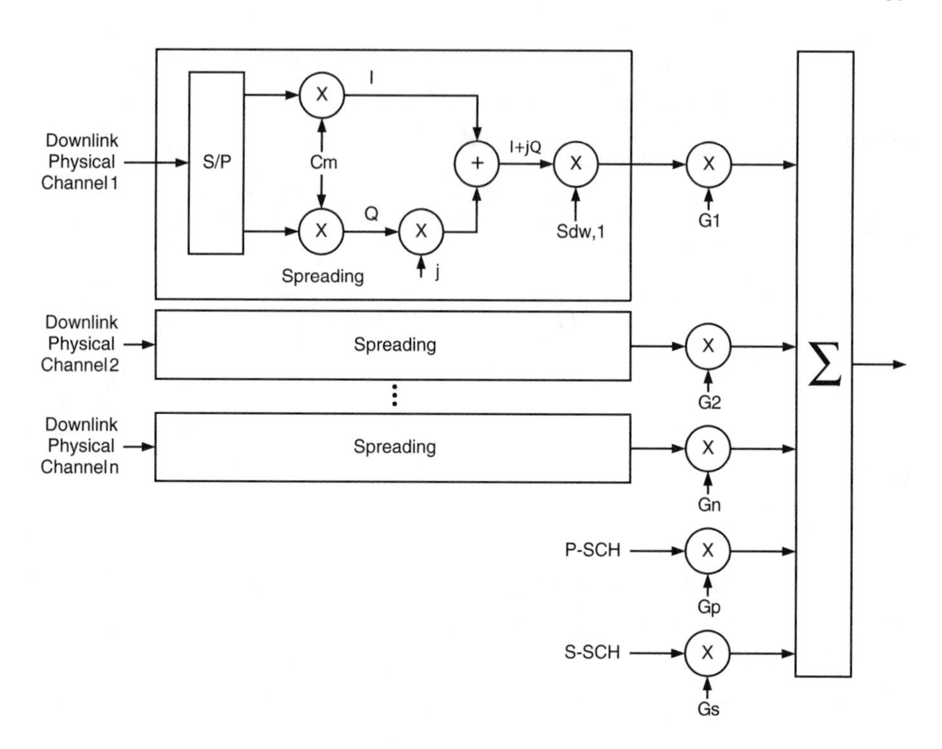

FIGURE 8.17
Downlink spreading.

S-SCH are multiplied by the respective gain factors G_p and G_s. Figure 8.17 illustrates the downlink spreading and combination process.

8.12 Spreading Codes

The subsections that follow describe the channelization and scrambling operations with a focus on UTRA FDD technology.

8.12.1 Channelization Codes

The channelization codes used in UTRA (all three technologies) are known as orthogonal variable spreading factor (OVSF) codes. Uplink and downlink channels make use of OVSF codes. The OVSF codes preserve the orthogonality between channels of different rates and spreading factors. They can be defined as

$$C_{1,0} = [1]$$

$$\begin{bmatrix} C_{2,0} \\ C_{2,1} \end{bmatrix} = \begin{bmatrix} C_{1,0} & C_{1,0} \\ C_{1,0} & -C_{1,0} \end{bmatrix} = \begin{bmatrix} 1 & 1 \\ 1 & -1 \end{bmatrix}$$

$$\begin{bmatrix} C_{2^{n+1},0} \\ C_{2^{n+1},1} \\ C_{2^{n+1},2} \\ C_{2^{n+1},3} \\ \cdots \\ C_{2^{n+1},2^{n+1}-2} \\ C_{2^{n+1},2^{n+1}-1} \end{bmatrix} = \begin{bmatrix} C_{2^n,0} & C_{2^n,0} \\ C_{2^n,0} & -C_{2^n,0} \\ C_{2^n,1} & C_{2^n,1} \\ C_{2^n,1} & -C_{2^n,1} \\ \cdots & \cdots \\ C_{2^n,2^n-1} & C_{2^n,2^n-1} \\ C_{2^n,2^n-1} & -C_{2^n,2^n-1} \end{bmatrix}$$

where $C_{\text{spreading\_factor},k}$ uniquely identifies the code and where $0 \leq k \leq$ spreading_factor $- 1$. The code tree of Figure 8.18 illustrates the OVSF codes. The leftmost value in each channelization code word corresponds to the chip first transmitted in time.

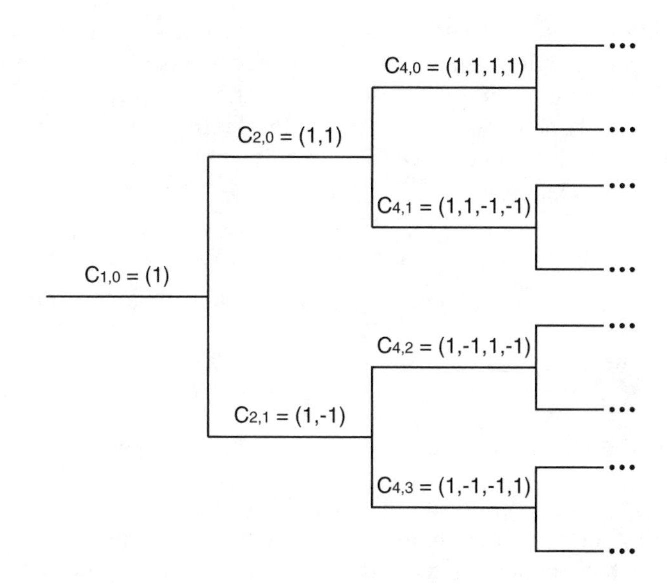

FIGURE 8.18
OVSF code tree.

8.12.2 Uplink Scrambling Codes

The uplink physical channels make use of complex-valued scrambling codes for the scrambling operation. Two types of scrambling codes may be used: long scrambling codes and short scrambling codes. The first alternative is implemented when a Rake receiver is used for the reception. The second alternative is used when advanced multiuser detectors or interference cancellation receivers are used for the reception. The latter alternative is preferable for easing the implementation of advanced receiver structures. There are $2^{[24]}$ long scrambling codes and $2^{[24]}$ short scrambling codes and they are assigned by higher layers.

The long scrambling codes are 38,400 chips long and are formed by means of long scrambling sequences. These scrambling sequences are constructed by appropriately combining two sequences, x and y, with the first generated by the primitive polynomial:

$$X^{25} + X^3 + 1$$

and the second generated by the primitive polynomial:

$$X^{25} + X^3 + X^2 + X + 1$$

The resulting sequences constitute segments of Gold sequences. The several scrambling code sequences are obtained by appropriately performing additions of shifted versions of the two sequences. The configuration of the uplink long scrambling code generator is depicted in Figure 8.19. The nth complex-valued long scrambling sequence $C_{\text{long},n}$ is given by

$$C_{\text{long},n}(i) = C_{\text{long},1,n}(i)\left[1 + (-1)^i\, jC_{\text{long},2,n}(2\lfloor i/2 \rfloor)\right]$$

where $i = 0, 1, \ldots, 2^{[25]} - 2$, $\lfloor . \rfloor$ denotes rounding to nearest lower integer, and

$$C_{\text{long},1,n}(i) = Z_n(i),\; C_{\text{long},2,n}(i) = Z_n[(i + 16777232)_{\text{mod } 2^{25}-1}]$$
$$Z_n(i) = +1, \quad \text{for } z_n(i) = 0, \quad \text{and } Z_n(i) = -1, \quad \text{for } z_n(i) = 1$$
$$z_n(i) = [x_n(i) + y(i)]_{\text{mod } 2}$$

$z_n(i)$, $x_n(i)$, and $y(i)$ denote the ith symbol of the sequences z_n, x_n, and y, with x and y constructed using the respective primitive polynomials defined above. The actual sequences depend on the chosen initial conditions.

The short scrambling codes are 256 chips long and are formed by means of short scrambling sequences. These scrambling sequences are constructed by appropriately combining three sequences, a, b, and c, with a generated by

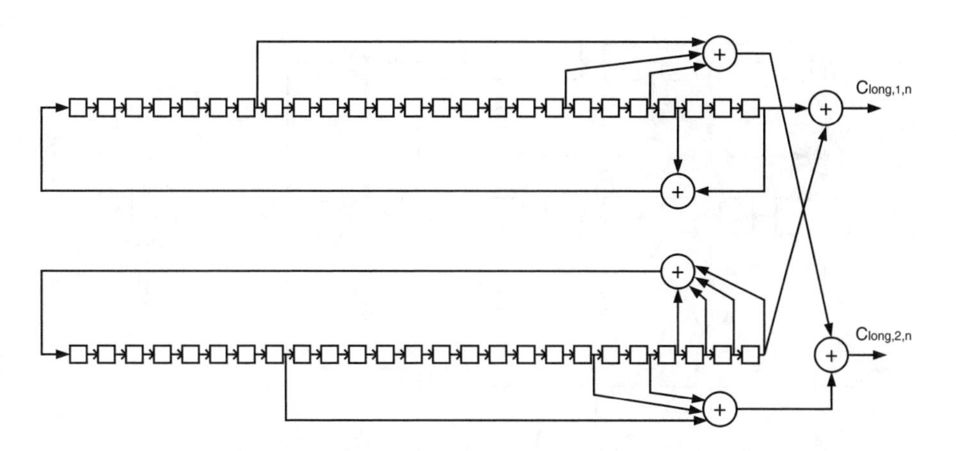

FIGURE 8.19
Uplink scrambling sequence generator.

the polynomial:

$$X^8 + X^5 + 3X^3 + X^2 + 2X + 1$$

b generated by the polynomial:

$$X^8 + X^7 + X^5 + X + 1$$

and c generated by the polynomial:

$$X^8 + X^7 + X^5 + X^4 + 1$$

The several scrambling code sequences are obtained by appropriately performing additions of shifted versions of the three sequences. The configuration of the uplink short scrambling code generator is depicted in Figure 8.20, where the mapper provides the mapping from $z_n\ (i) = 0, 1, 2,$ and 3 onto the output binary sequences $(C_{\text{short},1,n}(i_{\text{mod }256}), C_{\text{short},2,n}(i_{\text{mod }256})) = (+1, +1), (-1, +1), (+1, -1)$, and $(-1, -1)$. The nth complex-valued long scrambling sequence $C_{\text{short},n}$ is given by

$$C_{\text{short},n}(i) = C_{\text{short},1,n}(i_{\text{mod }256})[1 + (-1)^i\ jC_{\text{short},2,n}\ (2\lfloor(i_{\text{mod }256})/2\rfloor)]$$

where $i = 0, 1, \ldots, 254$, $\lfloor.\rfloor$ denotes rounding to nearest lower integer, and

$$z_n\ (i) = a\ (i) + 2b\ (i) + 2c\ (i)$$

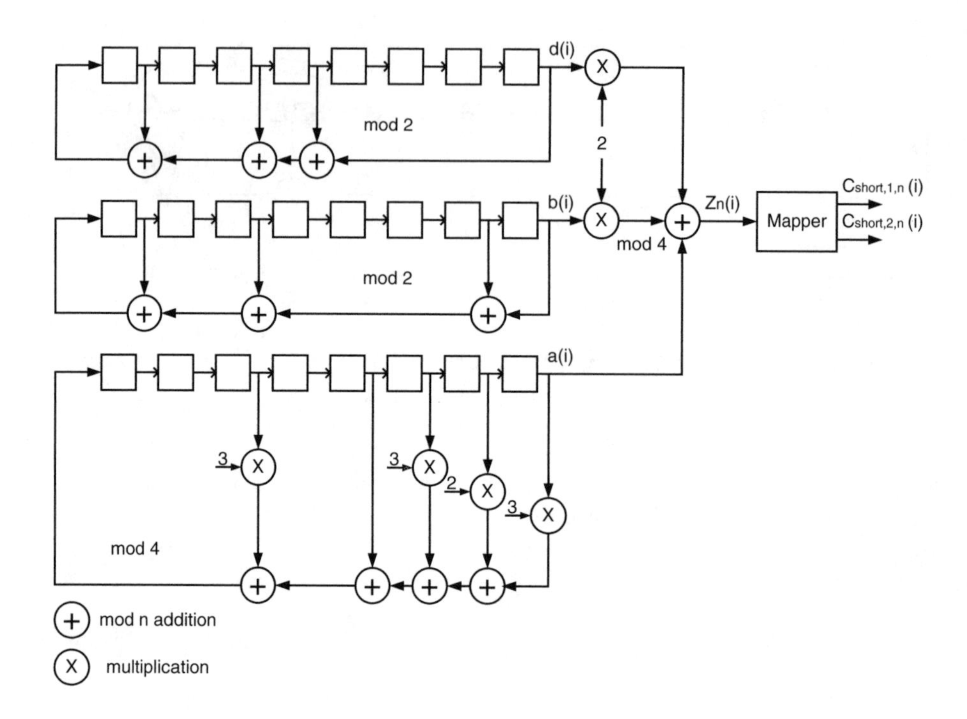

FIGURE 8.20
Uplink short sequence generator.

where $a(i)$, $b(i)$, and $c(i)$ denotes the ith symbol of the sequences a, b, and c, which are constructed using the respective primitive polynomials defined above.

Some of the scrambling sequences for the different physical channels are defined as follows. DPCCH and DPDCH may use either long or short scrambling codes. The nth uplink long scrambling code $S_{\text{long},n}(i)$ is given by

$$S_{\text{long},n}(i) = C_{\text{long},n}(i)$$

The nth uplink short scrambling code $S_{\text{short},n}(i)$ is given by

$$S_{\text{short},n}(i) = C_{\text{short},n}(i)$$

In both cases $i = 0, 1, \ldots, 38399$. The message part of PRACH uses a 10-ms scrambling code chosen among 8192 scrambling codes. The nth scrambling code in this case is given by

$$S_n(i) = C_{\text{long},n}(i + 4096)$$

where $i = 0, 1, \ldots, 38{,}399$. The message part of PCPCH may use one of 32,768 scrambling codes. These codes are 10 ms long and are divided into 512 groups each of which contains 64 codes, each group being allocated to a different cell. Either long or short codes may be used. The nth uplink long scrambling code $S_{\text{long},n}(i)$ is given by

$$S_{\text{long},n}(i) = C_{\text{long},n}(i)$$

The nth uplink short scrambling code $S_{\text{short},n}(i)$ is given by

$$S_{\text{short},n}(i) = C_{\text{short},n}(i)$$

In both cases $i = 0, 1, \ldots, 38{,}399$, and $n = 8192, 8193, \ldots, 40{,}959$.

8.12.3 Downlink Scrambling Codes

A considerable number of scrambling codes ($2^{18} - 1 = 262,143$) can be generated for the downlink, although not all of them are used. These scrambling codes are split into 512 sets, each of which contains one PSC and 15 SSCs. A one-to-one mapping between each PSC and 15 SSCs is provided. The set of PSCs is further split into 64 scrambling code groups.

The construction of the code sequences is based on the combination of two real sequences, x and y, into a complex sequence. One of the sequences, x, is constructed using the primitive polynomial

$$X^{18} + X^7 + 1$$

whereas the other, y, is constructed using the primitive polynomial:

$$X^{18} + X^{10} + X^7 + X^5 + 1$$

The resulting sequences constitute segments of Gold sequences and the resulting scrambling codes are repeated at every 10 ms. The several scrambling code sequences are obtained by appropriately performing additions of shifted versions of the two sequences. The configuration of the downlink scrambling code generator is depicted in Figure 8.21. The nth complex scrambling code sequence $S_n(i)$ is given by

$$S_n(i) = Z_n(i) + j\, Z_n[(i + 131072)_{\text{mod } 2^{18}-1}]$$

where $i = 0, 1, \ldots, 38{,}399$, and

$$Z_n(i) = +1, \quad \text{for } z_n(i) = 0, \quad \text{and } Z_n(i) = -1, \quad \text{for } z_n(i) = 1$$

$$z_n(i) = x(i+n)_{\text{mod } 2^{18}-1} + y(i)_{\text{mod } 2}$$

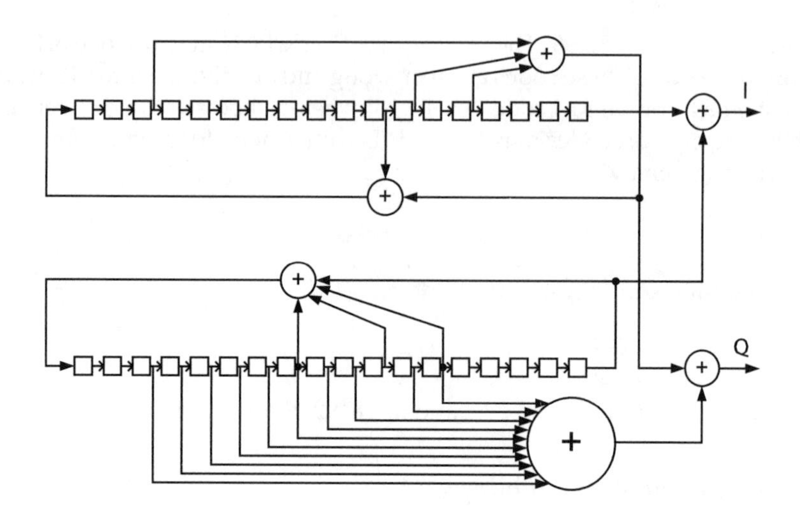

FIGURE 8.21
Downlink scrambling code generator.

where $z_n(i)$, $x(i)$, and $y(i)$ denote the ith symbol of the sequences z_n, x, and y, with x and y constructed using the respective primitive polynomials defined above. The actual sequences depend on the chosen initial conditions. Each cell is allocated one, and only one, PSC. In particular, the physical channels P-CCPCH, P-CPICH, PICH, AICH, AP-AICH, CD/CA-ICH, CSICH, and S-CCPCH carrying PCH use the PSC allocated to the corresponding cell. The remaining downlink physical channels of the cell can use either PSC or one SSC of the set associated with the PSC of the cell. It is possible to have a mixture of PSC and SSC for one CCTrCH.

There are also the synchronization codes. In particular, there are one primary synchronization code and 16 secondary synchronization codes. The primary synchronization code is constructed as a hierarchical Golay sequence and is chosen to present good aperiodic autocorrelation properties. It is a complex-valued sequence, with identical real and imaginary components. It is given by

$$(1 + j) \times \langle a, a, a, -a, -a, a, -a, -a, a, a, a, -a, a, -a, a, a \rangle$$

where

$$a = \langle 1, 1, 1, 1, 1, 1, -1, -1, 1, -1, 1, -1, 1, -1, -1, 1 \rangle$$

Each of the 16 secondary synchronization codes is complex-valued, with identical real and imaginary components. The kth secondary synchronization code, $k = 1, 2, \ldots, 16$, is given by

$$(1 + j) \times \langle h_m(0) \times z(0), \ldots, h_m(255) \times z(255) \rangle$$

where $m = 16(k - 1)$, $h_m(i)$ and $z(i)$ denote the ith symbol in the sequence h_m and z; h_m corresponds to the sequence given by the mth row of the Hadamard matrix, defined as

$$H_0 = [1]$$
$$H_n = \begin{bmatrix} H_{n-1} & H_{n-1} \\ H_{n-1} & -H_{n-1} \end{bmatrix}$$

and

$$z = \langle b, b, b, -b, b, b, -b, -b, b, -b, b, -b, -b, b, -b, -b \rangle$$

with

$$b = \langle 1, 1, 1, 1, 1, 1, -1, -1, -1, 1, -1, 1, -1, 1, 1, -1 \rangle$$

8.13 UTRA Procedures

This section describes some of the procedures used in UTRA. Although most of the descriptions here concern UTRA FDD, equivalent procedures are also found in UTRA TDD. Some of the procedures that are specific to UTRA TDD are also described.

8.13.1 Cell Search

Cell search procedure is carried out by the UE. It consists in searching for a cell and determining the downlink scrambling code and common channel frame synchronization of that cell. Typically, three steps are followed in the cell search procedure:

1. *Slot Synchronization.* The UE uses the primary synchronization code of SCH, which is common to all cells, to acquire slot synchronization to a cell. A matched filter may be used for this purpose and the slot timing can be obtained by detecting the peaks at the filter output.

2. *Frame Synchronization and Code-Group Identification*. The UE uses the secondary synchronization code to acquire frame synchronization and identify the code group of the cell found in the previous step. This is accomplished by correlating all possible SSC sequences with the received signal and identifying the maximum correlation output. This is achievable because the cyclic shifts of the sequences are unique.
3. *Scrambling Code Identification*. The UE determines the PSC used by the cell found in the first step. This is accomplished by means of a symbol-by-symbol correlation over the CPICH with all codes within the code group identified in the second step. The P-CCPCH can now be detected and system- and cell-specific BCH information be read.

If information on which scrambling codes to search for is provided, the last two steps can be simplified.

8.13.2 Common Physical Channel Synchronization

The radio frame timing of all common physical channels are related to the radio frame timing of P-CCPCH, which is determined in the cell search procedure. Therefore, after cell search, all common physical channels can have their radio frame timing determined.

8.13.3 Radio Link Establishment and Monitoring

A Node B radio link set can be in one of three possible states: initial state, in-sync state, and out-of-sync state. The corresponding state transition diagram is illustrated in Figure 8.22. The establishment of a radio link varies depending on whether or not a radio link already exists. If no radio link exists, the following procedure is followed:

- The radio link sets that are to be set up are considered to be in the initial state. Transmission of downlink DPCCH is started and, if any data are to be transmitted, transmission of downlink DPDCH is also started.
- UE uses P-CCPCH timing and timing offset information obtained from UTRAN to establish downlink chip and frame synchronization of DPCCH.
- Transmission of uplink DPCCH is started after higher layers consider the downlink physical channel established and the activation time signaled to UE (if any) has been reached.
- Uplink chip and frame synchronization is established by UTRAN. After a number of successive in-sync indications are received from

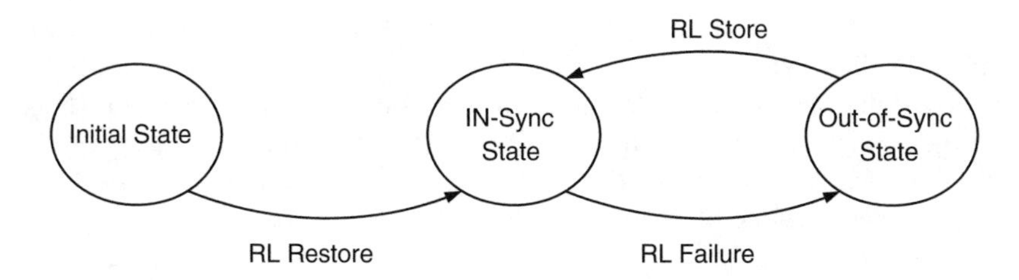

FIGURE 8.22
State diagram for Node B radio link set.

Layer 1, Node B triggers an RL restore procedure indicating which radio link set has obtained synchronization. The corresponding radio link set then is considered to be in the in-sync state.

If a radio link already exists, a similar procedure is followed, but with the difference that if a radio link is added to an existing radio link, this radio link set is considered to be in the state the radio link set was prior to the addition of the radio link.

After the radio links are established, a link-monitoring procedure is followed by UE and Node B so that failure/restore management is carried out. In case of failure, Node B triggers an RL failure procedure indicating that the corresponding radio link set is considered to be in the out-of sync state.

Downlink radio links are monitored by the UE. In a steady-state condition, an out-of-sync condition shall be reported by the UE if either of the following criteria are fulfilled:

- The quality of DPCCH, estimated over the previous 160-ms period, is worse than a certain threshold.
- All transport blocks with a CRC attached, received over the previous 160-ms period, and the 20 most recently received blocks with a CRC attached present incorrect CRC.

In a steady-state condition, an in-sync condition shall be reported by the UE if both of the following criteria are fulfilled:

- The quality of DPCCH, estimated over the previous 160-ms period, is acceptable.
- At least one transport block with a CRC attached is received in a TTI ending in the current frame with correct CRC.

Uplink radio links are monitored by Node B. The synchronization status of all radio link sets shall be checked by Layer 1 in Node B in every frame. Primitives are used to indicate such status to the RL failure/restored triggering functions. Only one synchronization status indication shall be given per radio link set. The criteria used to indicate in-sync or out-of-sync conditions are not subject to specification.

8.13.4 Uplink DPCCH and DPDCH Reception

Upon reception of the uplink dedicated channels, namely, DPDCH and DPCCH, the following tasks are performed by the receiver of the base station:

- Despreading of the DPCCH and buffering of the DPDCH using the maximum bit rate (smallest spreading factor)
- Estimation of the channel from the pilot bits received on the DPCCH (every slot)
- Estimation of the signal-to-interference ratio from the pilot bits (every slot)
- Transmission of the TPC command in the downlink direction to control the uplink transmission power (every slot)
- Decoding of the TPC bits to adjust the downlink power for the respective connection (every slot)
- Decoding of the FBI bits and adjustment of the diversity antenna phases, or phases and amplitudes, according to the transmission diversity mode (over two or four slots)
- Decoding of the TFCI information from the DPCCH to detect the bit rate and channel decoding parameters for DPDCH (every 10-ms frame)
- Decoding of the DPDCH according to the TTI (10, 20, 40, or 80 ms)

The reception in the downlink direction includes the functions as described before, but some peculiarities are noted:

- The dedicated physical channels, DPDCH and DPCCH, have a constant spreading factor, but DSCH has a varying spreading factor.
- The FBI bits do not appear in the downlink direction.
- A common pilot channel is used in addition to the pilot bits on DPCCH.
- In case of transmission diversity, the UE receives pilot patterns from the two antennas, and the channel estimation is performed with these two patterns.

8.13.5 Uplink Power Control

Uplink inner-loop power control is used for the uplink physical channels. PRACH during random-access procedure and PCPCH during CPCH access procedure use specific power control algorithms. This is detailed later in this chapter. The power control procedure used for PCPCH is similar to that used for DPCCH/DPDCH.

The DPCCH initial transmit power and the relative transmit power offset between DPCCH and DPDCHs are set by higher layers. Subsequent adjustments of power levels through the power control procedure affect both channels equally so that the relative transmit power between these channels is maintained. The UE transmit power is adjusted to maintain the received uplink signal-to-interference ratio (SIR) at Node B above a SIR target (SIR_{target}). For such a purpose, the cell in the active set (serving cells) estimates the SIR of the received DPCH ($SIR_{estimate}$). TPC commands are then generated by the serving cells and transmitted once per slot, as follows:

- If $SIR_{estimate} > SIR_{target}$, then the TPC is set to 0.
- If $SIR_{estimate} < SIR_{target}$, then the TPC command is set to 1.

The UE receives one or more TPC commands per slot and derives a single TPC command (TPC_cmd) per slot. The transmit power is then adjusted with a step of $\Delta_{adjust} = \Delta_{TPC} \times \text{TPC\_cmd}$, in decibels, where $\Delta_{TPC}-$, the step size, in decibels, is a Layer 1 parameter (1 or 2 dB).

8.13.6 Downlink Power Control

Downlink channels have their transmit power determined by the network. On the other hand, some rules exist concerning the ratio of the transmit power between the different downlink channels, as briefly described next.

DPCCH/DPDCH

DPCCH and DPDCH undergo the same power control procedure, and the relative power between these two channels, as determined by the network, is not affected by the power control algorithm.

In the power control process, the UE does not make any assumptions about how the downlink power control is set by UTRAN. The UE assists UTRAN in this process by assessing the downlink SIR and recommending increase or decrease in the transmitted power. The SIR is assessed by means of the pilot bits received within DPCCH at the UE. A TPC command is generated and sent in the first available TPC field in the uplink DPCCH in each slot. In fact, depending on the mode of operation, which is set by UTRAN, either a unique TPC command in each slot is sent or the same TPC command is sent over three slots. More specifically, the aim is to keep the received SIR above a SIR target (SIR_{target}). A higher-layer outer loop adjusts SIR_{target} independently for each connection. The UE estimates the received downlink DPCCH/DPDCH power as well as the received interference and an estimated SIR ($SIR_{estimate}$) is determined. TPC commands are then generated as follows:

- If $SIR_{estimate} > SIR_{target}$, then the TPC is set to 0, requesting a transmit power decrease.
- If $SIR_{estimate} < SIR_{target}$, then the TPC command is set to 1, requesting a transmit power increase.

Upon receiving the TPC command(s), the downlink DPCCH/DPDCH power is adjusted accordingly. In the case of a unique TPC command, the power is updated at every slot. In the case of three TPC commands, an estimate of the TPC commands is carried out over three slots and the power is updated accordingly at every three slots. The downlink power is then adjusted to a new power $P(k)$. This is obtained as a function of the current power $P(k-1)$, of the kth power adjustment due to the inner loop power control $P_{TPC}(k)$, and of a correction $P_{bal}(k)$. Such a correction is obtained according to the downlink power control procedure for balance radio link powers toward a common reference power. The power control function is given by

$$P(k) = P(k-1) + P_{TPC}(k) + P_{bal}(k)$$

where all the elements are in decibels. The correction power is

$$P_{bal}(i) = \text{sign}\{(1-r)[P_{REF} - P(i)]\} \times \min\{|(1-r)[P_{REF} - P(i)]|, P_{bal,max}\}$$

where $0 \leq r \leq 1$ is a convergence coefficient, P_{REF} is a reference transmission power in dBm (signaled by higher layers), $P_{bal,max}$ is the maximum power change limit for radio link power balancing control (signaled by higher layers and set to be multiple of the power control step size Δ_{TPC}) and $\text{sgn}(x)$ is the signal function ($= -1, 0, +1$ if $x < 0, x = 0, x > 0$, respectively). The actual transmission power must be set to as close as possible to $P(i)$. $P_{TPC}(k)$ assumes the values $+\Delta_{TPC}$, 0, or $-\Delta_{TPC}$ depending on the current estimated TPC, on $P_{TPC}(k)$ averaged over a certain window size, and on Δ_{TPC}. The power control step size Δ_{TPC} can take on the following values: 0.5, 1, 1.5, and 2 dB.

PDSCH

The PDSCH power control can be performed by one of the following options: inner-loop power control based on the power control commands sent by the UE on the uplink DPCCH or slow power control.

AICH, PICH, and CSICH

Higher layers inform the UE about the relative power of AICH, PICH, and CSICH compared to the P-CPICH transmit power. The power of these channels are measured as the power per transmitted acquisition indicator (for AICH), as the power over the paging indicators (for PICH), and as the power per transmitted status indicator (for CSICH).

S-CCPCH

The power of the data field power and that of the TFCI and pilot fields may be offset and the offset may vary in time.

8.13.7 Paging Procedure

Once registered within a network, an UE is allotted a paging group, for which paging indicators are dedicated. These paging indicators are periodically transmitted on the PICH to indicate the presence of the paging message belonging to that paging group. After detecting a PI, the UE shall decode the next PCH frame appearing on the S-CCPCH. The paging message appears on the S-CCPCH 7680 chips after the end of transmission of paging indicators on PICH. Note that the frequency with which the PIs are transmitted has a direct impact on the UE battery life. This is because to detect these PIs the UE must leave the save battery mode (sleep mode).

8.13.8 Random-Access Procedure

A random-access procedure is initiated by the UE upon request originated from the MAC sublayer. Before such a procedure can be initiated, several pieces of information shall be available to Layer 1 from RRC. These include, among others:

- Preamble scrambling code
- Message duration (10 or 20 ms)
- Set of available RACH subchannels (out of 12 RACH subchannels) for each access service class
- Set of available signatures for each access service class
- Power-ramping factor
- Number of preamble retransmissions
- Initial preamble power, the power offset between the preamble power
- Random-access message power

At the initiation of the random-access procedure, the following information shall be available to Layer 1 from MAC:

- Transport format for the message part of PRACH
- Access service class of the PRACH transmission
- Transport block set (data) to be transmitted

The following main steps comprise the random-access procedure.

1. One uplink access slot corresponding to the set of available RACH subchannels is randomly selected.
2. A signature from the set of available signatures is randomly selected.
3. A preamble using the selected uplink access slot, signature, and preamble transmission power is transmitted.
4. AP-AICH is monitored to detect the acquisition indicator (AI).
5. Detection of AI. If no positive or negative AI is detected, then following steps are carried out.
   a. Another uplink access slot corresponding to the set of available RACH subchannels is randomly selected.
   b. Another signature from the set of available signatures is randomly selected.
   c. The transmission power by the power ramp step is increased. If the maximum allowable power is exceeded by 6 dB, a Layer 1 status "No Ack on AICH" is passed to the MAC layer and the random-access procedure is terminated.
   d. Whether or not the number of retransmissions has been reached is verified. In the positive case, a Layer 1 status "No Ack on AICH" is passed to the MAC layer and the random-access procedure is terminated. In the negative case, repeat from step 3.
6. Detection of AI. If a negative AI is detected, then a Layer 1 status "NAck on AICH received" is passed to the MAC layer and the random-access procedure is terminated.
7. Detection of AI. If a positive AI is detected, the random-access message is transmitted three or four uplink access slots after the uplink access slot of the transmitted preamble, depending on the AICH transmission timing parameter.
8. Successful exit. A Layer 1 status "RACH message transmitted" is passed to the MAC layer and the random-access procedure is terminated.

8.13.9 CPCH Access Procedure

A CPCH access procedure is initiated by the UE upon request originated from MAC sublayer. Before such a procedure can be initiated several items of information shall be available to Layer 1 from RRC. Such information, available in the system information message for each PCPCH in a CPCH set allocated to a cell, includes:

- Uplink Access Preamble (AP) scrambling code
- Uplink Preamble signature set
- Access preamble slot subchannels group
- AP-AICH preamble channelization code
- Uplink collision detection (CD) preamble scrambling code
- CD preamble signature
- CD preamble slot subchannels group
- CD-AICH preamble channelization code
- CPCH uplink scrambling code
- DL-DPCCH channelization code

Some physical layer parameters are made available by the RRC and MAC layers:

- Maximum number of retransmitted preambles
- Initial open-loop power level, power step size
- CPCH transmission timing parameter
- Length of power control preamble (0 or 8 slots)
- Number of frames for the transmission of start of message indicator in DL-DPCCH for CPCH
- Set of transport format parameters
- Transport format of the message part
- Data to be transmitted

The performance of CPCH access is improved with the use of CSICH. CSICH is a separate downlink channel used to indicate the occupation status of different PCPCHs. The use of CISCH avoids unnecessary access attempts when all PCPCHs are occupied. A further improvement can be achieved with the activation of the channel assignment (CA) functionality. In that case, besides the occupation indication of different PCPCHs, the CSICH also conveys the CA message, which informs the maximum data rate of each PCPCH.

The CA message is transmitted in parallel with the CD message. The following main steps comprise the CPCH access procedure, in which case the CA functionality is assumed to be active.

1. The status indicators of CSICH are detected. If the maximum available data rate is less than the requested data rate, the access attempt is aborted and a failure message is sent to the MAC layer. The availability of each PCPCH is retained.
2. CPCH-AP signature from the set of available signatures is randomly selected.
3. An uplink access slot from the available CPCH-AP access slots is randomly selected.
4. The AP using the selected uplink access slot, signature, and preamble transmission power is transmitted.
5. AP-AICH is monitored to detect the acquisition indicator (AI).
6. Detection of AI. If no positive or negative AI is detected, the UE tests the value of the most recent transmission of the status indicator corresponding to the PCPCH selected immediately before the AP transmission. If it indicates "not available," the access attempt is aborted and a failure message is sent to the MAC layer. Otherwise, the following steps are carried out:
   a. The next slot available in the subchannel group used is selected. (A minimum separation of three or four access slots from the last transmission must exist, depending on the transmission timing parameter.)
   b. The transmission power by the specified power step is increased.
   c. Whether or not the number of retransmissions has been reached is verified. In the positive case, a Layer 1 failure message is passed to the MAC layer and the CPCH access procedure is terminated.
7. Detection of AI. If a negative AI is detected, then a Layer 1 failure message is passed to the MAC layer and the CPCH access procedure is terminated.
8. Detection of AI. If a positive AI is detected the UE randomly selects one CD signature and one CD access slot and transmits a CD preamble. It then waits for the CD/CA-ICH and the CA message from Node B.
9. Monitoring of CD/CA-ICH. If the UE does not receive a CD/CA-ICH in the designated slot or if it receives a CD/CA-ICH in the designated slot with a signature that does not match the signature

used in the CD preamble, then a Layer 1 failure message is passed to the MAC layer and the CPCH access procedure is terminated.

10. Monitoring of CD/CA-ICH. If the UE receives a CD/CA-ICH in the designated slot with a matching signature and a CA message indicating one of the PCPCHs known to be free, the UE transmits the power control preamble followed by the message portion of the burst. If the CA message indicates a PCPCH known to be busy, then a Layer 1 failure message is passed to the MAC layer and the CPCH access procedure is terminated.
11. Detection of Start of Message Indicator. The UE monitors a number of frames indicated by higher layers in DL-DPCCH for CPCH, to detect the start of message indicator, a known sequence repeated on a frame-by-frame basis.
12. Detection of Start of Message Indicator. If the start of message indicator is not detected, then a Layer 1 failure message is passed to the MAC layer and the CPCH access procedure is terminated.
13. Detection of Start of Message Indicator. If the start of message indicator is detected, then a continuous transmission of packed data is carried out.
14. Inner-Loop Power Control. During CPCH Packet Data transmission, uplink PCPCH and DL-DPCCH are inner-loop-power-controlled by UE and UTRAN.
15. Detection of Emergency Stop Command. If an emergency stop command sent by UTRAN is detected, then a Layer 1 failure message is passed to the MAC layer and the CPCH access procedure is terminated.
16. Detection of DL-DPCCH Loss. If loss of DL-DPCCH is detected, then the UE halts the CPCH transmission, a Layer 1 failure message is passed to the MAC layer, and the CPCH access procedure is terminated.
17. Successful Exit. To indicate end of transmission, several empty frames, with the number set by higher layers, are sent.

8.13.10 Transmit Diversity

Two transmit diversity modes are defined in UTRA: open-loop mode (OLM) and closed-loop mode (CLM). In OLM, the transmission is independent of a feedback information from the UE. In CLM, the FBI message available on the uplink DPCCH is determined so that transmission can be adequately adjusted at UTRAN to maximize the UE received power.

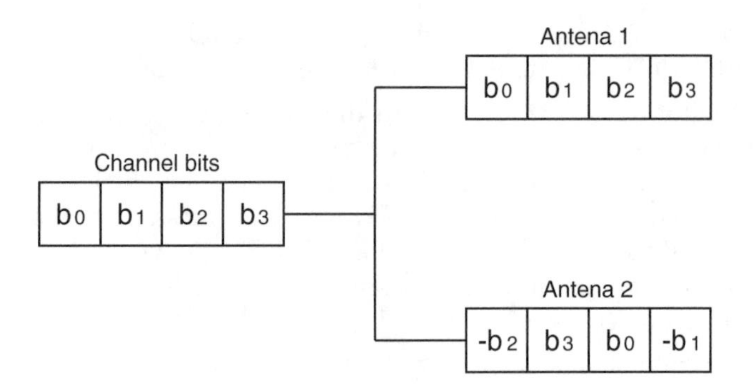

FIGURE 8.23
STTD encoding.

Open Loop Mode

Two diversity techniques are defined in OLM: space time block coding based transmit diversity (STTD) and time-switched transmit diversity (TSTD). In STTD, the encoding is applied on blocks of four consecutive channel bits. A generic block diagram for STTD encoder is shown in Figure 8.23. STTD encoding is optional in UTRAN and STTD support is mandatory at the UE.

In TSTD, the slots may hop from antenna 1 to antenna 2. For example, TSTD can be implemented with the even-numbered slots transmitted on antenna 1 and the odd numbered slots on antenna 2. TSTD is optional in UTRAN and TSTD support is mandatory in the UE.

Closed-Loop Mode

The general block diagram to support CLM transmit diversity is shown in Figure 8.24. In CLM, the UE uses CPICH to estimate the channels received from each antenna. Such an estimate is performed once every slot and is used to generate control information, which is fed back to UTRAN. Feedback signaling message bits are then transmitted on the portion of FBI field of uplink DPCCH slots. UTRAN processes such a message transmission to adjust the transmission adequately to maximize the UE received power. The update rate is 1500 Hz and the feedback bit rate is 1500 bit/s. Two diversity modes are defined in CLM: Mode 1 and Mode 2.

In Mode 1, orthogonal dedicated pilot symbols in DPCCH are sent on both antennas. The UE estimates the optimum phase adjustment for antenna 2. In this case, antenna 1 maintains the same phase while the phase of antenna 2 is modified according to the UE request. The adjustment of antenna 2 is based on the sliding average over two consecutive feedback commands (1 bit per command). Hence, four different settings ($\pm\pi, \pm\pi/2$) are possible.

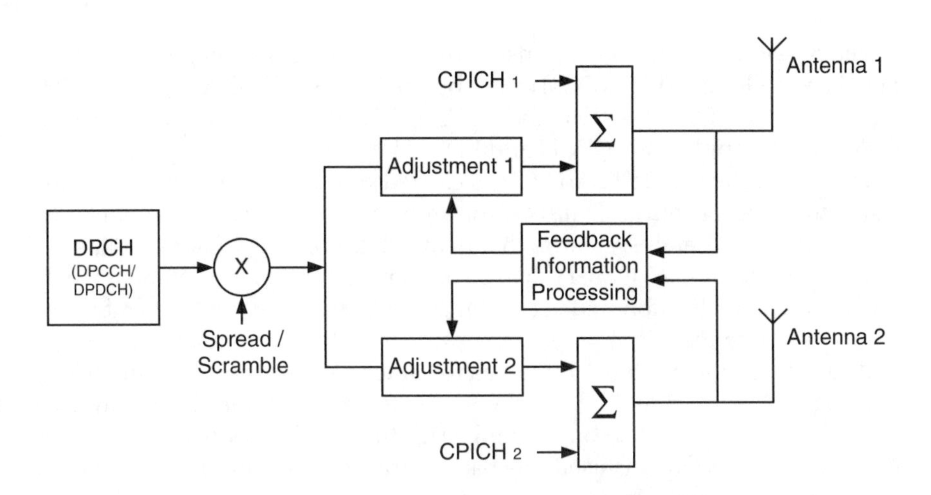

FIGURE 8.24
Closed-loop mode transmit diversity.

In Mode 2, the same dedicated pilot symbols in DPCCH are sent on both antennas. The UE estimates the optimum phase as well as the amplitude adjustments. The sliding average in this case is carried out over four consecutive feedback commands (1 bit per command); 1 bit is used for amplitude adjustment, whereas 3 bits are used for phase adjustment. The amplitudes can be adjusted to 0.2 and 0.8 or to 0.8 and 0.2, for antennas 1 and 2, respectively. The phase difference between the two antennas can be set to 8 possible values $(\pm\pi, \pm\pi/4, \pm\pi/2, \pm 3\pi/4)$.

Transmit Diversity

The application of transmit diversity follows the criteria described next.

- Simultaneous use of STTD and CLM on the same physical channel is not possible.
- The application of transmit diversity to P-CCPCH and SCH is compulsory if it is used on any other downlink channel.
- The transmit diversity mode used for a PDSCH shall be the same as that for DPCH associated with this PDSCH.
- The transmit mode (OLM or CLM) on the associated DPCH may not change during the duration of the PDSCH frame and within the slot prior to the PDSCH frame. Within CLM, however, a change between the two modes, Mode 1 and Mode 2, is allowed.

The possible application of transmit diversity to the several physical channels is as follows. STTD (OLM) may be applied for P-CCPCH, S-CCPCH, DPCH, PICH, PDSCH, AICH, and CSICH. TSTD may applied to SCH only. CLM may be applied for DPCH and PDSCH.

The application of TSTD to SCH is carried out by transmitting the even-numbered slots of both PSC and SSC through antenna 1 and the odd-numbered slots of both PSC and SSC through antenna 2. If OLM or CLM is applied to any downlink channel, CPICH shall be transmitted from both antennas using the same channelization and scrambling codes. In this case, the symbol sequence for antenna 1 is $A = 1 + j$, the same for all 10 subslots, within a slot, and for all 15 slots within a frame; as for antenna 2 the symbol sequence is $+A, -A, -A, +A, +A, -A, -A, +A + A, -A$, for one slot and the negative of it for the next slot, and so on, starting with this symbol sequence for slot 0. If no transmit diversity is used, the pattern transmitted through antenna 1 is replicated for antenna 2.

In UTRA TDD, two other transmit diversity schemes are supported as follows. For dedicated physical channels, switched transmitter diversity (STD) and transmit adaptive antennas (TxAA) are applicable.

8.13.11 Handover Procedure

Intramode handover, intermode handover, and intersystem handover can be supported within UTRA.

Intramode Handover

Intramode handover comprises the handover procedures within each UTRA mode. In UTRA FDD, soft handover, softer handover, and hard handover are supported. UTRA TDD does not support soft handover (or macrodiversity).

Intramode handover is heavily based on the energy-per-chip (E_c) to total noise power (N_0) estimate. Specifically for UTRA FDD, the required quantities can be measured by the UE using the CPICH. The pilot symbols are used to estimate the received signal code power (RSCP) (received power on one code after despreading). The wideband channel is used to estimate the received signal strength (RSSI). The ratio between the received signal code power and the total power received in the channel bandwidth (E_c/N_o) is obtained as RSCP/RSSI.

The relative timing between cells is also an important parameter to be obtained to adjust the transmission timing for coherent combining in the Rake receiver. A bad estimate of such a parameter may render difficult the combination of signals from different base stations as well as the power control operation in soft handover. The downlink timing can be adjusted in steps of 256 chips and is carried out under RNC command. Within a 10-ms window, the relative timing between cells can be found from PSC phase (note that the code period is 10 ms). For a timing uncertainty larger than this, the system frame number (SFN) needs to be extracted from the P-CCPCH. Hard handovers do not require such accurate (chip-level) timing information.

The actual handover algorithms are left as an implementation issue.

Intermode Handover

Intermode handover comprises handover between the UTRA modes (FDD and TDD). For example, a dual-mode FDD-TDD UE operating on FDD mode measures the power level from the TDD cells within the area. This is carried out through the TDD CCPCH bursts, which are sent twice during the 10-ms TDD frame. Because the TDD cells within the same coverage area are synchronized, finding one slot with the reference midamble implies that other TDD cells bear the same burst timing with reference power.

Intersystem Handover

Intersystem handover comprises handover between the different cellular technologies (e.g., GSM, cdma2000). This type of handover requires that the UE receive information from the target system within compressed frames to allow measurements from other frequencies. Note that handover in such a case is always of the hard handover type.

8.13.12 Timing Advance

Timing advance is an UTRA TDD feature. Timing advance can be used to minimize interference between adjacent time slots in large cells. For example, in UTRA TDD-3.84 the guard period is 25 μs (96 chips), which yields a cell of 3.75 km. If larger cells are required, then a timing advancement scheme may be implemented to align the separate transmission instants in the base station receiver. In such cases, a 6-bit number, with an accuracy of 4 chips (1.042 μs) is used. The required timing advance is estimated by the base station, and the UE, under command from higher layers, adjusts its transmission accordingly. The application of the timing advance scheme may extend the cell radius to 9.2 km.

8.13.13 Dynamic Channel Allocation

Dynamic channel allocation (DCA) is an UTRA TDD feature. As already mentioned, the resource units in UTRA TDD constitute the channel frequency, time slot, and code. The allocation of resources in UTRA TDD may be carried out on a dynamic basis with the support of the UE and the base station through periodic signal monitoring and reporting. Two types of DCA are supported: slow DCA and fast DCA.

Slow DCA

The allocation of resources to the cells is defined as slow DCA. The slow DCA can be terminated at any network entity above the base stations forming the seamless coverage area (having the same uplink/downlink partitioning). This means that slow DCA is terminated at the RNC. The allocation of resources is then carried out by means of negotiation between adjacent cells, to speed up the handover process and to minimize interference. The slow DCA algorithm allocates the resource units in a cell-related preference list, which is then used by the fast DCA algorithm. The list is updated on a frame-by-frame basis.

Fast DCA

The allocation of resources to bearer services is defined as fast DCA. The fast DCA is always terminated at the base station. The fast DCA uses the required resources available within the cell-related preference list, as previously provided by slow DCA. Multirate services are supported by aggregating codes (multicode) or time slots (multislot) or codes and time slots. The allocation of resources may vary depending on different strategies. It may be possible to allocate the resources based on the least interference condition; or to aggregate several time slots for diversity purposes; or vary time slot, code, and frequency according to a predetermined scheme also for diversity purposes. The number of allocated codes varies in accordance with the channel characteristics, environment, and system implementation. The allocation strategy is also dependent on the services. For real-time services, resources must be allocated for the duration of the call, but the resources may vary to comply with the allocation criteria. For non-real-time services, the resources are allocated for the period of transmission of the data and the best-effort strategy is used.

8.14 Interference Issues

If UTRA TDD and UTRA FDD use separate frequency bands, the two technologies may coexist with aggregated benefits. The coexistence of both technologies, however, must be carefully planned to avoid mutual interference. Interference is necessarily a problem in case the spectrum allocations of these technologies are contiguous. In this scenario, co-siting UTRA FDD and UTRA TDD base stations seems not to be technically attractive because a guard band would be necessary to minimize mutual interference. Interference between the mobile stations of one technology to the base stations of another and between mobile stations of different technologies is also possible, and this may be minimized with intersystem and interfrequency handovers.

Interference within UTRA FDD basically follows the same principles as already explored in other CDMA systems. In UTRA TDD, on the other hand, the sharing of the same frequency by both uplink and downlink renders the interference issue more complex. Implicitly, TDD systems must work synchronously. Therefore, neighboring base stations of different operators may interfere with each other if internetwork synchronization is not worked out. Moreover, because of the asymmetric nature of uplink and downlink, a different degree of asymmetry between uplink and downlink in adjacent cells, even if the respective base stations are synchronized, will cause interference. The use of DCA helps to reduce the interference in this case. In summary, frame-level synchronization between base stations within the same system (operator) is required. Moreover, frame-level synchronization between base stations of different systems (operators) within neighboring area is recommended. More than an optional feature, DCA is a necessity in TDD systems.

8.15 Summary

The radio interface specifications for both UTRA FDD and UTRA TDD have been developed with the strong objective of harmonization of these two components to achieve maximum commonality. Here, important physical parameters and higher-layer protocols are common to both technologies. On the other hand, because of the peculiarities of one or another technology, with the physical layers having different parameters to control, in an actual implementation the algorithms for both receiver and radio resource management differ between these two technologies. In the UTRA TDD base station, advanced receivers are necessary, whereas in the mobile station the solution for the receiver depends on the details of the performance requirements.

The CN specifications are based on an evolved GSM-MAP architecture and capabilities are included so that operation with an evolved ANSI-41 based CN is possible. The radio interfaces are defined to accommodate a wide range of services including speech, data, and multimedia, with these being simultaneously used by a subscriber and multiplexed on a single carrier. Both UTRA FDD and UTRA TDD can provide data rate services with similar QoS. UTRA TDD cells, on the other hand, can be smaller than UTRA FDD cells for the same data rate services because of the TDMA duty cycle of

UTRA TDD. However, this can be overcome if the timing advance feature is implemented.

References

1. ITU-T Recommendation Q.1701: Framework for IMT-2000 Networks, March 1999.
2. Supplement to ITU-T Recommendation Q.1701: Framework for IMT-2000 Networks—Roadmap to IMT-2000 Recommendations, Standards and Technical Specifications, June 2000.
3. Holma, H. and Toskala, A., Eds., *WCDMA for UMTS—Radio Access for Third Generation Mobile Communications*, John Wiley & Sons, Chichester, U.K., 2000.
4. ITU-R Recommendation M.1035: Framework for the Radio Interface(s) and Radio Sub-system Functionality for International Mobile Telecommunications-2000 (IMT-2000), 1994.
5. ITU-T Recommendation Q.1711: Network Functional Model for IMT-2000, March 1999.
6. Patel, P. and Dennett S., The 3GPP and 3GPP2 movements toward an all-IP mobile network, *IEEE Personal Commun.*, 62–64, August 2000.
7. 3GPP TR 23.922, Architecture for an All IP Network, December 1999.
8. 3GPP TS 25.211 v3.5.0 (2000-12), Physical Channels and Mapping of Transport Channels onto Physical Channels (FDD).
9. 3GPP TS 25.212 v3.5.0 (2000-12), Multiplexing and Channel Coding (FDD).
10. 3GPP TS 25.211 v3.4.0 (2000-12), Spreading and Modulation (FDD).
11. 3GPP TS 25.211 v3.5.0 (2000-12), Physical Layer Procedures (FDD).
12. CWTS TS C101 v3.1.1 (2000-9), Physical Layer—General Description.
13. CWTS TS C102 v3.3.0 (2000-9), Physical Channels and Mapping of Transport Channels onto Physical Channels.
14. CWTS TS C103 v2.2.0 (1999-10), Multiplexing and Channel Coding.
15. CWTS TS C102 v3.0.0 (1999-10), Physical Layer Procedures.
16. CWTS TS C102 v3.0.0 (1999-10), MAC Protocol Specification
17. CWTS TS C102 v2.1.0 (1999-10), RLC Protocol Specification.
18. ETSI TS C125 321 v3.6.0 (2000-12), MAC Protocol Specification.

9

cdma2000

9.1 Introduction

The IMT-2000 CDMA multicarrier radio interface is referred to as cdma2000. It is a wideband spread spectrum radio interface designed to meet the requirements of 3G wireless systems as well as the requirements of 3G evolution of the 2G TIA/EIA-95-B family standards. cdma2000 provides full backward compatibility with TIA/EIA-95-B. Backward compatibility permits cdma2000 infrastructure to support TIA/EIA-95-B mobile stations and allows cdma2000 mobile stations to operate in TIA/EIA-95-B systems. cdma2000 proposes backward compatibility with cdmaOne to provide a smooth transition from 2G to 3G networks. One important aspect of backward compatibility is the ability to support an overlay of cdma2000 and cdmaOne networks in the same spectrum.

A wide range of data rates is supported that accommodates the various wireless services. These applications range from plain quality voice and low-rate packet data services to high-quality voice and high-rate data services, with voice and data services provided on a nonconcurrent or concurrent basis, as required.

cdma2000 accomplishes the wideband transmission requirements in two different ways. The forward transmission may utilize either direct spread (DS) technology or multicarrier (MC) technology. The reverse link, on the other hand, always uses DS technology. The MC implementation of the cdma2000 forward link facilitates cdmaOne and cdma2000 overlay design. In an MC implementation, an $N\times1.25$ MHz cdma2000 system ($N=1$, 3, 6, 9, or 12) can overlay N contiguous cdmaOne carriers, where N is the spreading rate (SR) number. SR 1 is referred to as $1\times$ and SR 3 is referred to as $3\times$, which are the two technologies defined in cdma2000 standards. The $1\times$ component includes enhancements for high-rate packet data access. In addition to the different SRs and to meet the different quality-of-service (QoS) requirements, cdma2000 incorporates a number of radio configurations (RCs). These RCs are specified in terms of the achievable rate transmissions, frame duration, and prespreading modulation schemes. The RCs are different for downlink and uplink. For SR 1 and 3, cdma2000 standards specify ten RCs, numbered sequentially from 1 to 10, for the downlink, and seven RCs, numbered sequentially from 1 to 7, for the uplink. Collectively, these RCs form the radio interface, which consists of the 1× and 3× components. Some of these RCs are designed to accomplish backward compatibility with cdmaOne. RCs from 1 through 9 for the forward link and RCs from 1 to 6 for the reverse link accommodate a wide range of services and data rates. RC 10 for the forward link and RC 7 for the reverse link are specific for high rate packet data access using a separate carrier.

Different sets of channels are defined for these two SRs. A set of channels transmitted between base station and mobile stations within a given frequency assignment is referred to as a CDMA channel. In the forward direction, this is known as the forward CDMA channel and in the reverse direction this is referred to as the reverse CDMA channel. In both directions, long code PN sequences are used, but with different purposes. In the forward link, the long code is used for scrambling on the forward CDMA channel, whereas in the reverse link it is used for spreading on the reverse CDMA channel. On both forward and reverse CDMA channels, it provides limited privacy, although on the latter it uniquely identifies a mobile station. In addition to different RCs used to achieve a wide range of transmission rates, Walsh functions with different lengths are used with the same purpose.

Certain forward-link channels, such as the pilot channel, may be shared by the overlay and underlay systems. The overlay implementation may take advantage of the reusability feature, in which case the cdmaOne network elements or entities can be reused and upgraded to accommodate the cdma2000 technology. System planning and optimization tools developed for the cdmaOne system can also be reused or upgraded for the cdma2000 system design. The cdma2000 radio transmission technology includes an $N = 1$ option that supports all the features and performance enhancements that are provided by higher-order options such as $N = 3, 6, 9$, or 12. Although it uses the same bandwidth, cdma2000 with $N = 1$ may support twice as many voice users as compared with cdmaOne. In addition, data services are also more efficiently treated in this new technology.

cdma2000 also supports the reuse of existing cdmaOne service standards, such as those concerning speech services, data services, short message services, and over-the-air provisioning and activation services. Full handovers of voice and data calls from one system to the other constitute one very interesting feature supported by cdmaOne and cdma2000. The ability to perform smooth handovers between cdmaOne and cdma2000 systems allows operators to gradually build up cdma2000 networks in areas where additional capacity and enhanced services are needed. The operators are able to maintain service continuity while offering advanced services in a selected portion of their service areas.

To be backward compatible with existing cdmaOne systems, the cdma2000 radio interface retains many of the attributes of the cdmaOne air interface design. Some of these air interface attributes are in support of the following:

- Synchronized base station operation to facilitate fast handovers between cdmaOne and cdma2000 networks
- Chip rates that are multiples of the cdmaOne chip rate to allow compatible frequency planning with cdmaOne and to simplify design of cdmaOne/cdma2000 dual-mode terminals
- Frame structure and numerology consistent with cdmaOne
- Code-multiplexed pilot signal from the base station to the mobile station to facilitate fast acquisitions and handovers between cdmaOne and cdma2000 users.

Many other features are incorporated that provide for flexibility and improve the performance.

The descriptions of the cdma2000 technology in this chapter are fully based on References 1 through 8. Note that in the cdma2000 technology two distinct groups of RCs can be formed: one for general-purpose RCs (RCs from 1 through 9 for the forward link and for RCs from 1 through 6 for the reverse link) and another for the high-rate packet data access RCs (RC 10 for the forward link and RC 7 for the reverse link). That for high-rate packet data access evolved from the high data rate (HDR) technology to the 1× evolved high-speed data only (1×EV-DO) design: both projects present almost indistinguishable characteristics.

The great majority of the sections in this chapter describe the RCs of the first group. A specific section is dedicated for the description of the RCs of the second group.

9.2 Network Architecture

In the development of cdma2000, the core network specifications are based on an evolved ANSI-41 and Internet protocol (IP) network. These specifications also include the necessary capabilities for operation with an evolved GSM-MAP-based core network. The cdma2000 Reference Model is illustrated in Figure 9.1. Figure 9.1 identifies the network entities (NEs), represented by

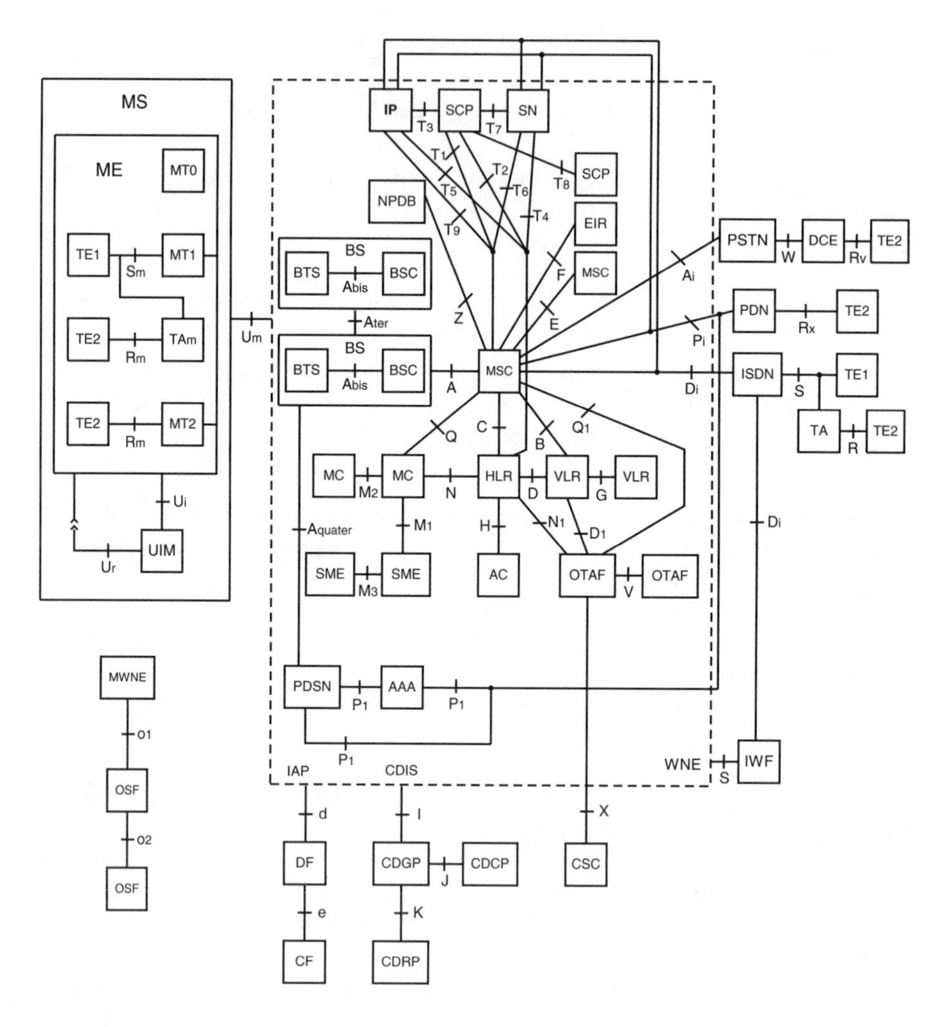

FIGURE 9.1
cdma2000 Wireless Network Reference Model.

squares and rectangles, and the associated reference points (RPs), represented by a line connecting the NEs.

An NE encompasses a group of functions implemented by a part of a physical device, or by a physical device, or by a number of distributed physical devices. There are three types of NEs:

1. *Specific Network Entity*, which is an individual instance of the network
2. *Collective Entity*, which contains encompassed NEs that are an instance of the collective

3. *Composite Entity,* which contains encompassed NEs that are part of the composite

An RP is a conceptual point that divides two groups of functions, not necessarily constituting a physical interface. An RP becomes a physical interface only when the NEs on either side of it are contained in different physical devices.

9.2.1 Network Entities

This subsection describes the various NEs of the reference model shown in Figure 9.1.

- *Authentication, Authorization, and Accounting (AAA).* The AAA provides IP functionalities to support authentication, authorization, and accounting. The AAA interacts with other AAA entities to perform AAA functions when the home AAA is outside the serving mobile network.
- *Authentication Center (AC).* The AC manages the authentication information related to the MS. The AC may or may not be located within and be indistinguishable from an HLR. An AC may serve more than one HLR.
- *Base Station (BS).* A BS provides the radio means for MSs to access the network services. It includes two other NEs, namely, BSC and BTS.
- *Base Station Controller (BSC).* The BSC provides control and management functions for one or more BTSs. The BSC exchanges messages with both the BTS and the MSC. Traffic and signaling information concerning call control, mobility management, and MS management may pass transparently through the BSC.
- *Base Transceiver System (BTS).* The BTS provides transmission capabilities across the Um RP. The BTS consists of radio devices, antennas, and equipment.
- *Call Data Collection Point (CDCP).* The CDCP collects the IS-124 format call detail information.
- *Call Data Generation Point (CDGP).* The CDGP provides call detail information to the CDCP in the IS-124 format. This may be the NE that converts call detail information from a proprietary format into the IS-124 format. All information from the CDGP to the CDCP must be in the IS-124 format.
- *Call Data Information Source (CDIS).* The CDIS can be the source of call detail information. This information may be in proprietary format. It is not required to be in IS-124 format.

- *Call Data Rating Point (CDRP).* The CDRP takes the unrated IS-124 format call detail information and applies the appropriate charge- and tax-related information. The charge and tax information is added using the IS-124 format.
- *Collection Function (CF)—[Intercept].* The CF collects intercepted communications for a lawfully authorized law enforcement agency. The CFs typically include:

  The ability to receive and process call contents information for each intercept subject

  The ability to receive information regarding each intercept subject
- *Customer Service Center (CSC).* The CSC is an entity where service provider representatives receive telephone calls from customers wishing to subscribe to initial wireless service or to request a change in the customer's existing service. The CSC interfaces proprietarily with the over-the-air service provisioning function (OTAF) to perform network and MS-related changes necessary to complete the service provisioning request.
- *Data Circuit Equipment (DCE).* The DCE provides a non-ISDN user–network interface.
- *Delivery Function (DF)—[Intercept].* The DF delivers intercepted communications to one or more collection functions. The DFs typically include:

  The ability to accept call contents for each intercept subject over one or more channels from each access function

  The ability to deliver call contents for each intercept subject over one or more channels to a collection function as authorized for each law enforcement agency

  The ability to accept information over one or more data channels and combine that information into a single data flow for each intercept subject

  The ability to filter or select information on an intercept subject before delivery to a collection function as authorized for a particular law enforcement agency

  The optional ability to detect audio in-band DTMF digits for translation and delivery to a collection function as authorized for a particular law enforcement agency

  The ability to duplicate and deliver information on the intercept subject to one or more collection functions as authorized for each law enforcement agency

  The ability to provide security to restrict access

- *Equipment Identity Register (EIR).* The EIR is the register to which user equipment identity may be assigned for record purposes.
- *Home Location Register (HLR).* The HLR is the location register to which a user identity is assigned for record purposes. These include subscriber information such as electronic serial number (ESN), mobile directory number (MDN), profile information, current location, and authorization period.
- *Integrated Services Digital Network (ISDN).* The ISDN is defined in accordance with the appropriate ANSI T1 standards.
- *Intelligent Peripheral (**IP**).* The **IP** performs specialized resource tasks such as playing announcements, collecting digits, performing speech-to-text or text-to-speech conversion, recording and storing voice messages, facsimile services, data services, etc. (This chapter uses bold notation **IP** for intelligent peripheral and the normal notation IP for Internet protocol.)
- *Intercept Access Point (IAP).* The IAP provides access to the communications to, or from, the equipment, facilities, or services of an intercept subject.
- *Interworking Function (IWF).* The IWF provides information conversion for one or more WNEs. An IWF may have an interface to a single WNE providing conversion services. An IWF may augment an identified interface between two WNEs, providing conversion services to both WNEs.
- *Managed Wireless Network Entity (MWNE).* The MWNE is a wireless entity within the collective entity or any specific NE with OS wireless management needs, including another OS.
- *Message Center (MC).* The MC stores and forwards short messages. The MC may also provide supplementary services for short message service.
- *Mobile Station (MS).* The MS is a wireless terminal used by subscribers to access network services over a radio interface. MSs include portable units, units installed in vehicles, and fixed location MSs. The MS is the interface equipment used to terminate the radio path at the subscriber.
- *Mobile Switching Center (MSC).* The MSC switches MS-originated or MS-terminated traffic. An MSC is usually connected to at least one base station. It may connect to the other public networks (PSTN, ISDN, etc.), other MSCs in the same network, or MSCs in different networks. The MSC may store information to support these capabilities.
- *Mobile Terminal 0 (MT0).* The MT0 is a self-contained data-capable MS termination that does not support an external interface.

- *Mobile Terminal 1 (MT1)*. The MT1 is an MS termination that provides an ISDN user–network interface.
- *Mobile Terminal 2 (MT2)*. The MT2 is an MS termination that provides a non-ISDN user–network interface.
- *Number Portability Database (NPDB)*. The NPDB provides portability information for portable directory numbers.
- *Operations Systems Function (OSF)*. The OSF is defined by the Telecommunications Management Network (TMN) OSF. These functions include the element management layer (EML), network management layer (NML), service management layer (SML), and business management layer (BML), functions that span all operations system functions.
- *Over-the-Air Service Provisioning Function (OTAF)*. The OTAF provides an interface to customer service centers (CSCs) to support service provisioning activities.
- *Packet Data Serving Node (PDSN)*. The PDSN provides IP functionality to the mobile network. A PDSN establishes, maintains, and terminates link layer sessions to the mobile station. A PDSN routes IP datagrams to the packet data network (PDN). A PDSN may act as a mobile IP foreign agent in the mobile network. It may interface with one or more base stations to provide the link layer session. A PDSN interacts with the AAA to provide IP authentication, authorization, and accounting support. A PDSN may interface to one or more IP networks either public or intranet to provide IP network access.
- *Packet Data Network (PDN)*. A PDN, such as the Internet, provides a packet data transport mechanism between processing network entities capable of using such services.
- *Public-Switched Telephone Network (PSTN)*. The PSTN is defined in accordance with the appropriate ANSI T1 standards.
- *Service Control Point (SCP)*. The SCP acts as a real-time database and transaction processing system that provides service control and service data functionality.
- *Service Node (SN)*. The SN provides service control, service data, specialized resources, and call control functions to support bearer-related services.
- *Short Message Entity (SME)*. The SME composes and decomposes short messages. An SME may, or may not be located within and be indistinguishable from an HLR, MC, VLR, MS, or MSC.
- *Terminal Adapter (TA)*. The TA converts signaling and user data between a non-ISDN and an ISDN interface.

- *Terminal Adapter m (TAm)*. The TAm converts signaling and user data between a non-ISDN and an ISDN interface.
- *Terminal Equipment 1 (TE1)*. The TE1 is a data terminal that provides an ISDN user–network interface.
- *Terminal Equipment 2 (TE2)*. The TE2 is a data terminal that provides a non-ISDN user–network interface.
- *User Identity Module (UIM)*. The UIM contains subscription information such as the NAM and may contain subscription feature information. The UIM can be integrated into any mobile terminal or it may be removable.
- *Visitor Location Register (VLR)*. The VLR is the location register other than the HLR used by an MSC to retrieve information for handling of calls to or from a visiting subscriber. The VLR may, or may not be located within, and be indistinguishable from an MSC. The VLR may serve more than one MSC.
- *Wireless Network Entity (WNE)*. The WNE is an NE in the wireless collective entity.

9.2.2 Reference Points

This subsection describes the various RPs of the Reference Model shown in Figure 9.1. The Um RP is the only RP that is by definition a physical interface. The other RPs are physical interfaces if NEs on either side of them are contained in different physical devices. An interface exists when two NEs are interconnected through exactly one RP.

- *RP A*: the interface between BSC and MSC
- *RP Ai*: the interface between **IP** and PSTN, plus the interface between MSC and PSTN, plus the interface between SN and PSTN
- *RP Abis*: the interface between BSC and BTS
- *RP Ater*: the BS–BS interface
- *RP Aquater*: the interface between PDSN and BS
- *RP B*: the interface between MSC and VLR
- *RP C*: the interface between MSC and HLR
- *RP D*: the interface between VLR and HLR
- *RP d*: the interface between IAP and DF
- *RP D 1*: the interface between OTAF and VLR
- *RP D i*: the interface between **IP** and ISDN, plus the interface between IWF and ISDN, plus the interface between MSC and ISDN, plus the interface between SN and ISDN

- *RP E*: the interface between MSC and MSC
- *RP e*: the interface between CF and DF
- *RP F*: the interface between MSC and EIR
- *RP G*: the interface between VLR and VLR
- *RP H*: the interface between HLR and AC
- *RP I*: the interface between CDIS and CDGP
- *RP J*: the interface between CDGP and CDCP
- *RP K*: the interface between CDGP and CDRP
- *RP L*: Reserved
- *RP M1*: the interface between SME and MC
- *RP2*: the MC–MC interface
- *RP M 3*: the SME–SME interface
- *RP N*: the interface between HLR and MC
- *RP N 1*: the interface between HLR and OTAF
- *RP O1*: the interface between MWNE and OSF
- *RP O2*: the OSF–OSF interface
- *RP Pi*: the interface between MSC, IWF, PDSN, AAA, and PDN; this RP is also the interface between PDSN and AAA
- *RP Q*: the interface between MC and MSC
- *RP Q 1*: the interface between MSC and OTAF
- *RP R*: the interface between TA and TE2
- *RP Rm*: the interface between TE2 and TAm plus the interface between TE2 and MT2
- *RP Rv*: the interface between DCE and TE2
- *RP Rx*: the interface between PPDN and TE2
- *RP S*: the interface between ISDN and TE1
- *RP m*: the interface between TE1 and MT1 plus the interface between TE1 and TAm
- *RP T1*: the interface between MSC and SCP
- *RP T2*: the interface between HLR and SCP
- *RP T3*: the interface between **IP** and SCP
- *RP T4*: the interface between HLR and SN
- *RP T5*: the interface between **IP** and MSC
- *RP T6*: the interface between MSC and SN
- *RP T7*: the interface between SCP and SN

- *RP T8*: the interface between SCP and SCP
- *RP T9*: the interface between HLR and **IP**
- *RP Ui*: the interface between integrated UIM and an MT
- *RP Um*: the interface between BS and MS, which corresponds to the air interface
- *RP Ur*: the interface between the removable-UIM and an MT
- *RP V*: the interface between OTAF and OTAF
- *RP W*: the interface between DCE and PSTN
- *RP X*: the interface between CSC and OTAF
- *RP Y*: the interface between WNE and IWF
- *RP Z*: the interface between MSC and NPDB

9.3 Radio Interface Protocol Architecture

Radio bearer services are handled by the radio interface protocols. cdma2000 air interface protocols comply with ISO/OSI Reference Model layering requirements. A general protocol model, as depicted in Figure 9.2, is defined for cdma2000 radio interfaces. The protocol architecture is modularly composed

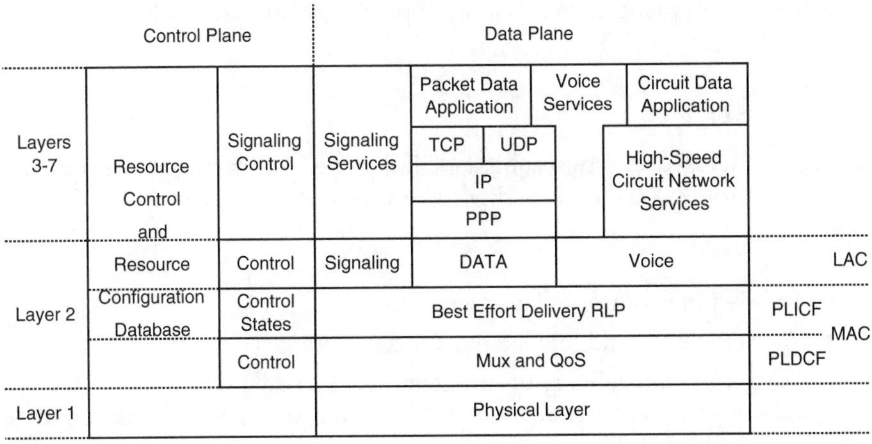

FIGURE 9.2
A general protocol model for cdma2000 radio interfaces.

of layers and planes that are logically independent of each other. Two planes are defined: the control plane and the data plane. The control plane is responsible for all cdma2000-specific control signaling (signaling protocol). The data plane is responsible for the transmission and reception of all user-related information (user traffic), such as coded voice in a voice call, or packets in an Internet connection.

cdma2000 radio transmission technology provides protocols and services that correspond to the bottom two layers of the ISO/OSI Reference Model, namely, Layer 1 and Layer 2. The upper layers, i.e., Layers 3 to 7, support applications and protocols that comply with IMT-2000 services and QoS requirements.

9.3.1 Upper Layers

In the data plane, Layers 3 to 7, the upper layers, provide signaling services, packet data applications, voice services, and circuit data applications. Signaling services provide signaling to support the mobile station operation. Packet data applications are supported by the following protocols: transmission control protocol (TCP), user datagram protocol (UDP), Internet protocol (IP), and point-to-point protocol (PPP). Voice services include conventional voice telephony services through PSTN accesses as well as voice Internet telephony. Circuit data applications are supported by high-speed circuit network services. These three applications encompass the conception of a generalized multimedia service model. Virtually any concurrent combination of voice, packet data, and high-speed packet data services is allowed.

In the control plane, Layers 3 to 7 support signaling protocols.

9.3.2 Layer 2

Layer 2 is divided into two sublayers: link access control (LAC) and media access control (MAC). LAC and MAC are designed to provide the following:

- A wide range of upper layer services
- High efficiency and low latency for data services
- Advanced QoS delivery of circuit and packed data services (e.g., limitations on acceptable delay, on bit error rate, on frame error rate, and combination of these)
- Advanced multimedia services (concurrent voice, packet data, and circuit data services with individual QoS requirements)

Link Access Control

The LAC sublayer supports high point-to-point transmission over the air. Services supported by LAC include signaling and, optionally, circuit data. A high degree of flexibility is accomplished, with encoded voice data transported in the form of packet data or circuit data traffic, as required. For voice services being transported directly by the physical layer, for backward compatibility purposes, LAC services are considered to be null. If a QoS higher than that provided by lower layers is required, LAC makes use of end-to-end reliable automatic repeat request (ARQ) protocols to guarantee error-free information delivery. Alternatively, if an adequate QoS is provided by these lower layers, then LAC may be omitted. Layer 3 and LAC (Layer 2) communicate with each other by means of service access points (SAPs). Service data units (SDUs) from Layer 3 are transferred to LAC through these SAPs. SDUs are then encapsulated into LAC protocol data units (LAC PDUs), which are subject to segmentation and reassembly. LAC PDUs or encapsulated PDU fragments and interface control information are transferred to MAC in the form of PDU control status blocks (PCSBs). PCSBs contain relevant information about LAC PDUs or LAC PDU fragments, CDMA system time boundaries (superframe, frame, time slots), error indicators, etc. All this process occurs in the data plane, with the control of it carried out by the control plane. The communication between LAC and MAC is also provided by SAPs.

Media Access Control

The MAC sublayer is composed of the physical-layer-independent convergence function (PLICF) and the physical-layer-dependent convergence function (PLDCF). MAC, by means of PLICF and PLDCF, supports three important tasks:[9]

1. *Best-Effort Delivery.* A reliable transmission over the air link is supported, with a best effort level of reliability provided by a radio link protocol (RLP).
2. *MAC Control States.* Sophisticated procedures for controlling the access of data services to the physical layer are used.
3. *Multiplexing and QoS Control.* Conflicting requests from competing services and prioritization of access requests are enforced by negotiated QoS levels.

A simple active/inactive two-state control mechanism, as provided by cdmaOne, cannot cope with the requirements for high-speed data services with many competing users. In such a simple mechanism, idle users in the active state may cause excessive interference. In the same way, the relatively long time and high system overhead required for a transition from the inactive state renders this scheme inefficient. cdmaOne circumvents this through a sophisticated MAC control mechanism. Such a mechanism is supported by a data services state diagram, in which multiple instances of the state machine are supported, one for each active packet or circuit data instance. Figure 9.3 illustrates the state transition diagram supported by MAC.

The null state is the default state in which an instance is encountered prior to the activation of a packet data service. After a packet service is invoked, initialization procedures are required and a transition to the initialization state occurs. If this is unsuccessful, a transition back to the null state occurs. In the successful event, the control hold state is reached. In this state, a dedicated control channel is established that provides means for MAC control

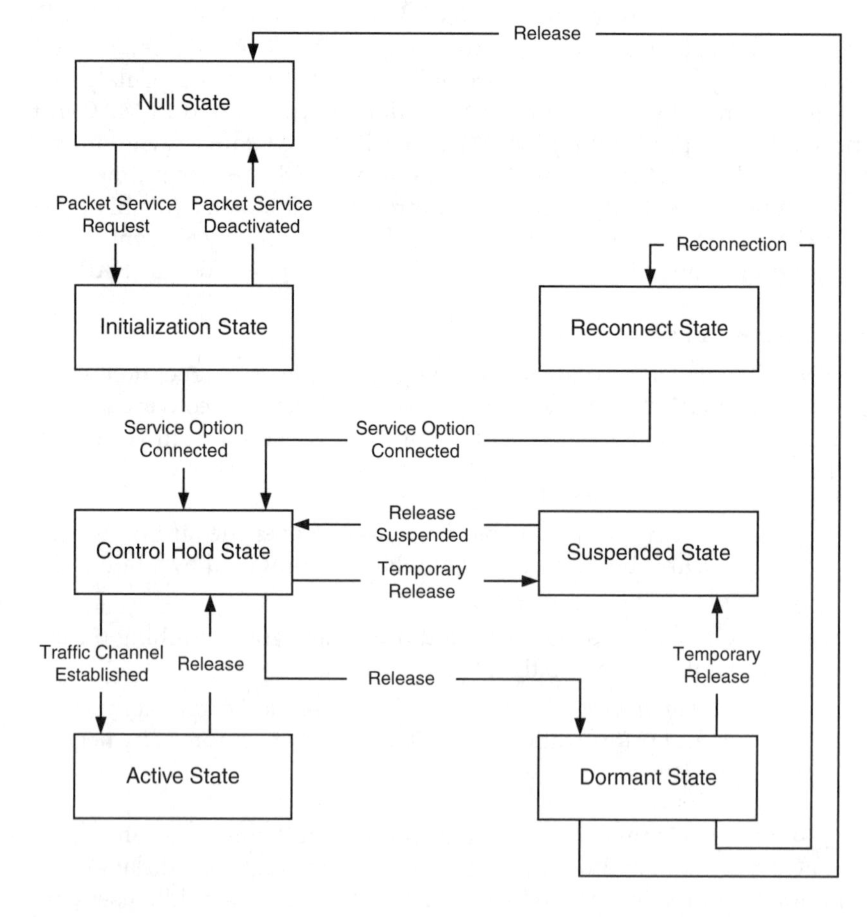

FIGURE 9.3
Data services state diagram.

commands to flow between user and base station with virtually no latency. High-speed data burst initiation and power control are examples of commands. The active state then is reached from the control hold state after the assignment of a traffic channel; the dedicated channel is kept. While in this state, if no data exchange occurs for a predefined period of time (T_{active}), the traffic channel is released and the instance returns to the control hold state. The dedicated control channel is kept for a fast traffic channel reassignment, if required. From the control hold state the suspended state is reached if no data exchange occurs for another predefined period of time (T_{hold}). In the suspended state, the control channel is released, but it may be quickly reassigned, if required, for the state information for RLP is kept. Therefore, a virtual active set is maintained by both base station and mobile station. This allows the base station or mobile station to choose the best base station to be used in the event that a packed data traffic for the user occurs. From the suspended state the return to the control hold state occurs after a control channel is established. If no data exchange occurs for a period of time exceeding a predefined period (T_{suspend}) the dormant state is attained. In the dormant state, no base station or MSC resources are maintained. On the other hand, a short data burst mode is provided to support the delivery of short messages. In this case, the overhead of a transition to the active state is avoided. The dormant state is also reached from the control hold state if a release message is sent but the PPP state is to be maintained. From the dormant state if data information is to be sent the move is to the reconnect state. From there a dedicated control channel is assigned and the control hold state is attained. From the dormant state, if PPP is terminated, the instance returns to the null state.

9.3.3 Layer 1

Layer 1 comprises the physical layer. The physical layer is responsible for the information transport between base station and mobile stations. At the transmission side, it transforms the information into an over-the-air waveform. At the reception side, it performs the inverse operation. A detailed description of the physical layer is provided in several sections of this chapter.

9.4 Logical Channels

A logical channel is an SAP providing a logical communication path between logical instances. In particular, logical channels convey information between Layer 3 and the LAC sublayer. Logical channels are defined in terms of their intended use. Logical channels may perform the following tasks:

- Convey information from one or several sources
- Deliver the information to a single target only or to multiple targets
- Carry signaling or user data information
- Transfer information in the forward direction or in the reverse direction

In cdma2000, logical channels are defined for the following purposes: synchronization, broadcast, general signaling, dedicated signaling, and access. Multiple instances of the same logical channel may be deployed.

A logical channel name consists of three lowercase letters followed by "ch" for channel. A hyphen is used after the first letter. The first letter indicates whether it is a forward (f) or reverse (r) channel; the second letter indicates whether it is a dedicated (d) or common (c) channel; and the third letter indicates whether it is a traffic (t) or signaling (s) channel. cdma2000 systems use the following types of logical channels to carry signaling information: f-csch/r-csch (forward and reverse common signaling channel, respectively) and f-dsch/r-dsch (forward and reverse dedicated signaling channel, respectively). The following is a description of the cdma2000 logical channels.

- *Forward Dedicated Traffic Channel (f-dtch) and Reverse Dedicated Traffic Channel (r-dtch).* f-dtch and r-dtch are point-to-point channels used to convey user data traffic in the forward and reverse directions, respectively. They are allocated on a permanent basis throughout the duration of the active state of a data service. They convey information to a single PLICF instance.
- *Forward Common Traffic (f-ctch) and Reverse Common Traffic Channel (r-ctch).* f-ctch and r-ctch are point-to-point channels used to convey short data bursts associated with the data service. They are allocated for use throughout the duration of the short bursts. Mobile stations or several PLICF instances may share these channels.
- *Forward Dedicated MAC Channel (f-dmch_control) and Reverse Dedicated MAC Channel (r-dmch_control).* f-dmch_control and the f-dmch_control are point-to-point channels used to convey MAC messages. They are allocated during the active state and control hold state of a data service. They convey information to a single PLICF instance.
- *Forward Common MAC Channel (f-cmch_control) and Reverse Common MAC Channel (r-cmch_control).* f-cmch_control is a point-to-multipoint channel used by the base station to convey MAC messages. r-cmch_control is a multipoint-to-point channel shared on a contention basis

by mobile stations used to convey MAC messages. MAC messages are conveyed in the dormant state or suspended state of a data service.

- *Dedicated Signaling Channel (dsch).* dsch conveys upper layer signaling data to a single PLICF instance.
- *Common Signaling Channel (csch).* csch conveys upper layer signaling data from several mobile stations or PLICF instances.

9.5 Physical Channels

A physical channel provides a physical communication path between instances. Physical channels are defined in terms of their radio characteristics such as frequency, coding, timing, power control policies, etc. In cdma2000, a physical channel has its name given in uppercase letters. The first letter indicates whether it is a Forward (F) or Reverse (R) channel, and this is followed by a hyphen. The remaining letters specify the functions of the channel, the last two always being "CH" for channel. A number of physical channels are defined within cdma2000 standards and this shall be listed in this section. The description below does not approach the specific use of the channel nor does it mention the spreading rate or radio configuration to which the channel belongs. This is a brief description: the details are left to be given later in this chapter.

- *Forward Pilot Channel (F-PICH).* The F-PICH is an unmodulated, direct-sequence spread spectrum signal transmitted continuously by each base station.
- *Transmit Diversity Pilot Channel (F-TDPICH).* The F-TDPICH is an unmodulated, direct-sequence spread spectrum signal transmitted continuously by a CDMA base station to support forward-link transmit diversity.
- *Auxiliary Pilot Channel (F-APICH).* The F-APICH is required for forward-link spot beam-forming networks using smart antennas.
- *Auxiliary Transmit Diversity Pilot Channel (F-ATDPICH).* The F-ATDPICH is a transmit diversity pilot channel associated with an auxiliary pilot channel.
- *Dedicated Auxiliary Pilot Channel (F-DAPICH).* The F-DAPICH is an unmodulated, direct sequence spread spectrum signal transmitted continuously by a CDMA base station. It is used with antenna beam-forming application and beam-steering techniques to improve system performance.

- *Sync Channel (F-SYNCH).* The F-SYNCH is a code channel that transports the synchronization message for initial time synchronization purposes.
- *Paging Channel (F-PCH).* The F-PCH is a code channel used for transmission of control information and pages.
- *Quick Paging Channel (F-QPCH).* The F-QPCH is an uncoded, spread, and On-Off-Keying (OOK) modulated spread spectrum signal. It instructs mobile stations operating in the slotted mode to receive either the forward common control channel or the paging channel starting in their next respective frames.
- *Broadcast Control Channel (F-BCCH).* The F-BCCH is a code channel used to convey broadcast overhead messages and short message service broadcast messages.
- *Forward Common Control Channel (F-CCCH).* The F-CCCH is used for communication of Layer 3 and MAC control messages from a base station to one or more mobile stations.
- *Common Assignment Channel (F-CACH).* The F-CACH is used by the base station to acknowledge a mobile station accessing the enhanced access channel. For the reservation access mode, it is used to transmit the address of a reverse common control channel and associated common power control subchannel.
- *Common Power Control Channel (F-CPCCH).* The F-CPCCH is used to transmit power control bits to multiple mobile stations.
- *Forward Fundamental Channel (F-FCH).* The F-FCH is a portion of a forward traffic channel that carries a combination of higher-level data and power control information.
- *Forward Supplemental Code Channel (F-SCCH).* The F-SCCH is a portion of the forward traffic channel that operates in conjunction with a forward fundamental channel to provide higher data rate services.
- *Forward Supplemental Channel (F-SCH).* The F-SCH is a portion the forward traffic channel that operates in conjunction with a forward fundamental channel or a forward dedicated control channel to provide higher data rate services.
- *Forward Dedicated Control Channel (F-DCCH).* The F-DCCH is a portion of the forward traffic channel used for the transmission of higher-level data, control information, and power control information.
- *Access Channel (R-ACH).* The R-ACH is used for short signaling message exchanges, such as call originations, responses to pages, and registrations.

- *Enhanced Access Channel (R-EACH).* The R-EACH is used for transmission of short messages, such as signaling, MAC messages, response to pages, and call originations.
- *Reverse Common Control Channel (R-CCCH).* The R-CCCH is a portion of a reverse CDMA channel used for the transmission of digital control information and data from one or more mobile stations to a base station.
- *Reverse Pilot Channel (R-PICH).* The R-PICH is an unmodulated, direct-sequence spread spectrum signal transmitted continuously by a CDMA mobile station.
- *Reverse Dedicated Control Channel (R-DCCH).* The R-DCCH is a portion of reverse traffic channel used for the transmission of higher-level data and control information from a mobile station to a base station while a call is in progress.
- *Reverse Fundamental Channel (R-FCH).* The R-FCH is a portion of a reverse traffic channel carrying higher-level data and control information from mobile station to a base station.
- *Reverse Supplemental Code Channel (R-SCCH).* The R-SCCH is a portion of the reverse traffic channel that operates in conjunction with the reverse fundamental channel to provide higher data rate services. Optionally, it may also operate with other reverse supplemental code channels for the same purpose.
- *Reverse Supplemental Channel (R-SCH).* The R-SCH is the reverse traffic channel that operates in conjunction with the reverse fundamental channel or the reverse dedicated control channel to provide higher data rate services.

9.6 Mapping of Channels

A logical channel can be mapped onto one or more physical channels. The mapping of channels occurs in the multiplex layer, which is a protocol layer situated between Layer 2 and Layer 1. A logical channel may have permanent and exclusive use of a physical channel (e.g., the synchronization channel). It may also have temporary, but still exclusive, use of a physical channel (e.g., successive r-csch access probe sequences can be sent on different physical access channels). Or it may share the physical channel with other logical channels (requiring a multiplex function to perform the mapping, possibly on a PDU-by-PDU basis). In certain cases, a logical channel can be mapped

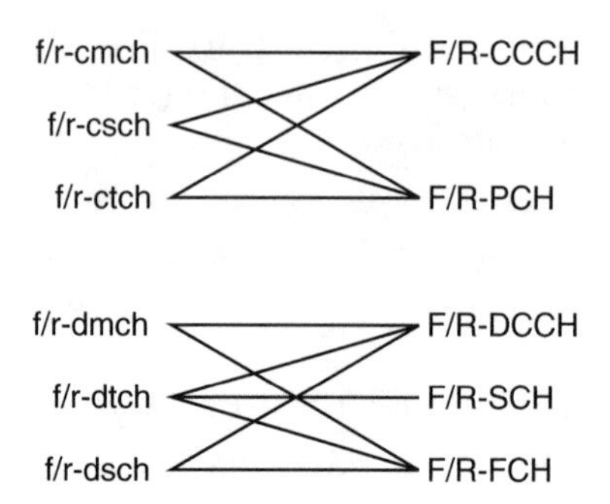

FIGURE 9.4
Possible mapping of logical channels and physical channels.

onto another logical channel to form one effective logical channel capable of carrying different traffic types (e.g., the broadcast channel and the general forward signaling channel are mapped onto one common logical channel carrying signaling information). Because a logical channel can carry only one PDU at a time, there must be a serialization process at Layer 3 to ensure a deterministic behavior. Figure 9.4 illustrates the possible mapping of logical channels and physical channels. Next illustrated is how some voice and data services may use these channels.[10]

For plain-quality voice services, upper layer signaling frames on f/r-dsch, RLP frames on f/r-dtch, MAC messages on f/r-dmch, voice frames on f/r-dtch, and power control information are multiplexed into F/R-FCH. The multiplexing may be carried out using the dim-and-burst or blank-and-burst mechanisms. For higher-quality voice services, upper layer signaling frames on f-dsch use the F/R-DCCH, whereas voice frames on f/r-dtch and power control information are multiplexed into F/R-FCH.

For packet data services, upper layer signaling frames on r/f-dsch, MAC messages on f/r-dmch, and data frames on f/r-dtch are multiplexed into F/R-FCH. In this case, the MAC control is performed on a centralized basis because F/R-FCH may be in soft handover. As an alternative to this operation mode, packet data services may be offered through F/R-DCCH, which may or may not be in soft handover. In this case, the MAC control can be performed either on a centralized or distributed basis.

For concurrent plain-quality voice service and data service, upper layer signaling frames on f/r-dsch, MAC messages on f/r-dmch, and low-rate RLP frames on f/r-dtch are multiplexed into dual-frame size (5 and 20 ms) F/R-FCH. RLP frames on f/r-dtch containing packet data are transmitted on the F/R-SCH. For concurrent higher-quality voice service and data service a similar mapping occurs, but upper layer signaling frames and MAC messages are multiplexed into F/R-DCCH.

9.7 Achievable Rates

A wide range of data rates is supported that accommodates various wireless services. These applications range from plain-quality voice and low-rate packet data services to high-quality voice and high-rate data services, with voice and data services provided on a nonconcurrent or concurrent basis, as required. To meet the QoS requirements of various wireless services, an adequate combination of SRs, RCs, FCH, and SCH is required.

As already mentioned, the possible SRs are $N = 1, 3, 6, 9$, or 12, corresponding to a bandwidth of $N \times 1.25 = 1.25$, 3.75, 7.5, 11.25, and 15 MHz, with SR 1 and SR 3 specified in the cdma2000 standards. The RCs are different for downlink and uplink. For SR 1 and 3, cdma2000 standards specify ten RCs, numbered sequentially from 1 to 10, for the downlink, and seven RCs, numbered sequentially from 1 to 7, for the uplink. RC 1 and RC 2 are backward compatible with cdmaOne. RC 1, 3, 4, 6, and 7 for the downlink and RC 1, 3, and 5 for the uplink are derived from Rate Set 1. RC 2, 5, 8, and 9 for the downlink and RC 2, 4, and 6 for the uplink are derived from Rate Set 2. Rate Set 1 yields transmission rates of 1.2, 2.4, 4.8, and 9.6 kbit/s, whereas Rate Set 2 yields transmission rates of 1.8, 3.6, 7.2, and 14.4 kbit/s. RC 10 for the downlink and RC 7 for the uplink are specific for high-rate packet data access and are detailed later in this chapter.

FCH complies with a channel structure similar to that defined in the TIA/EIA-95-B standard. Such a structure allows for variable-rate transmissions with blind rate detection. The standard 20-ms frame duration is specified but a 5-ms frame option is also permitted. The latter option is used to accommodate low-latency transmission of signaling messages over FCH in case such a channel is used for this purpose.

SCH conveys data traffic in circuit or packet mode. Its transmission rate and its duration are scheduled by the base station and a wide range of data rates is supported. Multiple SCHs may be aggregated to support multiple data streams with different QoS requirements.

In both downlink and uplink, FCHs and SCHs are code-multiplexed and transmitted in parallel. Therefore, independent coding, interleaving, and power control mechanisms can be applied to these channels. In this case, different QoS requirements can be accommodated over these two physical

TABLE 9.1

Achievable Data Rates

Spreading Rate	Bandwidth (MHz)	Reverse (kbit/s)	Forward (MHz)
1	1.25	1.2–307.2	1.2–307.2
3	3.75	1.2–1036.8	1.2–1036.8
6	7.5	1.2–2073.6	1.2–2073.6
9	11.25	1.2–2073.6	1.2–2457.6
12	15	1.2–2073.6	1.2–2457.6

channels. Note that this can be used, for example, to transmit voice and data, each complying with its own QoS requisite.

The achievable transmission rates of cdma2000 are shown in Table 9.1.

In addition to different RCs used to achieve a wide range of transmission rates, Walsh codes with different lengths are also used. A Walsh function W_m^M corresponds to a Walsh code of length M that is serially constructed from the nth row of an $M \times M$ Hadamard matrix. In such a matrix, the zeroth row is given by the Walsh function zero, the first row is given by the Walsh function 1, and so on. For SR 1 and for SR 3, the Walsh function spreading sequence runs at a rate of 1.2288 and 3.6864 Mchip/s, respectively. Therefore, they repeat with the respective periods of $N/1.2288$ μs and $N/3.6864$ μs.

SR 1 and SR 3 use different long code sequences. For SR 1 the long code sequence runs at 1.2288 Mchip/s, whereas for SR 3 its rate is 3.6864 Mchip/s. The long code generator for SR 1 is implemented with delay elements whose outputs are conveniently fed back and modulo-2-combined in accordance with its generator polynomial. An n-bit long code mask (a 42-bit binary number) creates a unique identity of the long code. Figure 9.5 sketches a long code generator for SR 1. The long code generator for SR 3 uses the SR 1 long code generator as the basis for the SR 3 long code generator. Figure 9.6 sketches the SR 3 long code generator.

9.8 Forward Link

This section describes the main features of the cdma2000 forward link.

9.8.1 General

cdma2000 forward link encompasses the following basic features:

- Walsh codes are used for channelization purposes. The length of the Walsh codes is different for different transmission rates.

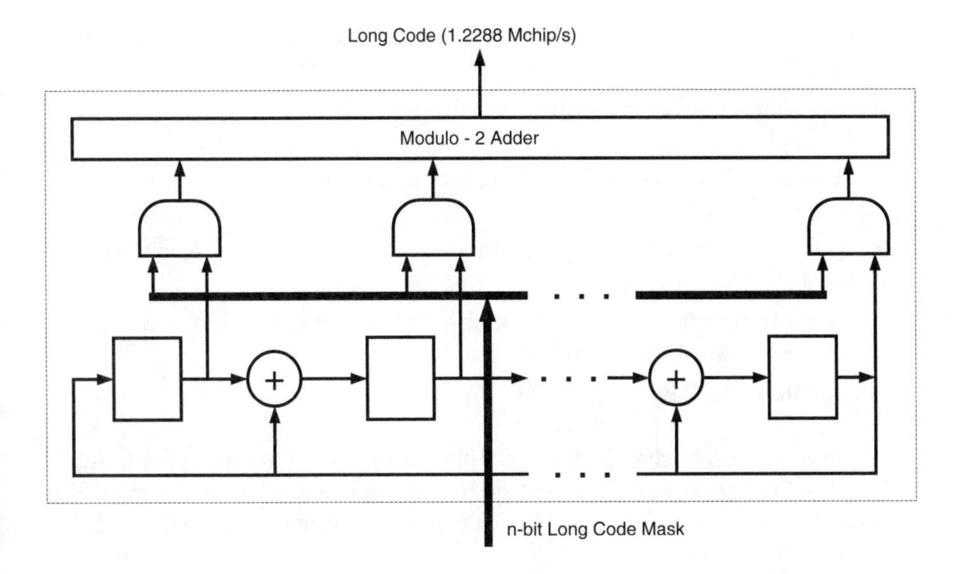

FIGURE 9.5
Long code generator for SR 1.

- Long code is used for scrambling on the forward CDMA channel.
- Nonorthogonal codes are also used for channelization purposes, in case the number of orthogonal Walsh codes is insufficient. These are quasi-orthogonal codes generated by appropriately masking orthogonal Walsh functions (codes) with quasi-orthogonal (masking) functions (QOFs).

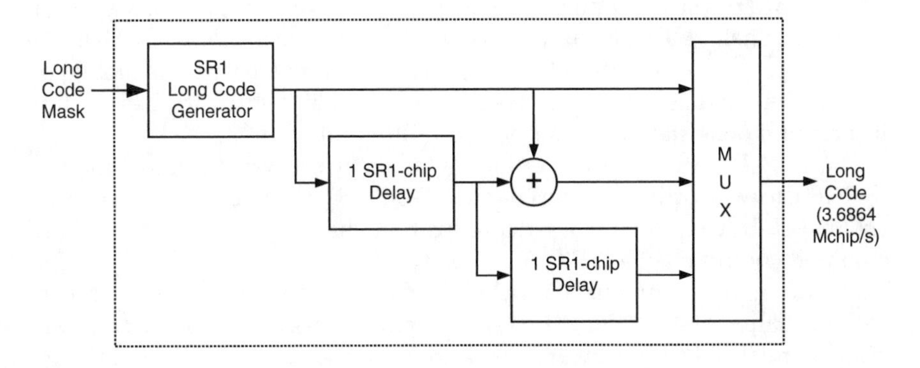

FIGURE 9.6
Long code generator for SR 3.

- QPSK modulation is used prior to the spreading operation. The use of QPSK doubles the number of Walsh codes available.
- Convolutional codes, with constraint length of 9, are used for voice and data.
- Turbo codes, with constraint length of 4, are used for data running on the supplemental channels.
- Closed-loop fast power control (up to 800 updates per seconds) is used.
- 20-ms frames are used for signaling and user information and 5-ms frames are used for control information.
- Transmit diversity is applicable.

Observe, however, that, in order for cdma2000 to be backward compatible with cdmaOne, some of the mentioned features are present only in certain SRs and RCs. For example, QPSK modulation does not appear in some RCs of SR 1.

9.8.2 Spreading Rate

Forward link achieves the wideband transmission by utilizing the direct spread technology or multicarrier approach. In an MC implementation, the modulation symbols after coding and interleaving are demultiplexed onto $N \times 1.25$ MHz carriers ($N = 1, 3, 6, 9$, or 12), each carrier with a 1.2288 Mchip/s rate. In the DS implementation, the modulation symbols after coding and interleaving are spread by a $N \times 1.2288$ Mchip/s sequence, with the spread signal being modulated onto a carrier.

9.8.3 Physical Channels

The forward physical channels can be divided into two groups: forward dedicated channels (F-DCHs) and Forward Common Channels (F-CCHs). The F-DCHs convey information from the base station to a particular mobile on a point-to-point basis. The F-CCHs convey information from the base station to a set of mobile stations on a point-to-multipoint basis.

The F-CCHs comprise the following channels: forward paging channel (F-PCH), forward quick paging channel (F-QPCH), forward broadcast control channel (F-BCCH), forward synchronization channel (F-SYNCH), forward common control channel (F-CCCH), forward pilot channels (F-PiCHs), forward common assignment channel (F-CACH), and forward common power control channel (F-CPCCH). The F-PiCHs encompass four channels: forward pilot channel (F-PCH), forward auxiliary pilot channel (F-APICH), forward transmit diversity pilot channel (F-TDPICH), and forward auxiliary diversity pilot (F-ATDPICH).

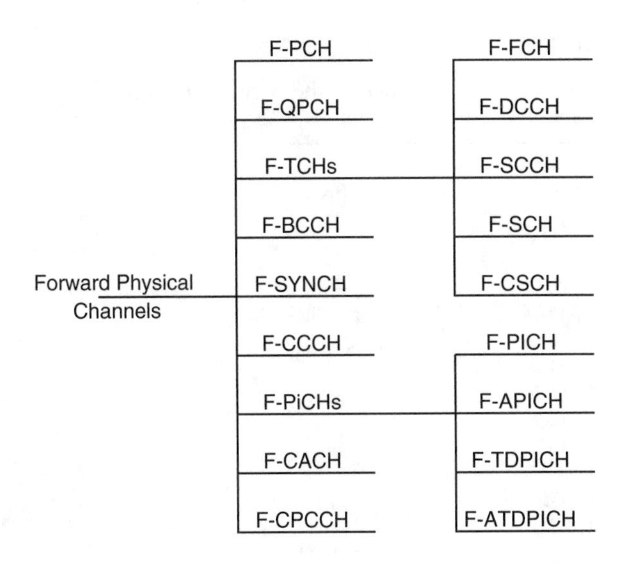

FIGURE 9.7
Forward physical channels.

The F-DCHs comprise the forward traffic channels (F-TCHs). The F-TCHs encompass four channels: forward fundamental channel (F-FCH), forward dedicated control channel (F-DCCH), forward supplemental code channel (F-SCCH), and forward supplemental channel (F-SCH).

Figure 9.7 illustrates the forward physical channels.

The use of some channels varies depending on the SR as well as on the RC. Table 9.2 indicates the range of valid channels for each channel type for SR 1 and SR 3. Note that SR 3 encompasses fewer channels than SR 1. In particular, F-TDPICH, F-ATDPICH, F-PCH, and F-SCCH appear in SR 1 but not in SR 3. Each of the code channels shown in Table 9.2 is spread by the appropriate Walsh function or quasi-orthogonal function. A quadrature spreading is then applied to each code channel, the quadrature pair of PN sequences running at 1.2288 Mchip/s. Frequency division multiplexing may be used within the base station for multiple forward CDMA channels. cdma2000 Revision A standards state that if a base station transmits the F-CCCH on a forward CDMA channel, the base station shall also transmit the F-BCCH on the same forward CDMA channel.

9.8.4 Radio Configuration

cdma2000 standards specify ten RCs, numbered sequentially from 1 to 10, for the forward traffic channels. Table 9.3 shows some radio characteristics of the RCs from 1 through 9 and SRs from 1 through 3 for the forward link.

TABLE 9.2

Range of Valid Channels in the Forward Link for SR 1 and SR 3

Channel Name	SR 1	SR 3
F-PICH	1	1
F-TDPICH	1	—
F-APICH	Not specified	Not specified
F-ATDPICH	Not specified	—
F-SYNCH	1	1
F-PCH	7	—
F-BCCH	8	8
F-QPCH	3	3
F-CPCCH	4	4
F-CACH	7	7
F-CCCH	7	7
F-DCCH	1 per F-TCH	1 per F-TCH
F-FCH	1 per F-TCH	1 per F-TCH
F-SCCH	7 per F-TCH RC 1, RC 2	—
F-SCH	2 per F-TCH RC 3, RC 4, RC 5	2 per F-TCH RC 3, RC 4, RC 5

In Table 9.3, TD stands for transmit diversity. For SR 1, the base station may support orthogonal transmit diversity (OTD) or space time spreading (STS) on the forward dedicated channels (F-DCCH, F-FCH, and F-SCH), and forward common channels (F-BCCCH, F-QPCH, F-CPCCH, F-CACH, and F-CCCH). For RC 3 through RC 9, the forward dedicated control channel and forward fundamental channel also allow a 9.6 kbit/s, 5-ms format.

As already mentioned RC 1, 3, 4, 6, and 7 are derived from Rate Set 1, whereas RC 2, 5, 8, and 9 are derived from Rate Set 2. A number of "must," "shall," and "may" recommendations concerning the use of different RCs and SRs appear in cdma2000 standards. In particular, in order for forward and reverse links to be compatible at the base station the RCs shown in Table 9.4 must be supported.

It is noteworthy that not all forward traffic channels are supported by the RCs. In particular, only F-FCH is supported by RCs from 1 through 9. F-DCCH and F-SCH are supported by RC 3 through RC 9. F-SCCH is supported only by RC 1 and RC 2.

9.8.5 Power Control

A fast forward power loop control mechanism for the forward link is incorporated into cdma2000 specifications. The algorithm provides control updates

TABLE 9.3

RC Characteristics for Forward Traffic Channel

RC	SR	Data Rates (kbit/s)	Coding Rate	Pre-Spreading Symbols	TD
1	1	1.2, 2.4, 4.8, 9.6	1/2	BPSK	—
2	1	1.8, 3.6, 7.2, 14.4	1/2	BPSK	—
3	1	1.2, 1.35, 1.5, 2.4, 2.7, 4.8, 9.6, 19.2, 38.4, 76.8, 153.6	1/4	QPSK	Allowed
4	1	1.2, 1.35, 1.5, 2.4, 2.7, 4.8, 9.6, 19.2, 38.4, 76.8, 153.6, 307.2	1/2	QPSK	Allowed
5	1	1.8, 3.6, 7.2, 14.4, 28.8, 57.6, 115.2, 230.4	1/4	QPSK	Allowed
6	3	1.2, 1.35, 1.5, 2.4, 2.7, 4.8, 9.6, 19.2, 38.4, 76.8, 153.6, 307.2	1/6	QPSK	—
7	3	1.2, 1.35, 1.5, 2.4, 2.7, 4.8, 9.6, 19.2, 38.4, 76.8, 153.6 307.2, 614.4	1/3	QPSK	—
8	3	1.8, 3.6, 7.2, 14.4 28.8, 57.6, 115.2 230.4, 460.8	1/4 (20 ms) or 1/3 (5 ms)	QPSK	—
9	3	1.8, 3.6, 7.2, 14.4, 28.8, 57.6, 115.2, 230.4, 259.2, 460.8, 518.4, 1,036.8	1/2 (20 ms) or 1/3 (5 ms)	QPSK	—

TABLE 9.4

Base Station RCs for Compatibility between Forward and Reverse Links

Channel Supported	Reverse Link	Forward Link
F/R-FCH	RC 1	RC 1
F/R-FCH	RC 2	RC 2
F/R-FCH or F/R-DCCH	RC 3	RC 3, RC 4, RC 6, or RC 7
F/R-FCH or F/R-DCCH	RC 4	RC 5, RC 8, or RC 9
F/R-FCH or F/R-DCCH	RC 5	RC 6 or RC 7
F/R-FCH or F/R-DCCH	RC 6	RC 8 or RC 9

at the rate of up to 800 updates per second. It is similar to the power control mechanism used for the reverse link in cdmaOne. The mobile station measures the forward link traffic channel power and instructs the base station to increase or decrease the power, as required. The power control command is time-multiplexed on the reverse pilot channel. Note that the traffic channel, in fact, comprises several channels. Therefore, as far as power control is concerned, several options can be exercised. In one of them, the higher-rate channel, chosen between F-FCH and F-SCH, is monitored for power control purposes. The gain setting for the other channel is then determined based on the power relation between the two channels. In another option, F-FCH and F-SCH are monitored separately, with independent quality targets. The power gain for each one of these channels is then applied independently, as required.

9.8.6 Transmit Diversity

cdma2000 provides means for TD. TD is used to improve the link performance. For SR 1, the base station may support orthogonal transmit diversity (OTD) or space time spreading (STS) on the forward dedicated channels (F-DCCH, F-FCH, and F-SCH), and forward common channels (F-BCCCH, F-QPCH, F-CPCCH, F-CACH, and F-CCCH). For SR 3, the MC approach already provides for TD. OTD and STS are optional forward-link transmission methods in which channel symbols are conveniently distributed between two antennas. These symbols are then spread by a Walsh code or quasi-orthogonal function and transmitted using the same carrier.

For SR 1, in case no TD is provided, the channel symbols are demultiplexed and fed into the forward transmission block, as shown in Figure 9.8.

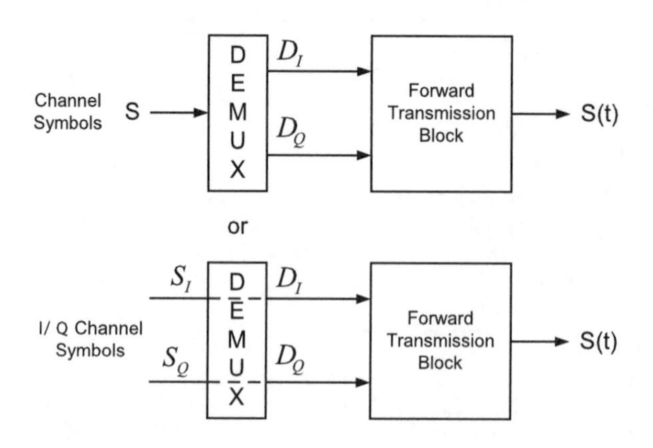

FIGURE 9.8
Forward transmission without diversity (SR 1).

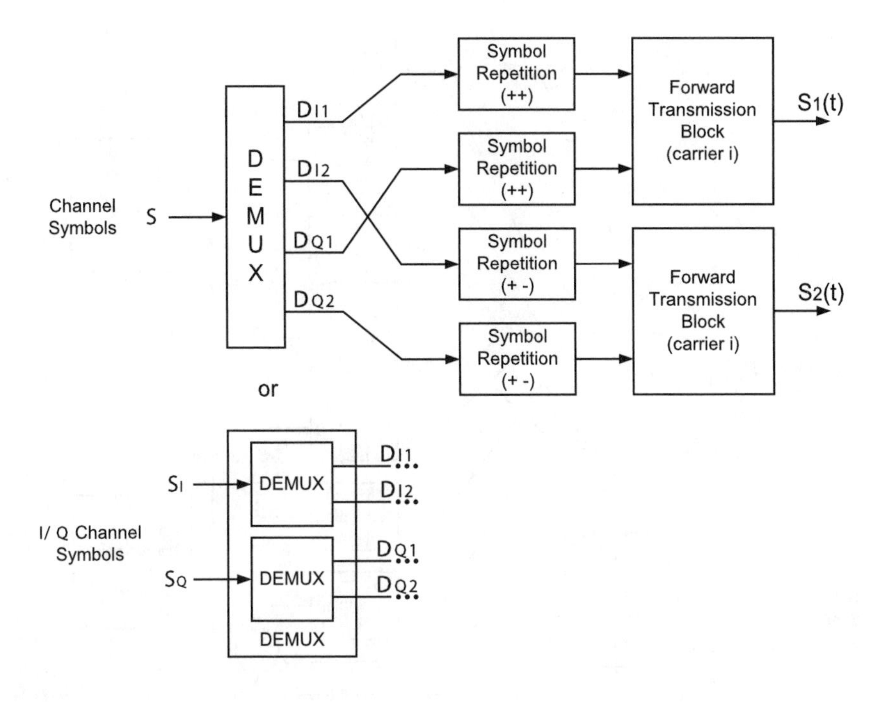

FIGURE 9.9
Forward OTD scheme (SR 1).

For the SR 1, if TD is provided, for both TD techniques (OTD and STS), encoding is applied on blocks of four consecutive channel bits. Such encoding encompasses bit repetition (2×), inversion of bit polarity, and, in the STS case, bit addition. The encoded bits are then I- and Q-mapped, spread, modulated, and transmitted by means of two antennas. Figure 9.9 illustrates the OTD process. Figure 9.10, below, sketches the STS scheme.

For SR 3, the serial coded information symbols are split into three parallel streams, each of which is spread by a Walsh code and a long PN sequence. Each information stream is then transmitted using a different carrier. Figure 9.11 shows the distribution of symbols for each transmission block.

The DEMUX functions in all transmission schemes distribute input symbols sequentially from the top to bottom output paths. Figures 9.8 through 9.11 show that two types of DEMUX are used. The first serves channel structures in which the symbols to be transmitted are not presented in the I/Q form. The second type is used for channel structures in which the symbols to be transmitted are already presented in the I/Q form. This depends on the type of channel to be transmitted and is explored in more depth later in this chapter.

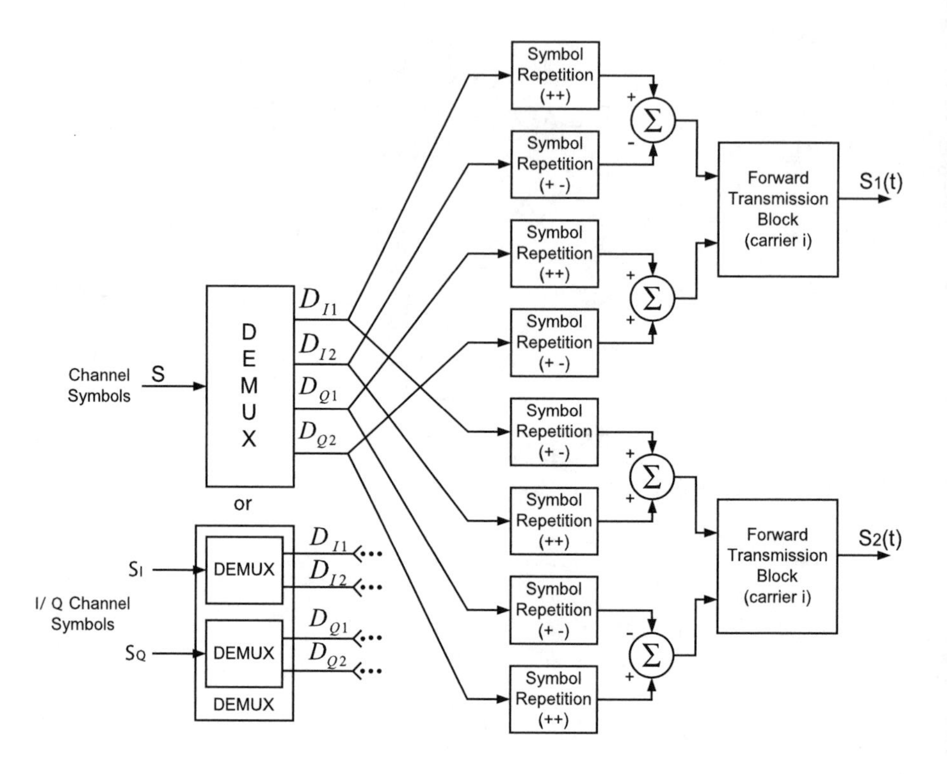

FIGURE 9.10
STS transmit diversity scheme (SR 1).

9.8.7 Transmission Block

The forward transmission block is shown in Figure 9.12. Channelization is provided by means of Walsh functions or quasi-orthogonal functions. Walsh codes are used preferentially, but if the forward link is running out of orthogonal codes quasi-orthogonal functions can be used to increase the number of usable codes. The code lengths vary to achieve different information rates. A further codification is included in the transmission chain with the in-phase and quadrature signal components able to experience a 90° rotation. If rotation is applied, an $I + jQ$ input signal is coded as $-Q + jI$. Another codification is implemented by means of a complex spreading. For an $I + jQ$ input signal, the in-phase arm of the complex multiplier yields $I \times C_{\text{short}I} - Q \times C_{\text{short}Q}$, whereas the quadrature arm of the complex multiplier gives $I \times C_{\text{short}Q} + Q \times C_{\text{short}I}$. $C_{\text{short}I}$ and $C_{\text{short}Q}$ are short in-phase and quadrature PN sequences. Note that both codes and both signals are present in both arms of the complex multiplier. Therefore, even if one of the signals is nil, as is the case of the cdmaOne signals, the quadrature spreading operation is still applied to the

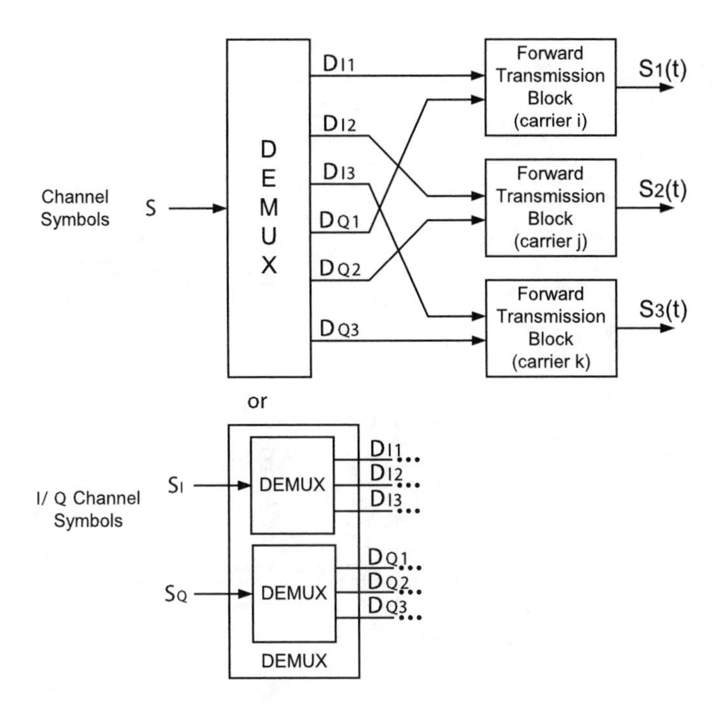

FIGURE 9.11
Forward MC scheme (SR 3)—forward link.

other signal. Baseband filtering follows and finally modulation is applied. For cdmaOne-compatible signals, a BPSK modulation is accomplished. For other signals, a QPSK modulation is implemented. Note, however, that the modulation block is exactly the same in both cases. The difference is obtained in the demultiplexing operation carried out previously. For cdmaOne signals, one of the DEMUX outputs is nil, in which case the complex multiplier yields the same symbol in both arms of the modulator. This leads to a BPSK modulation. For a normal operation of the DEMUX, with the signals sequentially presented at its outputs, each two information bits are then mapped onto one QPSK symbol. The use of a QPSK modulator doubles the number of Walsh codes available.

For cdmaOne applications, or if no diversity is applied, only one transmission block is used. For diversity applications in SR 1, two transmission blocks are used, both with the same carrier. For SR 3 applications, three transmission blocks are utilized, each of which uses a different carrier frequency. In all cases, two PN sequences, $C_{\text{short}I}$ and $C_{\text{short}Q}$, at a chip rate of 1.2288 Mchip/s, are used to modulate the in-phase (cosine) and quadrature (sine) RF carriers, respectively. Two pulse shaping filters of the finite impulse response (FIR)

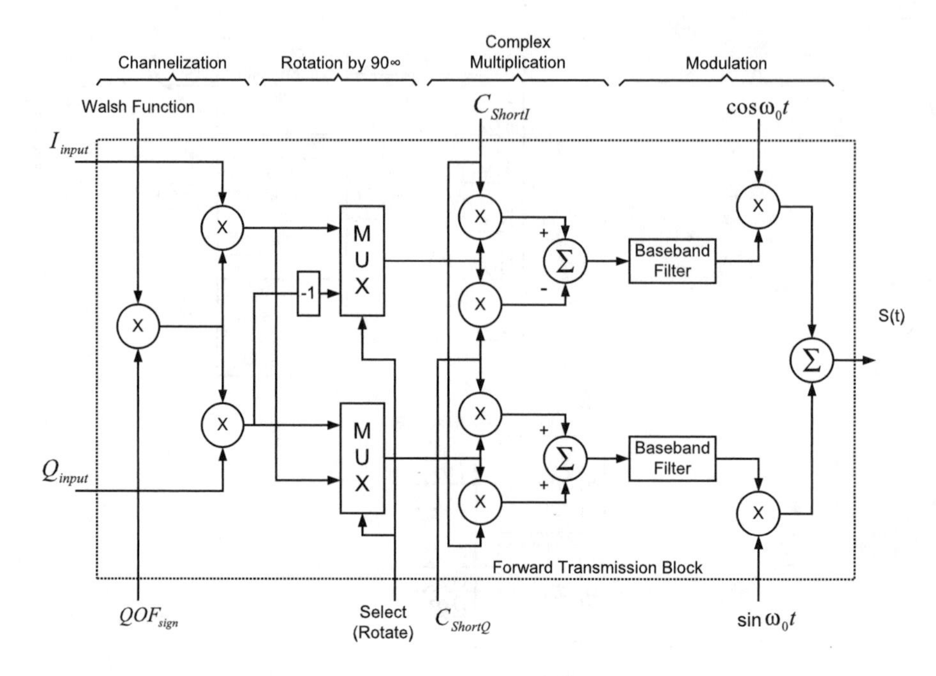

FIGURE 9.12
Forward link transmission block.

type, one at the in-phase branch and the other at the quadrature branch, are used to confine the radiated spectrum within the 1.25 bandwidth. The resulting spreading leads to $N \times 1.25$ MHz, where N is the SR number. The in-phase PN code, $C_{\text{short}I}$, and quadrature PN code, $C_{\text{short}Q}$, are generated by 15-stage linear feedback shift registers. They run at 1.2288 Mchip/s and have their characteristic polynomials given as

$$C_{\text{short}I} = 1 + x^5 + x^7 + x^8 + x^9 + x^{13} + x^{15} \tag{9.1}$$

$$C_{\text{short}Q} = 1 + x^3 + x^4 + x^5 + x^6 + x^{10} + x^{11} + x^{12} + x^{15} \tag{9.2}$$

The pilot PN sequence period is $2^{15}/1{,}228{,}800 = 26.666\ldots$ ms, with exactly 75 pilot PN sequence repetitions occurring every 2 s.

The long PN code, C_{long}, is generated by 42-stage linear feedback shift registers running at 1.2288 Mchip/s. Its characteristic polynomial is

$$C_{\text{long}} = 1 + x + x^2 + x^3 + x^5 + x^6 + x^7 + x^{10} + x^{16} + x^{17} + x^{18} + x^{19} + x^{21} + x^{22} + x^{25} + x^{26} + x^{27} + x^{31} + x^{33} + x^{35} + x^{42} \tag{9.3}$$

TABLE 9.5

Walsh Functions for the Reverse CDMA Channels for RC 3 through RC 6

Channel Type	Walsh Function
Reverse pilot channel	W_0^{32}
Enhanced access channel	W_2^8
Reverse common control channel	W_2^8
Reverse dedicated control channel	W_8^{16}
Reverse fundamental channel	W_4^{16}
Reverse supplemental channel 1	W_1^2 or W_2^4
Reverse supplemental channel 2	W_2^4 or W_6^8

9.9 Reverse Link

This section describes the main features of the cdma2000 reverse link.

9.9.1 General

cdma2000 forward link encompasses the following basic features:

- *Pilot-based coherent detection.* The reverse link incorporates a pilot channel. The pilot channel is used with the aim of providing a coherence phase for the base station to perform coherent demodulation. It is also used as a reference signal for power control purposes, rendering power control algorithms independent of the FCH data rate. In this case, power loop control latency is reduced and can be performed without the establishment of an explicit rate.
- *Code multiplexing.* Channels in the reverse link are primarily code-multiplexed with an independent gain setting for each channel. In this case, multiple services can be supported more efficiently, with QoS requirements set independently for optimal allocation of reverse-link resources. Walsh codes are used for channelization purposes. The length of the Walsh codes is different for different transmission rates. Table 9.5 shows the Walsh functions used on the reverse CDMA channels for RC 3 through RC 6.
- *Long code.* Long code is used for spreading on the reverse CDMA channel. It uniquely identifies a mobile station.

- *Continuous waveform.* A continuous waveform is used for all data rates. This minimizes interference to biomedical devices (e.g., hearing aids, pacemakers, etc.) and allows a range increase at lower transmission rates.
- *Convolutional codes,* with constraint length of 9, are used for voice and data.
- *Turbo codes,* with constraint length of 4, are used for data running on the supplemental channels.
- *20-ms frames* are used for signaling and user information and 5-ms frames are used for control information.

Observe, however, that in order for cdma2000 to be backward compatible with cdmaOne, some of the mentioned features are present only in certain SRs and RCs.

9.9.2 Spreading Rate

The reverse-link transmission achieves the wideband transmission by utilizing the direct spread approach. In such a case, the modulation symbols after coding and interleaving are spread over an $N \times 1.25$ MHz band ($N = 1, 3, 6, 9$, or 12). Each system spreads the modulated symbols by means of PN sequences running at $N \times 1.2288$ Mchip/s, with the spread signal modulated onto a carrier.

9.9.3 Physical Channels

Reverse physical channels can be divided into two groups: reverse dedicated channels (R-DCHs) and reverse common channels (R-CCHs). The R-DCHs convey information from a particular mobile station to the base station on a point-to-point basis. The R-CCHs convey information from multiple mobile stations to the base station on a multipoint-to-point basis.

The R-CCHs comprise the following channels: reverse common control channel (R-CCCH), reverse enhanced access channel (R-EACH), and reverse access channel (R-ACH). The R-DCHs comprise the following channels: reverse fundamental channel (R-FCH), reverse dedicated control channel (R-DCCH), reverse supplemental code channel (R-SCCH), reverse supplemental channel (R-SCH), and reverse power control subchannel (R-PCSCH). In addition, a reverse pilot channel (R-PICH) is included for channels operating in RCs other than 1 and 2. Traffic-carrying operation for RC 3 to RC 6 is performed by means of the following channels: R-FCH, R-SCH, R-DCCH, and R-PCSCH. For RC 1 and RC 2 this is performed by the following channels: R-FCH and R-SCCH. Common control operation makes use of the

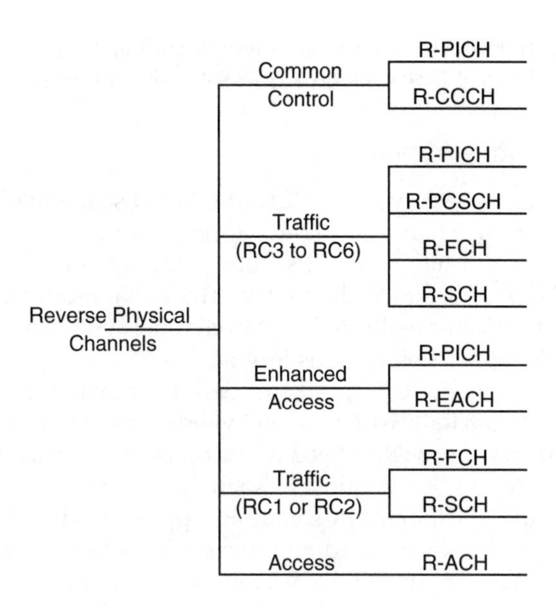

FIGURE 9.13
Reverse physical channels.

following channels: R-CCCH and R-PICH. Enhanced access operation is carried out through the following channels: R-EACH and R-PICH. Access operation is performed by R-ACH. Figure 9.13 illustrates the reverse physical channels.

The use of some channels varies depending on the SR as well as on the RC. Table 9.6 indicates the range of valid channels for each channel type for SR 1

TABLE 9.6

Range of Valid Channels in the Reverse Link for SR 1 and SR 3

Channel Name	SR 1	SR 3
R-PICH	1	1
R–ACH	1	—
R–EACH	1	—
R–CCCH	1	1
R–DCCH	1	1
R–FCH	1	1
R–SCCH	7 RC 1, RC 2	—
R–SCH	2 RC 3, RC 4	2

and SR 3. Note that SR 3 encompasses fewer channels than SR 1. In particular, R-ACH, R-EACH, and R-SCCH appear in SR 1 but not in SR 3.

9.9.4 Radio Configuration

cdma2000 standards specify seven RCs, numbered sequentially from 1 to 7, for the reverse traffic channels. Table 9.7 shows some radio characteristics of the RCs from 1 through 6 and SRs from 1 through 3 for the reverse link. In Table 9.7, "64-ary" refers to the 64-ary orthogonal modulation. For RC 3 through RC 6, the reverse dedicated control channel and reverse fundamental channel also allow a 9.6 kbit/s, 5-ms format.

As already mentioned, RC 1, 3, and 5 are derived from Rate Set 1 and RC 2, 4, and 6 are derived from Rate Set 2. Rate Set 1 yields transmission rates of 1.2, 2.4, 4.8, and 9.6 kbit/s, whereas Rate Set 2 yields transmission rates of 1.8, 3.6, 7.2, and 14.4 kbit/s. A number of "must," "shall," and "may" recommendations concerning the use of different RCs and SRs appear in cdma2000 standards. In particular, in order for forward and reverse links to be compatible at the base station the RCs shown in Table 9.8 must be supported.

It is noteworthy that not all forward traffic channels are supported by the RCs. In particular, only R-FCH is supported by RCs from 1 through 6. R-DCCH and R-SCH are supported by RC 3 through RC 6. R-SCCH is supported only by RC 1 and RC 2.

TABLE 9.7

RC Characteristics for Reverse Traffic Channel

RC	SR	Data Rates (kbit/s)	Coding Rate	Pre-Spreading Symbols	Pilot
1	1	1.2, 2.4, 4.8, 9.6	1/3	64-ary	—
2	1	1.8, 3.6, 7.2, 14.4	1/2	64-ary	—
3	1	1.2, 1.35, 1.5, 2.4, 2.7, 4.8, 9.6, 19.2, 38.4, 76.8, 153.6	1/4	BPSK	Inserted
3	1	307.2	1/2	BPSK	Inserted
4	1	1.8, 3.6, 7.2, 14.4, 28.8, 57.6, 115.2, 230.4	1/4	BPSK	Inserted
5	3	1.2, 1.35, 1.5, 2.4, 2.7, 4.8, 9.6, 19.2, 38.4, 76.8, 153.6	1/4	QPSK	Inserted
5	3	307.2, 614.4	1/3	BPSK	Inserted
6	3	1.8, 3.6, 7.2, 14.4, 28.8, 57.6, 115.2, 230.4, 460.8	1/4	BPSK	Inserted
6	3	1,036.8	1/2	BPSK	Inserted

TABLE 9.8

Mobile Station RCs for Compatibility between Forward and Reverse Links

Channel Supported	Forward Link	Reverse Link
F/R-FCH	RC 1	RC 1
F/R-FCH	RC 2	RC 2
F/R-FCH or F/R-DCCH	RC 3 or RC 4	RC 3
F/R-FCH or F/R-DCCH	RC 5	RC 4
F/R-FCH or F/R-DCCH	RC 6 or RC 7	RC 3 or RC 5
F/R-FCH or F/R-DCCH	RC 8 or RC 9	RC 4 or RC 6

9.9.5 Transmission Block

The reverse transmission block for RC 1 and RC 2 (SR 1) is the same as that for cdmaOne and is shown in Figure 9.14. The blocks in Figure 9.14 have been extensively explored in Chapter 5 and are not further described here. For RC 3 and RC 4 (SR 1) and for RC 5 and RC 6 the reverse transmission block is shown in Figure 9.15. Note in Figure 9.15 that channels are code-multiplexed with independent gain settings for each channel. In this case, multiple services can be supported more efficiently, with QoS requirements set independently for optimal allocation of reverse-link resources. The difference between the transmission block for RC 3 and RC 4 (SR 1) and RC 5 and RC 6 (SR 3) is that all the prebaseband and prefilter operations occur at the chip rate of 1.2288 Mchip/s for the first case and 3.6864 Mchip/s for the second case. For SR 1, the in-phase PN code, $C_{\text{short}I}$, and quadrature PN code, $C_{\text{short}Q}$, are generated by 15-stage linear feedback shift registers. They run at 1.2288 Mchip/s and

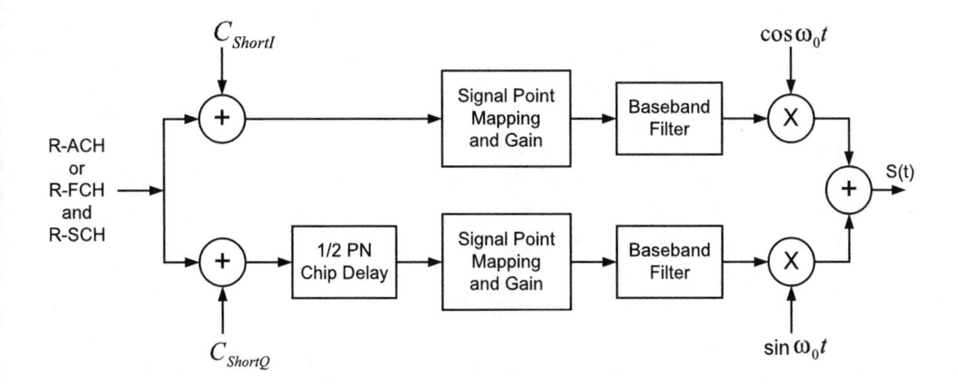

FIGURE 9.14
Reverse link transmission block for RC 1 and RC 2 (SR 1).

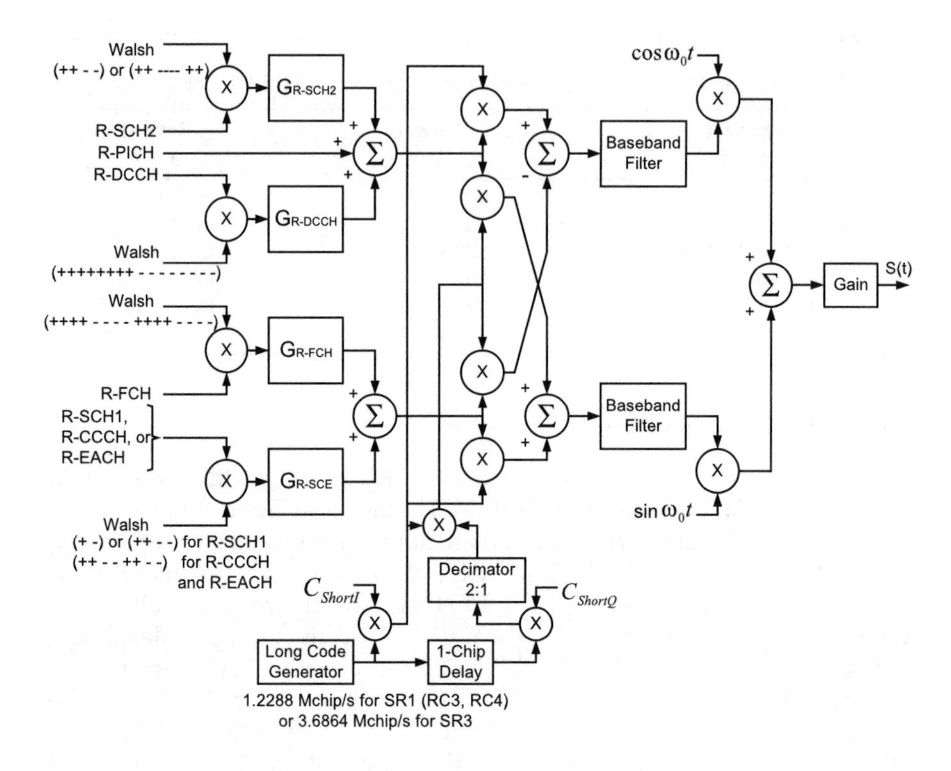

FIGURE 9.15
Reverse link transmission block for RC 3 and RC 4 (SR 1) and for RC 5 and RC 6 (SR 3).

have their characteristic polynomials given as

$$C_{\text{short}I} = 1 + x^5 + x^7 + x^8 + x^9 + x^{13} + x^{15} \tag{9.4}$$

$$C_{\text{short}Q} = 1 + x^3 + x^4 + x^5 + x^6 + x^{10} + x^{11} + x^{12} + x^{15} \tag{9.5}$$

The pilot PN sequence period is $2^{15}/1{,}228{,}800 = 26.666\ldots$ ms, with exactly 75 pilot PN sequence repetitions occurring every 2 s. For SR 3, the I and Q PN sequences are truncated sequences of a maximal length linear feedback shift register based on the characteristic polynomial:

$$C_{\text{short}} = 1 + x^3 + x^5 + x^9 + x^{20} \tag{9.6}$$

The I and Q PN sequences are formed using this $2^{20} - 1$ length maximal length sequence but truncated after 3×2^{15} chips. The I PN sequence has its starting position such that the first 20 chips, in hexadecimal form, are 80,114.

The Q PN sequence has its starting position equal to that of the I PN sequence delayed by $2^{19} - 1$ chips. The chip rate is 3.6864 Mchip/s and the PN sequence period is obtained as $3 \times 2^{15}/3,686,400 = 26.666\ldots$ ms. Here, again, exactly 75 PN sequence repetitions occur every 2 s. The long PN code, C_{long}, is generated by 42-stage linear feedback shift registers with its characteristic polynomial given as

$$C_{\text{long}} = 1 + x + x^2 + x^3 + x^5 + x^6 + x^7 + x^{10} + x^{16} + x^{17} + x^{18} + x^{19} + x^{21} + x^{22} + x^{25} + x^{26} + x^{27} + x^{31} + x^{33} + x^{35} + x^{42} \qquad (9.7)$$

9.10 Forward Physical Channels

This section describes the main characteristics of the forward physical channels. As already mentioned, the forward physical channels can be grouped into forward link common channels and forward link dedicated channels. The following channels belong to the first group: F-PCH, F-QPCH, F-BCCH), F-SYNCH, F-CCCH, F-PiCHs, F-CACH, and F-CPCCH. The F-PiCHs encompass five channels: F-PCH, F-APICH, F-TDPICH, F-ATDPICH, and F-DAPICH. The second group contains the forward traffic channels, namely: F-FCH, F-DCCH, F-SCCH, F-SCH.

The five forward pilot channels are unmodulated, spread spectrum signals used by mobile stations for synchronization, phase reference, and demodulation purposes. They carry neither payload information nor signaling information, essentially consisting of short PN chips. They define the cell boundary and can be used as beacon signals to facilitate rapid pilot searches as well as to perform handoff measurements. All base stations use the same short PN code and are distinguished by the different offsets of that code.

A forward traffic channel comprises one or more code channels used to transport user and signaling traffic from the base station to the mobile station. The number of code channels composing a forward traffic channel depends on the type of traffic to be carried and on the desirable rate, these varying according to the SR and RC. For SR 1 and for RCs from 1 through 9 there may be one F-FCH and one F-DCCH. In addition, and for RC 1 and RC 2 only, there may be as many as seven F-SCCH. Still, for SR 1, and only for RC 3, RC 4, and RC 5 there may be as many as two F-SCH. For SR 3, there may be one F-FCH, one F-DCCH, and as many as two F-SCH. These are code channels received in parallel by the mobile station. The power levels and the QoS requirements of theses channels are set differently for each channel, as required. Their Walsh codes are not predetermined and are assigned on a demand basis.

In the description of the channels, a signal point mapping block is present in all channel structures. The signal point mapping block maps the binary levels 0 and 1 onto +1 and −1, respectively.

9.10.1 Forward Pilot Channel

The F-PICH is an unmodulated, direct-sequence spread spectrum signal transmitted continuously by each base station, unless the base station is classified as a *hopping pilot beacon* base station. The F-PICH prior to Walsh spreading contains a sequence of zeros. Such a sequence is combined with the Walsh code 0, length 64 (W_0^{64}), which also encompasses a sequence of zeros. The F-PICH allows a mobile station to acquire the timing of the forward CDMA channel, provides a phase reference for coherent demodulation, and provides means for signal strength comparisons between base stations for handoff purposes. Only one F-PICH is used per forward CDMA channel for both SR 1 and SR 3. Figure 9.16 depicts the F-PICH structure for both SR 1 and SR 3. The outputs S_I and S_Q shown in Figure 9.16 constitute the inputs of the DEMUX blocks shown in Figure 9.8, for SR 1, and Figure 9.11, for SR 3.

9.10.2 Forward Transmit Diversity Pilot Channel

The F-TDPICH is an unmodulated, direct-sequence spread spectrum signal transmitted continuously by a CDMA base station. It is used to support forward-link transmit diversity. F-PICH and F-TDPICH provide phase references for coherent demodulation of those forward-link CDMA channels deploying transmit diversity. The transmission of F-TDPICH does not imply a decrease of the transmit power of F-PICH. On the contrary, the base station should continue to use sufficient power on the F-PICH to ensure that a mobile station is able to acquire and estimate the forward CDMA channel without

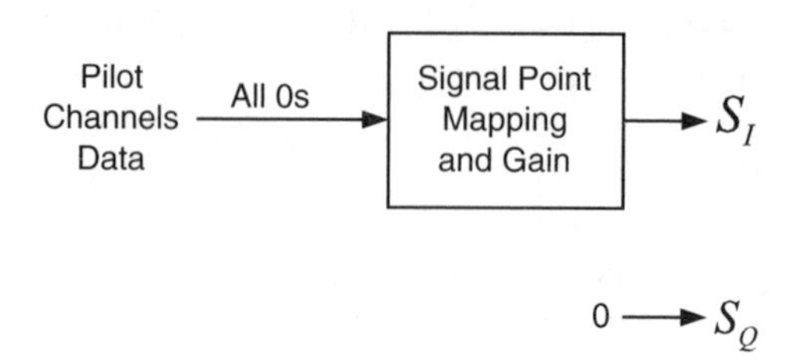

FIGURE 9.16
Forward pilot channels structure.

using energy from the F-TDPICH. F-TDPICH is transmitted with Walsh code 16, length 128 (W_{16}^{128}). Only one F-TDPICH is used per forward CDMA channel, with this channel provided in SR 1 and not in SR 3. Its configuration is the same as that shown in Figure 9.16.

9.10.3 Forward Auxiliary Pilot Channel

The F-APICH is used for forward-link spot beam-forming purposes in networks with smart antennas. The utilization of F-APICH provides for high data rate applications in specific locations. It is used as a phase reference for coherent demodulation of those forward-link CDMA channels associated with it. Zero or more F-APICHs can be transmitted by the base station on an active forward CDMA channel. An F-APICH can be shared by a number of distinct mobiles in the same spot beam. The locations served by F-APICHs may vary, as required. Spot beams can be used to increase coverage of a particular geographic point or to increase capacity of hot spots. Systems making use of such an option must provide for separate forward-link channels for the specific area. F-APICHs are code-multiplexed with other forward-link channels. This obviously reduces the number of Walsh codes available for traffic. To reduce this effect, long Walsh codes are used for these channels. The F-APICH is transmitted with Walsh code n, length N (W_n^N), where $N \leq 512$ and $1 \leq n \leq N - 1$. The Walsh code number and Walsh code length are determined by the base station. This channel is used in SR 1 and in SR 3, with the number of them per forward CDMA channel not specified. Its configuration is the same as that shown in Figure 9.16.

9.10.4 Forward Auxiliary Transmit Diversity Pilot Channel

The F-ATDPICH is a transmit diversity pilot channel associated with an F-APICH. F-ATDPICH and F-APICH provide phase references for coherent demodulation of those forward-link CDMA channels associated with the F-APICH. F-ATDPICH is transmitted with Walsh code $n + N/2$, length N ($W_{n+N/2}^N$), where $N \leq 512$ and $1 \leq n \leq N - 1$. The Walsh code number and Walsh code length are determined by the base station. This channel is used in SR 1 and not in SR 3, with the number per forward CDMA channel not specified. Its configuration is the same as that shown in Figure 9.16.

9.10.5 Forward Dedicated Auxiliary Pilot Channel

The F-DAPICH is an optional auxiliary pilot channel used on a dedicated basis for a given mobile station. It is an unmodulated, direct-sequence spread spectrum signal transmitted continuously by a CDMA base station. F-DAPICH is code-multiplexed with other forward-link channels. Its Walsh code number and the corresponding Walsh code length are determined by the base station. F-DAPICH is employed aiming at antenna beam-forming applications and beam-steering techniques to increase the coverage or date rate for a particular mobile station. Note that F-DAPICH cannot be considered a common channel. This channel is used for periodic channel estimations so that the forward-link antenna pattern can be adequately adjusted for better performance.

9.10.6 Forward Synchronization Channel

The F-SYNCH is a code channel conveying the synchronization message. Such a message is used by the mobile station to acquire initial time synchronization. F-SYNCH is implemented in cdma2000 as it is in cdmaOne. F-SYNCH is a low-powered, low-rate channel (1.2 kbit/s) that contains a single, repeating message referred to as the sync channel message. This message is continuously broadcast by the cell and contains parameters, such as system identification number, network identification number, cell or sector Short PN offset, system time, long code state, and paging channel data rate. This channel is transmitted with Walsh code 32, length 64 (W_{32}^{64}) for both SR 1 and SR 3, one per forward CDMA channel. The F-SYNCH structure is depicted in Figure 9.17.

9.10.7 Forward Paging Channel

The F-PCH is a code channel used for transmission of control information and pages from a base station to the mobile stations. It conveys system overhead

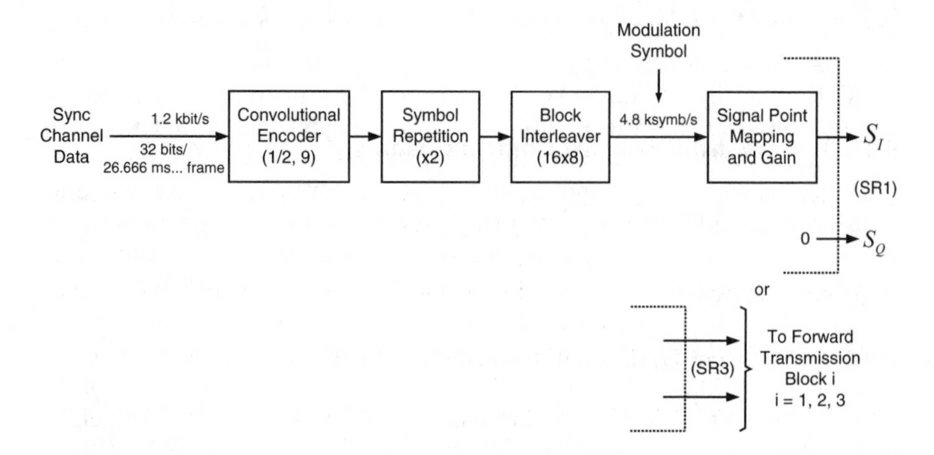

FIGURE 9.17
Forward synchronization channel structure.

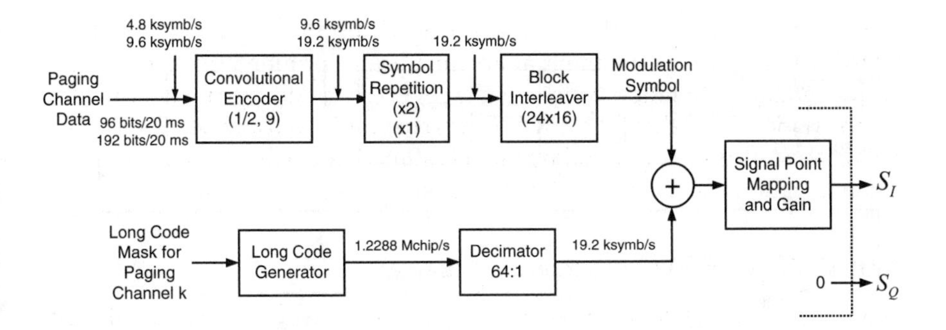

FIGURE 9.18
Forward paging channel structure.

information and mobile station specific messages. It is identical to the paging channel of cdmaOne. F-PCH transmits in the slotted mode, each slot with 80 ms of duration. Mobile stations, on the other hand, may operate in either the slotted mode or nonslotted mode. Paging and control messages for a mobile station operating in the nonslotted mode can be conveyed in any of the F-PCH slots. Therefore, the nonslotted mode of operation requires the mobile station to monitor all the slots. The slotted mode of operation requires the assignment of a specific slot to the mobile station; this feature is used to save battery. There may be as many as seven F-PCHs per forward CDMA in SR 1. SR 3 does provide for F-PCH. The primary F-PCH is assigned Walsh code number 1, length 64 (W_1^{64}), with the remaining F-PCHs of the same length and numbered sequentially from 2 to 7 (W_{2-7}^{64}). These channels operate at full rate (9.6 kbit/s) and at half rate (4.8 kbit/s). The F-PCH is illustrated in Figure 9.18.

9.10.8 Forward Broadcast Control Channel

The F-BCCH is a code channel used for transmission of control information from a base station to the mobile stations. It conveys broadcast overhead messages and short message service broadcast messages. (Mobile specific messages are not sent on this channel, but on the F-CCCH.) There may be as many as eight F-BCCHs per forward CDMA in both SR 1 and SR 3. The specific Walsh code used is determined by the base station and such information is conveyed by the F-SYNCH. In both SR 1 and SR 3, 744 bits are transmitted in slots of 40, 80, or 160 ms. The 744 bits together with 16 quality indicator bits and eight encoder tail bits lead to data rates of, respectively, 19.2, 9.6, and 4.8 kbit/s. Different Walsh codes are used for the different F-BCCH structures. The F-BCCH structure is illustrated in Figure 9.19.

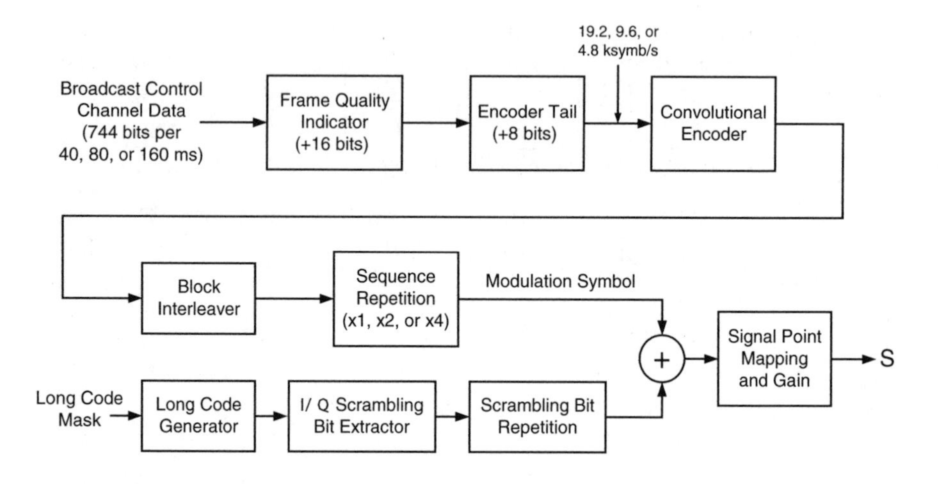

FIGURE 9.19
Forward broadcast control channel structure.

F-BCCH for SR 1

The long code generator for the SR 1 F-BCCH operates with a chip rate of 1.2288 Mchip/s. The I/Q Scrambling Bit Extractor block extracts the I and Q pairs at a rate given by the modulation symbol rate divided by twice the scrambling bit repetition factor. The scrambling repetition factor, in the scrambling repetition bit block, is equal to 1 for the non-TD mode and 2 for the TD mode. Two operation options can be found for the F-BCCH, depending on the convolutional encoder used. One of the options uses a 1/4-rate convolutional encoder with constraint length of 9. The other option uses a 1/2-rate convolutional encoder with constraint length of 9. In the first case, the block interleaver is of 3,072 symbols, whereas in the second case the block interleaver is of 1,535 symbols. The modulation symbol rates (rate after the block interleaver) are, respectively, 76.8 and 38.4 ksymb/s. The Walsh codes in the respective cases are numbered n with lengths 32 (W_n^{32}) and 64 (W_n^{64}).

F-BCCH for SR 3

The long code generator for the SR 3 F-BCCH operates with a chip rate of 3.6864 Mchip/s. The I/Q scrambling bit extractor block extracts the I and Q pairs at a rate given by the modulation symbol rate divided by the scrambling bit repetition factor multiplied by 6. The scrambling repetition factor, in the scrambling repetition bit block, is equal to 3. A 1/3-rate convolutional encoder with constraint length of 9 is used, in which case the block interleaver operates with 2,304 symbols. The modulation symbol rate (rate after the block interleaver) is, therefore, 57.6 ksymb/s. The Walsh codes are numbered n with lengths 128 (W_n^{128}).

9.10.9 Forward Quick Paging Channel

The F-QPCH is an uncoded, spread, and on-off-keying modulated spread spectrum signal used in support of the operation of F-PCH and F-CCCH. It is sent by the base station to inform mobile stations operating in the slotted mode whether to receive the F-PCH or the F-CCCH starting in their respective next frames. The use of F-QPCH reduces the time a mobile station needs to process received data, resulting in increased battery life. This is because the mobile does not have to activate its processors to understand the messages of the channel. Indicators are used to facilitate the task. These indicators are recognized by threshold-based detection. Therefore, if there is no new message for the mobile station in the F-PCH or in the F-CCCH, it does not have to activate its processors to decode the message in the assigned slot. Data rates of 4.8 and 2.4 ksymb/s can be used. Slots of 80 ms are specified to convey two indicators per mobile in each slot. The resulting indicator rates are, respectively, 9.6 and 4.8 ksymb/s. The F-QPCH slots are aligned to initiate 20 ms before the start of the zero-offset pilot PN sequence. In SR 1, the symbols are repeated two or four times to yield a constant rate of 19.2 ksymb/s. In SR 3, the repetition factors are, respectively, 3 and 6, leading to a transmission rate of 28.8 ksymb/s. One of the indicators, the paging indicator, serves the purpose of instructing a slotted mode mobile station to monitor the F-PCH or the F-CCCH starting in the next frame. The other indicator, the configuration change indicator, serves the purpose of instructing a slotted-mode mobile station to monitor the F-PCH, the F-CCCH, and the F-BCCH, after an idle handoff has been performed. This is carried out to determine whether the mobile station should update its stored parameters, in case the cell configuration parameters have changed. There may be up to three F-QPCHs per forward CDMA both in SR 1 and SR 3. These channels are assigned the Walsh codes numbered 48, 80, and 112, and length 128 (W_{48}^{128}, W_{80}^{128}, and W_{112}^{128}, respectively). The F-QPCH structure is illustrated in Figure 9.20.

9.10.10 Forward Common Control Channel

The F-CCCH conveys Layer 3 and MAC control messages from a base station to one or more mobile stations. The coding parameters are identical to those of F-PCH. It essentially replaces the F-PCHs for higher data rate configurations carrying mobile station specific messages. Therefore, F-CCCHs are effectively paging channels optimized for packet services, in which case F-PCHs are not

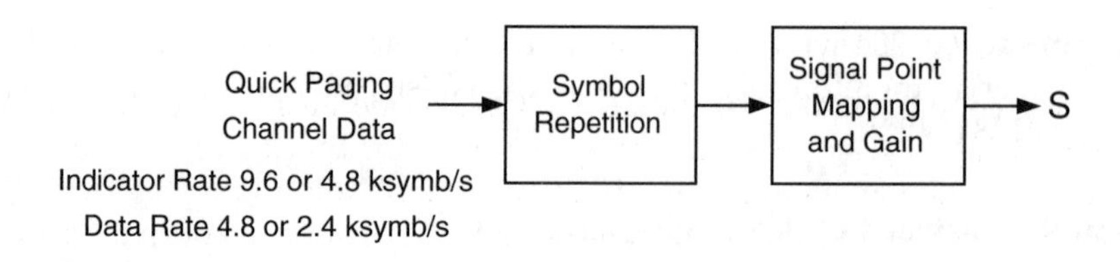

FIGURE 9.20
Forward quick paging channel structure.

used. An F-CCCH contains slots of 80-ms duration accommodating 20-, 10-, or 5-ms frames. Paging and control messages for a mobile station operating in the nonslotted mode can be conveyed in any of the F-CCCH slots. Therefore, the nonslotted mode of operation requires the mobile station to monitor all the slots. The slotted mode of operation requires the assignment of a specific slot to the mobile station, a feature used to save battery. Although the data rate of the F-CCCHs may vary from frame to frame, for any given frame transmitted to the mobile station the data rate of that frame is previously known to that mobile station. There may be as many as seven F-CCCHs per forward CDMA in both SR 1 and SR 3. The specific Walsh code used is determined by the base station and such information is conveyed by the F-SYNCH. In both SR 1 and SR 3, three data rates are possible: 9.6, 19.2, and 38.4 kbit/s. The F-CCCH structure is illustrated in Figure 9.21.

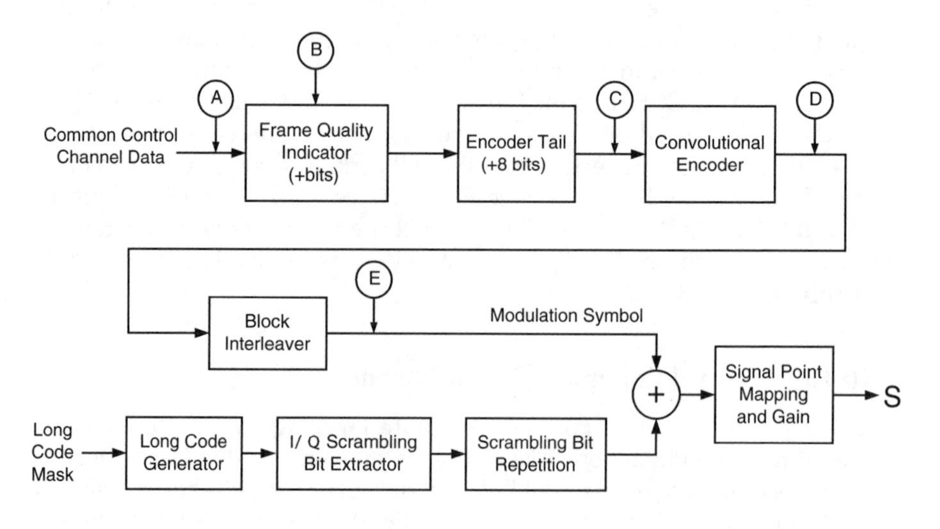

FIGURE 9.21
Forward common control channel structure.

F-CCCH for SR 1

The long code generator for the SR 1 F-CCCH operates with a chip rate of 1.2288 Mchip/s. The I/Q scrambling bit extractor block extracts the I and Q pairs at a rate given by the modulation symbol rate divided by twice the scrambling bit repetition factor. The scrambling repetition factor, in the scrambling repetition bit block, is equal to one for the non-TD mode and two for the TD mode. Two operation options can be found for the F-BCCH, depending on the convolutional encoder used. One of the options uses a 1/4-rate convolutional encoder with constraint length of 9. The other option uses a 1/2-rate convolutional encoder with constraint length of 9. The Walsh codes in the respective cases for the respective transmission rates are W_n^{16}, W_n^{32}, and W_n^{64}, and W_n^{32}, W_n^{64}, and W_n^{128}. The various parameters for the points (A, B, C, D, E) shown in Figure 9.21 are specified in Table 9.9.

F-CCCH for SR 3

The long code generator for the SR 3 F-CCCH operates with a chip rate of 3.6864 Mchip/s. The I/Q scrambling bit extractor block extracts the I and Q pairs at a rate given by the modulation symbol rate divided by the scrambling bit repetition factor multiplied by 6. The scrambling repetition factor,

TABLE 9.9

Forward Common Control Channel Parameters

Configuration	A (bits/ms)	B (bits)	C (kbit/s)	D (symbols)	E (ksymb/s)
SR1	172/5	12	38.4	768	153.6
1/4 rate	172/10	12	19.2	768	76.8
	360/10	16	38.4	1536	153.6
	172/20	12	9.6	768	38.4
	360/20	16	19.2	1536	76.8
	744/20	16	38.4	3072	153.6
SR1	172/5	12	38.4	384	76.8
1/2 rate	172/10	12	19.2	384	38.4
	360/10	16	38.4	768	76.8
	172/20	12	9.6	384	19.2
	360/20	16	19.2	768	38.4
	744/20	16	38.4	1536	76.8
SR3	172/5	12	38.4	576	115.2
	172/10	12	19.2	576	57.6
	360/10	16	38.4	1152	115.2
	172/20	12	9.6	576	28.8
	360/20	16	19.2	1152	57.6
	744/20	16	38.4	2304	115.2

in the scrambling repetition bit block, is equal to 3. A 1/3-rate convolutional encoder with constraint length of 9 is used. The Walsh codes for the three transmission rates are, respectively, W_n^{64}, W_n^{128}, and W_n^{256}. The various parameters for the points (A, B, C, D, E) shown in Figure 9.21 are specified in Table 9.9.

9.10.11 Forward Common Assignment Channel

The F-CACH is used by the base station to acknowledge a mobile station accessing the R-EACH. In the reservation access mode, it is used to convey the address of an R-CCCH and the associated R-CPCSCH. This is the case in which the mobile station requests a channel for longer messaging. The mobile station then is informed of R-CCCH on the F-CACH. Concomitantly, an R-CPCSCH is also assigned for closed-loop power control purposes. The F-CACH provides rapid reverse-link channel assignments to support random-access packet data transmission. The base station may choose not to support F-CACHs, in which case F-BCCHs may be used instead. There may be as many as seven F-CACHs per forward CDMA in both SR 1 and SR 3. The 32 channel bits per 5 ms frame together with eight quality indicator bits and eight encoder tail bits lead to a data rate of 9.6 kbit/s. The F-CACH structure is illustrated in Figure 9.22. The signal point mapping block in this case maps the binary levels 0 and 1 onto +1 and −1, respectively, in the presence of a message, or onto 0, in the absence of a message.

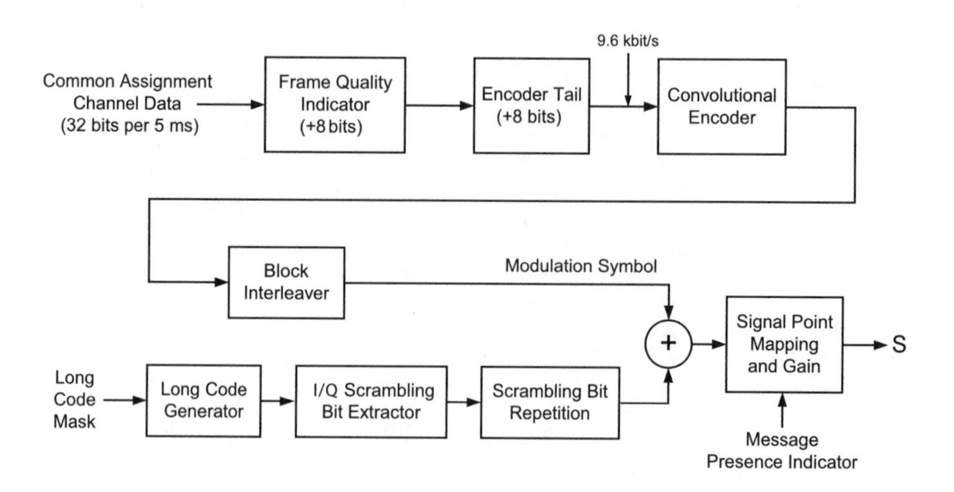

FIGURE 9.22
Forward common assignment channel structure.

F-CACH for SR 1

The long code generator for the SR 1 F-CACH operates with a chip rate of 1.2288 Mchip/s. The I/Q scrambling bit extractor block extracts the I and Q pairs at a rate given by the modulation symbol rate divided by twice the scrambling bit repetition factor. The scrambling repetition factor, in the scrambling repetition bit block, is equal to one for the non-TD mode and two for the TD mode. Two operation options can be found for the F-BCCH, depending on the convolutional encoder used. One of the options uses a 1/4-rate convolutional encoder with constraint length of 9. The other option uses a 1/2-rate convolutional encoder with constraint length of 9. In the first case, the block interleaver is of 192 symbols, whereas in the second case the block interleaver is of 96 symbols. The modulation symbol rates (rate after the block interleaver) are, respectively, 38.4 and 19.2 ksymb/s. The Walsh codes in the respective cases are W_{32}^{64} and W_{32}^{128}.

F-CACH for SR 3

The long code generator for the SR 3 F-BCCH operates with a chip rate of 3.6864 Mchip/s. The I/Q scrambling bit extractor block extracts the I and Q pairs at a rate given by the modulation symbol rate divided by the scrambling bit repetition factor multiplied by 6. The scrambling repetition factor, in the scrambling repetition bit block, is equal to 3. A 1/3-rate convolutional encoder with constraint length of 9 is used, in which case the block interleaver operates with 144 symbols. The modulation symbol rates (rate after the block interleaver) is, therefore, 28.8 ksymb/s. The Walsh code is W_{32}^{256}.

9.10.12 Forward Common Power Control Channel

The F-CPCCH conveys power control bits (PCBs) to multiple mobile stations operating in one of the following modes: power controlled access mode (PCAM), reservation access mode (RAM), or designated access mode (DAM). In PCAM, the mobile station accesses the R-EACH to transmit an enhanced access preamble, an enhanced access header, and enhanced access data in the enhanced access probe using closed-loop power control. In RAM, the mobile station accesses R-EACH and R-CCCH. On R-EACH, it transmits an enhanced access preamble and an enhanced access header in the enhanced access probe. On R-CCCH, it transmits the enhanced access data using closed-loop power control. In DAM, the mobile station responds to requests received on F-CCCH. Each PCB, known as common power control subchannel, consists of one common power control bit. These PCBs are used to adjust the power levels of R-CCCH and R-EACH. The base station may support operation on one to four F-CPCCHs. The PCBs (subchannels) are time-multiplexed on the F-CPCCH. Each subchannel controls an R-CCCH or an R-EACH. The

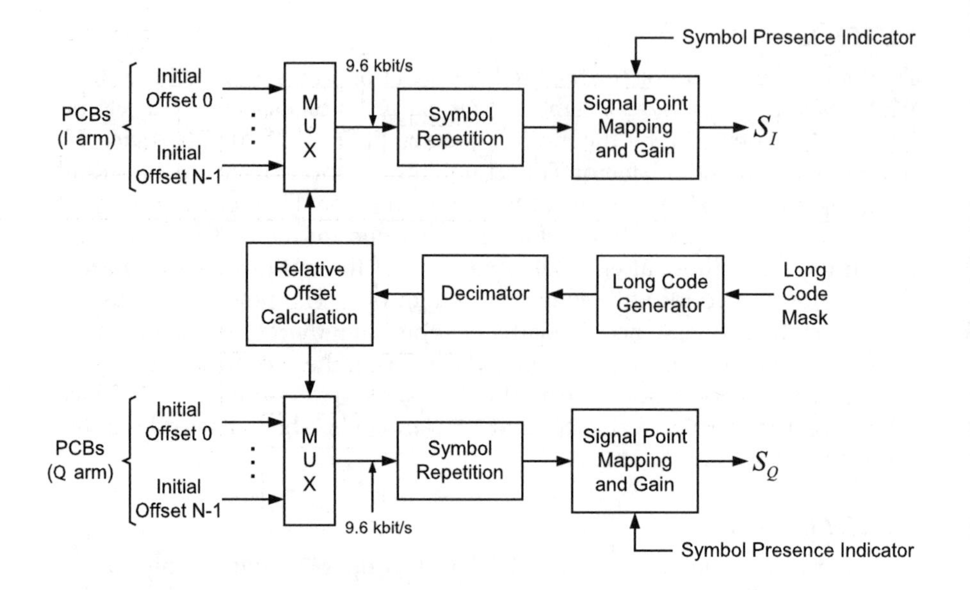

FIGURE 9.23
Forward common power control channel parameters.

F-CPCCH structure is depicted in Figure 9.23. The output data rate of the MUX block in the I arm and in the Q arm is constant and equal to 9.6 kbit/s. Three update rates are possible: 800, 400, and 200 bit/s. Given the 9.6 kbit/s fixed rate, the number of multiplexed subchannels is, respectively, 12, 24, and 48.

F-CPCCH for SR 1

The long code generator for the SR 1 F-CPCCH operates with a chip rate of 1.2288 Mchip/s. The scrambling repetition factor, in the scrambling repetition bit block, is equal to one for the non-TD mode and two for the TD mode yielding a symbol rate of 9.6 and 19.2 ksymb/s, respectively. The Walsh codes in the respective cases are W_{32}^{64} and W_{32}^{128}.

F-BCCH for SR 3

The long code generator for the SR 3 F-BCCH operates with a chip rate of 3.6864 Mchip/s. The scrambling repetition factor, in the scrambling repetition bit block, is equal to three, yielding a symbol rate of 28.8 ksymb/s. The Walsh code W_{32}^{256}.

9.10.13 Forward Fundamental Channel and Forward Supplemental Code Channel

F-FCH and F-SCCH operate jointly as specified in RC 1 and RC 2 of SR 1. Such a combination provides higher data rate services and backward compatibility

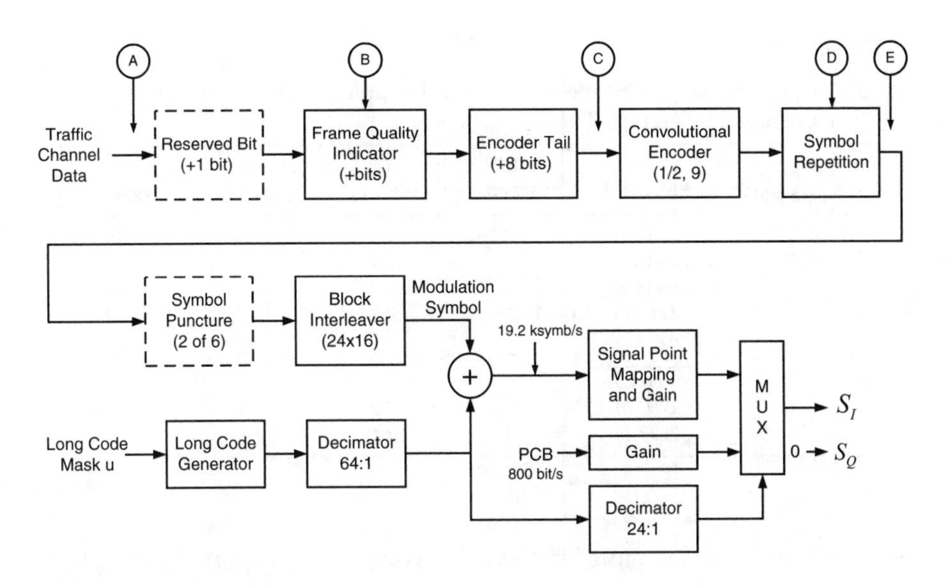

FIGURE 9.24
Forward fundamental channel and forward supplemental code control channel structure for RC 1 and RC 2 of SR 1.

with cdmaOne. RC 1 and RC 2, respectively, support Rate Set 1 and Rate Set 2 of cdmaOne. One F-FCH and up to seven F-SCCH can be used simultaneously for a forward traffic channel. These channels transmit at variable rates the changes occurring on a frame-by-frame basis, in which case the receiver is required to provide for rate detection. The basic transmission rates are 1.2, 2.4, 4.8, and 9.6 kbit/s for RC 1, and 1.8, 3.6, 7.2, and 14.4 kbit/s for RC 2. Figure 9.24 illustrates the F-FCH and F-SCCH channel structure. The dashed-line boxes in Figure 9.24 indicate the boxes present in RC 2 (but not in RC 1). The various parameters for the points (A, B, C, D, E) shown in Figure 9.24 are specified in Table 9.10. In both configurations, the modulation symbol rate is 19.2 ksymb/s. In the same way, the PCB rate is 800 bit/s. Note, however, that PCBs are not punctured in for F-SCCHs, but only for F-FCHs.

9.10.14 Forward Fundamental Channel and Forward Supplemental Channel

F-FCH and F-SCH operate jointly as specified in RC 3, RC 4, and RC 5, for SR 1, and in RC 6, RC 7, RC 8, and RC 9, for SR 3. These channels use frame structures in multiples of 20 ms. A 5-ms frame can also be utilized but only by F-FCH (not by F-SCH). In the same way, 40- and the 80-ms frames are used only for F-SCHs. The 5-ms structure is mainly used for signal carrying purposes, whereas the 20-ms structure is mostly utilized to convey user

TABLE 9.10

Forward Fundamental Channel and Forward Supplemental Code Control Channel Parameters for RC 1 and RC 2 of SR 1

Configuration	A (bits/ms)	B (bits)	C (kbit/s)	D (factor)	E (ksymb/s)
RC 1	16/20	0	1.2	× 8	19.2
(SR 1)	40/20	0	2.4	× 4	19.2
	80/20	8	4.8	× 2	19.2
	172/20	12	9.6	× 1	19.2
RC 2	21/20	6	1.8	× 8	28.8
(SR 1)	55/20	8	3.6	× 4	28.8
	125/20	10	7.2	× 2	28.8
	267/20	12	14.4	× 1	28.8

information. Note, therefore, that an F-SCH does not transport signaling traffic. These channels use orthogonal variable length Walsh codes with their lengths ranging from 2 to 128, depending on the desired data rate and on the QoS requirements. One F-FCH and up to two F-SCHs can be used simultaneously for a forward traffic channel. These channels transmit at variable rates the changes occurring on a frame-by-frame basis, in which case the receiver is required to provide for rate detection. The QoS target can be set individually for each channel, as required. When using 20-ms frames, configurations and parameters for the F-SCHs are the same as those for the F-FCH. In general, F-SCH carries data rates higher than F-FCH. This channel is transmitted at variable rates on a frame-by-frame basis. Figure 9.25 illustrates the F-FCH and F-SCH channel structure. The various parameters for the points (A, B, C, D, E, F, G, H, I) shown in Figure 9.25 are specified in Table 9.11. A convolutional encoder or turbo encoder is used depending on the number of encoder input bits. Above a certain value, turbo encoding is used; otherwise, convolutional encoding with a constraint length of 9 is used. RC 6 does not employ turbo encoding. The dashed-line box in Figure 9.25 indicates the box present only in RC 5 and RC 9. In Table 9.11, n is the length of the frame in multiples of 20 ms. In Figure 9.25, the long code generator runs at a chip rate of 1.2288 Mchip/s for RC 3, RC 4, and RC 5 for SR 1, and at chip rate of 3.6864 Mchip/s for RC 6, RC 7, RC 8, and RC 9 for SR 3. For RC 3, RC 4, and RC 5, the I/Q scrambling bit extractor block extracts the I and Q pairs at a rate given by the modulation symbol rate divided by twice the scrambling bit repetition factor. In the same way, the scrambling repetition factor, in the scrambling repetition bit block, is equal to one for the non-TD mode and two for the TD mode. For RC 6, RC 7, RC 8, and RC 9, the I/Q scrambling bit extractor block extracts

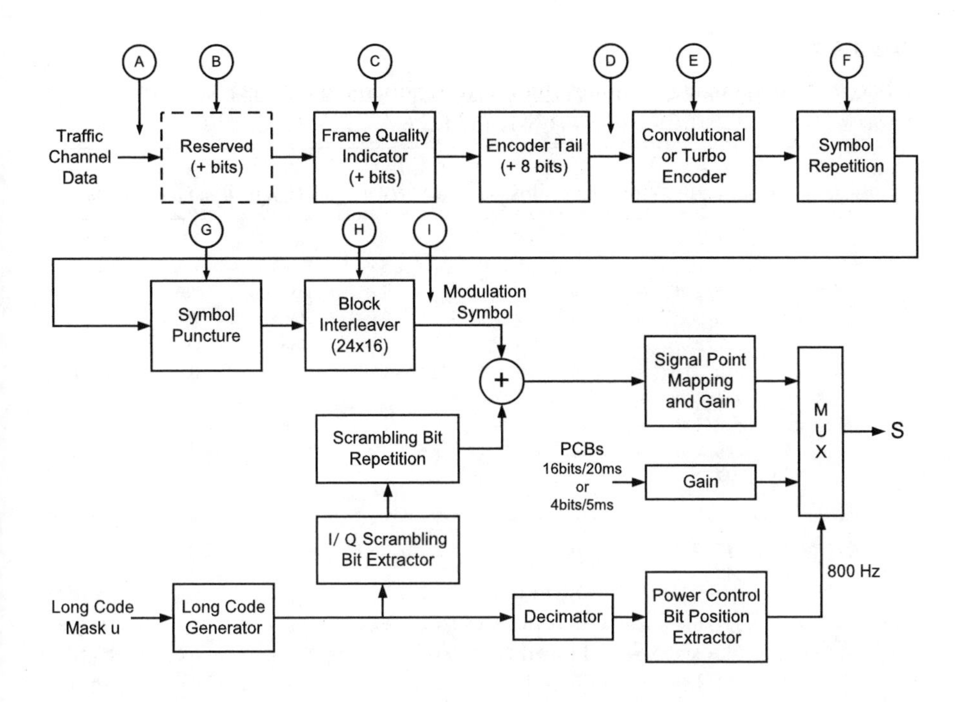

FIGURE 9.25
Forward fundamental channel and forward supplemental code channel structure for RC 3, RC 4, RC 5 of SR 1, and RC 6, RC 7, RC 8, and RC 9 of SR 3.

the I and Q pairs at a rate given by the modulation symbol rate divided by the scrambling bit repetition factor multiplied by 6. In the same way, the scrambling repetition factor, in the scrambling repetition bit block, is equal to 3.

9.10.15 Forward Dedicated Control Channel

The F-DCCH is a portion of a forward traffic channel used for the transmission of higher-level data, control information, and power control information. It operates in RC 3, RC 4, and RC 5 for SR 1, and in RC 6, RC 7, RC 8, and RC 9 for SR 3, supporting data rates from 1.05 to 14.4 ksymb/s, depending on the RC. There may be one F-DCCH per forward traffic channel. It uses frames of 5 or 20 ms for any of the RCs and Walsh codes W_n^{64}, W_n^{128}, W_n^{64}, W_n^{128}, W_n^{256}, W_n^{128}, W_n^{256} for RC 3 to RC 9, respectively. The Walsh code channel number for the F-DCCH is determined by the base station. Figure 9.26 illustrates the F-DCCH channel structure. The dashed-line box in Figure 9.26 indicates the box present only in RC 5 and RC 9. The various parameters for the points

TABLE 9.11

Forward Fundamental Channel and Forward Supplemental Code Control Channel Parameters for RC 3, RC 4, RC 5 of SR 1, and RC 6, RC 7, RC 8, and RC 9 of SR 3

Configuration	A (bits/ms)	B (bits)	C (bits)	D (kbit/s)	E (rate)	F (factor)	G (deletion)	H (symbols)	I (ksymb/s)
RC3	24/5	—	16	9.6	1/4	1×	None	192	38.4
(SR1)	16/20	—	6	1.5	1/4	8×	1 of 5	768	38.4
	40/20*n*	—	6	2.7/*n*	1/4	4×	1 of 9	768	38.4/*n*
	80/20*n*	—	8	4.8/*n*	1/4	2×	None	768	38.4/*n*
	172/20*n*	—	12	9.6/*n*	1/4	1×	None	768	38.4/*n*
	360/20*n*	—	16	19.2/*n*	1/4	1×	None	1,536	76.8/*n*
	744/20*n*	—	16	38.4/*n*	1/4	1×	None	3,072	153.6/*n*
	1,512/20*n*	—	16	76.8/*n*	1/4	1×	None	6,144	307.2/*n*
	3,048/20*n*	—	16	153.6/*n*	1/4	1×	None	12,288	614.4/*n*
	1 to 3,047/20*n*	—	—	—	—	—	—	—	—
RC4	24/5	—	16	9.6	1/2	1×	None	96	19.2
(SR1)	16/20	—	6	1.5	1/2	8×	1 of 5	384	19.2
	40/20*n*	—	6	2.7/*n*	1/2	4×	1 of 9	384	19.2/*n*
	80/20*n*	—	8	4.8/*n*	1/2	2×	None	384	19.2/*n*
	172/20*n*	—	12	9.6/*n*	1/2	1×	None	384	19.2/*n*
	360/20*n*	—	16	19.2/*n*	1/2	1×	None	768	38.4/*n*
	744/20*n*	—	16	38.4/*n*	1/2	1×	None	1,536	76.8/*n*
	1,512/20*n*	—	16	76.8/*n*	1/2	1×	None	3,072	153.6/*n*
	3,048/20*n*	—	16	153.6/*n*	1/2	1×	None	6,144	307.2/*n*
	6,120/20*n*	—	16	307.2/*n*	1/2	1×	None	12,288	614.4/*n*
	1 to 6119/20*n*	—	—	—	—	—	—	—	—
RC5	24/5	0	16	9.6	1/4	1×	None	192	38.4
(SR1)	21/20	1	6	1.8	1/4	8×	4 of 12	768	38.4
	55/20*n*	1	8	3.6/*n*	1/4	4×	4 of 12	768	38.4/*n*
	125/20*n*	1	10	7.2/*n*	1/4	2×	4 of 12	768	38.4/*n*
	267/20*n*	1	12	14.4/*n*	1/4	1×	4 of 12	768	38.4/*n*
	552/20*n*	0	16	28.8/*n*	1/4	1×	4 of 12	1,536	76.8/*n*
	1,128/20*n*	0	16	57.6/*n*	1/4	1×	4 of 12	3,072	153.6/*n*
	2,280/20*n*	0	16	115.2/*n*	1/4	1×	4 of 12	6,144	307.2/*n*
	4,584/20*n*	0	16	230.4/*n*	1/4	1×	4 of 12	12,288	614.4/*n*
	1 to 3,048/20*n*	—	—	—	—	—	—	—	—
RC6	24/5	—	16	9.6	1/6	1×	None	288	57.6
(SR3)	16/20	—	6	1.5	1/6	8×	1 of 5	1,152	57.6
	40/20*n*	—	6	2.7/*n*	1/6	4×	1 of 9	1,152	57.6/*n*
	80/20*n*	—	8	4.8/*n*	1/6	2×	None	1,152	57.6/*n*
	172/20*n*	—	12	9.6/*n*	1/6	1×	None	1,152	57.6/*n*
	360/20*n*	—	16	19.2/*n*	1/6	1×	None	2,304	115.2/*n*
	744/20*n*	—	16	38.4/*n*	1/6	1×	None	4,608	230.4/*n*
	1,512/20*n*	—	16	76.8/*n*	1/6	1×	None	9,216	460.8/*n*
	3,048/20*n*	—	16	153.6/*n*	1/6	1×	None	18,432	921.6/*n*
	6,120/20*n*	—	16	307.2/*n*	1/6	1×	None	36,864	1,843.2/*n*
	1 to 6119/20*n*	—	—	—	—	—	—	—	—

TABLE 9.11

Continued

Configuration	A (bits/ms)	B (bits)	C (bits)	D (kbit/s)	E (rate)	F (factor)	G (deletion)	H (symbols)	I (ksymb/s)
RC7	24/5	—	16	9.6	1/3	1×	None	144	28.8
(SR3)	16/20	—	6	1.5	1/3	8×	1 of 5	576	28.8
	$40/20n$	—	6	$2.7/n$	1/3	4×	1 of 9	576	$28.8/n$
	$80/20n$	—	8	$4.8/n$	1/3	2×	None	576	$28.8/n$
	$172/20n$	—	12	$9.6/n$	1/3	1×	None	576	$28.8/n$
	$360/20n$	—	16	$19.2/n$	1/3	1×	None	1,152	$57.6/n$
	$744/20n$	—	16	$38.4/n$	1/3	1×	None	2,304	$115.2/n$
	$1{,}512/20n$	—	16	$76.8/n$	1/3	1×	None	4,608	$230.4/n$
	$3{,}048/20n$	—	16	$153.6/n$	1/3	1×	None	9,216	$460.8/n$
	$6{,}120/20n$	—	16	$307.2/n$	1/3	1×	None	18,432	$921.6/n$
	$12{,}264/20n$	—	16	$614.4/n$	1/3	1×	None	36,864	$1{,}843.2/n$
	1 to $12{,}263/20n$	—	—	—	—	—	—	—	—
RC8	24/5	0	16	9.6	1/3	2×	—	288	57.6
(SR3)	21/20	1	6	1.8	1/4	8×	—	1,152	57.6
	$55/20n$	1	8	$3.6/n$	1/4	4×	—	1,152	$57.6/n$
	$125/20n$	1	10	$7.2/n$	1/4	2×	—	1,152	$57.6/n$
	$267/20n$	1	12	$14.4/n$	1/4	1×	—	1,152	$57.6/n$
	$552/20n$	0	16	$28.8/n$	1/4	1×	—	2,304	$115.2/n$
	$1{,}128/20n$	0	16	$57.6/n$	1/4	1×	—	4,608	$230.4/n$
	$2{,}280/20n$	0	16	$115.2/n$	1/4	1×	—	9,216	$460.8/n$
	$4{,}584/20n$	0	16	$230.4/n$	1/4	1×	—	18,432	$921.6/n$
	$9{,}192/20n$	0	16	$460.8/n$	1/4	1×	—	36,864	$1{,}843.2/n$
	1 to $9{,}191/20n$	—	—	—	—	—	—	—	—
RC9	24/5	0	16	9.6	1/3	1×	None	144	28.8
(SR3)	21/20	1	6	1.8	1/2	8×	None	576	28.8
	$55/20n$	1	8	$3.6/n$	1/2	4×	None	576	$28.8/n$
	$125/20n$	1	10	$7.2/n$	1/2	2×	None	576	$28.8/n$
	$267/20n$	1	12	$14.4/n$	1/2	1×	None	576	$28.8/n$
	$552/20n$	0	16	$28.8/n$	1/2	1×	None	1,152	$57.6/n$
	$1{,}128/20n$	0	16	$57.6/n$	1/2	1×	None	2,304	$115.2/n$
	$2{,}280/20n$	0	16	$115.2/n$	1/2	1×	None	4,608	$230.4/n$
	$4{,}584/20n$	0	16	$230.4/n$	1/2	1×	None	9,216	$460.8/n$
	$9{,}192/20n$	0	16	$460.8/n$	1/2	1×	None	18,432	$921.6/n$
	$20{,}712/20n$	0	16	$1{,}036.8/n$	1/2	1×	2 of 18	36,864	$1{,}843.2/n$
	1 to $18{,}408/20n$	—	—	—	—	—	—	—	—

(A, B, C, D, E, F, G, H) shown in Figure 9.26 are specified in Table 9.12. In Figure 9.26, the long code generator runs at a chip rate of 1.2288 Mchip/s for RC 3, RC 4, and RC 5 for SR 1, and at a chip rate of 3.6864 Mchip/s for RC 6, RC 7, RC 8, and RC 9 for SR 3. For RC 3, RC 4, and RC 5, the I/Q scrambling bit extractor block extracts the I and Q pairs at a rate given by the modulation symbol rate

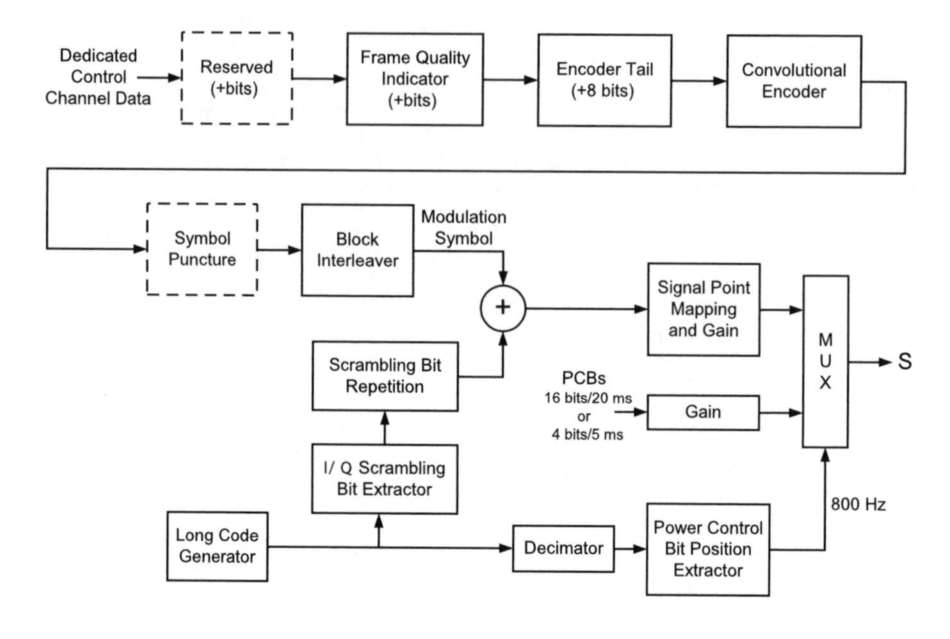

FIGURE 9.26
Forward dedicated control channel structure for RC 3, RC 4, RC 5 of SR 1, and RC 6, RC 7, RC 8, and RC 9 of SR 3.

divided by twice the scrambling bit repetition factor. In the same way, the scrambling repetition factor, in the scrambling repetition bit block, is equal to one for the non-TD mode and two for the TD mode. For RC 6, RC 7, RC 8, and RC 9, the I/Q scrambling bit extractor block extracts the I and Q pairs at a rate given by the modulation symbol rate divided by the scrambling bit repetition factor multiplied by 6. In the same way, the scrambling repetition factor, in the scrambling repetition bit block, is equal to 3.

9.11 Reverse Physical Channels

This section describes the main characteristics of the reverse physical channels. As already mentioned, the reverse physical channels can be divided into two groups: reverse dedicated channels (R-DCHs) and reverse common channels (R-CCHs). The R-DCHs convey information from a particular mobile station to the base station on a point-to-point basis. The R-CCHs convey information from multiple mobile stations to the base station on a

TABLE 9.12

Forward Dedicated Control Channel Parameters for RC 3, RC 4, RC 5 of SR 1, and RC 6, RC 7, RC 8, and RC 9 of SR 3

Configuration	A (bits/ms)	B (bits)	C (bits)	D (kbit/s)	E (rate)	F (factor)	G (deletion)	H (symbols)	I (ksymb/s)
RC3	24/5	—	16	9.6	—	—	—	192	38.4
(SR1)	172/20	—	12	9.6	—	—	—	768	38.4
	1 to 171/20	—	12 or 16	1.05 – 9.55	—	—	—	768	38.4
RC4	24/5	—	16	9.6	—	—	—	96	19.2
(SR1)	172/20	—	12	9.6	—	—	—	384	19.2
	1 to 171/20	—	12 or 16	1.05 – 9.55	—	—	—	384	19.2
RC5	24/5	0	16	9.6	—	—	None	192	38.4
(SR1)	267/20	1	12	14.4	—	—	4 of 12	768	38.4
	1 to 268/20	0	12 or 16	1.05 – 14.4	—	—	—	768	38.4
RC6	24/5	—	16	9.6	—	—	—	288	57.6
(SR3)	172/20	—	12	9.6	—	—	—	1152	57.6
	1 to 171/20	—	12 or 16	1.05 – 9.55	—	—	—	1152	57.6
RC7	24/5	—	16	9.6	—	—	—	144	28.8
(SR3)	172/20	—	12	9.6	—	—	—	576	28.8
	1 to 171/20	—	12 or 16	1.05 – 9.55	—	—	—	576	28.8
RC8	24/5	0	16	9.6	1/3	2×	—	288	57.6
(SR3)	267/20	1	12	14.4	1/4	1×	—	1152	57.6
	1 to 268/20	0	12 or 16	1.05 – 14.4	1/4	—	—	1152	57.6
RC9	24/5	0	16	9.6	1/3	—	—	144	28.8
(SR3)	267/20	1	12	14.4	1/2	—	—	576	28.8
	1 to 268/20	0	12 or 16	1.05 – 14.4	1/2	—	—	576	28.8

multipoint-to-point basis. The R-CCHs comprise the following channels: reverse common control channel (R-CCCH), reverse enhanced access channel (R-EACH), and reverse access channel (R-ACH). The R-DCHs comprise the following channels: reverse fundamental channel (R-FCH), reverse dedicated control channel (R-DCCH), reverse supplemental code channel (R-SCCH), reverse supplemental channel (R-SCH), and reverse power control subchannel (R-PCSCH). In addition, a reverse pilot channel (R-PICH) is included for channels operating in RCs other than RC 1 and RC 2.

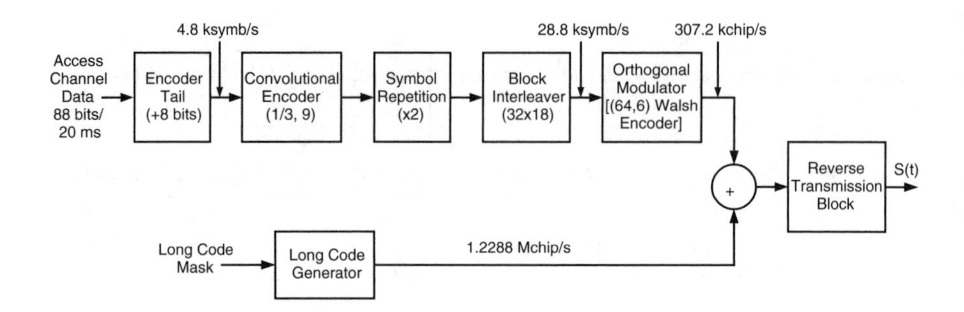

FIGURE 9.27
Structure for the reverse access channel for SR 1.

9.11.1 Reverse Access Channel

The R-ACH is used to convey short signaling exchanges related to Layer 3 and MAC messages. These messages are due to, for example, call originations, responses to pages, and registrations. There is one R-ACH per reverse CDMA channel, with this R-ACH used in RC 1 and RC 2 for SR 1. It operates on a slotted random-access basis, with multiple access provided by means of the slotted-ALOHA algorithm. To allow for backward compatibility with cdmaOne, R-ACH is identical to the access channel specified in cdmaOne. Thus, because cdmaOne does not support a pilot channel in the reverse link, R-PICH is not used to support R-ACH. The R-ACH structure is illustrated in Figure 9.27.

9.11.2 Reverse Enhanced Access Channel

The R-EACH, like the R-ACH, is used to convey short signaling exchanges related to Layer 3 and MAC messages. These messages are due to, for example, call originations, responses to pages, and registrations. In addition, it can be used to transmit moderate-sized data packets. R-EACH replaces R-ACH for RC 3, RC 4, RC 5, and RC 6, with one R-EACH provided per reverse CDMA channel. Access to R-EACH is provided by means of random-access protocols. An access probe in this channel comprises an enhanced access preamble (EAP), an enhanced access header (EAH), and the enhanced access data (EAD). Depending on how the access probe is transmitted, three modes of operation are defined: basic access mode (BAM), power-controlled access mode (PCAM), and reservation access mode (RAM). In BAM, the access probe comprises EAP and EAD. In PCAM, the access probe consists of three elements: EAP, EAH, and EAD. In RAM, the access probe encompasses EPA and EAH. To facilitate the detection process at the base station, R-PICH is

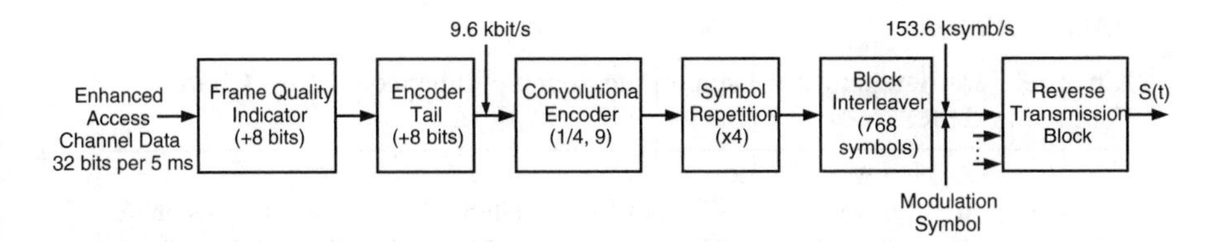

FIGURE 9.28
Channel structure for the header on the reverse enhanced access channel for SR 1 and SR 3.

transmitted during the enhanced access channel probe. The channel structure of the header on the R-EACH is depicted in Figure 9.28. The channel structure for the data on the R-EACH is shown in Figure 9.29. The various parameters for the points (A, B, C, D, E, F) shown in Figure 9.29 are specified in Table 9.13. R-EACH uses Walsh code W_2^8.

9.11.3 Reverse Common Control Channel

The R-CCCH is used for transmission of control information and data from one or more mobile stations. Typically, an R-CCCH is set up after permission to transmit is obtained as a result of a request to transmit sent on R-EACH. R-CCCH can operate in two modes: reservation access mode (RAM) or designated access mode (DAM). In RAM, the mobile station accesses R-EACH and R-CCCH. On R-EACH, it transmits an enhanced access preamble and an enhanced access header in the enhanced access probe. On R-CCCH, it transmits the enhanced access data using closed-loop power control. In DAM, the mobile station responds to requests received on F-CCCH. R-CCCH can be power controlled in RAM or in DAM and may support soft handoff in RAM. Each R-CCCH is associated with a single F-CCCH, and can be used for signaling and user data if reverse traffic channels are not in use. Like R-EACH, R-CCCH

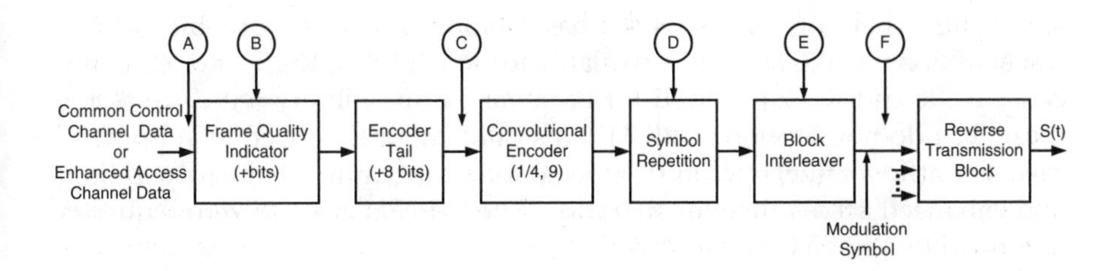

FIGURE 9.29
Channel structure for the data on the reverse enhanced access channel for SR 1 and SR 3.

TABLE 9.13

Channel Parameters for the Data on the Reverse Enhanced Access Channel for SR 1 and SR 3

Configuration	A (bits/ms)	B (bits)	C (kbit/s)	D (deletion)	E (symbols)	F (ksymb/s)
SR1	172/5	12	38.4	1×	768	153.6
	360/10	16	38.4	1×	1536	153.6
	172/10	12	19.2	2×	1536	153.6
	744/20	16	38.4	1×	3072	153.6
	360/20	16	19.2	2×	3072	153.6
	172/20	12	9.6	4×	3072	153.6
SR3	172/5	12	38.4	1×	768	153.6
	360/10	16	38.4	1×	1536	153.6
	172/10	12	19.2	2×	1536	153.6
	744/20	16	38.4	1×	3072	153.6
	360/20	16	19.2	2×	3072	153.6
	172/20	12	9.6	4×	3072	153.6

uses a random-access protocol to transmit probes. To facilitate the detection process at the base station, R-PICH is transmitted during the enhanced access channel probe. The channel structure for the data on the R-EACH is shown in Figure 9.29. The various parameters for the points (A, B, C, D, E, F) shown in Figure 9.29 are specified in Table 9.13. R-EACH uses Walsh code W_2^8.

9.11.4 Reverse Pilot Channel and Reverse Power Control Subchannel

The R-PICH consists of an unmodulated, direct-sequence spread spectrum signal transmitted continuously by a CDMA mobile station. It provides a phase reference for coherent demodulation and may provide a means for signal strength measurement. The R-PICH is used for initial acquisition, time tracking, Rake-receiver coherent reference recovery, and power control measurements. Therefore, it assists the base station in detecting mobile station transmissions. R-PICH is only available for RC 3, RC 4, RC 5, and RC 6. Because RC 1 and RC 2 are used for backward compatibility with cdmaOne, these RCs do not support R-PICH. This pilot channel is used in support of reverse traffic channel operation, reverse common control channel operation, and enhanced access channel operation. The R-PICH is an all-zero sequence identified by the Walsh code W_0^{32}.

The R-PCSCH consists of power control bits (PCBs) indicating the quality of the forward link. It is used by the mobile station to assist the base station in controlling the power of the forward traffic channels of RC 3 through RC 9. Power control information runs at a rate of 800 bit/s.

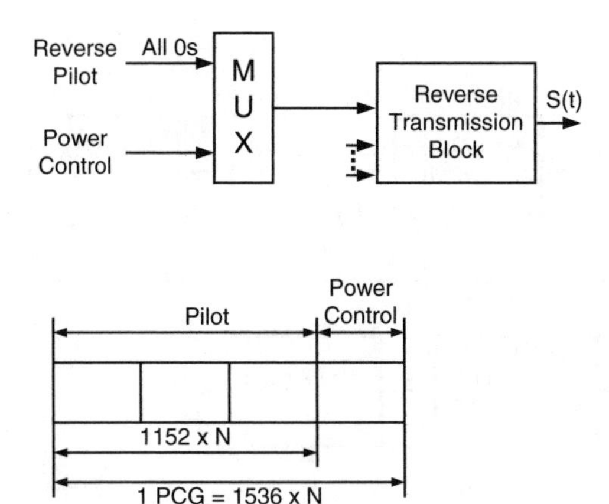

FIGURE 9.30
Channel structure for the reverse pilot channel and reverse power control subchannel.

R-PICH and R-PCSCH are time-multiplexed, and the total time of R-PICH and R-PCSCH represents the time of one power control group (PCG). One PCG contains $1536 \times N$ chips, where N is the SR number. One fourth of one PCG ($384 \times N$ chips) is composed of PCBs, whereas three fourths ($1152 \times N$ chips) contains pilot channel bits. All PN chips sent on the R-PICH within one PCG are transmitted at the same power level. R-PICH can be transmitted with the gated transmission mode enabled or disabled. When disabled, the mobile station shall transmit R-PCSCH in every PCG. When enabled, the mobile station shall transmit R-PICH only in specific PCGs. The structures of R-PICH and R-PCSCH are illustrated in Figure 9.30.

9.11.5 Reverse Fundamental Channel and Reverse Supplemental Code Channel

R-FCH and R-SCCH operate jointly as specified in RC 1 and RC 2 of SR 1. Such a combination provides higher data rate services and backward compatibility with cdmaOne. RC 1 and RC 2, respectively, support Rate Set 1 and Rate Set 2 of cdmaOne. One R-FCH and as many as seven R-SCCH can be used simultaneously for a reverse traffic channel. These channels transmit at variable rates the changes occurring on a frame-by-frame basis. Therefore, the receiver is required to provide for rate detection. The basic transmission rates are 1.2, 2.4, 4.8, and 9.6 kbit/s for RC 1, and 1.8, 3.6, 7.2, and 14.4 kbit/s for RC 2. Figure 9.31 illustrates the R-FCH and R-SCCH channel structure. The dashed-line boxes in Figure 9.31 indicate the boxes present in RC 2 (but not in RC 1). The various parameters for the points (A, B, C, D, E, F) shown in Figure 9.31 are specified in Table 9.14.

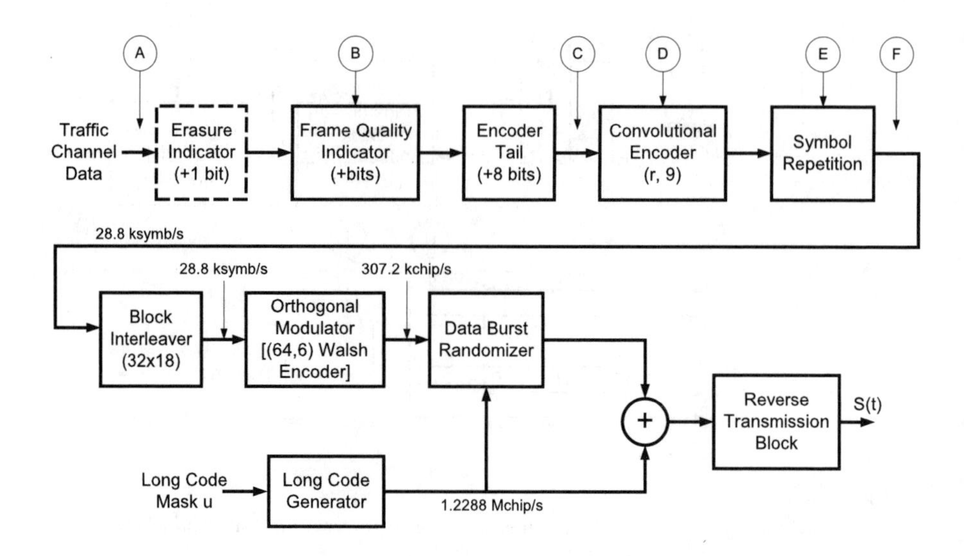

FIGURE 9.31
Reverse fundamental channel and reverse supplemental code control channel structure for RC 1 and RC 2 of SR 1.

9.11.6 Reverse Fundamental Channel and Reverse Supplemental Channel

R-FCH and R-SCH operate jointly as specified in RC 3 and RC 4 for SR 1, and in RC 5 and RC 6 for SR 3. These channels use frame structures in multiples of 20 ms. A 5-ms frame can also be utilized but only by R-FCH (not by R-SCH).The 5-ms structure is mainly used for signaling carrying purposes, whereas the 20-ms structure is mostly utilized to convey user information. Note, therefore,

TABLE 9.14
Reverse Fundamental Channel and Reverse Supplemental Code Control Channel Parameters for RC 1 and RC 2 of SR 1.

Configuration	A (bits/ms)	B (bits)	C (kbit/s)	D (rate)	E (factor)	F (ksymb/s)
RC1	16/20	0	1.2	1/3	8×	28.8
(SR1)	40/20	0	2.4	1/3	4×	28.8
	80/20	8	4.8	1/3	2×	28.8
	172/20	12	9.6	1/3	1×	28.8
RC2	21/20	6	1.8	1/2	8×	28.8
(SR1)	55/20	8	3.6	1/2	4×	28.8
	125/20	10	7.2	1/2	2×	28.8
	267/20	12	14.4	1/2	1×	28.8

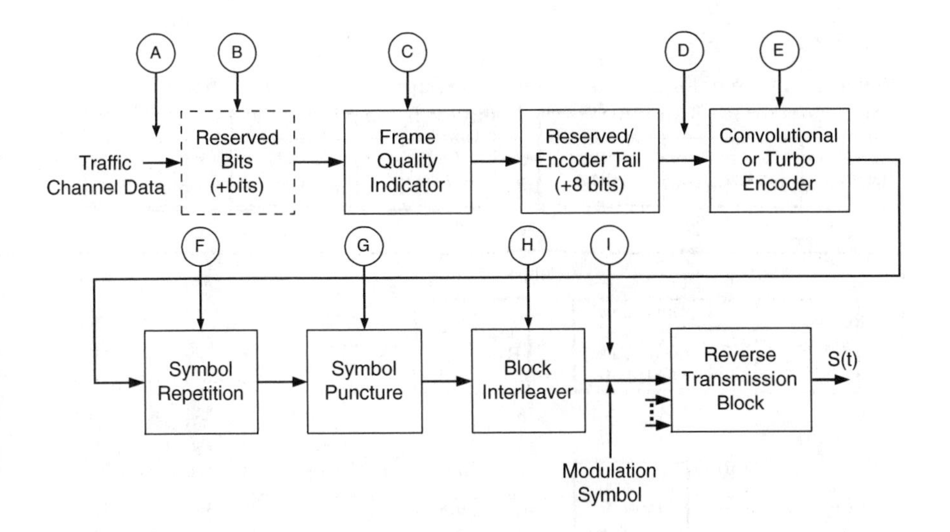

FIGURE 9.32
Reverse fundamental channel and reverse supplemental code channel structure for RC 3 and RC 4 of SR 1, and RC 5 and RC 6 of SR 3.

that an F-SCH does not transport signaling traffic. These channels use orthogonal variable length Walsh codes with their lengths ranging from 2 to 128, depending on the desired data rate and on the QoS requirements. One R-FCH and up to two R-SCHs can be used simultaneously for a reverse traffic channel. These channels transmit at variable rates the changes occurring on a frame-by-frame basis. Therefore, the receiver is required to provide for rate detection. The QoS target can be set individually for each channel, as required. Figure 9.32 illustrates the R-FCH and R-SCH channel structure. The various parameters for the points (A, B, C, D, E, F, G, H, I) shown in Figure 9.32 are specified in Table 9.15. A convolutional encoder or turbo encoder is used depending on the number of encoder input bits. Above a certain value turbo encoding is used; otherwise, convolutional encoding with constraint length of 9 is used. The dashed-line box in Figure 9.32 indicates the box present only in RC 4 and RC 6. In Table 9.15, n is the length of the frame in multiples of 20 ms.

9.11.7 Reverse Dedicated Control Channel

The R-DCCH is used for transmission of higher-level data and control information while a call is in progress. It is supported by RCs other than RC 1 and RC 2. The R-DCCH structure is illustrated in Figure 9.33. The various parameters for the points (A, B, C, D, E, F, G) shown in Figure 9.33 are specified in Table 9.16.

TABLE 9.15

Reverse Fundamental Channel and Reverse Supplemental Code Channel Parameters for RC 3 and RC 4 of SR 1, and RC 5 and RC 6 of SR 3

Configuration	A (bits/ms)	B (bits)	C (bits)	D (kbit/s)	E (deletion)	F (symbols)	G (ksymb/s)
RC3	24/5	—	16	9.6	—	384	76.8
(SR1)	172/20	—	12	9.6	—	1536	76.8
	1 to 171/20	—	12 or 16	1.05–9.55	—	1536	76.8
RC4	24/5	0	16	9.6	None	384	76.8
(SR1)	267/20	1	12	14.4	8 of 24	1536	76.8
	1 to 268/20	0	12 or 16	1.05–14.4	—	1536	76.8
RC5	24/5	—	16	9.6	—	384	76.8
(SR3)	172/20	—	12	9.6	—	1536	76.8
	1 to 171/20	—	12 or 16	1.05–9.55	—	1536	76.8
RC6	24/5	0	16	9.6	None	384	76.8
(SR3)	267/20	1	12	14.4	8 of 24	1536	76.8
	1 to 268/20	0	12 or 16	1.05–14.4	—	1536	76.8

9.12 High-Rate Packet Data Access

cdma 2000 1× systems use a dedicated RF channel for high-rate packet data services. Such a solution evolved from the high data rate (HDR) technology (see Chapter 6) to the 1× evolved high-speed data only (1×EV-DO) design, both projects presenting almost indistinguishable characteristics. The cdma2000 high-rate packet data air interface (also known as 1×EV-DO) is optimized for non-real-time, high-speed packet data services and has been included in the cdma2000 specifications to increase its data transmission capability. The cdma2000 high-rate packet data access (HRPDA) feature completes the set of cdma2000 RCs already explored in this chapter. In particular, HRPDA constitutes the RC 10 in the forward link and RC 7 in the reverse link.

In HRPDA, the access terminal (user terminal) and the access network (base station) jointly determine the highest rate a subscriber can support at any instant. This is accomplished by means of the deployment of a combination of techniques based on channel measurement, channel control, and interference suppression and mitigation. In particular, each access terminal assesses the quality of the signals (carrier-to-interference ratio, or CIR) received from neighboring access networks. The best access network is chosen

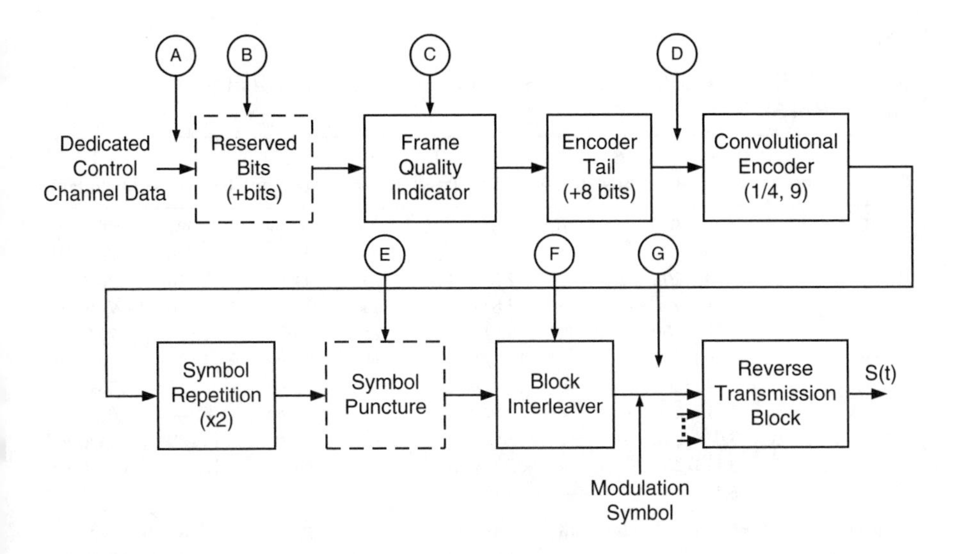

FIGURE 9.33
Reverse dedicated control channel structure for RC 3 and RC 4 of SR 1, and RC 5 and RC 6 of SR 3.

and its CIR, or equivalently the supportable data rate value, is conveyed to the corresponding access network. Note that this may change continuously because of the nonstationary condition of the wireless channel (motion of the user, change of the propagation conditions, etc.). Therefore, the monitoring of the CIRs of the various access networks and the selection of the best access network must be performed continuously, this having a direct implication on the supportable data rate. Packet transmissions on the forward link, then, occur with data rates and packet durations that vary in accordance with the user channel conditions. Hence, it may be said that the HRPDA forward link supports dynamic data rates. Packet transmissions on the reverse link, in turn, occur with variable rates but fixed packet duration. The HRPDA reverse link does not support dynamic data rates.

In the descriptions that follow, and differently from what appears in the HRPDA documents, the same convention as that used to name the cdma2000 physical channels has been adopted.

9.12.1 Forward Link—General

The forward CDMA channel, using a dedicated 1× RF channel for data transmission purposes, comprises the following time-multiplexed channels:

TABLE 9.16

Reverse Dedicated Control Channel Structure for RC 3 and RC 4 of SR 1, and RC 5 and RC 6 of SR 3

Configuration	A (bits/ms)	B (bits)	C (bits)	D (kbit/s)	E (rate)	F (factor)	G (deletion)	H (symbols)	I (ksymb/s)
RC3	24/5	—	16	9.6	1/4	2×	None	384	76.8
(SR1)	16/20	—	6	1.5	1/4	16×	1 of 5	1,536	76.8
	40/20n	—	6	2.7/n	1/4	8×	1 of 9	1,536	76.8/n
	80/20n	—	8	4.8/n	1/4	4×	None	1,536	76.8/n
	172/20n	—	12	9.6/n	1/4	2×	None	1,536	76.8/n
	360/20n	—	16	19.2/n	1/4	1×	None	1,536	76.8/n
	744/20n	—	16	38.4/n	1/4	1×	None	3,072	153.6/n
	1,512/20n	—	16	76.8/n	1/4	1×	None	6,144	307.2/n
	3,048/20n	—	16	153.6/n	1/4	1×	None	12,288	614.4/n
	6,120/20n	—	16	307.2/n	1/4	1×	None	12,288	614.4/n
	1 to 6,119/20n	—	—	—	—	—	—	—	—
RC4	24/5	0	16	9.6	1/4	2×	None	384	76.8
(SR1)	21/20	1	6	1.8	1/4	16×	8 of 24	1,536	76.8
	55/20n	1	8	3.6/n	1/4	8×	8 of 24	1,536	76.8/n
	125/20n	1	10	7.2/n	1/4	4×	8 of 24	1,536	76.8/n
	267/20n	1	12	14.4/n	1/4	2×	8 of 24	1,536	76.8/n
	552/20n	0	16	28.8/n	1/4	1×	4 of 12	1,536	76.8/n
	1,128/20n	0	16	57.6/n	1/4	1×	4 of 12	3,072	153.6/n
	2,280/20n	0	16	115.2/n	1/4	1×	4 of 12	6,144	307.2/n
	4,584/20n	0	16	230.4/n	1.4	1×	4 of 12	12,288	614.4/n
	1 to 3,048/20n	—	—	—	—	—	—	—	—
RC5	24/5	—	16	9.6	1/4	2×	None	384	76.8
(SR3)	16/20	—	6	1.5	1/4	16×	1 of 5	1,536	76.8
	40/20n	—	6	2.7/n	1/4	8×	1 of 9	1,536	76.8/n
	80/20n	—	8	4.8/n	1/4	4×	None	1,536	76.8/n
	172/20n	—	12	9.6/n	1/4	2×	None	1,536	76.8/n
	360/20n	—	16	19.2/n	1/4	1×	None	1,536	76.8/n
	744/20n	—	16	38.4/n	1/4	1×	None	3,072	153.6/n
	1,512/20n	—	16	76.8/n	1/4	1×	None	6,144	307.2/n
	3,048/20n	—	16	153.6/n	1/4	1×	None	12,288	614.4/n
	6,120/20n	—	16	307.2/n	1/3	1×	None	18,432	921.6/n
	12,264/20n	—	16	614.4/n	1/3	1×	None	36,864	1843.2/n
	1 to 12,263/20n	—	—	—	—	—	—	—	—
RC6	24/5	0	16	9.6	1/4	2×	None	384	76.8
(SR3)	21/20	1	6	1.8	1/4	16×	8 of 24	1,536	76.8
	55/20n	1	8	3.6/n	1/4	8×	8 of 24	1,536	76.8/n
	125/20n	1	10	7.2/n	1/4	4×	8 of 24	1,536	76.8/n
	267/20n	1	12	14.4/n	1/4	2×	8 of 24	1,536	76.8/n
	552/20n	0	16	28.8/n	1/4	1×	None	2,304	115.2/n
	1,128/20n	0	16	57.6/n	1/4	1×	None	4,608	230.4/n

TABLE 9.16

Continued

Configuration	A (bits/ms)	B (bits)	C (bits)	D (kbit/s)	E (rate)	F (factor)	G (deletion)	H (symbols)	I (ksymb/s)
	$2{,}280/20n$	0	16	$115.2/n$	1/4	1×	None	9,216	$460.8/n$
	$4{,}584/20n$	0	16	$230.4/n$	1/4	1×	None	18,432	$921.6/n$
	$9{,}192/20n$	0	16	$460.8/n$	1/4	1×	None	36,864	$1843.2/n$
	$20{,}712/20n$	0	16	$1036.8/n$	1/2	1×	2 of 18	36,864	$1843.2/n$
	1 to $18{,}408/20n$	—	—	—	—	—	—	—	—

- *Pilot Channel (F-PICH)*. The F-PICH is used for synchronization purposes at the access terminal.
- *Forward Medium Access Control (MAC) Channel (F-MACCH)*. The F-MACCH consists of two subchannels:

  Reverse Power Control Channel (F-RPCCH). The F-RPCCH is used for power control purposes.

  Reverse Activity Channel (F-RACH). The F-RACH conveys a reverse-link activity bit stream.
- *Control Channel (F-CCH)*. The F-CCH conveys control messages as well as user traffic.
- *Forward Traffic Channel (F-TCH)*. The F-TCH carries user data at variable rate.

Each channel is further code-division-multiplexed into quadrature Walsh channels. Each slot in the forward link contains 2048 chips and, at a chip rate of 1.2288 Mchip/s, the corresponding slot duration is $2048 \div 1228.8 = 1.66\ldots$ ms. Within each slot, F-PICH, F-MACCH, and F-CCH or F-TCH are time-multiplexed and transmitted at the same power level. The data within F-TCH and F-CCH are encoded into blocks, which in the HRPDA documents are called physical layer packets. After the encoding operation, the following signal processing techniques are sequentially applied: scrambling, interleaving, modulation, sequence repetition, and puncturing. The resulting sequences are then demultiplexed into 16 pairs of parallel streams, a pair constituted by the in-phase (I) and quadrature (Q) branches. These parallel streams are covered with a distinct 16-ary Walsh function at a chip rate to yield Walsh symbols at 76.8 ksymb/s. These symbols are then added to compose a single I stream and a single Q stream running at a chip rate of 1.2288 Mchip/s. Preamble, pilot channel, MAC channels, and traffic channels are then time-multiplexed and quadrature-spread, modulated, and transmitted. The physical layer packets can be transmitted in 1 through 16 slots. If more

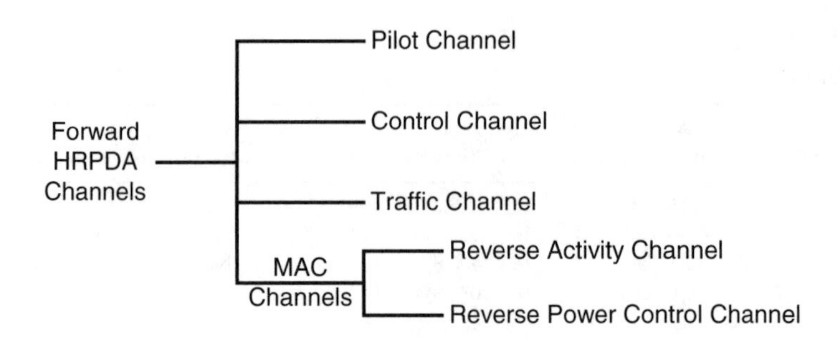

FIGURE 9.34
HRPDA forward-link channels.

than one slot is required, a four-slot interlacing scheme is used. In this case, the slots for a given packet are separated by three slots, which are used for other packets.

9.12.2 Forward-Link Channels

The forward-link channels are depicted in Figure 9.34. These channels are time-multiplexed and each channel is further decomposed into code division multiplexed quadrature Walsh channels.

Pilot Channel

The F-PICH is an unmodulated signal used by the access terminal for synchronization, initial acquisition, phase recovery, timing recovery, and maximal ratio combining. It is also used for predicting the received CIR. The F-PICH consists of all-0 symbols and is assigned the Walsh cover 0. It is transmitted only on the I branch.

Forward MAC Channel

The F-MACCH is composed of Walsh channels. They are orthogonally covered and BPSK-modulated on either I or Q phase of the carrier. The Walsh channels are identified by a MACIndex, the index identifying a unique 64-ary Walsh cover and a unique modulation phase. For a given MACIndex i, the corresponding Walsh functions are

$$W^{64}_{i=2} \text{ for } i = 0; 2; \ldots ; 62$$
$$W^{64}_{(i-1)=2+32} \text{ for } i = 1; 3; \ldots ; 63$$

The even-numbered MACIndex identifies the MAC channels to be assigned to the I branch, whereas the odd-numbered MACIndex identifies the MAC channels to be assigned to the Q branch. The symbols for the Walsh covers are transmitted four times per slot, with each burst containing 64 chips. The bursts immediately precede or follow the pilot bursts of each slot.

Reverse Power Control Channel. The F-RPCCH is used for the transmission of the reverse power control (RPC) bit stream directed to a given access terminal with an open connection. It is assigned to one of the MAC channels available. The F-RPCCH is transmitted at a rate of 600 bit/s, with each RPC symbol transmitted four times per slot, in bursts of 64 chips.

Reverse Activity Channel. The F-RACH is used for the transmission of the reverse activity bit (RAB) stream. It uses the MAC channel with MACIndex 4 and operates at a rate of 600/RABLength bit/s. RABLength is a field in the public data of the traffic channel assignment of the route update protocol.

Control Channel

The F-CCH is used to transmit broadcast messages as well as access-terminal-directed messages. The messages are transmitted at a rate of 76.8 or 38.4 ksymb/s. Its modulation characteristics are the same as those for the F-TCH at the corresponding rates. The F-CCH has a preamble that is covered by a biorthogonal cover sequence with MACIndex 2 or 3, used, respectively, for rates 76.8 and 38.4 kbit/s.

Forward Traffic Channel

The F-TCH is used to convey physical layer packets. It is a packet-based, variable-rate channel operating at rates ranging from 38.4 to 2457.6 kbit/s. It makes use of a preamble, consisting of all-0 symbols transmitted on the I branch. The preamble sequence is covered by a 32-chip biorthogonal sequence, which is specified in terms of the 32-ary Walsh functions and their bit-by-bit complements. These Walsh functions are given, respectively, as

$$W^{32}_{i=2} \text{ for } i = 0; 2; \ldots; 62$$
$$\overline{W^{32}_{(i-1)=2}} \text{ for } i = 1; 3; \ldots; 63$$

The physical layer packets transmitted on the F-TCH are encoded with code rates of 1/3 or 1/5 with 6 bits of the TAIL field discarded. A sequence of signal processing techniques, such as turbo encoding, turbo interleaving, scrambling, and channel interleaving, is applied before modulation. Different modulation schemes, such as QPSK, 8-PSK, and 16-QAM, are used depending on the data rate. After modulation, sequence repetition and symbols puncturing are applied. The I and Q streams are then demultiplexed into 16 parallel streams. The individual streams at the demultiplexer output are assigned to 1 of 16 distinct Walsh channels $W_k^{16}, k = 0; 1; \ldots; 15$. The modulated symbols on each branch are scaled to maintain a constant transmit power, independently of the data rate. The scaled Walsh chips are summed on a chip-by-chip basis.

9.12.3 Forward-Link Quadrature Spreading

The spreading sequence used for quadrature spreading has a length of 2^{15} PN chips (32,768 chips). The characteristic polynomials for the I and Q branches are given by

$$C_{\text{short}I} = 1 + x^2 + x^6 + x^7 + x^8 + x^{10} + x^{15}$$
$$C_{\text{short}Q} = 1 + x^3 + x^4 + x^5 + x^9 + x^{10} + x^{11} + x^{12} + x^{15}$$

9.12.4 Forward-Link Data Rates and Modulation Parameters

The date rates supported by the forward traffic channel are shown in Table 9.17. As already mentioned, the data rates supported by the control channel are 76.8 or 38.4 kbit/s. Table 9.17 also shows the modulation parameters for the forward link.

9.12.5 Forward-Link Transmission

Figure 9.35 summarizes the transmission chain of the HRPDA forward link.

9.12.6 Reverse Link—General

The reverse CDMA channel, using a dedicated 1× RF channel for data transmission purposes, comprises the following channels:

- *Reverse Traffic Channel (R-TCH).* The R-TCH is used by the access terminal to transmit user-specific traffic or signaling information. It consists of four channels:

  Pilot Channel (R-TPICH). The R-TPICH is used by the access network for coherent demodulation of the R-TDCH.

  Reverse Rate Indicator Channel (R-RRICH). The R-RRICH is used to indicate the data transmission rate on the R-TDCH.

  Data Rate Control (R-DRCCH). The R-DRCCH is used by the access terminal to indicate to the access network the supportable F-TCH data rate. It also informs the best-serving sector on the forward CDMA channel.

TABLE 9.17

RC Characteristics for Forward Traffic Channel and Control Channel.

Data Rates (kbit/s)	Bits	Slots	Coding Rate	Modulation Scheme	Preamble Pilot MAC Data (chips)
38.4	1024	16	1/5	QPSK	1,024 3,072 4,096 24,576
76.8	1024	8	1/5	QPSK	512 1,536 2,048 12,288
153.6	1024	4	1/5	QPSK	256 768 1,024 6,144
307.2	1024	2	1/5	QPSK	128 384 512 3,072
614.4	1024	1	1/3	QPSK	64 192 256 1,536
307.2	2048	4	1/3	QPSK	128 768 1,024 6,272
614.4	2048	2	1/3	QPSK	64 384 512 3,136
1228.8	2048	1	1/3	QPSK	64 192 256 1,536
921.6	3072	2	1/3	8-PSK	64 384 512 3,136

TABLE 9.17

Continued

Data Rates (kbit/s)	Bits	Slots	Coding Rate	Modulation Scheme	Preamble Pilot MAC Data (chips)
1843.2	3072	1	1/3	8-PSK	64 192 256 1,536
1228.8	4096	2	1/3	16-QAM	64 384 512 3,136
2457.6	4096	1	1/3	16-QAM	64 192 256 1,536

Acknowledgment Channel (R-ACKCH). The R-ACKCH is used to inform the access network about the success of the reception of the data packet transmitted on the F-TCH.

Data Channel (R-TDCH). The R-TDCH is used by the access terminal for the actual transmission of data.

- *Access Channel (R-ACH).* The R-ACH is used by the access terminal to initiate communication with the access network. It is also used to respond to an access terminal directed message. The R-ACH consists of two channels:

Pilot Channel (R-APICH). The R-APICH is used by the access network for coherent demodulation of the R-ADCH.

Data Channel (R-ADCH). The R-ADCH conveys data concerning user access.

The R-RRICH symbols are time-division-multiplexed with the R-TPICH; the resulting channel is still referred to as pilot channel. The pilot channel, the R-DRCCH, the R-ACKCH, and the R-TDCH are orthogonally spread by Walsh functions of length 4, 8, or 16. Each R-TCH is identified by a user long code. In the same way, the R-ACH for each sector is identified by a distinct access channel long code. The frame duration for the R-ACH and R-TCH is 26.66... ms, with each frame consisting of 16 slots. Each slot contains 2048 chips and has a duration of 1.66... ms.

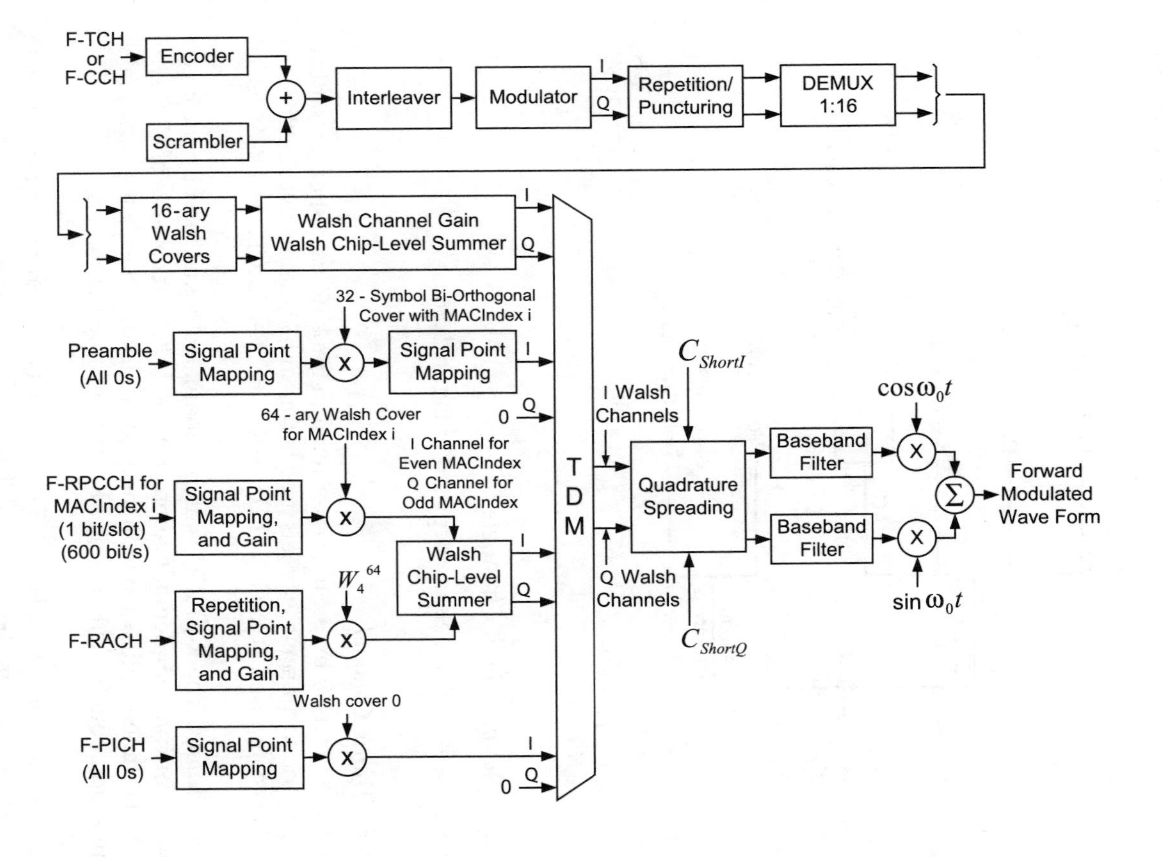

FIGURE 9.35
Transmission chain of the HRPDA forward link.

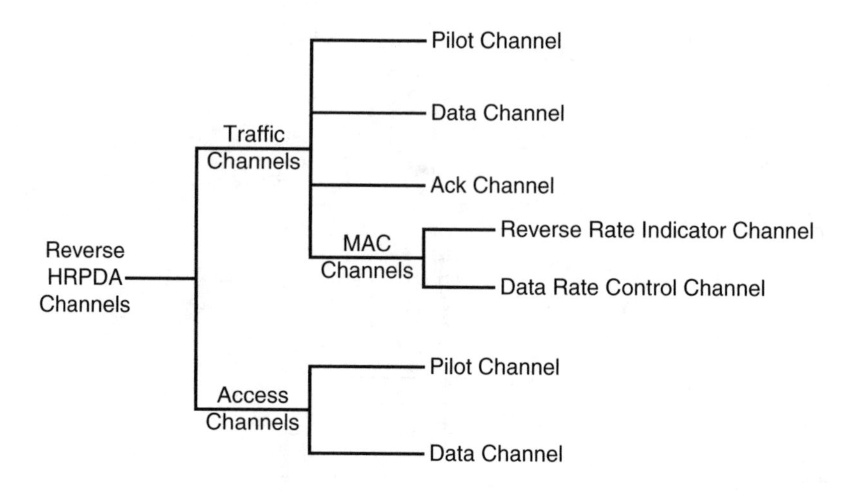

FIGURE 9.36
HRPDA reverse channels.

9.12.7 Reverse-Link Channels

The reverse-link channels are depicted in Figure 9.36.

Reverse Traffic Channel

The R-TCH is used by the access terminal to transmit user-specific traffic to the network. In the same way, it is used to transmit signaling information. The R-TCH is composed of the R-TPICH, R-RRICH, R-DRCCH, and R-ACKCH.

Pilot Channel. The R-TPICH is an unmodulated signal used by the access network for coherent demodulation of the R-TDCH. It consists of all-0 symbols and is time-multiplexed with the R-RRICH, both sharing the same Walsh channel and both being transmitted at the same power.

Reverse Rate Indicator Channel. The R-RRICH is used to indicate the data transmission rate on the R-TDCH. The RRI symbol indicating such a rate contains 3 bits and is transmitted at a rate of one symbol per 16-slot physical layer packet. Each symbol is encoded into a 7-bit code word and each code word is repeated 37 times. After the repetition process, the last 3 bits are discarded. Such operations yield 256 binary symbols per physical layer packet. This is then time-multiplexed with the R-TPICH and spread with the 16-chip Walsh function W_0^{16}, yielding 256 RRI chips per slot. These RRI chips occupy the first 256 chip positions of every slot and are transmitted on the I channel.

Data Rate Control Channel. The R-DRCCH is used by the access terminal to indicate to the access network the supportable F-TCH data rate. It also informs the selected serving sector on the forward CDMA channel. Four bits are used to indicate the requested F-TCH data rate. The DRC values are transmitted at a rate of 600/DRCLength DRC values per second. The DRCLength parameter is conveyed by the forward traffic channel MAC protocol. The DRC values are block-encoded into 8-bit biorthogonal code words, and each code word is transmitted twice per slot. Each bit of the repeated code word is spread by the Walsh function W_i^8, $i = 0; 1; \ldots; 7$, where i is the DRCCover. Each resulting Walsh chip is further spread by the Walsh function W_8^{16} and the R-DRCCH is transmitted on the Q channel.

Acknowledgment Channel. The R-ACKCH is used to inform the access network about the success of the reception of the data packet transmitted on the F-TCH. An R-ACKCH is transmitted by the access terminal whenever an F-TCH directed to it is detected. The R-ACKCH is BPSK-modulated, a bit 0 indicating a successful reception and a bit 1 indicating an unsuccessful reception. The success of a reception is assessed by means of the FCS (frame check sum) mechanism. The R-ACKCH transmitted in response to an F-TCH occurring on time slot n of the forward channel occupies the time slot $n+3$ on the reverse channel. It is identified by the Walsh function W_4^8 and transmitted on the I channel.

Data Channel. The R-TDCH is used by the access terminal for the actual transmission of data. Transmissions of the R-TDCH within a frame occur at slot FrameOffset; the FrameOffset is public data of the reverse traffic channel MAC protocol. The data transmitted on the R-TCH are encoded, block interleaved, sequence repeated, and orthogonally spread by Walsh function W_2^4.

Access Channel

The R-ACH is used by the access terminal to initiate communication with the access network. It is also used to respond to an access terminal–directed message. The R-ACH consists of the R-APICH and R-ADCH. Access is carried out by means of access probes. An access probe consists of a preamble followed by a number of access channel physical layer packets. The preamble transmission makes use of the R-APICH only. On the other hand, during the access channel physical layer transmission, both R-APICH and R-ADCH are transmitted. During the preamble portion of the probe, the power of the R-APICH is higher than it is during the data portion of the probe. The physical layer packets on the access channel are transmitted at a fixed rate of 9.6 kbit/s.

Pilot Channel. The R-APICH is used by the access network for coherent demodulation of the R-ADCH. It is an unmodulated signal consisting of all-0 symbols. It is transmitted continuously during the access channel transmission. The R-APICH uses the Walsh function W_0^{16} and is transmitted on the I channel.

Data Channel. The R-ADCH conveys data concerning user access. Every access probe is composed of one or more physical layer packets transmitted on the R-ADCH. These packets are transmitted at a fixed rate of 9.6 kbit/s on the Q channel. The R-ADCH uses the Walsh function W_2^4.

9.12.8 Reverse-Link Quadrature Spreading

After the orthogonal spreading, the R-ACKCH, R-DRCH, and R-TDCH chip sequences are given a gain relative to that of the R-TPICH. A similar process occurs with the R-ADCH, with its gain relative to that of R-APICH. An adequate combination of these channels forms the I channel and the Q channel sequences, and these sequences are quadrature spread. The quadrature spreading occurs at a rate of 1.2288 Mchip/s and is equivalent to a complex multiplication operation of these I and Q channels and I and Q PN sequences. These PN sequences, in turn, are obtained as follows. The I PN sequence is the result of the multiplication of the I long-code PN sequence and the I access terminal common short PN sequence. The Q PN sequence is obtained by the multiplication of the decimated Q long-code PN sequence and the Q access terminal common short PN sequence. The access terminal common short-code PN sequences have a period of 2^{15} chips (32,768 chips), with its I and Q components based on the following respective characteristic polynomials:

$$
\begin{aligned}
C_{\text{short}I} &= 1 + x^5 + x^7 + x^8 + x^9 + x^{13} + x^{15} \\
C_{\text{short}Q} &= 1 + x^3 + x^4 + x^5 + x^6 + x^{10} + x^{11} + x^{12} + x^{15}
\end{aligned}
$$

The long PN code, C_{long}, is generated by 42-stage linear feedback shift registers running at 1.2288 Mchip/s. Its characteristic polynomial is given as

$$
\begin{aligned}
C_{\text{long}} = {} & 1 + x + x^2 + x^3 + x^5 + x^6 + x^7 + x^{10} + x^{16} + x^{17} + x^{18} \\
& + x^{19} + x^{21} + x^{22} + x^{25} + x^{26} + x^{27} + x^{31} + x^{33} + x^{35} + x^{42}
\end{aligned}
$$

Its I and Q components are generated by a modulo-2 inner product of the 42-bit state vector of the sequence generator and two 42-bit masks. These masks vary depending on the channel on which the access terminal is transmitting.

TABLE 9.18

RC Characteristics for Reverse Traffic Channel and Access Channel

Data Rates (kbit/s)	9.6	19.2	38.4	76.8	153.6
Reverse rate index	1	2	3	4	5
Bits	256	512	1024	2048	4096
Duration (ms)	26.66...	26.66...	26.66...	26.66...	26.66...
Code rate	1/4	1/4	1/4	1/4	1/2
Symbols	1024	2048	4096	8192	8192
Symbol rates (ksymb/s)	38.4	76.8	153.6	307.2	307.2
Repetition	×8	×4	×2	×1	×1
Modulation symbol rates (ksymb/s)	307.2	307.2	307.2	307.2	307.2
Modulation scheme	BPSK	BPSK	BPSK	BPSK	BPSK
PN chips per bit	128	64	32	16	8

At the beginning of every period of the short codes, the long-code generator is reloaded with the hexadecimal value 0 × 24B91BFD3A8. Note, therefore, that the long codes have the same period as the short codes.

9.12.9 Reverse-Link Data Rates and Modulation Parameters

The date rates supported by the reverse traffic channel are shown in Table 9.18. As already mentioned, the access channel operates at a fixed data rate of 9.6 kbit/s. Table 9.18 also shows the modulation parameters for the reverse link.

9.12.10 Reverse-Link Transmission

Figure 9.37 summarizes the transmission chain of the HRPDA reverse traffic channel. Figure 9.38 summarizes the transmission chain of the HRPDA reverse access channel.

9.12.11 Open-Loop Power Control Operation

The open-loop operation is based on the power of the received F-PICH. The power of the R-ADCH relative to that of the R-APICH depends on the transmission rate and is obtained by means of public data of the access channel MAC protocol. The power of R-APICH during the preamble portion of the access probe increases in subsequent probes, as already explained. The initial

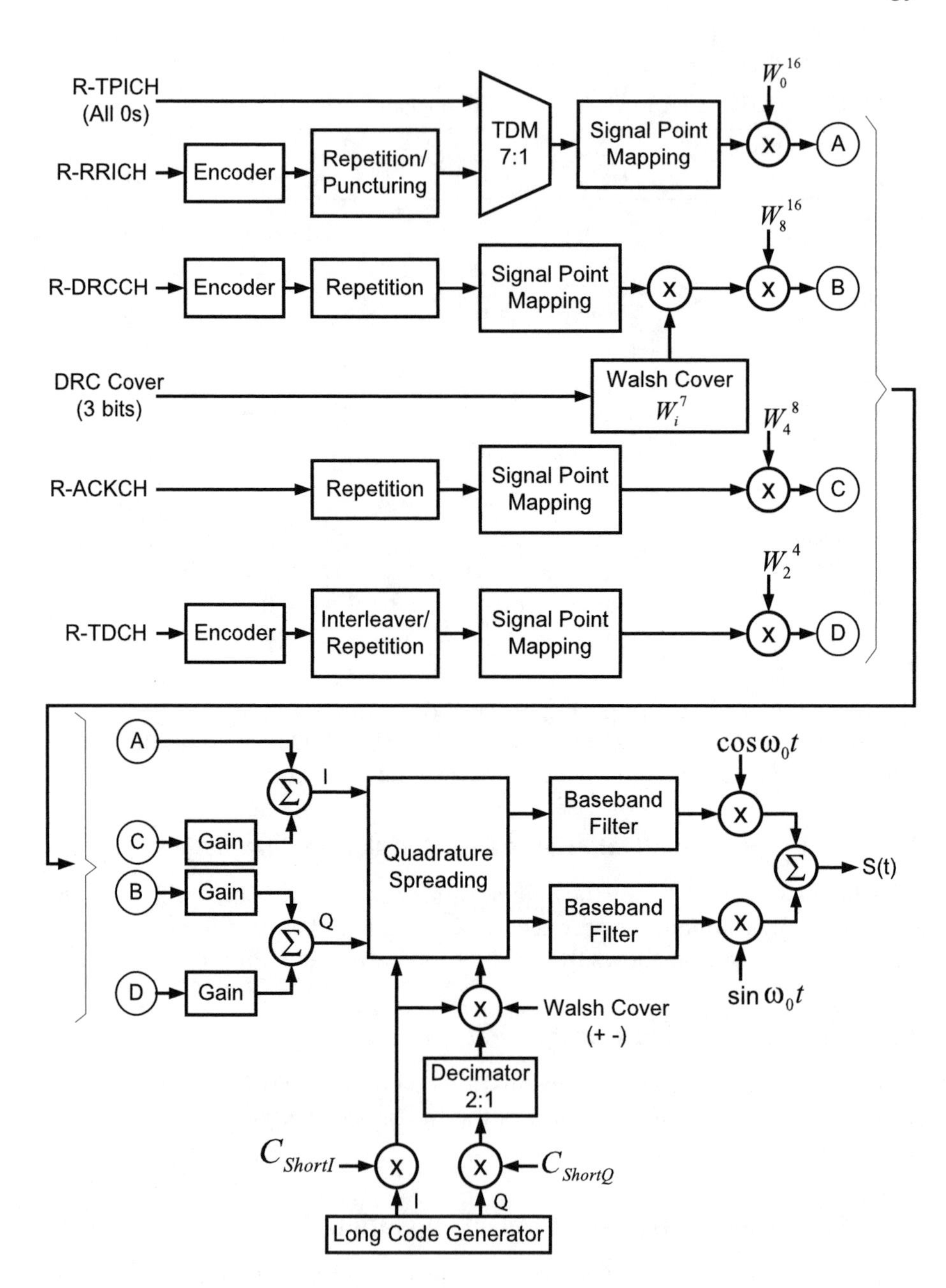

FIGURE 9.37
Transmission chain of the HRPDA reverse traffic channel.

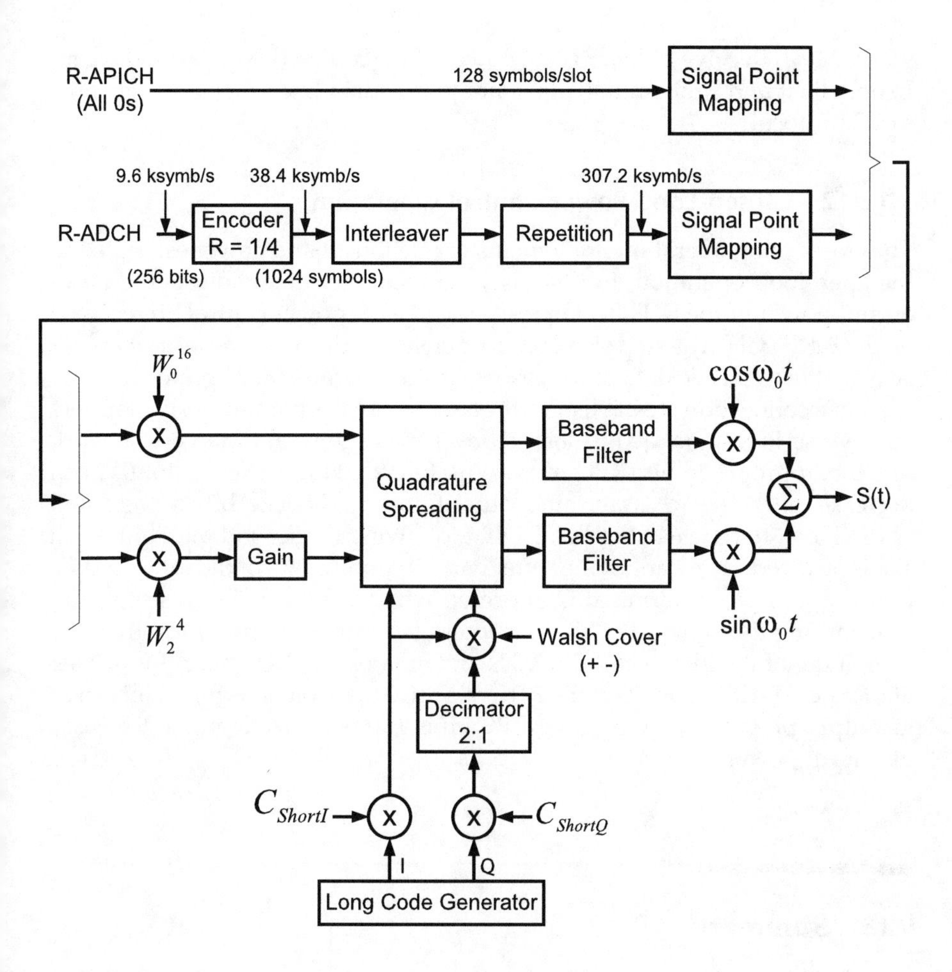

FIGURE 9.38
Transmission chain of the HRPDA reverse access channel.

power of the R-TPICH is set to be equal to that of the R-APICH at the end of the last access probe minus the difference in the forward-link mean received signal power from the end of the last access probe to the start of the reverse traffic channel transmission. The subsequent adjustments of the R-TPICH are obtained by means of the closed-loop power control operation. The initial power of the R-APICH, X_0, is set as

$$\begin{aligned} X_0 = &- \text{Mean Received Power (dBm)} \\ &+ \text{OpenLoopAdjust} \\ &+ \text{ProbeInitialAdjust} \end{aligned}$$

where Mean Received Power is the power of the received F-PICH, and Open-LoopAdjust and ProbeInitialAdjust are public data from the access channel MAC protocol.

9.12.12 Closed-Loop Power Control Operation

The closed-loop operation implements corrections to the power estimated in the open-loop operation. In this case, the access terminal adjusts the mean output power of the R-TPICH in response to each power control bit received on the F-RPCCH. The step size used to increase or decrease the output power is available as public data of the reverse traffic channel MAC protocol.

After a connection is established, the access network continuously monitors the reverse-link signal quality. Based on such monitoring, the access network sets the appropriate bit (RPC bit) on the F-RPCCH to increase (bit 0) or to decrease (bit 1) the access terminal output power. The RPC bit is considered as received after the 64-chip MAC burst following the second pilot burst of a slot is received. By means of the softer handoff public data of the route update protocol, the access terminal is informed whether two different sectors are transmitting the same RPC bit. The access terminal then provides diversity combining of the identical F-RPCCHs and obtains one RPC bit from each set of identical F-RPCCHs. If all the RPC bits are 0, the access terminal increases its output power. If any resulting RPC bit is 1, the access terminal decreases its output power.

9.13 Summary

cdma2000 is a wideband spread spectrum radio interface designed to meet the requirements of the 3G wireless systems as well as the requirements of the 3G evolution of the 2G TIA/EIA-95-B family standards. Full backward compatibility with TIA/EIA-95-B is provided that permits cdma2000 infrastructure to support TIA/EIA-95-B mobile stations. Backward compatibility allows cdma2000 mobile stations to operate in TIA/EIA-95-B systems.

cdma2000 accomplishes the wideband transmission requirements in two different ways. The forward transmission may utilize either direct spread technology or multicarrier technology. The reverse link always uses DS technology. The MC implementation of the cdma2000 forward link facilitates the cdmaOne and cdma2000 overlay design. In an MC implementation, an $N \times 1.25$ MHz cdma2000 system ($N = 1, 3, 6, 9$, or 12) can overlay N contiguous cdmaOne carriers, where N is the spreading rates (SR) number. In addition to the different SRs and to meet the different QoS requirements, cdma2000 incorporates a number of radio configurations (RCs). These RCs are specified in terms of the achievable rate transmissions, frame duration, and prespreading modulation schemes.

SR 1 is referred to as 1× and SR 3 is referred to as 3×. For SR 1 and SR 3, cdma2000 standards specify ten RCs, numbered sequentially from 1 to 10, for the downlink, and seven RCs, numbered sequentially from 1 to 7, for the uplink. Collectively, these RCs form the radio interface, which consists of the 1× and 3× components. RCs from 1 through 9 for the forward link and RCs from 1 to 6 for the reverse link accommodate a wide range of services and data rates. RC 10 for the forward link and RC 7 for the reverse link are specific for high-rate packet data access using a separate carrier. That for high-rate packet data access evolved from the high data rate (HDR) technology to the 1× evolved high-speed data-only (1×EV-DO) design; both projects present almost indistinguishable characteristics.

References

1. 3GPP2 S.R0005-A, Network Reference Model for cdma2000 Spread Spectrum Systems, December 13, 1999.
2. 3GPP2 S.R0003, 3GPP2 System Capability Guide, Release A, January 20, 2000.
3. 3GPP2 C.S0001-A, Introduction to cdma2000 Standards for Spread Spectrum Systems, Release A, June 9, 2000.
4. 3GPP2 C.S0004-A, Signaling Link Access Control (LAC) Specification for cdma2000 Spread Spectrum Systems, Release A, June 9, 2000.
5. 3GPP2 C.S0002-A, Physical Layer Standard for cdma2000 Spread Spectrum Systems, Release A, June 9, 2000.
6. 3GPP2 C.S0003-A, Medium Access Control (MAC) Standard for cdma2000 Spread Spectrum Systems, Release A, June 9, 2000.
7. 3GPP2 C.S0024-Version 2.0, High Rate Packet Data Air Interface Specification, October 27, 2000.
8. Document 8F/334-E, Proposed Revision Material for Recommendation ITU-R M.1457, June 19, 2001.
9. Knisely, D. N, Li, Q. and S. Ramesh, N. S., cdma2000: a third-generation radio transmission technology, *Bell Labs Tech*, J., 63–78, July–September 1998.
10. Garg, V. K., *IS-95 CDMA and cdma2000 Cellular/PCS Systems Implementation*, Prentice-Hall, Upper Saddle River, NJ, 2000.

Part V
Appendices

Appendix A

Open Systems Interconnection

The Open Systems Interconnection (OSI) Reference Model was formulated by the International Standards Organization (ISO) in the early 1980s with computer networks as its target. The usefulness of such a model was soon recognized by communications engineers and its application extended to other complex communications networks. The model simplifies the design of complex networks by means of the use of a modular and structured approach in which network functions are identified that can be implemented independently of one another. In the OSI Reference Model, network operability is partitioned into seven layers and protocols implement the functionality assigned to each layer. Each layer provides services to the layer above it and uses the services from the layer below it. At the transmitting side, each layer adds its own header to the message received from the layer above it and delivers the composite message to the layer below it. On the receiving side, each layer removes the corresponding header from the message and delivers it to the layer above it.

Figure A.1 illustrates the OSI Reference Model in which the communication between two hosts (Host A and Host B) are depicted. In Figure A.1, the solid lines indicate a direct communication, whereas the broken lines indicate an indirect (virtual) communication. In such a reference model, the communications paths are composed of nodes connected by links. The seven layers of the OSI Reference Model are briefly described as follows:

- *Physical Layer (Layer 1)*. The physical layer handles the transmission of raw bits. It specifies the necessary procedures for the transfer of bits from one place in a network to another.
- *Data Link Layer (Layer 2)*. The data link layer aggregates a stream of bits into a frame.
- *Network Layer (Layer 3)*. The network layer provides means for routing among nodes within a packet-switched network.

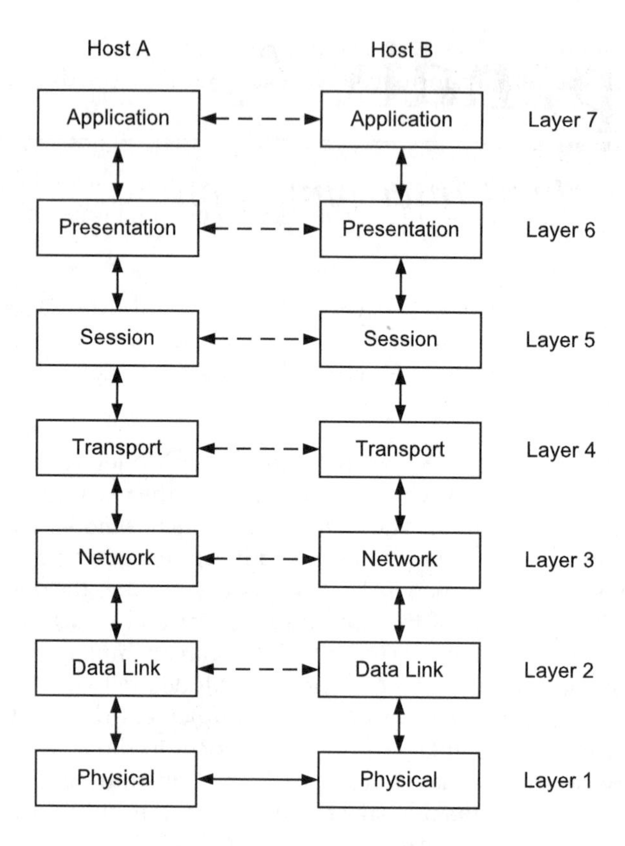

FIGURE A.1
The OSI/ISO Reference Model.

- *Transport Layer (Layer 4).* The transport layer establishes a process-to-process channel. It is responsible for the establishment, maintenance, and discontinuation of communications between two hosts, as well as end-to-end service quality. In the transmitting host, the following tasks are performed: reception and interpretation of user information delivered from the session layer; translation of the name of the destination (receiving) host to a network address; addition of its header information including the network address; delivery of the composite information (aggregate) to the network layer. In the receiving host, the transport layer accepts data from the network layer, verifies its consistency, and delivers it to the session layer.
- *Session Layer (Layer 5).* The session layer is responsible for creating and managing a name space to link different transports, which are considered as part of a single application.

- *Presentation Layer (Layer 6).* The presentation layer is responsible for providing a common format of data exchange between different types of peers.
- *Application Layer (Layer 7).* The application layer implements the functionality of the application. It is related to the purpose of the communication. Examples of application include e-mail, home banking, file transfer, and others.

Note that Layer 1, Layer 2, Layer 3, and Layer 4 (lower layers) provide means for transferring information between hosts. Layer 5, Layer 6, and Layer 7 (higher layers) provide means for user's interaction with a particular information service.

Appendix B

Signaling System Number 7

With the increase of complexity of modern networks, the adoption of a decentralized approach, and the necessity of integration of network elements, a clear separation between control information and user information for better efficiency became indispensable. Modern communication systems make use of distinct network elements and paths for control and user information applications. Within a communication network, the exchange of control information between network elements is performed by a signaling system.

Signaling System Number 7 (SS7) emerged as an international standard and gained worldwide acceptance. SS7 conforms to a layered model that parallels the OSI Reference Model. In SS7 terminology, these layers are called parts. The SS7 is responsible for the control of the fixed network as well as the mobile network. Figure B.1 illustrates the SS7 parts and the corresponding OSI layers.

- *Message Transfer Part Level 1 (MTP 1).* The MTP 1 defines the physical, electrical, and functional characteristics of a signaling digital link.
- *Message Transfer Part Level 2 (MTP 2).* The MTP 2 provides means for a correct point-to-point transmission of a message through the signaling link. It implements flow control, message sequence validation, and error check.
- *Message Transfer Part Level 3 (MTP 3).* The MTP 3 is responsible for the routing of messages between the network signaling points.
- *Telephone User Part (TUP).* The TUP is used to support analog circuit switched calls. It supports the connection between an MSC and a PSTN.
- *ISDN TUP (ISUP).* The ISUP supports both ISDN and non-ISDN calls in a digital network. It supports the connection between the MSC and an ISDN.

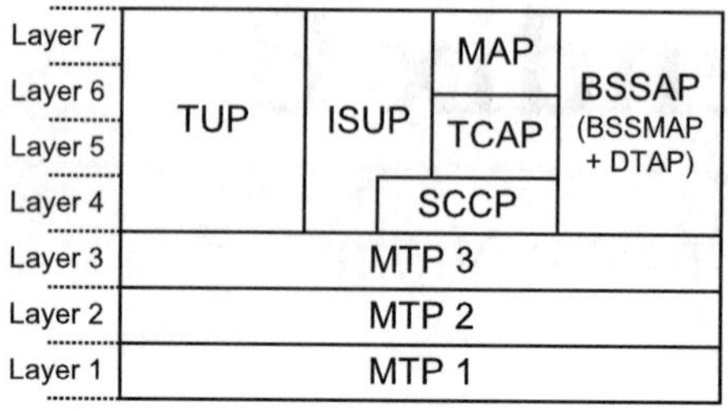

FIGURE B.1
SS7 and the corresponding OSI layers.

- *Signaling Connection Control Part (SCCP).* The SCCP supports connection- or connectionless-oriented network services. It is also responsible for the global title translation (GTT) service, which translates a global title (e.g., a 0-800, 1-800) into a destination point and a subsystem number.
- *Transaction Capabilities Application Part (TCAP).* The TCAP provides a set of protocols and services used by an application process.
- *Base Station System Management Application Part (BSSMAP).* The BSSMAP interfaces BSC with MSC.
- *Direct Transfer Application Part (DTAP).* The DTAP is used for message transfer between MSC and MS.
- *BSSMAP and DTAP.* These fuse into the BSSAP.
- *Mobile Application Part (MAP).* The MAP is the highest SS7 layer. It supports the connection between MSC, HLR, VLR, and some other network elements. MAP has been added to SS7 to perform control operations unique to mobile communication networks, specifically to GSM. A similar entity has also been incorporated into ANSI-41, for American cellular systems. The interfaces of MAP and ANSI-41 are detailed in Chapters 4 and 5, respectively.

Appendix C

Spread Spectrum

Spread spectrum is defined as a communication technique in which the intended signal is spread over a bandwidth in excess of the minimum bandwidth required to transmit the signal. This is accomplished by the use of a wideband encoding signal at the transmitter, which operates in synchronism with the receiver, where the encoding signal is also known. By allowing the intended signal to occupy a bandwidth far in excess of the minimum bandwidth required to transmit it, the signal will have a noiselike appearance.

The spread spectrum function constitutes an additional block included in a conventional communication system. It implements a second-level, an additional, modulation. Therefore, generating a spread spectrum signal involves two steps: first, the carrier is modulated by the baseband digital information with rate $R_b = 1/T_b$, where R_b is the information rate and T_b is the baseband (information) symbol duration; second, the modulated signal is used to modulate a wideband function with rate $R_c = 1/T_c$, where R_c is known as chip rate and T_c is the chip duration. After the first modulation process, the resulting signal is still a narrowband signal. After the second modulation process, the resulting signal is a wideband signal. The desired wideband signal arrives at the receiver together with other wideband signals, interference, and noise. If the set of spreading functions is chosen so that low cross-correlation among them exists, only the desired signal remains after the correlation process (see next section). Other waveforms are not correlated and will be spread, appearing as noise to the modulator. The correlated signal is then a band-pass signal, whereas the noise component is a wideband signal. Thus, by applying the composite signal (wanted band-pass signal plus wideband noise) to a band-pass filter, with a bandwidth just large enough to accommodate the wanted signal, the noise component is made band-pass. In this case, the filtered noise appears with much less power. Consequently, a jammer or interference power density in the information bandwidth of the received signal will be effective only if its total power is increased by the same amount as the bandwidth expansion of the signal.

Two main spread spectrum techniques are used: direct sequence spread spectrum and frequency hopping spread spectrum. This appendix introduces some very basic principles of spread spectrum.

C.1 Correlation

The correlation function quantifies the degree of similarity between two functions. Let $x(t)$ and $y(t)$ be two nonperiodic waveforms with finite energy. The cross-correlation function $R_{x,y}(\tau)$ is given by

$$R_{x,y}(\tau) = \int_{-\infty}^{\infty} x(t)\, y(t-\tau)\, dt$$

If $x(t)$ and $y(t)$ are periodic waveforms, with period T, then

$$R_{x,y}(\tau) = \frac{1}{T}\int_{0}^{T} x(t)\, y(t-\tau)\, dt$$

The autocorrelation $R_z(\tau)$ for either type of waveform ($z(t) = x(t)$, or $z(t) = y(t)$) is defined as

$$R_z(\tau) = R_{z,z}(\tau) = \int_{-\infty}^{\infty} z(t)\, z(t-\tau)\, dt$$

Assume that $z(t)$ is a binary waveform defined as

$$z(t) = \sum_{k=-\infty}^{\infty} Z_k \widehat{Z}\left(t - \frac{k}{W}\right)$$

where $Z_k \in \{+1, -1\}$, $\widehat{Z}(t)$ is the pulse shape, and $1/W$ is the duration of the pulse. Then

$$R_{x,y}(\tau) = \frac{W}{K}\int_{0}^{K/W} x(t)\, y(t-\tau)\, dt$$

where K is the number of pulses composing the period of the sequence. By using the above definitions, it can be shown that

$$R_{x,y}(\tau) = \frac{W}{K}\sum_{k=-\infty}^{\infty} R_{X,Y}(k)\, R_{\widehat{X},\widehat{Y}}\left(\tau - \frac{k}{W}\right)$$

where $R_{X,Y}(k)$ is the cross-correlation function of the two periodic binary sequences X_k and Y_k, and $R_{\hat{X},\hat{Y}}(\tau)$ is the nonperiodic cross-correlation function for the basic waveforms $\widehat{X}(t)$ and $\widehat{Y}(t)$. The cross-correlation between X_k and Y_k is defined as

$$R_{X,Y}(k) = \sum_{i=0}^{K-1} X_i Y_{i-k}$$

Note that the correlation between two sequences yields the difference between the number of bit-by-bit position agreements minus the number of disagreements. The autocorrelation function $R_Z(k)$ of the sequence Z_i is defined as

$$R_Z(k) = R_{Z,Z}(k) = \sum_{i=0}^{K-1} Z_i Z_{i-k}$$

The autocorrelation function for $z(t)$ is

$$R_z(\tau) = R_{z,z}(\tau) = \frac{W}{K} \sum_{k=-\infty}^{\infty} R_Z(k) R_{\hat{Z}}\left(\tau - \frac{k}{W}\right)$$

Assume that the pulse shape $\widehat{Z}(t)$ is a rectangle with amplitude equal to 1 and duration P. The autocorrelation function $R_{\hat{Z}}(\tau)$ is a triangle as shown in Figure C.1. Now, for a sequence of K binary symbols, in which the number of +1s and −1s differ by one, the autocorrelation is K, for $k = 0$ and -1 for $k \neq 0$. In such a case, the autocorrelation function for the complete waveform is as shown in Figure C.2.

Two real-valued waveforms are said to be orthogonal if $R_{x,y}(0) = 0$. In the same way, two sequences are said to be orthogonal if $R_{X,Y}(0) = 0$.

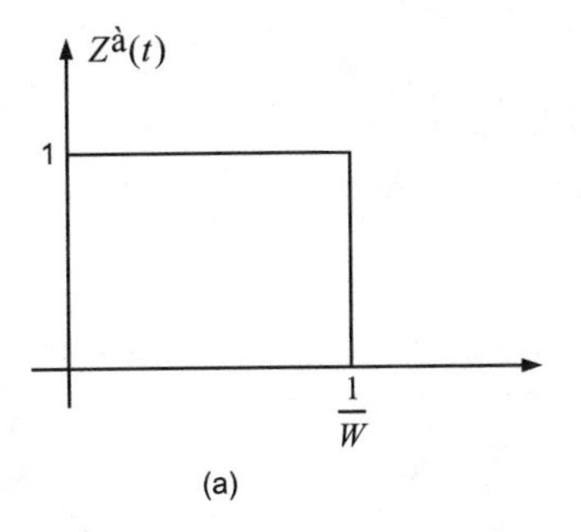

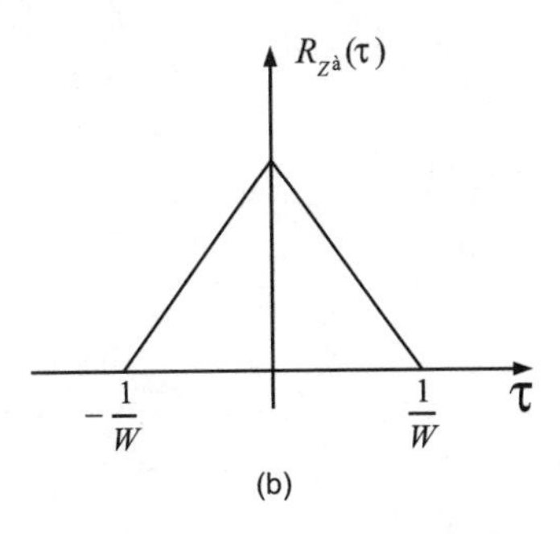

FIGURE C.1
A rectangular pulse shape (a) and its autocorrelation function (b).

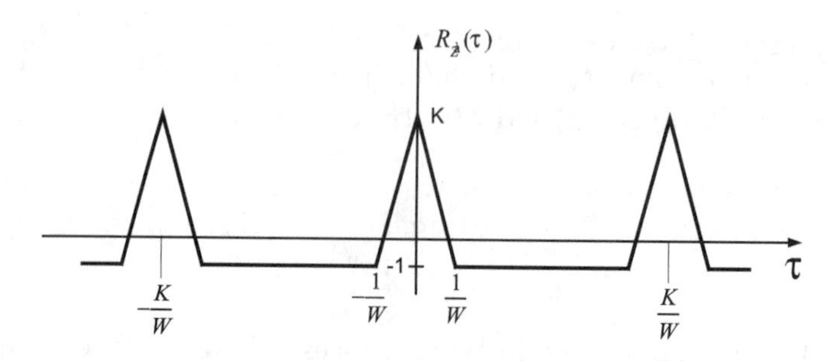

FIGURE C.2
Autocorrelation function of a real signal waveform.

C.2 Pseudonoise Sequences

Pseudonoise (PN) or pseudorandom sequences are used for two main purposes: data scrambling and spread spectrum modulation. PN sequences provide data scrambling by giving the data sequence a noiselike appearance. The use of PN sequence for spread spectrum modulation is briefly explained later in this appendix. Note, in the scrambling operation (as well as in the modulation operation), that both transmitter and receiver must work with exactly the same PN sequence. Therefore, the sequence cannot be completely random; otherwise, the original data would never be recovered from the scrambled data.

PN sequences are generated by means of linear feedback shift registers. For an n-stage linear feedback shift register, the maximum number of different combinations of n binary digits is 2^n. This implies that the period of the PN sequence cannot exceed 2^n. On the other hand, an all-zero state leads the shift register to remain in this null state forever. Therefore, the null state must be avoided, which makes the number of possible states of an n-stage shift register equal to $2^n - 1$. A sequence with a period equal to $2^n - 1$ is known as maximal length sequence or m-sequence or PN sequence. The following main properties characterize the m-sequences:

- *Balance Property*. Within a complete period of the sequence, the number of 1s and 0s differs from each other by at most 1.
- *Correlation Property*. By comparing a complete sequence with any shifted version of it, within the sequence period, the number of agreements minus the number of disagreements is always −1. That is, for a PN sequence Z_i, $R_Z(k) = -1$, for $k \neq 0$ and $R_Z(k) = 2^n - 1$, for $k = 0$.

C.3 Walsh Codes

The Walsh sequence can be generated by means of the Rademacher functions or by the Hadamard matrices. The Hadamard matrix is defined as

$$H_0 = [1]$$
$$H_n = \begin{bmatrix} H_{n-1} & H_{n-1} \\ H_{n-1} & -H_{n-1} \end{bmatrix}$$

The Walsh sequences are indexed by the row of the matrix. Hence, W_i corresponds to the sequence in row i of the Hadamard matrix. An example of the Hadamard matrix for $n = 2$ is shown as follows:

$$H_2 = \begin{bmatrix} 1 & 1 & 1 & 1 \\ 1 & -1 & 1 & -1 \\ 1 & 1 & -1 & -1 \\ 1 & -1 & -1 & 1 \end{bmatrix}$$

Thus, $W_1 = 1, -1, 1, -1$. The Walsh sequences are orthogonal to each other when correlated over their period.

C.4 Orthogonal Variable Spreading Factor Codes

Channelization in multirate CDMA systems can be provided by orthogonal variable spreading factor (OVSF) codes. Uplink and downlink channels make use of OVSF codes. The OVSF codes preserve the orthogonality between channels of different rates and spreading factors. They can be defined as

$$C_{1,0} = [1]$$
$$\begin{bmatrix} C_{2,0} \\ C_{2,1} \end{bmatrix} = \begin{bmatrix} C_{1,0} & C_{1,0} \\ C_{1,0} & -C_{1,0} \end{bmatrix} = \begin{bmatrix} 1 & 1 \\ 1 & -1 \end{bmatrix}$$

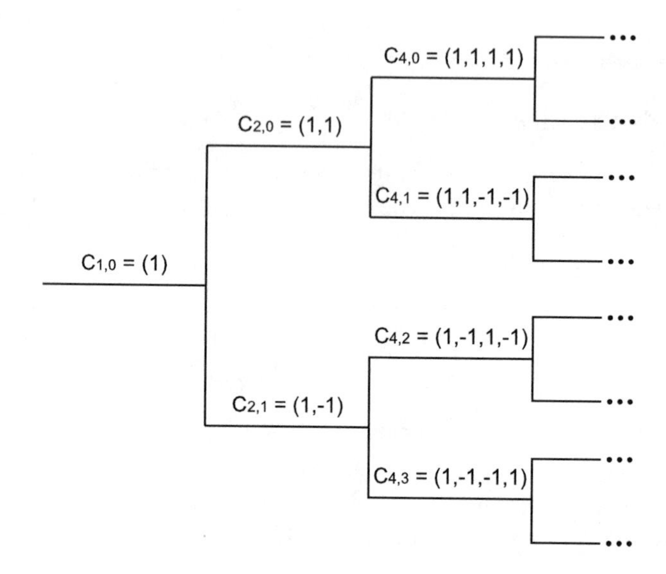

FIGURE C.3
OVSF code tree.

$$\begin{bmatrix} C_{2^{n+1},0} \\ C_{2^{n+1},1} \\ C_{2^{n+1},2} \\ C_{2^{n+1},3} \\ \cdots \\ C_{2^{n+1},2^{n+1}-2} \\ C_{2^{n+1},2^{n+1}-1} \end{bmatrix} = \begin{bmatrix} C_{2^n,0} & C_{2^n,0} \\ C_{2^n,0} & -C_{2^n,0} \\ C_{2^n,1} & C_{2^n,1} \\ C_{2^n,1} & -C_{2^n,1} \\ \cdots & \cdots \\ C_{2^n,2^n-1} & C_{2^n,2^n-1} \\ C_{2^n,2^n-1} & -C_{2^n,2^n-1} \end{bmatrix}$$

where $C_{\text{spreading\_factor},k}$ uniquely identifies the code and where $0 \le k \le$ spreading_factor $- 1$. The code tree of Figure C.3 illustrates the OVSF codes. The leftmost value in each channelization code word corresponds to the chip first transmitted in time.

C.5 Rake Receiver

In a multipath propagation environment, the received signal contains replicas of attenuated and delayed version of the transmitted signal. Assume that the signal is pseudorandom with a correlation width of $1/W$, as depicted in Figure C.2. Within $\pm 1/W$, by cross-correlating the received signal with the

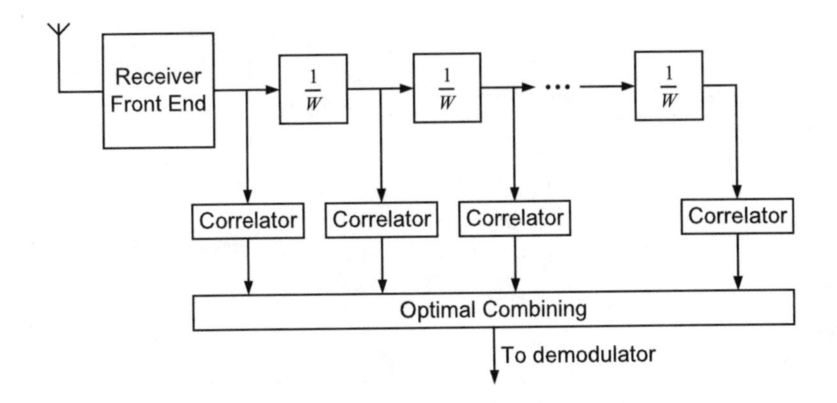

FIGURE C.4
Basic structure of a Rake receiver.

local replica of the signal, shifted by k/W, k an integer, the path with delay k/W, if any, is selected.

A Rake receiver is a tapped delay line, with delays equal to $1/W$, and with the output of each delay element connected to cross-correlators. These delay elements combined with the cross-correlators collect the signal energy from all the signal paths that fall within the span of the delay line and combine them optimally. A basic structure of the Rake receiver is shown in Figure C.4.

C.6 Processing Gain

Processing gain G is defined as the ratio between the bandwidth W of the spread signal and the bandwidth w of the unspread signal, i.e.,

$$G = \frac{W}{w}$$

which represents the gain achieved by processing a spread spectrum signal over an unspread signal. It can be obtained by the difference in decibels between the output signal-to-noise ratio (SNR_o, SNR of the spread information) and the input signal-to-noise ratio (SNR_i, SNR of the unspread information), i.e.,

$$10 \log G = \mathrm{SNR}_o - \mathrm{SNR}_i$$

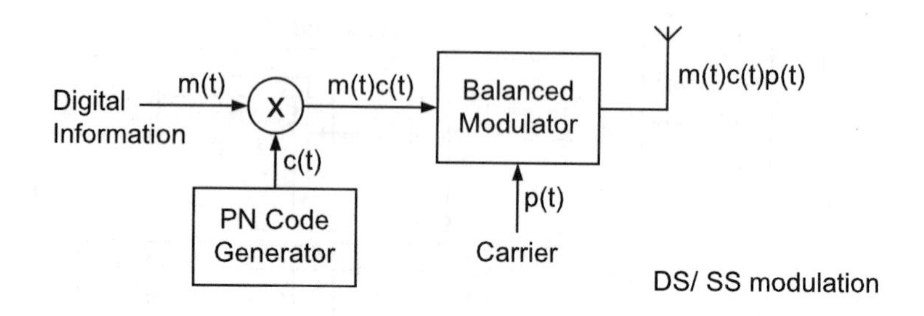

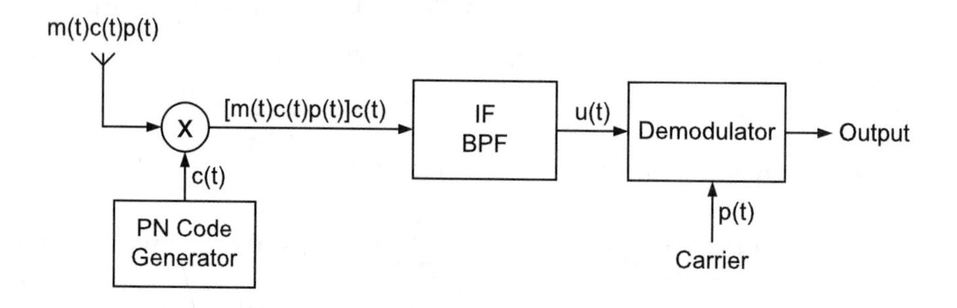

FIGURE C.5
A simplified model of a DS/SS system.

C.7 Direct Sequence Spread Spectrum

Direct sequence (DS), direct spread (DS), or pseudonoise (PN) spread spectrum (SS) uses a PN sequence to modulate a carrier (or a modulated carrier). In principle, any modulation technique such as AM (pulse), FM, or PM can be used. However, the most widespread form is the binary phase shift keying (BPSK) modulation. A simplified model of a DS/SS system is shown in Figure C.5. The information signal $m\,(t)$ and the spreading sequence $c\,(t)$ are combined before phase-modulating the carrier $p\,(t)$. The PN code $c\,(t)$ is a pseudorandom binary sequence with a chip rate R_c much larger than the information rate R_b. The multiplication of the PN code by the information signal results in a frequency spectrum with a bandwidth approximately equal to that of the PN sequence. The resultant signal is then modulated onto a carrier and transmitted. At the receiver, which is assumed to operate in synchronism with the transmitter, an exact replica of the PN code is used to unspread the received signal.

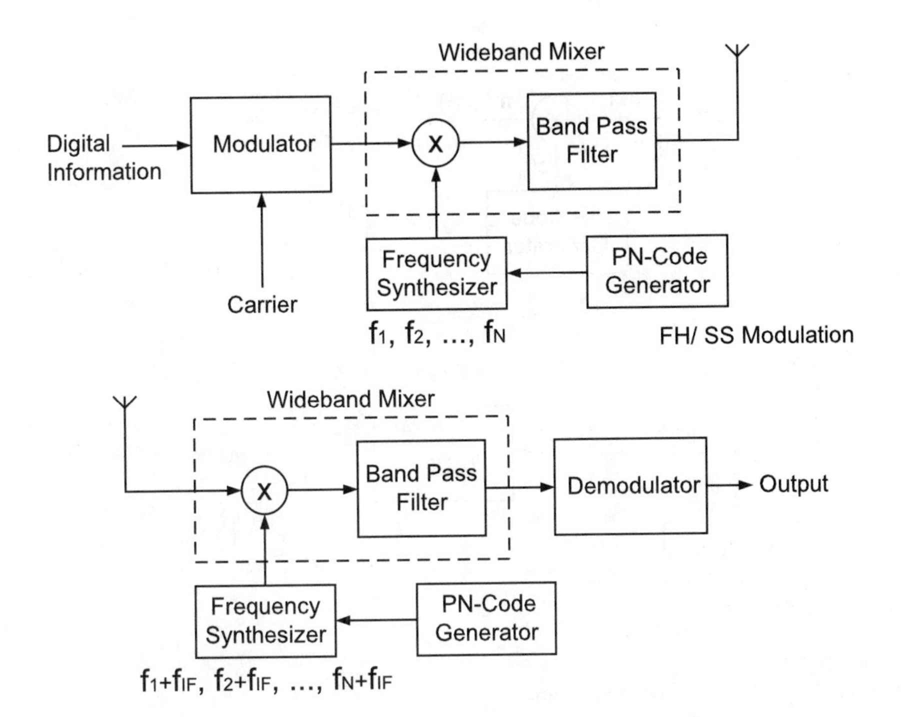

FIGURE C.6
A simplified model of an FH/SS system.

C.8 Frequency Hopping Spread Spectrum

Frequency hopping (FH) is a spread spectrum (SS) technique in which the carrier is allowed to hop from one frequency to another in a sequence dictated by a PN code. A simplified model of an FH/SS system is shown in Figure C.6. At any given time instant, the output of the FH transmitter is a single frequency. Over a period of time, its output signal spectrum is ideally rectangular, with the same amount of power in every frequency. In practice, however, the hopping process may generate spurious frequencies causing interchannel interference (cross talk). At the receiver, which is assumed to operate synchronously with the transmitter, the transmitted signal is mixed with a locally generated replica of the transmitter frequency sequence, offset by the intermediate frequency f_{IF}.

There two basic FH systems: slow FH (SFH) and fast FH (FFH). In SFH systems, several symbols of information are transmitted on each frequency hop, where each symbol is a chip. In FFH, several hops occur during the transmission of one symbol, where the chip is characterized by a hop.

Appendix D

Positioning of the Interferers in a Microcellular Grid

The number N of cells per cluster in a square cellular grid is given as $N = i^2 + j^2$, where $\{i, j\} \in Z$, Z defining the integer set. The clusters in a square grid can be of the collinear type or noncollinear type, as follows.

- Collinear type:

  $\sqrt{N} \in Z$, i.e., $i = 0$ or $j = 0$ (perfect square clusters), or

  $i = j$, i.e., $N = 2x^2$, where $x \in Z$.

- Noncollinear type:

  N is even, or

  N is odd and nonprime, or

  N is prime.

Note that there four groups of clusters: collinear, even noncollinear, odd nonprime noncollinear, and prime noncollinear. For each of these groups, the distance between an interferer positioned at the co-cell of the Lth co-cell layer and another at the target cell, with the target cell taken as the reference cell, is calculated differently. Such a distance is defined here as n_L, which is normalized with respect to the cell radius, i.e., n_L is given in number of cell radii. The relations that follow show the distance between the target cell and its interferers obtained for the mobile stations positioned for the worst-case condition. These are determined for each of the groups defined above and for the uplink and the downlink. The minimum distance between collinear co-cells is represented by the variable d, given in number of units of the cell radius. In the formulations that follow, the variable k is defined differently than it in the main text. Here, k is such that $k = 1, 2, 3, 4 \ldots$ (integer).

D.1 Collinear Type

In the collinear type, choose $m = i$ or $m = j$, whichever is non-nil. Then $d = 2m$.

- Uplink

$$n_k = 2mk - 1$$

- Downlink

$$n_k = 2mk$$

for both conditions, that is, at the vicinities of the base station and away from it.

D.2 Even Noncollinear Type

In the even noncollinear type, $d = N$.

- Uplink

$$n_k = 2Nk - 1$$

- Downlink

$$n_k = 2Nk$$

for both conditions, that is, at the vicinities of the base station and away from it.

D.3 Odd Nonprime Noncollinear Type

In the odd nonprime noncollinear type, N' is determined such that $N' = N/m$, where $m \in Z$. Then,

$$n_k = mn_{k(N')}$$

where $n_{k(N')}$ is the distance between the interferers at the kth layer and the target cell for reuse pattern N'.

D.4 Prime Noncollinear Type

In the prime noncollinear, $d = 2N$. We define p as a parameter that designates the prime number immediately prior to N, and that satisfies the relation $p = i^2 + j^2$. The only exception to this is when $N = 5$, for which $p = 3$.

- Uplink

$$\begin{aligned} n_{1+3(k-1)} &= p + 2N(k-1) \\ n_{2+3(k-1)} &= 2N - p + 2N(k-1) \\ n_{3+3(k-1)} &= 2Nk - 1 \end{aligned}$$

- Downlink

$$n_k = 2Nk$$

for both conditions, that is, at the vicinities of the base station and away from it. For the close-to-the-border condition:

$$\begin{aligned} n_k &= 2Nk \\ \overline{n}_{1+2(k-1)} &= p + 2N(k-1) \\ \overline{n}_{2+2(k-1)} &= 2N - p + 2N(k-1) \end{aligned}$$

Index

D

E

F

G

H

I

L

M

N

O

P

Q

R

S

T

U

V

W